TRACING COSMIC EVOLUTION WITH GALAXY CLUSTERS

COVER ILLUSTRATION:

Composite BIK image of the Lynx Field overlaid with X-ray contours from a Chandra Deep pointing.

Credit to P. Rosati and A. Stanford. For more details see Stanford, S.A. et al, 2001, Astrophysical Journal **552**, 504.

A SERIES OF BOOKS ON RECENT DEVELOPMENTS IN ASTRONOMY AND ASTROPHYSICS

Publisher

THE ASTRONOMICAL SOCIETY OF THE PACIFIC
390 Ashton Avenue, San Francisco, California, USA 94112-1722
Phone: (415) 337-1100 **E-Mail: catalog@astrosociety.org**
Fax: (415) 337-5205 **Web Site: www.astrosociety.org**

A listing of all other ASP Conference Series Volumes and IAU Volumes published by the ASP is cited at the back of this volume

ASTRONOMICAL SOCIETY OF THE PACIFIC
CONFERENCE SERIES

Volume 268

TRACING COSMIC EVOLUTION WITH GALAXY CLUSTERS

Proceedings of the Sesto-2001 Workshop held in
Sesto Pusteria, Alto Adige/Südtirol, Italy
3-6 July 2001

Edited by

Stefano Borgani
INFN and Dipartimento di Astronomia, Università di Trieste, Trieste, Italy

Marino Mezzetti
Dipartimento di Astronomia, Università di Trieste, Trieste, Italy

and

Riccardo Valdarnini
International School for Advanced Studies (ISAS-SISSA), Trieste, Italy

Library of Congress Cataloging in Publication Data
Main entry under title

Card Number: 2002104267
ISBN: 1-58381-108-7

ASP Conference Series - First Edition

Printed in United States of America by Sheridan Books, Chelsea, Michigan

To our friend

and colleague

Giuliano Giuricin

June 15, 1950 - May 14, 2001

Table of Contents

Dedication v

Table of contents vii

Preface xv

Conference photograph xvi

Conference participants xvii

Clusters at High Redshift

Multi-wavelength Surveys for Distant Clusters 3
M. Postman (Invited Speaker)

The Cluster Environments of Radio Loud Quasars 15
J. M. Barr, M. N. Bremer, J. C. Baker and M. D. Lehnert

The Evolution of the Cluster Environments of Radio Sources at $z < 1.8$ 19
P. Best, M. Lehnert, H. Röttgering and G. Miley

A Search for High Redshift Clusters Associated with Radio Galaxies at 2 < z < 4 23
J. D. Kurk, B. P. Venemans, H. J. A. Röttgering, G. K. Miley and L. Pentericci

The Population of Galaxies in the Forming Cluster around the Radio Galaxy MRC 1138-262 at z=2.2 27
L. Pentericci, J. D. Kurk, H. J. A. Röttgering, G. K. Miley and B. P. Venemans

Evolution of the Cluster Population with the ROSAT Deep Cluster Survey 31
S. Borgani and P. Rosati

X-Ray Temperature and Morphology of z>0.8 Clusters of Galaxies . . . 35
B. Holden, S. A. Stanford, P. Rosati, P. Tozzi, G. Squires, S. Borgani and P. Eisenhardt

Tracing Cosmic Evolution with an *XMM* Serendipitous Cluster Survey . 43
A. K. Romer, P. T. P. Viana, C. A. Collins, A. R. Liddle, R. G. Mann and R. C. Nichol

Brightest Cluster Galaxies and the Las Campanas Distant Cluster Survey 47
D. Zaritsky, A. E. Nelson, L. Simard, A. H. Gonzalez and J. J. Dalcanton

Large-Scale Structure with Clusters

Large-Scale Structure Studies with Clusters of Galaxies 57
R. C. Nichol (Invited Speaker)

The REFLEX Cluster Survey: Probing the Mass Distribution in the Universe 69
L. Guzzo, H. Böhringer, C. A. Collins, P. Schuecker and the REFLEX Team

Clustering and Substructuring of REFLEX Clusters of Galaxies 77
P. Schuecker, H. Böhringer, L. Guzzo and C. A. Collins

The Large-Scale Environment of Dynamical Young Clusters of Galaxies 81
M. Plionis and S. Basilakos

Clustering Properties of Galaxy Clusters in Present and Future Surveys 86
L. Moscardini, S. Matarrese, P. Andreani, M. Bartelmann and H. Mo

The $\sigma_8(\Omega_M)$ Constraint from X-Ray Clusters Revised 90
E. Pierpaoli, D. Scott and M. White

Using Supercluster Geometry as a Cosmological Probe 94
S. Basilakos, V. Kolokotronis and M. Plionis

Clusters with SZ

Searching and Studying Clusters with the SZ Effect 101
J. G. Bartlett (Invited Speaker)

Masses of High-Redshift Clusters via SZ Effect Observations 113
L. Grego, M. Joy, J. E. Carlstrom, S. LaRoque, D. Nagai, K. Dawson, H. Ebeling, E. D. Reese and W. L. Holzapfel

SZ Surveys with the Arcminute MicroKelvin Imager 121
R. Kneissl and M. E. Jones

Internal Structure and Dynamics of Clusters

Tracing the Cluster Internal Dynamics with Member Galaxies 127
A. Biviano (Invited Speaker)

Segregation of Cluster Galaxies, and Cluster Substructure 139
P. Katgert, A. Biviano and T. Thomas

Merging Clusters in the Shapley Concentration 143
S. Bardelli, E. Zucca, T. Venturi, S. DeGrandi, S. Ettori and S. Molendi

Mass–to–Light Ratio of Galaxy Systems 147
M. Girardi, P. Manzato, M. Mezzetti, G. Giuricin and F. Limboz

Dark Matter in the Center of Galaxies and Galaxy Clusters: Ruling out the CDM Scenario? 151
E. D'Onghia, G. Chincarini, C. Firmani and D. Marchesini

ICM - Observations

X–ray Observations of the ICM Physics 157
M. Arnaud (Invited Speaker)

X-ray Statistical Properties of the Central Cool Component in Clusters of Galaxies 169
Y. Ikebe

A Pilot Study of the Tully-Fisher Relation in Cl0024+1654 at z=0.4 . . 173
A. J. Metevier, D. C. Koo and L. Simard

Cooling Flows 177
A. C. Fabian (Invited Speaker)

The New Emerging Model for the Structure of Cooling Cores in Clusters of Galaxies 187
H. Böhringer, K. Matsushita, Y. Ikebe and E. Churazov

The X-Ray Virial Relations for Relaxed Lensing Clusters Observed with Chandra 195
S. W. Allen, R. W. Schmidt and A. C. Fabian

Discovery of Holes in the Core of the Distant Galaxy Cluster RBS797 in a CHANDRA Observation 199
E. De Filippis, S. Schindler, A. Castillo-Morales, A. Schwope and J. Wambsganss

The Coma Cluster of Galaxies Observed by XMM-Newton 203
U. G. Briel, J. P. Henry, M. Arnaud and D. Neumann

Magnetic Fields in Clusters of Galaxies 207
F. Govoni, G. Giovannini, L. Feretti, D. Dallacasa and G. B. Taylor

Theory and Simulations

Frequently Asked Questions on the Luminosity–Temperature Relation in Groups and Clusters 213
F. Governato (Invited Speaker)

Simulations of Galaxy Clusters Observed by Chandra 220
A. Gardini

Evolution of Cluster Scaling Relations 224
P. Thomas

The Effect of Heating and Cooling on Scaling Laws of X-Ray Clusters . 228
O. Muanwong

Radiative Cooling in SPH Hydrodynamical Simulations of X-Ray Clusters 232
R. Valdarnini

Group Scaling Relations From a Cosmological Hydrodynamic Simulation: No Pre-heating Required? 236
R. Davé, N. Katz, L. Hernquist and D. Weinberg

The Thermal State of the ICM Probed with the SZ Effect 240
A. Cavaliere (Invited Speaker), A. Lapi and N. Menci

The Heating of the ICM: Energy Crisis and Viable Solutions 248
P. Tozzi

Blowing Bubbles in the ICM 253
V. Quilis, R. Bower and M. Balogh

Making the Connection: Feedback, X-rays and Galaxy Formation 257
R. G. Bower

Galaxy Population in Clusters

Formation and Evolution of Galaxies in Clusters 265
P. G. van Dokkum (Invited Speaker)

A Wide Field Survey of the Distant Rich Cluster Cl0024+1654 277
T. Treu, R. S. Ellis, P. Trivedi, J.-P. Kneib, A. Dressler, A. Oemler, P. Natarajan and I. R. Smail

Tracing Cosmic Evolution with the Coma Cluster 285
B. M. Poggianti, T. J. Bridges, B. Mobasher and D. Carter

Brightest Cluster Galaxy Formation in Low-Redshift Galaxy Clusters . 289
W. A. Barkhouse, H. K. C. Yee and O. López-Cruz

NIR and Optical Structural Parameters of Galaxies in the Cluster AC118 at z=0.31 293
G. Busarello, P. Merluzzi, M. Massarotti, F. La Barbera, M. Capaccioli and G. Theureau

The Dark Side of Star Formation in Galaxy Clusters: Spectroscopic Follow-up of Clusters Observed with ISOCAM 297
D. Fadda and P.-A. Duc

History of Mass Assembly and Star Formation in Clusters 301
T. Kodama, I. Smail, F. Nakata, S. Okamura and R. G. Bower

The Butcher-Oemler Effect in High Redshift X-ray Selected Clusters . . 305
D. A. Wake, C. A. Collins, B. W. Fairley, L. R. Jones, D. J. Burke, R. C. Nichol, K. A. Romer

Conference Summary

Conference Summary 311
R. S. Ellis (Invited Speaker)

Poster Session

A Dynamical Study of the Hot Galaxy Cluster 1E0657-56 329
R. Barrena, A. Biviano, M. Ramella, E. E. Falco and S. Seitz

New 2D Galaxy Systems within PDCS Fields 331
R. Barrena, E. Mediavilla, M. Ramella, M. Nonino, A. Biviano and W. Boschin

Deep Photographic Cluster Survey — DPCS 333
O. Baumann, H.-J. Röser, C. Tschamber, M. R. S. Hawkins and H. MacGillivray

Deprojection of Luminosity Functions of Galaxies in the Coma Cluster . 335
M. Beijersbergen, W. E. Schaap and J. M. van der Hulst

The Central X-Ray Structure of the Merging Cluster A3921 337
E. Belsole, J.-L. Sauvageot and H. Bourdin

Comparing the Field and Cluster Kormendy Relations 339
K. Blindert and H. K. C. Yee

A Serendipitous Cluster Survey in Chandra Archival Data 341
W. Boschin

The Properties of BCGs in X-Ray Selected Clusters 343
S. Brough, C. A. Collins, D. J. Burke, P. D. Lynam and R. G. Mann

The Cluster of Galaxies Abell 970 345
H. V. Capelato, L. Sodré Jr., D. Proust, G. B. Lima Neto, H. Cuevas, H. Quintana and P. Fouqué

Testing Cosmological Models with Negative Pressure 347
A. Cappi

Comparing the Star Formation in the Coma Cluster and the Field . . . 349
F. J. Castander, R. C. Nichol, A. Merrelli and the SDSS Collaboration

Distribution of Baryonic and Non-Baryonic Matter in Clusters of Galaxies 351
A. Castillo-Morales and S. Schindler

The TexOx Survey of Radio-Selected Galaxy Clusters 353
S. Croft, S. Rawlings, G. J. Hill, P. L. Gay and J. R. Tufts

Galaxy Morphological Segregation in Clusters: Local vs Global Conditions 355
M. Domínguez, H. Muriel and D. G. Lambas

The Atacama Submillimeter Telescope Experiment (ASTE) and Observations of Galaxy Clusters with Submillimeter Waves 357
H. Ezawa

Alignment of Main Axes of Galaxy Clusters 359
A. Faltenbacher and S. Gottlöber

WINGS: a Wide–field Imaging Nearby Galaxy clusters Survey 361
G. Fasano, D. Bettoni, C. Marmo, E. Pignatelli, B. M. Poggianti, M. Moles and P. Kjærgaard

New Evidences of Merging for the Galaxy Cluster Abell 521 363
C. Ferrari, S. Maurogordato, C. Benoist, E. Slezak, M. Arnaud and A. Cappi

Comparison of Subclustering Frequency Determined from 2D and 3D Data 365
P. Flin and J. Krywult

The Morphology-σ Relation in Hickson Compact Groups 367
P. Focardi and B. Kelm

ASCA Study of Dynamical Evolution in Nearby Clusters 369
T. Furusho

The Northern Sky Optical Cluster Survey 371
R. R. Gal and the DPOSS Team

Reacceleration of Relativistic Electrons in Cooling Flows: the Case of Perseus Cluster 373
M. Gitti, G. Brunetti, G. Setti

Statistics of Nearby Galaxy Groups from Volume Limited Catalogues: Algorithm and First Results. 375
G. Giudice

Constraints on $(\Omega_{\mathrm{m}}, \Omega_{\lambda})$ from Strong Lensing Clusters 377
G. Golse, J.-P. Kneib and G. Soucail

Tracing Cosmic Evolution with the Las Campanas Distant Cluster Survey 379
A. H. Gonzalez, J. J. Dalcanton, A. E. Nelson, L. Simard, D. Zaritsky and R. H. Wechsler

Mixed Dimentional Clustering Statistics 381
N. J. Horleston and S. Phillipps

The Energy Spectrum of Relativistic Electrons in the Coma Galaxy Cluster 383
C.-Y. Hwang and M.-Y. Tsay

Chandra Observation of the High-Redshift Galaxy Cluster MS 1054−0321 385
T. E. Jeltema, C. R. Canizares, M. W. Bautz, M. R. Malm, M. Donahue and G. P. Garmire

Searching for High-Redshift Clusters with *Planck Surveyor* 387
S. T. Kay

Far-Infrared Studies of Cluster Evolution with SIRTF 389
D. M. Kelly and G. H. Rieke

Statistics of Galaxies in Clusters 391
B. Kelm and P. Focardi

Non–Gaussian Morphology on Large Scales: Minkowski Functionals of the REFLEX Clusters of Galaxies 393
M. Kerscher, K. Mecke, P. Schuecker, H. Böhringer, L. Guzzo, S. De Grandi, C. A. Collins, S. Schindler and R. Cruddace

The Alignment Effect of Brightest Cluster Galaxies in the SDSS 395
R. S. J. Kim, J. Annis, M. A. Strauss, R. H. Lupton

Galaxy Cluster Mass Function Evolution Caused by Galaxy Mergings . 397
V. M. Kontorovich and S. I. Shelyag

Is Abell 970 a Recently Merged Cluster? 399
G. B. Lima Neto, H. V. Capelato and L. Sodré, Jr.

The BMW Deep X–Ray Cluster Survey 401
M. Longhetti, A. Moretti, L. Guzzo, S. Campana, M. R. Panzera, G. Tagliaferri, I. Dell'Antonio and D. Lazzati

Wide-Field Multicolour Imaging of Selected *REFLEX* Clusters 403
P. Lynam, H. Böhringer, P. Schuecker, T. Reiprich, E. Molinari, L. Guzzo, L. Barone, C. Collins, S. Schindler and D. Neumann

XMM–NEWTON Observation of Abell 1835 : Temperature, Mass and Gas Mass Fraction Profiles 405
S. Majerowicz, D. M. Neumann and T. H. Reiprich

Soft Cores in Late-Type Dwarf and LSB Galaxies from Hα Observations 407
D. Marchesini, E. D'Onghia, G. Chincarini, C. Firmani, P. Conconi, E. Molinari and A. Zacchei

3D Identification and Reconstruction of z $\sim$ 1 Clusters: Prospects for the DEEP2 Redshift Survey 409
C. Marinoni, M. Davis, J. Newman, A. Coil

The Colour-Magnitude Relation and the Age of Galaxies in the Cluster AC118 at z=0.31 413
P. Merluzzi, G. Busarello, M. Massarotti and F. La Barbera

A Deep Keck Spectral Survey of Cl0024+1654 at $z = 0.4$ 415
A. J. Metevier, D. C. Koo, L. Simard and G. D. Wirth

The Chemical Evolution of the ICM 417
A. Moretti, C. Chiosi and L. Portinari

The Radio Butcher-Oemler Effect: The Evolution of the Radio-selected Galaxy Population within Rich Clusters over the Last 5 Gyr . . . 419
G. E. Morrison, F. N. Owen, M. J. Ledlow, W. C. Keel, J. M. Hill and W. Voges

Galaxy Population in the Cluster of Galaxies around the Radio Galaxy 3C324 at $z = 1.2$ 421
F. Nakata, M. Kajisawa, T. Yamada, T. Kodama, K. Shimasaku, I. Tanaka, M. Doi, H. Furusawa, M. Hamabe, M.Iye, M. Kimura, Y. Komiyama, S. Miyazaki, S. Okamura, M.Ouchi, T.Sasaki, M. Sekigushi, M. Yagi and N. Yasuda

Cluster Structure in a Large Number of *ROSAT* and *ASCA* Clusters . . 423
N. Ota and K. Mitsuda

Deep ROSAT-HRI Observation of the cD Galaxy NGC 1399 in the Fornax Cluster: Morphology and Dynamical Status of the X-Ray Halo . . 425
M. Paolillo, G. Peres, G. Fabbiano and D.-W. Kim

The Luminosity Function of Clusters of Galaxies 427
I. Parolin, A. Moretti, E. Molinari and G. Chincarini

Recovering the Temperature Structure of Cooling–Flow Clusters 429
F. Pizzolato

ISOPHOT Results on Late-Type Virgo Cluster Galaxies 431
C. C. Popescu, R. J. Tuffs, H. J. Völk, D. Pierini and B. F. Madore

An XMM-Newton Mosaic Observation of A2163 433
G. W. Pratt, M. Arnaud and N. Aghanim

VIMOS IFU as a Tool to Observe the Cores of High z Clusters 435
D. Rizzo, on behalf of the VIRMOS Consortium

The Origin of the Density Profiles of Halos 437
E. Salvador-Solé and A. Manrique

The Structure of the Virgo Cluster from HI Data 439
T. Sanchis, J. M. Solanes and E. Salvador-Solé

The Nature of The Dwarf Population in Abell 868 441
R. M. Smith, P. J. Boyce, S. Phillipps, J. Bryn Jones, S. P. Driver and W. J. Couch

PCA of Cluster Surface Brightness Profiles 443
L. Spurgeon, S. Dos Santos, G. C. Stewart

Indices for SSPs with Enhanced Mix of α-Elements 445
R. Tantalo

Cluster Galaxy Populations from $z = 0.33-0.83$: Characterizing Evolution with CL1358+62 447
K. V. Tran, G. D. Illingworth and M. Franx

ISOPHOT Observations of Late-Type Virgo Cluster Galaxies 449
R. J. Tuffs, C. C. Popescu, H. J. Völk, D. Pierini, H. Hippelein, K. Leech, L. Metcalfe, I. Heinrichsen and C. Xu

Detection of Non-Random Patterns in Large-Scale Structure 451
R. Valdarnini

A Dynamical Modeling Technique for Clusters of Galaxies 453
E. Van Hese and H. Dejonghe

Radio Emission from the A3571 Cluster Complex in the Shapley Concentration 455
T. Venturi, S. Bardelli, M. Zagaria, I. Prandoni and R. Morganti

How to Pick up Interesting Clusters in X-Ray Surveys 457
P. T. P. Viana, A. R. Liddle, A. K. Romer and R. G. Mann

The L_x - σ Relation For Poor Groups of Galaxies 459
M. E. Zimer, J. S. Mulchaey and A. I. Zabludoff

Large-Scale Distribution and Spectral Properties of Galaxies in the Shapley Concentration 463
E. Zucca, S. Bardelli and A. Baldi

Author index 465

PREFACE

These proceedings are the written record of the Sesto–2001 Workshop on "Tracing Cosmic Evolution with Galaxy Clusters", which took place in Sesto Pusteria (Alto Adige/Südtirol, Italy) between the 3rd and the 6th of July 2001. This workshop, which follows the tradition of the meetings organized in Sesto during the 90s, has been organized by the Department of Astronomy of the University of Trieste (DAUT), the International School for Advanced Studies (SISSA–ISAS, Trieste), and the National Institute for Astrophysics (INAF) in Trieste.

The workshop has been dedicated to our friend and colleague Giuliano Giuricin, one of the "historical" organizers of the Sesto meetings, who suddently died few months ago, leaving a terrible void in his family, among his friends and in the scientific community.

The statistics of the workshop sound quite successful, with about 200 participants, 75 talks and about 100 posters: 10 review speakers gave overviews of the present state of the field and pointed out where significant developements are likely to occur; 12 targetted talks presented the most exciting results on cluster studies of the last year or two.

The aim of the Sesto–2001 Workshop was to provide a status-of-art overview on multiwavelength studies of galaxy clusters, from X–ray to UV, optical, near– and far-IR, sub-mm and radio bands. Two complementary aspects of cluster studies was emphasized. From the one hand, clusters of galaxies have been presented as powerful tracers of the structure of the Universe over the largest today accessible scales and over a sizeable fraction of the cosmological past light–cone. From the other hand clusters are powerful astrophysical laboratories, where to study the history of cosmic baryons and the processes of galaxy formation and evolution.

We believe that the conference was organized in the right moment. While the full exploitation of data from the past generation of X–ray satellites is coming to completion, Chandra and Newton–XMM are providing a significant change of perspective in our view of the complex physics taking place in clusters. Data from the Sloan Digital Sky Survey promise to provide the largest statistical sample of clusters in the nearby and moderately distant Universe. At the same time, detailed optical observations from 8m-class telescopes are allowing to trace the galaxy population in clusters out to $z \sim 1$ with unprecedented details.

We would like to take the opportunity to thank all the other members of the Organising Committee for their invaluable help in shaping a scientifically sound meeting: Hans Böhringer (MPE, Garching, Germany), Richard Bower (Durham, UK), Chris Collins (Liverpool, UK), Luigi Guzzo (Merate, Italy), and Piero Rosati (ESO, Garching, Germany). A special gratitude is due to our conference secretaries, Simonetta Fabrizio, Teresa Grione and Gabriella Schiulaz, for their efficient admnistrative skills, and to Benedikta Brunner (Tourist Office, Sesto Pusteria) and Wolf Holzer (Haus Sexten, Sesto Pusteria), for the efforts they spent to make the meeting running smoothly.

Stefano Borgani (INFN and DAUT, Trieste)
Marino Mezzetti (DAUT, Trieste)
Riccardo Valdarnini (SISSA, Trieste)

CONFERENCE PHOTOGRAPH

Conference participants

Steve ALLEN, Institute of Astronomy, University of Cambridge, Cambridge, UK ⟨ swa@ast.cam.ac.uk ⟩

Susan AMES , Radioastronomy Institute, Bonn University, Bonn, Germany ⟨ sames@astro.uni-bonn.de ⟩

James ANNIS, Experimental Astrophysics Group, Fermilab, Batavia, USA ⟨ annis@fnal.gov ⟩

Monique ARNAUD, Service d'Astrophysique, CEA Saclay, Gif sur Yvette, France ⟨ marnaud@discovery.saclay.cea.fr ⟩

Sandro BARDELLI, Osservatorio Astronomico di Bologna, Bologna, Italy ⟨ bardelli@bo.astro.it ⟩

Wayne BARKHOUSE, Department of Astronomy & Astrophysics, University of Toronto, Toronto, Canada ⟨ barkhous@astro.utoronto.ca ⟩

Jordi BARR, Astronomy Group, H.H. Wills Physics Laboratory, University of Bristol, Bristol, UK ⟨ j.barr@bristol.ac.uk ⟩

Rafael BARRENA, Instituto de Astrofisica de Canarias, La Laguna/Tenerife, Spain ⟨ rbarrena@ll.iac.es ⟩

James BARTLETT, Laboratoire d'Astrophysique, Observatoire Midi-Pyrenees, Toulouse, France ⟨ jim.bartlett@ast.obs-mip.fr ⟩

Spyros BASILAKOS, Astrophysics, Blackett Laboratory, Imperial College, London, UK ⟨ s.basilakos@ic.ac.uk ⟩

Marco BEIJERSBERGEN, Astronomy, Kapteyn Institute, Groningen, The Netherlands ⟨ beijersb@rendo.dekooi.nl, beijersb@astro.rug.nl ⟩

Elena BELSOLE, Service d'astrophysique/DAPNIA/DSM, CEA-Saclay, Gif sur Yvette, France ⟨ ebelsole@cea.fr ⟩

Philip BEST, Institute for Astronomy, Royal Observatory, Edinburgh, UK ⟨ pnb@roe.ac.uk ⟩

Andrea BIVIANO, Osservatorio Astronomico di Trieste, Trieste, Italy ⟨ biviano@ts.astro.it ⟩

Kris BLINDERT, Department of Astronomy and Astrophysics, University of Toronto, Toronto, Canada ⟨ blindert@astro.utoronto.ca ⟩

Hans BÖHRINGER, Max-Planck-Institut für extraterrestische Physik, Garching, Germany ⟨ hxb@mpe.mpg.de ⟩

Silvio BONOMETTO, Dipartimento di Fisica, Università di Milano-Bicocca, Milano, Italy ⟨ bonometto@mib.infn.it ⟩

Stefano BORGANI, INFN and Dipartimento di Astronomia, Università di Trieste, Trieste, Italy ⟨ borgani@ts.astro.it ⟩

Walter BOSCHIN, Dipartimento di Astronomia, Università di Trieste, Trieste, Italy ⟨ boschin@ts.astro.it ⟩

Richard BOWER, Dept. of Physics, University of Durham, Durham, UK ⟨ R.G.Bower@durham.ac.uk ⟩

Malcolm BREMER, Physics Department, Bristol University, Bristol, UK ⟨ m.bremer@bris.ac.uk ⟩

Sarah BRIDLE, Laboratoire d'Astrophysique, Observatoire Midi-Pyrenees, Toulouse, France ⟨ sarah@ast.obs-mip.fr ⟩

Ulrich G. BRIEL, Max-Planck-Institut für extraterrestrische Physik, Garching, Germany ⟨ ugb@xray.mpe.mpg.de, ugb@mpe.mpg.de ⟩

Sarah BROUGH, Astrophysics Research Institute, Liverpool John Moores University, Birkenhead, UK ⟨ sb@astro.livjm.ac.uk ⟩

Giovanni BUSARELLO, Osservatorio Astronomico di Capodimonte, Napoli, Italy ⟨ gianni@na.astro.it ⟩

Hugo Vicente CAPELATO, Divisao de Astrofisica, Instituto Nacional de Pesquisas Espaciais, Sao Jose dos Campos, Brazil ⟨ hugo@das.inpe.br ⟩

Alberto CAPPI, Osservatorio Astronomico di Bologna, Bologna, Italy ⟨ cappi@astbo3.bo.astro.it ⟩

Francisco Javier CASTANDER, Department of Astronomy, Yale University, New Haven, USA ⟨ fjc@astro.yale.edu, fjc@das.uchile.cl ⟩

Africa CASTILLO MORALES, Astrophysical Research Institute, Liverpool John Moores University, Birkenhead, UK ⟨ acm@astro.livjm.ac.uk ⟩

Alfonso CAVALIERE, Astrofisica, Dipartimento di Fisica, 2^a Università di Roma, Roma, Italy ⟨ cavaliere@roma2.infn.it ⟩

Guido CHINCARINI, Dipartimento di Fisica, Università degli Studi di Milano-Bicocca, Milano, Italy ⟨ guido@merate.mi.astro.it ⟩

Christopher COLLINS, Astrophysics Research Institute, Liverpool John Moores University, Birkenhead, UK ⟨ cac@astro.livjm.ac.uk ⟩

Nathan COURTNEY, University of Durham, Durham, UK ⟨ N.J.D.Courtney@durham.ac.uk ⟩

Steve CROFT, Department of Astrophysics, Oxford University, Oxford, UK ⟨ sdc@astro.ox.ac.uk ⟩

Romeel DAVÈ, Steward Observatory, Tucson, USA ⟨ rad@as.arizona.edu ⟩

Betty DE FILIPPIS, Astrophysics Research Institute, Liverpool John Moores University, Birkenhead, UK ⟨ bdf@astro.livjm.ac.uk ⟩

Sabrina DE GRANDI, Osservatorio Astronomico di Brera, Merate, Italy ⟨ degrandi@merate.mi.astro.it ⟩

Ricardo DEMARCO, ESO-European Southern Observatory, Garching bei München, Germany ⟨ rdemarco@eso.org ⟩

Elena D'ONGHIA, Osservatorio Astronomico di Brera, Merate, Italy ⟨ elena@merate.mi.astro.it ⟩

Mariano Javier DOMINGUEZ ROMERO, Grupo IATE, Observatorio Astronomico Cordoba, Cordoba, Argentina ⟨ mardom@mail.oac.uncor.edu ⟩

Harald EBELING, Institute for Astronomy, University of Hawaii, Honolulu, USA ⟨ ebeling@ifa.hawaii.edu ⟩

Richard S. ELLIS, Astronomy Department, California Institute of Technology, Pasadena, USA ⟨rse@astro.caltech.edu⟩

Stefano ETTORI, Insitute of Astronomy, University of Cambridge, Cambridge, UK ⟨settori@ast.cam.ac.uk⟩

Hajime EZAWA, Nobeyama Radio Observatory, National Astronomical Observatory of Japan, Nagano, Japan ⟨ezawa@nro.nao.ac.jp⟩

Andy FABIAN, Institute of Astronomy, University of Cambridge, Cambridge, UK ⟨acf@ast.cam.ac.uk⟩

Dario FADDA, Instituto de Astrofisica (IAC), La Laguna, Tenerife, Spain ⟨fadda@ll.iac.es⟩

Andreas FALTENBACHER, Astrophysikalisches Institut Potsdam, Potsdam, Germany ⟨afaltenbacher@aip.de⟩

Giovanni FASANO, Osservatorio Astronomico di Padova, Padova, Italy ⟨fasano@pd.astro.it⟩

Chiara FERRARI, CERGA, Observatoire de la Cote d'Azur, Nice, France ⟨ferrari@obs-nice.fr⟩

Piotr FLIN, Institute of Physics, Pedagogical University, Kielce, Poland ⟨sfflin@cyf-kr.edu.pl⟩

Paola FOCARDI, Dipartimento di Astronomia, Università di Bologna, Bologna, Italy ⟨pfocardi@astbo3.bo.astro.it⟩

Tae FURUSHO, Department of Physics, Tokyo Metropolitan University, Hachioji, Japan ⟨furusho@phys.metro-u.ac.jp⟩

Roy R. GAL, Department of Physics and Astronomy, Johns Hopkins University, Baltimore, USA ⟨rrg@pha.jhu.edu⟩

Alessandro GARDINI, Dipartimento di Astronomia, Università di Padova, Padova, Italy ⟨gardini@pd.astro.it⟩

Fabio GASTALDELLO, "G.Occhialini" Institute of Cosmic Physics, Milano, Italy ⟨gasta@ifctr.mi.cnr.it⟩

Joern GEISBUESCH, Cavendish Laboratory, Department of Physics, University of Cambridge, Cambridge, UK ⟨jg284@cam.ac.uk⟩

Simona GHIZZARDI, Istituto di Fisica Cosmica - CNR, Milano, Italy ⟨simona@ifctr.mi.cnr.it⟩

Roberto GILLI, Dept. of Physics and Astronomy, Johns Hopkins University, Baltimore, USA ⟨gilli@arcetri.astro.it⟩

Gabriele GIOVANNINI, Dipartimento di Fisica, Università di Bologna, Bologna, Italy ⟨ggiovann@ira.bo.cnr.it⟩

Marisa GIRARDI, Dipartimento di Astronomia, Università di Trieste, Trieste, Italy ⟨girardi@ts.astro.it⟩

Myriam GITTI, Dipartimento di Astronomia, Università di Bologna, Bologna, Italy ⟨mgitti@ira.bo.cnr.it⟩

Giuseppe GIUDICE, Dipartimento di Progettazione e gestione industriale, Università Federico II di Napoli, Napoli, Italy ⟨giudice@unina.it⟩

Mike GLADDERS, Dept. of Astronomy and Astrophysics, University of Toronto, Toronto, Canada ⟨ gladders@astro.utoronto.ca ⟩

Oleg GNEDIN, Institute of Astronomy, Cambridge University, Cambridge, UK ⟨ ognedin@ast.cam.ac.uk ⟩

Ghislain GOLSE, Laboratoire d'Astrophysique, Observatoire Midi-Pyrenees, Toulouse, France ⟨ ghislain.golse@ast.obs-mip.fr ⟩

Anthony H. GONZALEZ, Harvard-Smithsonian Center for Astrophysics; now: Department of Astronomy, University of Florida, Bryant Space Science Center, Gainesville, USA ⟨ anthony@astro.ufl.edu ⟩

Fabio GOVERNATO, Osservatorio Astronomico di Brera, Merate, Italy ⟨ fabio@merate.mi.astro.it ⟩

Federica GOVONI, Istituto di Radioastronomia, CNR Bologna, Bologna, Italy ⟨ fgovoni@ira.bo.cnr.it ⟩

Keith James Bruno GRAINGE, Cavendish Astrophysics, Cambridge University, Cambridge, UK ⟨ kjbg1@mrao.cam.ac.uk ⟩

Laura GREGO, Harvard-Smithsonian Center for Astrophysics, Cambridge, USA ⟨ lgrego@cfa.harvard.edu ⟩

Luigi GUZZO, Osservatorio Astronomico di Brera, Merate, Italy ⟨ guzzo@merate.mi.astro.it ⟩

Brad HOLDEN, University of California, Davis and Insitute for Geophysics and Planetary Physics, Livermore, USA ⟨ bholden@igpp.ucllnl.org ⟩

Nathan James HORLESTON, Wills Physics Laboratory, University of Bristol, Bristol, UK ⟨ nathan.horleston@bris.ac.uk ⟩

Chorng-Yuan HWANG, Graduate Institute of Astronomy, National Central University, Chung-Li, Taiwan ⟨ hwangcy@astro.ncu.edu.tw ⟩

Yasushi IKEBE, Extraterrestrische Physik, Max-Planck-Institut, Garching, Germany ⟨ ikebe@mpe.mpg.de ⟩

Tesla JELTEMA, Center for Space Research, M.I.T., Cambridge, USA ⟨ tesla@space.mit.edu ⟩

Peter KATGERT, Leiden Observatory, University of Leiden, Leiden, The Netherlands ⟨ katgert@strw.leidenuniv.nl ⟩

Scott T. KAY, Astronomy Centre, University of Sussex, Brighton, UK ⟨ S.T.Kay@sussex.ac.uk ⟩

Douglas KELLY, Steward Observatory, University of Arizona, Tucson, USA ⟨ dkelly@as.arizona.edu ⟩

Birgit KELM, Dipartimento di Astronomia, Università di Bologna, Bologna, Italy ⟨ kelm@astbo3.bo.astro.it ⟩

Daniel D. KELSON, The Observatories, Carnegie Institution of Washington, Pasadena, USA ⟨ kelson@ociw.edu ⟩

Martin KERSCHER, Sektion Physik, Ludwig-Maximilians-Universität, München, Germany ⟨ kerscher@theorie.physik.uni-muenchen.de ⟩

Rita S. J. KIM, Dept. of Physics and Astronomy, Johns Hopkins University, Baltimore, USA ⟨ rita@pha.jhu.edu ⟩

Ruediger KNEISSL, Astrophysics Group, Cavendish Laboratory, University of Cambridge, Cambridge, UK ⟨ rkneissl@mrao.cam.ac.uk ⟩

Taddy KODAMA, Department of Astronomy, University of Tokyo, Bunkyo-ku, Japan ⟨ kodama@astron.s.u-tokyo.ac.jp ⟩

David KOO, UCO/Lick Observatory, University of California, Santa Cruz, USA ⟨ koo@ucolick.org ⟩

Janusz KRYWULT, Institute of Physics, Pedagogical University, Kielce, Poland ⟨ krywult@pu.kielce.pl ⟩

Jaron KURK, Wiskunde en Natuurwetenschappen, Sterrewacht, Leiden, The Netherlands ⟨ kurk@strw.leidenuniv.nl ⟩

Francesco LA BARBERA, Osservatorio Astronomico di Capodimonte, Napoli, Italy ⟨ labarber@na.astro.it ⟩

Katy LANCASTER, AP Group, Cavendish Laboratory, University of Cambridge, Cambridge, UK ⟨ k.lancaster@mrao.cam.ac.uk ⟩

Gastao B. LIMA NETO, Departamento de Astronomia, Instituto Astronomico e Geofisico - USP, Sao Paulo, Brazil ⟨ gastao@iagusp.usp.br ⟩

Catarina LOBO, Centro de Astrofisica da Universidade do Porto, Porto, Portugal ⟨ lobo@astro.up.pt ⟩

Paul David LYNAM, X-ray Astronomy, M-P-I für Extraterrestrische Physik, Garching, Germany ⟨ pdl@xray.mpe.mpg.de ⟩

Andrea Valerio MACCIÒ, Dipartimento di fisica "G. Occhialini", Università Milano Bicocca, Milano, Italy ⟨ maccio@mib.infn.it ⟩

Manuela MAGLIOCCHETTI, Astrophysics, S.I.S.S.A., Trieste, Italy ⟨ maglio@sissa.it ⟩

Sebastien MAJEROWICZ, DAPNIA/SAp, CEA, Gif-sur-Yvette, France ⟨ smajerowicz@cea.fr ⟩

Danilo MARCHESINI, Osservatorio Astronomico di Merate-Brera, Merate, Italy ⟨ danilo@merate.mi.astro.it ⟩

Fabio MARDIROSSIAN, Dipartimento di Astronomia, Università di Trieste, Trieste, Italy ⟨ mardiros@ts.astro.it ⟩

Christian MARINONI, Astronomy Department, Berkeley University; now: Laboratoire d'Astrophysique de Marseille, Marseille, France ⟨ marinoni@astrsp-mrs.fr ⟩

Michele MASSAROTTI, Osservatorio Astronomico di Capodimonte, Napoli, Italy ⟨ michele@na.astro.it ⟩

Chiara MASTROPIETRO, Osservatorio Astronomico di Merate-Brera, Merate, Italy ⟨ chiara@merate.mi.astro.it ⟩

Kyoko MATSUSHITA, Max-Planck Institute für Extraterrestrische Phisik, Garching, Germany ⟨ matusita@xray.mpe.mpg.de ⟩

Hiroshi MATSUO, Advanced Technology Center, National Astronomical Observatory of Japan, Tokyo, Japan ⟨ h.matsuo@nao.ac.jp ⟩

Sophie MAUROGORDATO, CERGA, Observatoire de la Cote d'Azur, Nice, France ⟨ maurogor@purcell.obs-nice.fr ⟩

Amata MERCURIO, Dipartimento di Astronomia, Università di Trieste, Trieste, Italy ⟨ mercurio@ts.astro.it ⟩

Paola MERLUZZI, Osservatorio Astronomico di Capodimonte, Napoli, Italy ⟨ merluzzi@na.astro.it ⟩

Anne METEVIER, UCO/Lick Observatory, University of California, Santa Cruz, USA ⟨ anne@ucolick.org ⟩

Marino MEZZETTI, Dipartimento di Astronomia, Università di Trieste, Trieste, Italy ⟨ mezzetti@ts.astro.it ⟩

Silvano MOLENDI, Istituto di Fisica Cosmica, CNR, Milano, Italy ⟨ silvano@ifctr.mi.cnr.it ⟩

Alessia MORETTI, Dipartimento di Astronomia, Università di Padova, Padova, Italy ⟨ moretti@pd.astro.it ⟩

Alberto MORETTI, Osservatorio Astronomico di Merate-Brera, Merate, Italy ⟨ moretti@merate.mi.astro.it ⟩

Glenn MORRISON, IPAC, Caltech, Pasadena, USA ⟨ morrison@ipac.caltech.edu ⟩

Lauro MOSCARDINI, Dipartimento di Astronomia, Università di Padova, Padova, Italy ⟨ moscardini@pd.astro.it ⟩

Orrarujee MUANWONG, Astronomy Centre, University of Sussex, Brighton, UK ⟨ muanwong@astr.cpes.susx.ac.uk ⟩

Volker MÜLLER, Astrophysical Institute Potsdam, Potsdam, Germany ⟨ vmueller@aip.de ⟩

Fumiaki NAKATA, Dept. of Astronomy, University of Tokyo, Tokio, Japan ⟨ nakata@astron.s.u-tokyo.ac.jp ⟩

Jeffrey NEWMAN, Dept. of Astronomy, University of California at Berkeley, Berkeley, USA ⟨ jnewman@astro.berkeley.edu ⟩

Robert NICHOL, Dept. of Physics, Carnegie Mellon, Pittsburgh, USA ⟨ nichol@cmu.edu ⟩

Colin NORMAN, Space Telescope Science Institute, Baltimore, USA ⟨ norman@stsci.edu ⟩

Naomi OTA, Dept. of Physics, Tokyo Metropolitan University, Hachiouji, Japan ⟨ naomi@phys.metro-u.ac.jp ⟩

Maurizio PAOLILLO, Dipartimento di Scienze Fisiche ed Astronomiche, Università di Palermo, Palermo, Italy ⟨ paolillo@astropa.unipa.it ⟩

Ilaria PAROLIN, Università degli Studi di Milano and Osservatorio Astronomico di Brera-Merate, Milano, Italy ⟨ parolin@merate.mi.astro.it ⟩

Kristian PEDERSEN, Astronomical Observatory, University of Copenhagen, Copenhagen, Denmark ⟨ kp@astro.ku.dk ⟩

Laura PENTERICCI, Max Plank Institute für Astronomie, Heidelberg, Germany ⟨laura@mpia-hd.mpg.de⟩

Elena PIERPAOLI, Dept. of Physics & Astronomy, UBC, Lawrenceville, USA ⟨elena@geop.ubc.ca⟩

Fabio PIZZOLATO, "G.Occhialini" Institute of Cosmic Physics, Milano, Italy ⟨fabio@ifctr.mi.cnr.it⟩

Manolis PLIONIS, Institute of Astronomy & Astrophysics, National Observatory of Athens, Athens, Greece ⟨plionis@sapfo.astro.noa.gr⟩

Bianca Maria POGGIANTI, Osservatorio Astronomico di Padova, Padova, Italy ⟨biancap@pd.astro.it⟩

Cristina C. POPESCU, The Observatories of the Carnegie Institution of Washington, Pasadena, USA ⟨popescu@ociw.edu⟩

Marc POSTMAN, Space Telescope Science Institute, Baltimore, USA ⟨postman@stsci.edu⟩

Gabriel PRATT, Service d'Astrophysique, CEA/Saclay, Gif-sur-Yvette, France ⟨gwp@discovery.saclay.cea.fr⟩

Vicent QUILIS, Department of Physics, Durham University, Durham, UK ⟨vicent.quilis@durham.ac.uk⟩

Jesper RASMUSSEN, Astronomical Observatory, Niels Bohr Institute, Copenhagen East, Denmark ⟨jr@astro.ku.dk⟩

Steve RAWLINGS, Dept. of Astrophysics, Oxford University, Oxford, UK ⟨s.rawlings1@physics.ox.ac.uk⟩

Joerg RETZLAFF, Max-Planck-Institut für extraterrestrische Physik, Garching, Germany ⟨jrl@mpe.mpg.de⟩

Davide RIZZO, Osservatorio Astronomico di Merate-Brera, Merate, Italy ⟨rizzo@merate.mi.astro.it⟩

Hermann-Josef RÖSER, Max Plank Institute fur Astronomie, Heidelberg, Germany ⟨roeser@mpia-hd.mpg.de⟩

Kathy ROMER, Physics Department, Carnegie Mellon University, Pittsburgh, USA ⟨romer@cmu.edu⟩

Piero ROSATI, ESO - European Southern Observatory, Garching, Germany ⟨prosati@eso.org⟩

José María ROZAS AMADOR, Dpt. d'Astronomia i Meteorologia, Universitat de Barcelona, Barcelona, Spain ⟨jrozas@am.ub.es⟩

Eduard SALVADOR-SOLÉ, Dept. Astronomia i Meteorologia, Fac. de Fisica, Univ. de Barcelona, Barcelona, Spain ⟨eduard@am.ub.es⟩

Teresa SANCHIS, Dept. Astronomia i Meteorologia, Fac. de Fisica, Univ. de Barcelona, Barcelona, Spain ⟨tsanchis@am.ub.es⟩

Sabine SCHINDLER, Astrophysics Research Institute, Liverpool John Moores University, Birkenhead, UK ⟨sas@staru1.livjm.ac.uk⟩

Robert SCHMIDT, Institute of Astronomy, University of Cambridge, Cambridge, UK ⟨rschmidt@ast.cam.ac.uk⟩

Peter SCHUECKER, Max-Planck Institut für extraterrestrische Physik, Garching, Germany ⟨ peters@mpe.mpg.de ⟩

Uros SELJAK, Dept. of Physics, Jadwin Hall, Princeton University, Princeton, USA ⟨ useljak@princeton.edu ⟩

Sergey SHELYAG, Radio Astronomical Institute, Kharkov, Ukraine ⟨ sergey_shelyag@mail.ru ⟩

Rodney SMITH, Department of Physics and Astronomy, University of Wales, College of Cardiff, Cardiff, UK ⟨ R.Smith@astro.cf.ac.uk ⟩

Louise SPURGEON, Department of Physics and Astronomy, University of Leicester, Leicester, UK ⟨ les@star.le.ac.uk ⟩

Gordon STEWART, Department of Physics and Astronomy, University of Leicester, Leicester, UK ⟨ gcs@star.le.ac.uk ⟩

Rosaria TANTALO, Dipartimento di Astronomia, Università di Padova, Padova, Italy ⟨ tantalo@pd.astro.it ⟩

Gilles Paul THEUREAU, LPCE, CNRS - Paris Observatory, Orleans, France ⟨ theureau@cnrs-orleans.fr ⟩

Peter THOMAS, Astronomy Centre, CPES, University of Sussex, Brighton, UK ⟨ p.a.thomas@sussex.ac.uk ⟩

Sune TOFT, Astronomical Observatory, Copenhagen University, Copenhagen, Denmark ⟨ toft@astro.ku.dk ⟩

Giuseppe TORMEN, Dipartimento di Astronomia, Università di Padova, Padova, Italy ⟨ tormen@pd.astro.it ⟩

Luca TORNATORE, Dipartimento di Astronomia, Università di Trieste, Trieste, Italy ⟨ tornatore@ts.astro.it ⟩

Paolo TOZZI, Osservatorio Astronomico di Trieste, Trieste, Italy ⟨ tozzi@ts.astro.it ⟩

Kim-Vy TRAN, Dept. of Astronomy & Astrophysics, UC Santa Cruz, Santa Cruz, USA ⟨ vy@ucolick.org ⟩

Tommaso TREU, Dept. of Astronomy, California Institute of Technology, Pasadena, USA ⟨ tt@astro.caltech.edu ⟩

Dario TREVESE, Dipartimento di Fisica, Università di Roma "La Sapienza", Roma, Italy ⟨ dario.trevese@roma1.infn.it ⟩

Richard J. TUFFS, Max Planck Institut für Kernphysik, Heidelberg, Germany ⟨ Richard.Tuffs@mpi-hd.mpg.de ⟩

Riccardo VALDARNINI, Astrophysics, S.I.S.S.A., Trieste, Italy ⟨ valda@sissa.it ⟩

Roeland P. VAN DER MAREL, Space Telescope Science Institute, Baltimore, USA ⟨ marel@stsci.edu ⟩

Rien VAN DE WEYGAERT, Kapteyn Astronomical Institute, University of Groningen, Groningen, The Netherlands ⟨ weygaert@astro.rug.nl ⟩

Pieter VAN DOKKUM, Astronomy Dept, Caltech, Pasadena, USA ⟨ pgd@astro.caltech.edu ⟩

Emmanuel VAN HESE, Dept. of Mathematical Physics and Astronomy, Ghent University, Ghent, Belgium ⟨emmanuel@heze.rug.ac.be⟩

Bram Pieter VENEMANS, Wiskunde en Natuurwetenschappen, Sterrewacht, Leiden, The Netherlands ⟨venemans@strw.leidenuniv.nl⟩

Licia VERDE, Dept. Astrophysical Sciences Peyton Hall, Princeton University, Princeton, USA ⟨lverde@astro.princeton.edu⟩

Pedro T. P. VIANA, Centro de Astrofisica da U.P., Porto, Portugal ⟨viana@astro.up.pt⟩

David WAKE, Astrophysics Research Institue, Liverpool John Moores University, Birkenhead, UK ⟨daw@astro.livjm.ac.uk⟩

Ann I. ZABLUDOFF, Dept. of Astronomy, University of Arizona, Tucson, USA ⟨azabludoff@as.arizona.edu⟩

Dennis ZARITSKY, Steward Observatory, University of Arizona, Tucson, USA ⟨dennis@ngala.as.arizona.edu⟩

Marc ZIMER, Max Plank Institute für Astronomie, Heidelberg, Germany ⟨zimer@mpia-hd.mpg.de⟩

Valentina ZITELLI, Osservatorio Astronomico di Bologna, Bologna, Italy ⟨zitelli@bo.astro.it⟩

Elena ZUCCA, Osservatorio Astronomico di Bologna, Bologna, Italy ⟨zucca@bo.astro.it⟩

CLUSTERS AT HIGH REDSHIFT

Tracing Cosmic Evolution with Galaxy Clusters
ASP Conference Series, Vol. 268, 2002
S. Borgani, M. Mezzetti, and R. Valdarnini, eds.

Multi-wavelength Surveys for Distant Clusters

Marc Postman

Space Telescope Science Institute, 3700 San Martin Drive, Baltimore, MD 21208, U.S.A.

Abstract. Given the importance of clusters to the fields of cosmology and galaxy evolution, it is critical to understand how the cluster detection process affects (biases) ones scientific conclusions derived from a given cluster sample. I review the astrophysics, algorithms, and observational constraints that must be considered when attempting to assess cluster selection biases and their impact on the scientific constraints derived from multi-wavelength cluster surveys. In particular, one can more accurately quantify and understand the selection biases and assess cluster evolution when using joint optical/NIR/x-ray surveys than can be achieved when employing any single cluster survey.

1. Introduction

It is well appreciated by this audience that clusters of galaxies are important as tracers of mass on intermediate and large scales and as laboratories for exploring galaxy evolution. Recently, the application of coordinated multi-wavelength observations has significantly enhanced our understanding of the physics of cluster formation and evolution. In particular, the use of multi-wavelength data

- enables studies of a broad range of physical processes which control cluster and cluster galaxy evolution,
- expands the accessible redshift range,
- enhances detection of distant large-scale structure (e.g., intercluster filaments) via joint passband detection methods, and
- minimizes false positive detections and improves ones understanding of sample selection biases.

Now that the era of objective cluster detection across wide wavelength ranges is upon us, it is critical to ask **"How does the cluster detection process affect (bias) the scientific conclusions?"** This is an important question because *different types of clusters* can be found by different algorithms and/or at different wavelengths. The different detection outcomes are a consequence of a cluster detection S/N ratio that is generally a function of redshift, survey passband, and the mathematical details and statistical properties of the detection algorithm applied.

The wavelength-dependent differences in the properties of clusters are directly related to the physical phenomena that dominate the clusters "visibility." In the x-ray passbands (0.1 – 10 keV), it is the hot intracluster medium (ICM) that is being detected. The ICM in clusters typically shines with a luminosity in the range $43 < \log(L_X) < 45.5$. A luminous, centrally condensed ICM usually indicates a cluster in a state of dynamical equilibrium with a fairly deep gravitational potential. In the optical and near-IR passbands (0.6 – 2.2μ), it is one or more of the following that is being detected: a large overdensity of galaxies, a population of early-type galaxies with a narrow optical/NIR color range, and/or the distortion of the images of background galaxies due to gravitational lensing by the cluster. A large galaxy overdensity does not necessarily imply a well-relaxed or massive cluster. Hence, optical/NIR based surveys may include systems of significantly lower mass and/or age than those found in x-ray based surveys. The spectral energy distribution of most clusters peak in the optical/NIR with luminosities in the range $44 < \log(L_{opt}) < 46.5$. The millimeter wavebands (200 – 300 GHz) are ideal for detection of the Sunyaev-Zeldovich effect. The detection of the SZ effect is independent of redshift in as much as the electron density in the ICM and the effective electron temperature is independent of redshift (i.e., if n_e and T_e are functions of time, then so will be the amplitude of the SZ decrement). SZ based surveys have great potential for detecting systems in a quite homogeneous manner out to $z > 1$. However, the presence of a significant ICM is obviously required. At the longest wavelengths (1.4 – 30 GHz), detection of radio loud cluster galaxies at very high redshifts and detection of galaxies with bent radio lobes (suggesting passage through an ICM) can be achieved. Because radio-loud cluster galaxies may not always be a by-product of cluster formation, radio-based cluster surveys have significant selection biases. They are, nonetheless, capable of detecting systems at the highest redshifts ($z > 2$).

2. Astronomical & Observational Considerations

In addition to a variety of astrophysical parameters that dictate the area, depth, and geometry of a distant cluster survey, there are also several important observational considerations that must be accounted for. An obvious one is the much greater availability of optical/NIR facilities relative to x-ray telescopes. This makes completing large area surveys in the optical/NIR much easier. Competing with this is the fact that the detection of the ICM is a relatively unambiguous indicator of the physical reality of the cluster candidate. The pros and cons of performing cluster surveys in x-ray, optical, and SZ bandpasses are summarized in the table below.

The lack of a significant background in the x-rays is a key advantage of looking for clusters in this wavelength range. Figure 1 shows a beautiful example of this in the form of Chandra/ACIS imaging of the Lynx supercluster complex at $z \sim 1.3$ (Stanford *et al.* 2001). However, such x-ray images require a large investment of telescope time (190 ksec in the example here). The same clusters can be easily detected in NIR images with $\frac{1}{10}$th the exposure time. In addition, the presence of a significant background in the optical/NIR when searching for clusters at intermediate and high redshifts has lead to the development of several smart algorithms which selectively suppress the background and, in doing so,

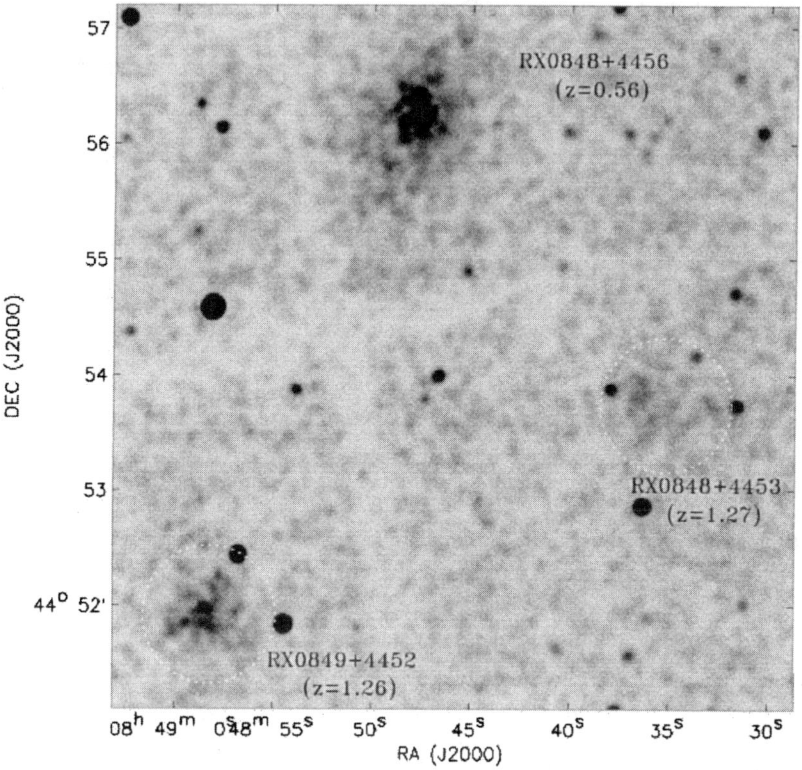

Figure 1. A 190 ksec Chandra/ACIS image of the Lynx supercluster at $z = 1.27$. Figure taken from Stanford *et al.* (2001).

yield relatively accurate ($|\delta z|/z \lesssim 0.1$) redshift estimates for the cluster candidates. Having these redshift estimates greatly accelerates the science that can be extracted from a survey conducted primarily[1] in two dimensions. Note that the $z = 1.27$ Lynx cluster in Figure 1, CL0848+4453, was initially discovered in a NIR image (Stanford *et al.* 1997).

Figure 2 summarizes the cluster survey area–depth parameter space covered to date (with a few near-future surveys included). This figure is modeled after that in Ebeling et al. (2001) but includes optical and NIR surveys as well. Survey flux limits have been converted to common cgs units for comparison across a broad wavelength range. Although the primary science motive for each listed survey is not always the study of cluster evolution, each one is well suited to the search for distant clusters. Within a given passband, the survey flux limit is an indicator of the effective redshift depth of the survey. However, in order to directly compare effective depths across a large spectral range one must also

[1] I say primarily because the use of color information can provide a quasi-third dimension that eliminates projection effects on scales larger than $\sim 10,000$ km s^{-1}.

understand how the cluster and galaxy luminosity functions vary with wavelength. Fortunately, this is relatively well known. The characteristic luminosity of clusters in the x-ray (0.5 – 2 keV), and the characteristic luminosity of cluster galaxies in the optical (0.8μ), and NIR (2.2μ) passbands are plotted for $z = 0.15$ and $z = 1$. Also shown are curves which indicate the areas and depths at which one would expect to find 10 and 100 clusters, respectively, with $L_x > 5 \times 10^{44}$ erg s^{-1} at any z (Ebeling *et al.* 2001).

X-ray Surveys	Optical/NIR Surveys	SZ (mm) Surveys
Detection of ICM is an unambiguous indicator of a real cluster	Large areas can be surveyed relatively quickly	Detection of decrement requires ICM, an unambiguous indicator of a real cluster
High contrast ($L_x \propto n_e^2$)	Many telescopes available	SZ signal is z independent (modulo ICM evolution)
Background is negligible	High QE detectors	Limited telescope access
Low (10%) spurious rate	Many algorithms produce cluster redshift estimate	Large area surveys are time intensive but upgrades to BIMA/OVRO could cut time by factor of 10
Limited telescope access	30% spurious rate (in 2D)	
Large area surveys are time intensive	Background is significant, need "smart" algorithms	
Requires optical/NIR follow-up for redshifts	Optical richness - cluster mass relation is noisy	

Recent deep optical/NIR surveys are now reaching areas in excess of 100 square degrees enabling the discovery of the richest clusters at $z \sim 1$ and the study of the evolution of the moments of the cluster distribution. In this regard, optical/NIR surveys still provide the largest numbers of distant ($z > 0.6$) cluster candidates. However, new untargetted SZ surveys and very wide area x-ray surveys (*e.g.*, MACS) provide substantial potential for the creation of distant cluster catalogs with very low contamination rates. The NOAO Deep Wide survey (Jannuzi et al. 2000) and the proposed *Primordial Explorer* (see http://prime.pha.jhu.edu) mission will provide superb datasets, enabling detection and study of clusters out to $z \sim 2$.

An optimal cluster survey strategy would be to take advantage of the benefits of each particular passband by observing a given region of sky in optical/NIR and x-rays. At the least, this will yield important constraints on differences in the selection functions at these wavelengths. At best, this will produce a comprehensive set of data with which to study cluster evolution. While this is currently difficult to do to sufficient depth to study distant clusters over a large area of sky, it is feasible over regions of a few square degrees. Indeed, several teams

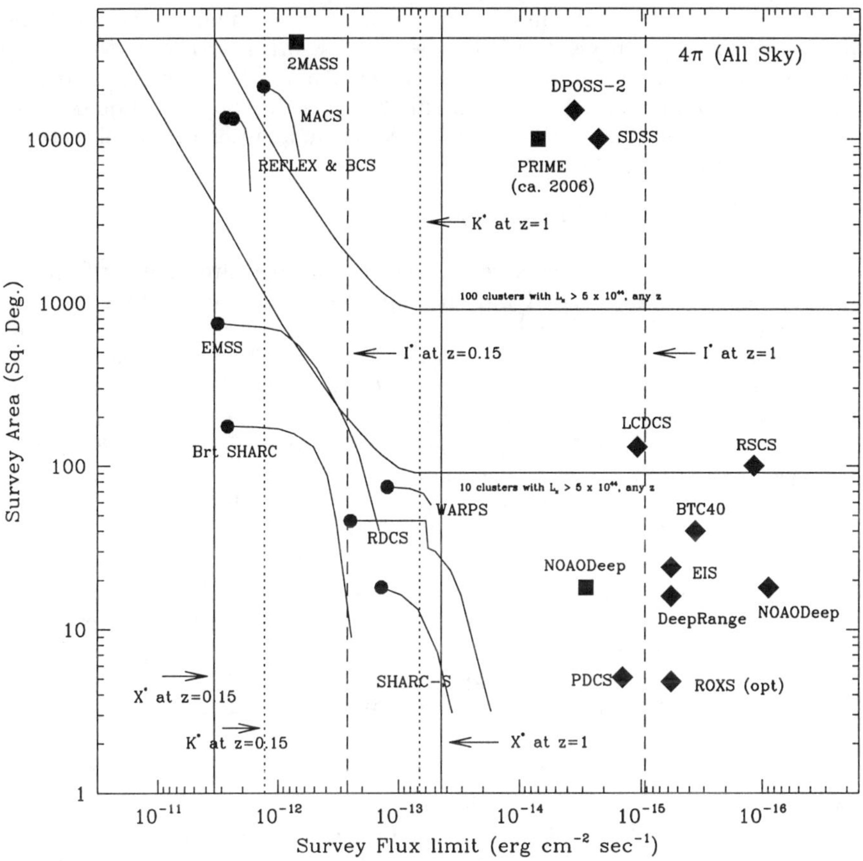

Figure 2. Survey flux limit versus survey area for a range of sky survey projects either completed or slated for completion in the near future. X-ray surveys are shown as large circles with trailing curves, to reflect the decreasing sensitivity of x-ray telescopes as targets move off-axis. NIR surveys are shown as squares, optical surveys are shown as diamonds. Vertical lines are the characteristic luminosities of clusters in the x-ray (0.5 – 2 keV) and of cluster galaxies in the optical (0.8μ) and NIR (2.2μ) passbands at $z = 0.15$ and $z = 1$. Data points in this figure are adopted from information provided in Postman *et al.* 1996; Postman *et al.* 1998; Jannuzi et al. 2000; de Propris et al. 1999; Voit & Donahue 1999; and Ebeling et al. 2001

have recently done precisely such studies either using existing x-ray data and re-imaging in the optical (Donahue *et al.* 2001) or using existing optical data and searching in x-ray archives for the corresponding high energy imaging (Holden *et al.* 1999). I review the conclusions reached from these two studies later in this talk to demonstrate the important constraints that multi-wavelength distant cluster surveys can yield. First, however, I will briefly review the cluster detection algorithms in use today, emphasizing some of the wavelength dependent pros and cons of each, as well as highlighting the importance of quantifying the selection function prior to attempting scientific analysis of any objectively derived cluster sample.

3. Cluster Detection Algorithms

The advancement of cluster detection algorithms has benefited greatly from the adaptation of well-known signal and image processing techniques to astronomical use. These include matched filters (*e.g.,* Postman *et al.* 1996, Kawasaki *et al.* 1998, Kepner *et al.* 1999, Lobo *et al.* 2000, Gladders & Yee 2000) and adaptive wavelet transforms (*e.g.,* Rosati *et al.* 1995). Other methods come to astronomy from the realm of multi-parametric statistical and clustering analysis such as the Voronoi Tessellation Technique (*e.g.,* Scharf *et al.* 1997, Ramella *et al.* 2001). One characteristic of distant clusters, almost regardless of observational bandpass, is that they are relatively low contrast features[2]. A key property of many of the above methods is, thus, to take the basic input data (*e.g.,* , an x-ray image or a galaxy catalog) and process it in a manner that significantly enhances the contrast. For example, the matched filter algorithm developed by Postman *et al.* (1996), and modified further by Kepner *et al.* (1999), enhances the contrast of a $z \sim 0.8$ cluster observed in the $I-$band sufficiently to turn a 2.4σ fluctuation in raw galaxy counts into a 4.5σ fluctuation in the matched filter signal. Similarly, the color cut used by Gladders & Yee (2000) in their red-sequence detection method can transform a 2σ fluctuation in galaxy counts (due to a $z \sim 1$ cluster) into a 5σ fluctuation. Most of the methods above can be applied at a wide range of redshifts and over a broad spectral range. Exceptions include methods which rely on cluster characteristics that span a limited spectral range (*e.g.,* the red galaxy sequence) or are best applied in the optical/NIR (*e.g.,* detection of gravitational lensing). Because each method comes with advantages and disadvantages, using multiple methods is desirable. Multiple algorithm application can assure high completeness across a broad range of cluster properties.

Indeed, the use of multiple algorithms highlights the importance of determining an accurate selection function. Figure 3, adopted from the work of Goto *et al.* 2001, shows an example of the differences in the richness and redshift distributions of cluster candidates identified using 4 different detection methods applied to identical data. The input data was the early release data from the Sloan Digital Sky Survey. The differences in the resulting cluster catalogs can be understood as a consequence of the different selection function each algorithm

[2]The contrast of a distant cluster in a survey is often low because the survey goal is to cover a modest to large area at the expense of depth. The contrast of the same cluster in a given bandpass can, of course, be increased by increasing the exposure time as in the Chandra/ACIS example in Figure 1.

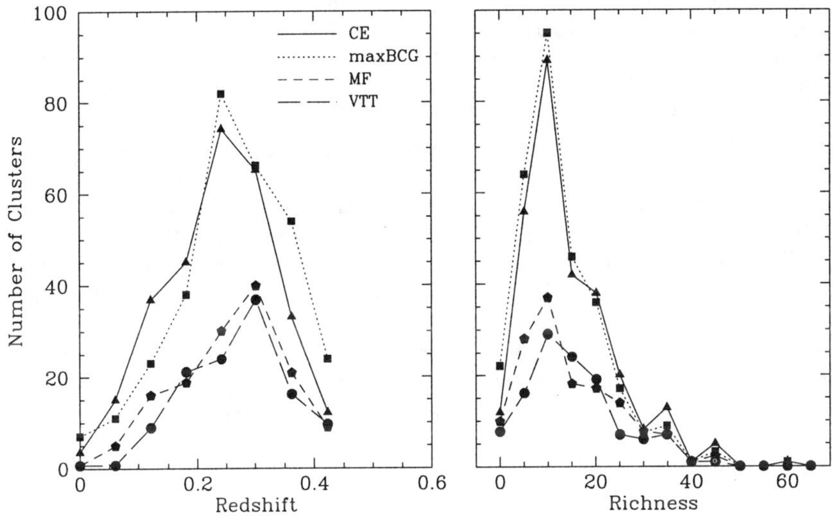

Figure 3. Number of clusters detected as functions of redshift and richness for 4 different search algorithms. The input data to each algorithm was identical and is taken from the SDSS early data release. This figure is adopted from the work of Goto *et al.* 2001. The 4 detection methods are: CE (cut & enhance algorithm which weights color information in a special way to increase the contrast of a cluster), maxBCG (searches for the red sequence and the BCG), MF (Matched filter algorithm), VTT (Voronoi Tessellation Technique). The histograms have NOT been corrected for the different selection functions intrinsic to each algorithm.

yields. Accurate interpretation of cluster data is thus critically dependent on accurate estimation of the relevant selection function.

An example of how cluster selection might affect scientific conclusions is a comparison of the recent work of Dressler *et al.* 1999 and Poggianti *et al.* 1999 (MORPHS) versus that of Balogh *et al.* 1999 and Ellingson *et al.* 2001 (CNOC). The CNOC study used 15 x-ray selected ($L_x > 2 \times 10^{44}$ erg s^{-1}) clusters in the redshift range $0.18 < z < 0.55$. These clusters are mostly rich, relaxed systems. The CNOC clusters also exhibit a prominent E/S0 red sequence in their color-magnitude diagrams. The 10 MORPHS clusters were primarily optically selected and lie in a similar redshift range, $0.37 < z < 0.56$. They span a wide range of mass and L_x, with 40% being classified as "irregular" in their overall morphology (as traced by the galaxy distribution). The CNOC clusters are reported to contain a low k+A galaxy fraction of 4.5% and after correction for scatter this value decreases to 1.5%. The field k+A fraction, for comparison, is 1 – 2%. In contrast, the MORPHS survey finds a significantly larger k+A fraction of $\sim 20\%$ and a significant excess of post-starburst galaxies. The MORPHS study also concludes that star formation in clusters is truncated relatively quickly (~ 1 Gyr). The CNOC study finds no excess starburst or post-starburst activity relative to the field and concludes that cluster galaxy star formation is truncated gradually over a 2 – 3 Gyr period. How can these two studies reach such different conclusions given the similarity in sample size and redshift range? The answer may, in part, lie in the different selection biases associated with optical vs. x-ray selection. X-ray luminous clusters at $z \lesssim 0.5$ tend to be dynamically older systems – as indicated by the existence of a well-established and centrally-compact intracluster medium. Their galaxy populations, at least within the central $500h^{-1}$ kpc, are dominated by early type galaxies with a narrow range in optical color. While such systems would easily be detected in optical surveys, systems that are dynamically younger may often not satisfy the minimum x-ray luminosity constraint and would thus be systematically excluded from an x-ray selected survey. Infall of late type galaxies into clusters may be more prominent in younger clusters and, hence, so to the presence of starburst and post-starburst objects. While one cannot attribute all differences between the CNOC and MORPHS survey results to the cluster selection process (*e.g.*, some of the differences are likely due to differences in the data analysis or data quality), it is clearly an important factor.

4. Case Studies

To better understand the consequences of using x-ray and optical selection criteria, as well as to improve constraints on cluster evolution in general, it is essential to study a common region of sky in multiple wavelengths. Two recent works, the CFHT Optical PDCS Survey (COPS; Holden *et al.* 1997, 1999, 2000; Adami *et al.* 2000) and the ROSAT Optical X-ray Survey (ROXS; Donahue *et al.* 2001), have done just this and have reached some interesting conclusions which are worth reviewing here.

In the COPS study, the ROSAT archive was searched for PSPC data with exposure times of 3 ksec or more and lying within 40 arcminutes of the centers of 31 optically selected clusters from the Palomar Distant Cluster Survey. The x-ray data were then processed using the SHARC source detection software

(Romer *et al.* 2000). A cumulative x-ray luminosity function (XLF) was derived from the resulting extended source catalog and compared with previous XLFs from Burns *et al.* (1996), Ebeling *et al.* (1997), Burke *et al.* (1997), and Rosati *et al.* (1998). Of the 31 PDCS clusters studied, 7 (23%) were detected in the x-ray data. The cumulative XLF of both the detections and the upper limits are consistent with those from previous x-ray selected samples. Holden *et al.* conclude that *optical selection does not appear to miss a significant number of x-ray emitting clusters within a given area* and that optical selection finds both intrinsically luminous and intrinsically faint ($L_x \leq 10^{43}$ erg s^{-1}) x-ray clusters. Of course, the errors on the derived XLF in the COPS study are substantial given the small sample. Further, while this effort was the first to systematically explore objectively derived optical and x-ray cluster samples it was not a double blind experiment – the existence of an optically selected cluster was the trigger for the ROSAT archive search.

The ROXS study takes the spirit of the COPS study and raises it one level higher by performing completely independent optical and x-ray cluster detection over a common 4.8 deg^2 area. The ROXS x-ray data are 23 deep ($T_{exp} > 8$ ksec) ROSAT PSPC pointings. Optical images in the V and I passbands were obtained for each ROSAT pointing. An adaptive wavelet algorithm (Rosati *et al.* 1995) was applied to the ROSAT images to detect extended sources. A matched filter algorithm was applied to the optical data to detect clusters and derive an optical richness estimate, Λ_{CL}, which corresponds to an effective optical luminosity. The availability of complete x-ray and optical data over the same sky area enabled the ROXS team to generate entirely independent cluster catalogs. The two catalogs were compared in a variety of ways and the following key results derived:

- 57 x-ray detections and 155 optical detections were found. 72% of the x-ray candidates are co-identified in the optical survey. Of the remaining 28%, one third are bona fide optically faint candidates. The rest are systems with poor x-ray flux measurements.

- Most $\Lambda_{CL} > 30$ optically selected clusters are real *if* the $L_x \propto \Lambda_{CL}^{\beta}$ relationship is steep. The best fit is $\beta = 3.8 \pm 0.8$.

- *There is no need to hypothesize a sizeable population of optically rich, x-ray faint clusters.* The observed cluster distribution is consistent with a steep $L_x - \Lambda_{CL}$ relation in which there is also significant scatter between x-ray and optical luminosity and a false positive rate (in the optical) of $\sim 25\%$.

Indeed, if the slope of the $L_x - \Lambda_{CL}^{\beta}$ relation is steeper than 2 and $\Lambda_{CL} \propto L_{Opt}$, then the M/L ratio of clusters will continue to increase with mass (at least into the moderate mass range of ROXS clusters). Specifically, if $L_x \propto M^2$ then $M/L_{Opt} \propto M^{1-2/\beta}$. Figure 4 shows this prediction overplotted on actual observational data obtained for low redshift clusters (Hradecky *et al.* 2000).

5. Goals for Multi-wavelength Distant Cluster Surveys

Multi-wavelength surveys for distant clusters have conclusively demonstrated their value in quantifying wavelength-dependent selection biases, establishing

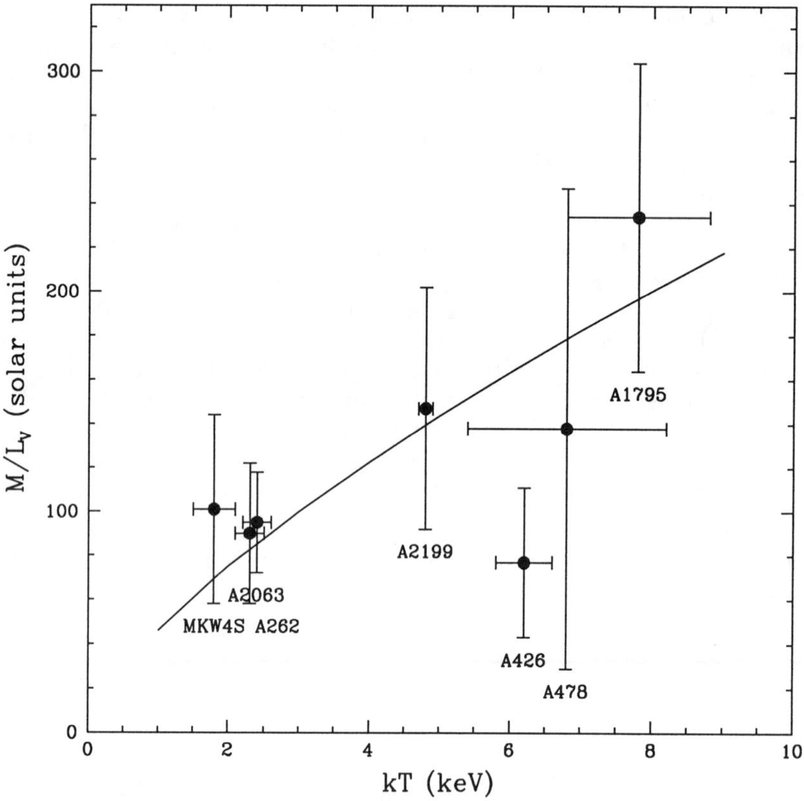

Figure 4. The optical M/L ratio of clusters as a function of the kinetic gas temperature of the ICM. Data points are from Hradecky *et al.* 2000. The solid curve is the prediction from the relationship $M/L_{Opt} \propto M^{1-2/\beta}$. A $L_x \propto T_x^{2.8}$ relation is adopted to convert luminosity to temperature.

fundamental relationships between global cluster properties, and for predicting and characterizing key evolutionary trends. The near term goals for the current and next generation of cluster surveys will likely focus on addressing the following key scientific questions:

- How do global cluster properties (mass, gas fraction, mass profile) depend on the size of the early type galaxy population, star formation rate, and internal kinematics. For example, *is the red sequence unique to virialized clusters?*

- How does clustering on large-scales evolve? Cluster surveys are now reaching volumes and depths that enable detection of several hundred clusters with $z > 0.6$. The utility of clusters as tracers of large-scale structure has been proven at low redshift and that same advantage can now be employed at $z \sim 1$.

- How do clusters themselves evolve? The assembly of a statistically complete sample of clusters in the range $1 < z < 2$ over an area of > 100 deg^2 is within reach of current technology (NIR mosaic cameras, more efficient millimeter band receivers for SZE detection). Coupled with spectroscopic, HST, and x-ray observations, such a large survey will yield precise constraints on cluster evolution.

To achieve these goals, equally dedicated efforts must be undertaken in the development and/or enhancement of cluster detection algorithms (*e.g.*, , see Nichol *et al.* 2001) and improved methods for comparing and analyzing independently derived cluster catalogs over a broad spectral range. This is especially important given the current or near future availability of homogeneous multi-wavelength surveys that have significant sky overlap (*e.g.*, 2MASS, SDSS, ROSAT/MACS, MAP, PRIME). The LSST (Tyson, Wittman, & Angel 2000) holds the potential for wide-area "mass"-selected cluster catalogs, substantially simplifying the scientific interpretation of the resulting cluster samples. In all these efforts, however, the question of understanding how the cluster detection process affects the scientific conclusions must always be in the fore.

References

Adami, C., Holden, B. P., Castander, F. J., Mazure, A., Nichol, R. C., Ulmer, M. P. 2000, A&A, 362, 825.

Balogh, M. L., Morris, S. L., Yee, H. K. C., Carlberg, R. G., Ellingson, E. 1999, ApJ, 527, 54.

Burke, D. J., Collins, C. A., Sharples, R. M., Romer, A. K., Holden, B. P., Nichol, R. C. 1997, ApJ, 488, L83.

Burns, J. O., Ledlow, M. J., Loken, C., Klypin, A., Voges, W., Bryan, G. L., Norman, M. L., White, R. A. 1996, ApJ, L49.

de Propris, R., Stanford, S. A., Eisenhardt, P. R., Dickinson, M., Elston, R. 1999, AJ, 118, 719.

Donahue, M., Mack, J., Scharf, C., Lee, P., Postman, M., Rosati, P., Dickinson, M., Voit, G. M., Stocke, J. T. 2001, ApJ, 552, L93.

Dressler, A., Smail, I., Poggianti, B. M., Butcher, H., Couch, W. J., Ellis, R. S., Oemler, A., Jr. 1999, ApJS, 122, 51.

Ebeling, H., Edge, A. C., Fabian, A. C., Allen, S. W., Crawford, C. S., Boehringer, H. 1997, ApJ, 479, L101.

Ebeling, H., Edge, A. C., & Henry, P. 2001, ApJ, 553, 668.

Ellingson, E., Lin, H., Yee, H. K. C., Carlberg, R. G. 2001, ApJ, 547, 609.

Gladders, M. D. & Yee, H. K. C. 2000, AJ, 120, 2148.

Goto, T. *et al.* 2001, AJ, submitted for publication.

Holden, B. P., Romer, A. K., Nichol, R. C., Ulmer, M. P. 1997, AJ, 114, 567.

Holden, B. P., Nichol, R. C., Romer, A. K., Metevier, A., Postman, M., Ulmer, M. P., Lubin, L. M. 1999, AJ, 118, 2002.

Holden, B. P., Adami, C., Nichol, R. C., Castander, F. J., Lubin, L., Romer, A. K., Mazure, A., Postman, M., Ulmer, M. P. 2000, AJ, 120, 23.

Hradecky, V., Jones, C., Donnelly, R. H., Djorgovski, S. G., Gal, R. R., Odewahn, S. C. 2000, ApJ, 543, 521.

Jannuzi, B. T., Dey, A., Tiede, G. P., Brown, M. J. I., NDWFS Team 2000, B.A.A.S., 197, 7709.

Kawasaki, W., Shimasaku, K., Doi, M., Okamura, S. 1998, A&AS, 130, 567.

Kepner, J., Fan, X., Bahcall, N., Gunn, J., Lupton, R., Xu, G. 1999, ApJ, 517, 78.

Lobo, C., Iovino, A., Lazzati, D., Chincarini, G. 2000, A&A, 360, 896.

Nichol, R. C. *et al.* 2001, astro-ph/0110230.

Poggianti, B. M., Smail, I., Dressler, A., Couch, W. J., Barger, A. J., Butcher, H., Ellis, R. S., Oemler, A., Jr. 1999, ApJ, 518, 576.

Postman, M., Lubin, L. M., Gunn, J. E., Oke, J. B., Hoessel, J. G., Schneider, D. P., Christensen, J. 1996, AJ, 111, 615.

Postman, M., Lauer, T. R., Szapudi, I., & Oegerle, W. 1998, ApJ, 506, 33.

Ramella, M., Boschin, W., Fadda, D., Nonino, M. 2001, A&A, 368, 776.

Romer, A. K., Nichol, R. C., Holden, B. P., Ulmer, M. P., Pildis, R. A., Merrelli, A. J., Adami, C., Burke, D. J., Collins, C. A., Metevier, A. J., Kron, R. G., Commons, K. 2000, ApJS, 126, 209.

Rosati, P., Della Ceca, R., Burg, R., Norman, C., Giacconi, R. 1995, ApJ, 445, L11.

Rosati, P., Della Ceca, R., Norman, C., Giacconi, R. 1998, ApJ, 492, L21.

Scharf, C. A., Jones, L. R., Ebeling, H., Perlman, E., Malkan, M., Wegner, G 1997, ApJ, 477, 79.

Stanford, S. A., Elston, R., Eisenhardt, P. R., Spinrad, H., Stern, D., Dey, A. 1997, AJ, 114, 2232.

Stanford, S. A., Holden, B., Rosati, P., Tozzi, P., Borgani, S., Eisenhardt, P. R., Spinrad, H. 2001, ApJ, 552, 504.

Tyson, J. A., Wittman, D., Angel, J. R. P., 2000, in proceedings of the Dark Matter 2000 conference (Santa Monica, February 2000) to be published by Springer, astro-ph/0005381.

Voit, M., & Donahue, M. 1999, private communication.

Tracing Cosmic Evolution with Galaxy Clusters
ASP Conference Series, Vol. 268, 2002
S. Borgani, M. Mezzetti, and R. Valdarnini, eds.

The cluster environments of radio loud quasars

J. M. Barr, M. N. Bremer

Department of Physics, University of Bristol, Tyndall Ave, Bristol BS8 1TL, U.K.

J. C. Baker

Astronomy Department, University of California, Berkeley, U.S.A.

M. D. Lehnert

Max-Planck-Institut für extraterrestrische Physik, Garching, Germany

Abstract.

We have carried out multi-colour imaging of the fields of a statistically complete sample of low-frequency selected radio loud quasars at $0.6 < z < 1.1$, in order to determine the characteristics of their environments. The largest radio sources are located in the field, and smaller steep-spectrum sources are more likely to be found in richer environments, from compact groups through to clusters. This radio-based selection (including source size) of high redshift groups and clusters is a highly efficient method of detecting rich environments at these redshifts. Although our single filter clustering measures agree with those of other workers, we show that these statistics cannot be used reliably on fields individually, colour information is required for this.

1. Introduction

Powerful radio sources at low redshift, (Cyg A, 3C295), are known to reside in massive elliptical galaxies such as those commonly found at the centre of clusters. AGN at low redshift have been shown to exist in environments of above average galactic density (Yee & Green 1987) and studies extending these measurements to higher redshift show no evolution in this trend (Ellingson, Yee, & Green 1991; Hill & Lilly 1991; Yee & Ellingson 1993; Wold et al. 2000). It is therefore reasonable to expect that at high redshift we may be able to find clusters by looking at the environments of AGN. This is further substantiated by theoretical arguments regarding the formation of *extended* lobe-dominated radio structures (*e.g.* Miley 1980). Such models require a confining medium to exist over hundreds of kpc, the scale lengths and pressures being comparable with those found in clusters of galaxies. An advantage of instigating a search in this manner is provided by the fact that a quasar will pinpoint the redshift of interest and through careful choice of filters facilitate the identification of associated galaxies by way of their colours.

These reasons make it possible to build a sample of distant groups and clusters from a study of radio loud quasars (RLQs), although care must be taken to understand the selection function. With reasonable estimates regarding the lifetime of the AGN and its effect on the cluster this can be achieved. In any case the opportunity to study clusters and their constituents is not lost.

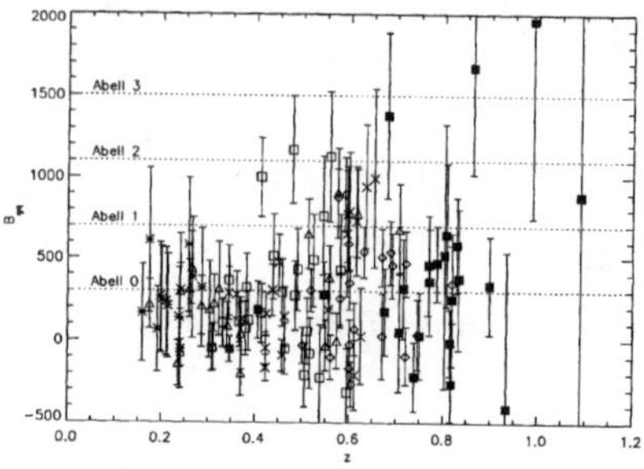

Figure 1. Values of the spatial cross correlation amplitude as a function of redshift for radio loud quasar fields. Filled squares, this work. Diamonds, Wold et al. (2000). Triangles, Yee & Ellingson (1993). Squares, Ellingson et al. (1991). Crosses, Yee & Green (1987). Asterisks, McClure & Dunlop (2000). The horizontal lines are the Abell classes in McClure & Dunlop.

A related benefit is that we may find and study small and compact groups at high redshift. Current optical and X-ray methods will be insensitive to such low mass/luminosity systems and the environments of RLQs may offer the best method of finding significant numbers of these systems at high redshift.

2. Sample and observations

The quasars used for this study are drawn from the Molonglo quasar sample (MQS, Kapahi et al. 1998) of low frequency selected radio loud quasars. This is a complete sample of 111 RLQs with $S_{408} > 0.95$ Jy in a contiguous area of southern sky ($-30° > \delta > -20°$, but excluding the R.A. range 14^h03^m - 20^h03^m). Within the range $0^h0^m <$ RA $< 14^h0^m$ there are 30 MQS objects which have $0.6 < z < 1.1$, 22 for which we have data. Eight sources with these criteria were *randomly* excluded due to observing time restrictions. This sample has many merits. Firstly it has a single selection criterion, that of low frequency radio luminosity. Thus it does not attempt to "match" catalogues of what may be intrinsically different objects. Secondly, of all objects in our chosen redshift range none are excluded based on optical or radio bias. The remaining objects can therefore be thought of as a *representative* sample. Also, the low frequency selection avoids preferentially selecting core dominated objects, ie: those which may be of intrinsically lower luminosity but have Doppler boosted emission due to their orientation to the line of sight. The MRQ is thus useful as a "complete and unbiased" sample on which to study the environmental properties of RLQs.

We have observed our objects in 2 or more filters which straddle the redshifted 4000Å break in order to maximise the contrast for early type galaxies. In several cases we also have near infrared (NIR) J and K' filters. The telescopes used were the CTIO 4m, the ESO 2.2m and 3.6m and the AAT.

3. Single filter analysis

Previous investigators have sought to quantify the excess of objects around quasars, (or radio galaxies, or any other object), by counting sources within a certain radius in a single filter (0.5 Mpc in most cases). We initially analyse clustering around quasars with reference to such a measure, the galaxy-quasar spatial cross correlation amplitude, B_{gq}. A detailed derivation can be found in Yee & Green (1984) and Longair & Seldner (1979).

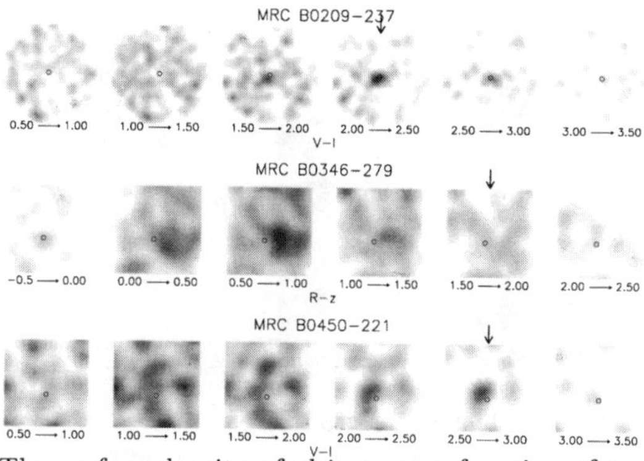

Figure 2. The surface density of objects as a function of position in the fields of three of our quasars, binned according to colour range. Cluster cores at the redshift of the quasar would produce peaks in the density maps indicated by the arrow. While all three fields produce high values of B_{gq}, we can see that in the second case the overdensity is at the wrong colour for the quasar redshift. The position of the quasar is marked by the circle.

The results of applying this method to our data can be seen in figure 1, as well as those for other investigators. Our results confirm that the quasars occupy a variety of environments and show no evidence for evolution to a redshift of 1.

There are, however, reasons why these methods are not necessarily the most accurate to use at high redshift. Firstly, they assume that the AGN is at the centre of any (roughly circular) overdensity. As we progress to higher redshifts, clusters are less likely to be in relaxed spheroidal systems and may exhibit significant substructure which can extend further than 0.5 Mpc from the quasar. Secondly, the number density of objects at the magnitude levels of interest displays variation on the scale-lengths of distant clusters making it difficult to estimate the background to subtract from any overdensity. Thirdly, even when an obvious overdensity is discovered, it may consist of objects which have the wrong colours to be galaxies at the redshift of the quasar (see fig. 2b). In this case these statistical analyses will not only be wrong, they may bias the average result systematically and in doing so render any measure of the average properties of a sample incorrect.

4. Multi-colour results

By choosing optical filters which straddle the 4000Å break, we can select the passively evolving elliptical galaxies at the redshift of the quasar. Figure 2. shows the density of objects around three of our quasars separated according to colour. We can identify agglomerations as having the potential to be at the quasars redshift by using these sort of plots. Further analysis looking at the colour dispersion or for a red sequence or using NIR filters can then quantify accurately which galaxies belong to the cluster core and which do not. Examples of these techniques are to be found in Baker et al. (2001) and Bremer et al. (2001). Because of the arguments regarding the formation of the radio lobes being dependent on a large scale medium, we would expect to see a correlation between the size of the source and the clustering around it. We see just such

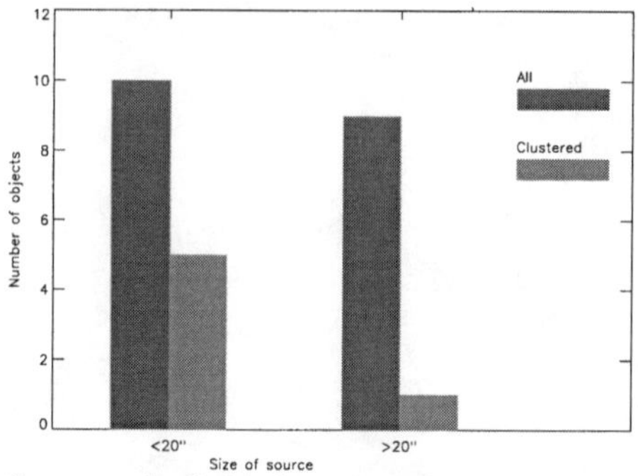

Figure 3. Our sample divided into small and large steep spectrum sources. Clustered environments are more often associated with smaller structures.

a correlation in figure 3. Our sample is separated into those sources with a radio size $< 20''$ and those with $> 20''$. Unresolved flat spectrum sources, ($\alpha < 0.6$), are rejected in order to prevent beamed sources with intrinsically lower luminosity from being included. The remaining sample of 19 steep spectrum objects is thus split approximately in half. If we loosely define our clusters as those with more than 10 associated core members, we see that five out of our six clusters are associated with smaller radio structures. We wouldn't expect to see a 1:1 relationship between size and clustering. The age of the source may have allowed it to expand despite a clustered environment, or conversely, perhaps we see a young source which is expanding unconfined. Projection effects may also hide the true size of the source.

We have found that RLQs exist in a variety of environments from the field through compact groups to rich clusters. In order to correctly quantify the environments of these sources, multicolour information must be used. There is evidence for a radio size - environment relation with smaller sources located in richer systems. By selecting small, powerful RLQs we can efficiently create a sample of distant groups and clusters. We also have the ability to "dial up" clusters at a specific redshift by looking at AGN.

References

Baker, J. C., et al., 2001, AJ, 121, 1821

Bremer, M. N., Baker, J. C., Lehnert, M. D. 2001, MNRAS, in press

Ellingson, E., Yee, H. K. C., Green, R. F. 1991, ApJ, 371, 49

Hill, G. J., Lilly, S. J. 1991, ApJ, 367, 1

Kapahi, V. K., et al., 1998, ApJS, 118, 327

Longair, M. S, Seldner, M. 1979, MNRAS, 189, 433

McClure, R. J., Dunlop, J. S. 2001, MNRAS, 321, 515

Miley, G. 1980, ARAA, 18, 165

Wold, M., Lacy, M., Lilje, P. B., Serjeant, S. 2000, MNRAS, 316, 267

Yee, H. K. C., Green, R. F. 1984, ApJ, 280, 79

Yee, H. K. C., Green, R. F. 1987, ApJ, 319, 28

Yee, H. K. C., Ellingson, E. 1993, ApJ, 411, 43

Tracing Cosmic Evolution with Galaxy Clusters
ASP Conference Series, Vol. 268, 2002
S. Borgani, M. Mezzetti, and R. Valdarnini, eds.

The evolution of the cluster environments of radio sources at $z < 1.8$

Philip Best

Institute for Astronomy, Royal Observatory, Blackford Hill, Edinburgh EH9 3HJ, UK

Matt Lehnert

MPE, Postfach 1312, 85741 Garching-bei-Munchen, Germany

Huub Röttgering, George Miley

Sterrewacht Leiden, Huygens Lab, Postbus 9513, 2300RA Leiden, NL

Abstract. An analysis of the cluster environments around distant radio galaxies is presented, in particular the results from new NTT deep optical–IR imaging of the fields of radio sources at $z \sim 1.6$. A net overdensity of K–band galaxies is found, together with a sharp peak in the angular cross–correlation amplitude, centred on the radio galaxies. This excess clustering is associated predominantly with red galaxies, with colours consistent with being old ellipticals at the radio source redshift. A large excess of such red galaxies is seen, particularly within 100 kpc of the radio source. These comprise amongst the most distant red sequences of cluster ellipticals yet discovered, that is, the highest redshift 'normal' clusters.

1. Introduction

The discovery and study of clusters at redshifts beyond one is of great importance for many aspects of structure formation and cosmology, as discussed throughout this volume. Unfortunately, identification of clusters at these redshifts is very challenging observationally: at X-ray wavelengths existing surveys suffer sensitivity limits and the field of view of the current generation of X-ray telescopes make them inefficient for wide-area surveys; at optical and near–infrared wavelengths distant clusters show only low contrast above the high background counts at faint magnitudes. Fortunately an alternative method of locating distant clusters does exist, using powerful radio sources as probes.

Radio sources are hosted by giant elliptical galaxies (e.g. McLure & Dunlop 2000, Best et al 1998) and are amongst the most massive galaxies known in the early Universe. At low redshifts they are typically found in galaxy groups, or poor clusters, but environmental richness appears to increase with redshift: at $z \sim 0.5$, the large amplitude of the galaxy cross–correlation function around radio galaxies and an Abell clustering classification indicate that about 40% of radio sources are located in clusters of Abell richness class 0 or greater (e.g. Hill & Lilly 1991). At $z \sim 1$ there is overwhelming evidence from observations at a wide variety of wavelengths that at least some powerful radio sources are located at the centres of clusters (e.g. Best 2000 and references therein). Best (2000) analysed UKIRT K–band data of a sample of 3CR radio galaxies at $z \sim 1$ and found a sharp peak in the cross–correlation function surrounding the radio galaxies, corresponding to a mean environmental richness of between Abell class 0 and 1. Colour–magnitude (C-M) relations showed red sequences around some (but not all) radio sources (see also Barr et al, this volume), but a detailed analysis of the C-M relations and the source–to–source variations was prohibited by the small field of view (70 by 70 arcsec) of the UKIRT frames.

At still higher redshifts, strong evidence is found for (proto-)clusters around radio galaxies at $z > 2$ (e.g. Pentericci et al, Kurk et al, this volume). At these redshifts, however, studies have been limited to objects selected by line emission: the sequence of red galaxies characteristic of low redshift clusters is not seen.

In this article the first results are presented of a very deep optical–IR imaging study of the environments of radio sources at $1.44 < z < 1.7$, in order to investigate their environments, to determine to what redshift the centres of clusters are dominated by a population of red galaxies, and ultimately to use this red sequence to investigate the evolution of the cluster ellipticals.

2. The fields of radio galaxies at $z \sim 1.6$

We have defined a complete subsample of 9 radio sources with redshifts $1.44 < z < 1.7$ and galactic latitudes $|b| > 20°$ from the equatorial sample of Best et al (1999; $S_{408\mathrm{MHz}} > 5\,\mathrm{Jy}$; $-30 < \delta < +10°$). Deep 5 by 5 arcminute R, J, K images of the fields of five of these sources were obtained in Sept 2000 and Aug 2001 using the NTT, reaching typical depths of $K \sim 20.5$, $J \sim 22.3$ and $R \sim 26$. Although final results await the completion of the imaging program, these multi–colour data provide some promising early results, including:

• *An excess of K–band counts.* The number counts in the infrared K–band have been determined as a function of magnitude in each field (see Table 1). These results show a clear excess of number counts fainter than $K \sim 18.5$. Note that the magnitude of the radio source host galaxy is typically $K \sim 18$.

• *A peak in the angular cross–correlation function around the radio sources.* For each field the angular cross–correlation function has been determined following the method outlined by Best (2000); cross–correlation statistics have been calculated both for all galaxy–galaxy pairs, and for only radio galaxy–galaxy pairs. Combining the data from the different fields, there is a clear excess of clustering around the radio host galaxies, as compared to the field in general (Figure 1), the mean cross–correlation amplitude indicating cluster environments of Abell richness class 0 to 1. To investigate how the cross–correlation amplitude varies with galaxy colour, the data were split into different colour bins. The clustering excess is predominantly associated with red galaxies; those galaxies with colours consistent with being cluster ellipticals are very strongly clustered around the radio galaxy, whilst blue galaxies show no excess clustering. There is some field-to-field variation; this will be investigated when the dataset is completed.

• *Overdensities of red galaxies.* In many of the fields there is a large overdensity of extremely red galaxies (see also Hall et al 2001), with colours consistent with passively evolving elliptical galaxies at the radio source redshift. There is

K	Raw Counts	Corrected Counts	Counts per mag. per sq. degree*	Error*	Literature Counts*	Excess Galaxies
15.0–15.5	3	3.0	308	177	218	0
15.5–16.0	5	5.0	513	229	620	0
16.0–16.5	7	7.0	719	271	960	−1
16.5–17.0	23	23.0	2362	492	1530	3
17.0–17.5	32	32.3	3320	587	2530	3
17.5–18.0	52	52.5	5396	751	5540	0
18.0–18.5	95	96.9	9958	1041	9470	2
18.5–19.0	128	130.6	13418	1215	10500	9
19.0–19.5	208	214.4	22029	1613	11600	34
19.5–20.0	276	293.6	30164	2068	17200	42
20.0–20.5	294	381.8	39226	3565	22700	54

Table 1. The K–band number counts combined across the radio galaxy fields. (*: counts per magnitude per square degree.)

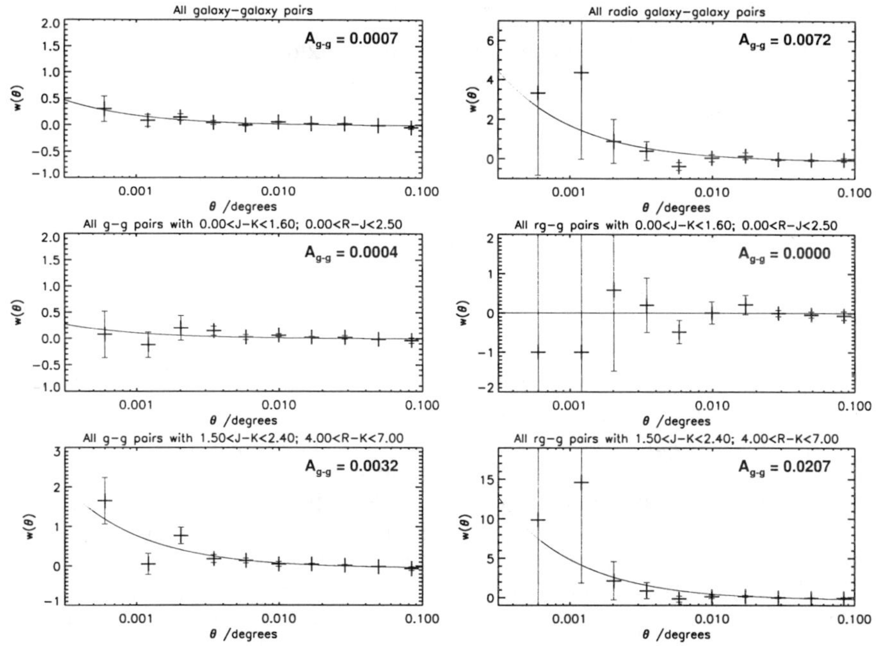

Figure 1. Angular cross–correlation analyses of the fields of the 3 radio galaxies observed in Sept 2000. The left–hand plots consider all galaxy-galaxy pairs, and the right–hand plots only radio galaxy–galaxy pairs. The upper plots are for all of the galaxies in the field, and show a clear excess of clustering around the radio galaxy. The middle plots show blue galaxies, demonstrating that these have no preferential clustering around the radio galaxy. The bottom plots are for galaxies with colours consistent with being old elliptical galaxies at the radio source redshift: these have a remarkably high cross–correlation amplitude.

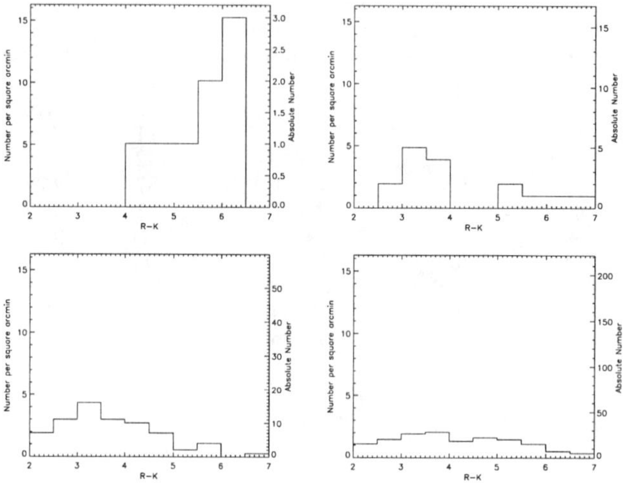

Figure 2. $R-K$ colour histograms for galaxies around 2025-155, as a function of distance from the radio source: Upper left – 0 to 100 kpc radius; Upper right – 100 to 250 kpc radius; Lower left – 250 to 500 kpc radius; Lower right – 500 to 1000 kpc radius. Note the strong excess of galaxies with very red colours, particularly in the inner 100 kpc.

especially strong clumping of red galaxies on a $\sim$ 100kpc scale. This can be seen, for example, in colour cuts as a function of radius (Figure 2).

3. Conclusions

We have presented some of the first results of a program of optical–IR imaging of the fields of radio galaxies at redshifts $z \sim 1.6$. Three different analyses all provide strong evidence for significant galaxy overdensities around some of these radio sources. Work to spectroscopically confirm these redshift and to study the properties of the cluster galaxies is continuing. The presence of the extremely red galaxies is particularly exciting, as it indicates that a substantial population of elliptical galaxies have already formed, and must have ceased star formation a significant time ($\gtrsim$ 1 Gyr) before redshift 1.5. This red sequence also makes these amongst the most distant 'normal' clusters known, and thus offers unique prospects for advancing studies of galaxy formation and evolution.

References

Best P.N., Longair M.S., Röttgering H.J.A., 1998, MNRAS, 295, 549
Best P.N., Röttgering H.J.A., Lehnert M.D., 1999, MNRAS, 310, 223
Best P.N., 2000, MNRAS, 317, 720
Hall P.B., et al., AJ, 121, 1840
Hill G.J. & Lilly S.J., 1991, ApJ, 367, 1
McLure R.J. & Dunlop J.S., MNRAS, 317, 249

Tracing Cosmic Evolution with Galaxy Clusters
ASP Conference Series, Vol. 268, 2002
S. Borgani, M. Mezzetti, and R. Valdarnini, eds.

A search for high redshift clusters associated with radio galaxies at 2 < z < 4

J.D. Kurk, B.P. Venemans, H.J.A. Röttgering, G.K. Miley

Sterrewacht Leiden, P.O. Box 9513, 2300 RA, Leiden, The Netherlands

L. Pentericci

Max-Planck-Institut für Astronomie, Königstuhl 17, D-69117, Heidelberg, Germany

Abstract. High redshift radio galaxies are amongst the most massive galaxies in the early Universe and have properties expected from central galaxies in forming clusters. We are carrying out an observational programme on the VLT to find and study galaxy proto clusters around radio galaxies at redshifts $2 < z < 4$. First, we use narrow band imaging to select candidate galaxies which show excess Lyα emission at redshifts similar to the central radio galaxy. Then, we use multi object spectroscopy to confirm the redshifts of these candidates and measure the velocity dispersion of the cluster members. Our goal is to observe a sample of $\sim$ 10 targets and investigate galaxy overdensities as a function of redshift. Here, we report on the current progress of the programme and show some preliminary results which include the discovery of a structure of galaxies at redshift 4.1.

1. Introduction

Searching for clusters of galaxies or their progenitors (proto clusters) at redshifts > 2 is important for constraining galaxy evolution, large scale structure formation and cosmological models (e.g. Giavalisco et al. 1998, Bahcall, Fan, & Cen 1997). Radio galaxies can act as beacons of these distant structures.

Luminous radio galaxies are the oldest and most massive known galaxies in the early Universe (De Breuck et al. 2000). In standard cosmological models massive galaxies form in the deepest of the first potential wells, in which - at either a later or earlier time depending on the galaxy formation model - other galaxies will form to assemble a cluster. There exists a diverse collection of evidence that high redshift radio galaxies (HzRGs) are actually residing in dense cluster environments, which includes: (i) the excess of companion galaxies around HzRGs (e.g. Röttgering et al. 1996), (ii) the extreme radio rotation measures of HzRGs indicating that they are embedded in hot magnetized gas (Carilli et al. 1997), and (iii) giant Lyα emitting gas halos surrounding HzRGs which can extent up to 200 kpc (e.g. Kurk et al. 2001).

In principle, using narrow band imaging in rest-frame Lyα star forming galaxies can be found effectively over a narrow redshift range. However, since there is only a small amount of dust needed to extinguish Lyα emission from recombining hydrogen ions, it was not clear whether Lyα photons could escape at all from star forming galaxies (Pritchet 1994). Now, using large telescopes several searches for Lyα emitters are underway and show good results (Rhoads et al. 2000, Stiavelli et al. 2001). With the presumption that powerful radio galaxies are located in overdense regions, we have started a search for high redshift clusters using the narrow band imaging technique to target Lyα emitting galaxies around radio galaxies at redshifts $2 < z < 4$.

2. The sample

¿From our collection of almost 150 radio galaxies at $z > 2$, we have chosen eight objects with redshifts suitable for Lyα imaging with the available narrow band VLT/FORS filters: four at $z \sim 2$ and three at $z \sim 3$, while for two more galaxies at $z = 4.1$ a custom filter was manufactured. These galaxies were also selected on their large radio, optical/IR continuum and Lyα luminosities, which are characteristics of massive galaxies. For each of the ten radio galaxy fields we (plan to) do spectroscopy of ~ 40 emitter candidates and aim to confirm > 15 emitters. With this sample, we can study what fraction of HzRGs are located in galaxy overdensities and how properties of these overdensities (e.g. velocity dispersions, sizes) and its cluster-galaxies (e.g. rotation curves, star formation rates) change with redshift. Simultaneously, we can study the giant gas halos associated with the radio galaxies.

Currently, imaging of five fields has been carried out, resulting in the detection of fifteen to sixty emitters in each field with Lyα rest frame equivalent width > 20 Å and significant signal-to-noise. Multi object spectroscopy has been completed for two fields: MRC 1138-262 and TN J1338-1942. The latter has the largest redshift of our sample and is described in the following section. The former was the subject of our pilot project and is described in detail by Pentericci et al. (this volume). Briefly, the existence of 14 Lyα emitters in this field was confirmed, with redshifts close to the radio galaxy but distributed in two groups (grey histogram in Fig. 1).

3. TN J1338-1942 at z = 4.11

About 25 candidates with above mentioned selection criteria were discovered in the field around TN J1338-1942. Subsequent spectroscopy of 22 of these led to the confirmation of 20 Lyα emitters at $z \sim 4.1$ (Fig. 2). Their spatial distribution is inhomogeneous: it is extended in the direction of the radio axis and Lyα halo of the radio galaxy (Fig. 2). Their velocity distribution (black histogram in Fig. 1) is narrower than that of MRC 1138-262 and has a dispersion ($\sigma_{\rm v}$) of ~ 350 km s^{-1} which implies a total virial mass ($M_{\rm vir}$) of 2×10^{14} $M_{\odot}$, assuming spherical symmetry with a virial radius (R) of 1.5 Mpc, using $M_{\rm vir} = 5\ R\ \sigma_{\rm v}^2\ /\ G$, where G is the gravitation constant. To determine whether we actually observe an overdensity here with respect to a field which does not contain a (proto) cluster, we have compared the number density of emitters around 1338-1942 with observations of blank fields by Cowie & Hu (1998) and Rhoads et al. (2000). The number density of emitters around 1338-1942 is five

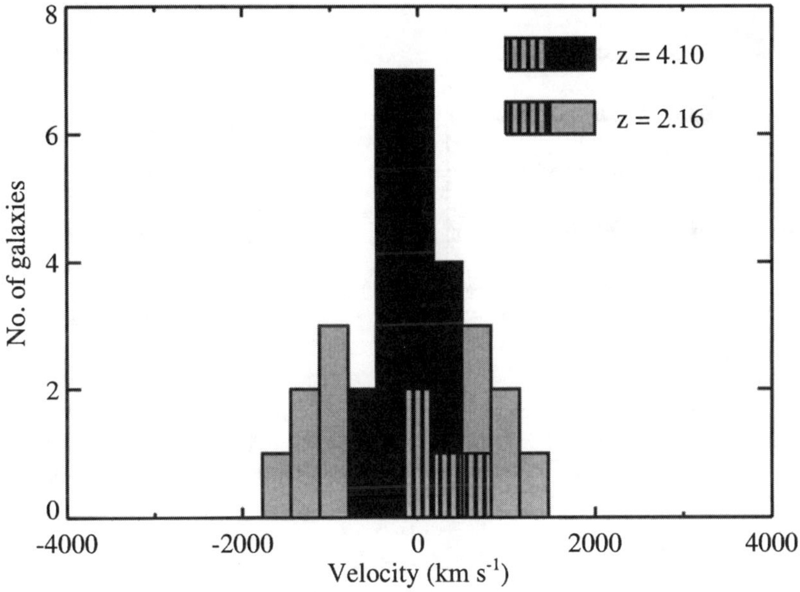

Figure 1. Velocity distributions of confirmed emitters around MRC 1138-262 at $z = 2.16$ (grey histogram) and TN J1338-1942 at $z = 4.11$ (black histogram). The emitters around 1138 are clustered in two groups, separated by ~ 1800 km s^{-1}, while the emitters around 1338 have approximately a Gaussian distribution around the radio galaxy.

times higher than the field density and comparable to the density of emitters around 1138-262.

4. Conclusions and outlook

Using narrow band imaging and subsequent multi object spectroscopy on the VLT, we have confirmed the existence of concentrations of galaxies around radio galaxies at $z = 2.2$ and 4.1. These observations are consistent with the idea that HzRGs are brightest cluster galaxies in forming clusters. Furthermore, it shows that narrow band imaging at the wavelength of redshifted Lyα is an effective method to find (proto) clusters at high redshift. We expect to uncover several more of these structures in the coming years. Obviously, these systems will be subject to further study at a multitude of wavelength bands, contributing important clues to outstanding questions in galaxy formation and cosmology.

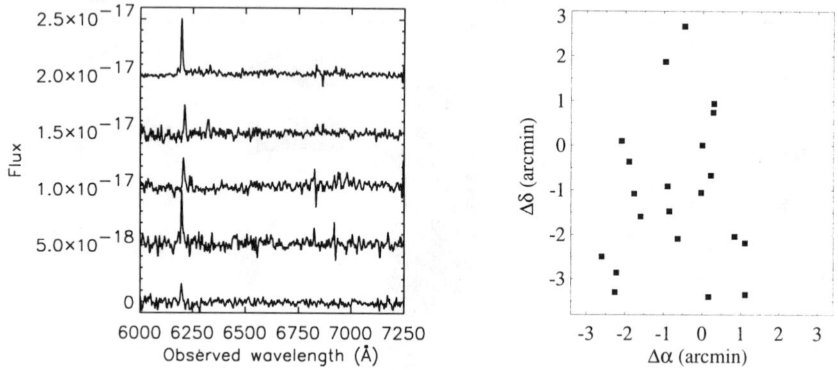

Figure 2. *Left* Spectra of five Lyα emitters found in the field of TN J1338-1942. *Right* Positions of the Lyα emitters around the radio galaxy TN J1338-1942. The radio galaxy itself is located at the origin.

5. References

Bahcall, N. A., Fan, X., Cen, R. 1997, ApJ, 485, L53

Carilli, C. L., Röttgering H. J. A., van Ojik, R., Miley, G. K., & van Breugel, W. J. M. 1997, ApJS, 109, 1

Cowie, L. L. & Hu, E. M. 1998, AJ, 115, 1319

De Breuck, C., van Breugel, W., Röttgering, H. J. A., & Miley, G. K. 2000, A&AS, 143, 303

Giavalisco, M., Steidel, C. C., Adelberger, K. L., Dickinson, M. E., Pettini, M., & Kellog, M. 1998, ApJ, 503, 543

Röttgering, H. J. A., West, M. J., Miley, G. K., & Chambers, K. C. 1996, A&A, 307, 376

Kurk, J. D. et al. 2000, A&A, 358, L1

Kurk, J. D., Röttgering, H. J. A., Miley, G. K., & Pentericci, L. 2001 in Rev Mex AA, Conf. Proc. 'Emission Lines from Jet Flows', ed. W. Henney, W. Steffen, L. Binette, & A. Raga, astro-ph/0102337

Pentericci, L. et al. 2000, A&A, 361, L25

Pritchet, C. 1994, PASP, 106, 1052

Rhoads, J. E., Malhotra, S., Dey, A., Stern, D., Spinrad, H., & Jannuzi, B. T. 2000, ApJ, 545, L85

Stiavelli, M., Scarlata, C., Panagia, N., Treu, T., Bertin, G., & Bertola, F. 2001, ApJ in press, astro-ph/0105503

Tracing Cosmic Evolution with Galaxy Clusters
ASP Conference Series, Vol. 268, 2002
S. Borgani, M. Mezzetti, and R. Valdarnini, eds.

The population of galaxies in the forming cluster around the radio galaxy MRC 1138-262 at z=2.2

L. Pentericci

Max-Planck-Institute für Astronomie, Königstuhl 17, D-69117, Heidelberg, Germany

J.D. Kurk, H.J.A. Röttgering, G.K. Miley, B.P. Venemans

Sterrewacht Leiden, P.O. Box 9513, 2300 RA, Leiden, The Netherlands

Abstract. We have recently discovered a forming cluster around the radio galaxy MRC 1138-262 at redshift 2.2. Besides the population of Lyα emitting galaxies that have been confirmed spectroscopically, we have detected many candidate Hα emitters that seem to have a different spatial distribution from the other galaxies: they are more clustered towards the center of the cluster and seem to be distributed along the same direction as the radio source. We present here the characteristics of the Lyα and Hα emitters and study the nature of these populations.

1. Introduction

One of the most interesting questions in modern astrophysics concerns the formation of structures in the early universe. The most luminous high redshift radio galaxies are unique probes of galaxy and cluster formation: they are amongst the oldest and most massive objects in the early universe and there is considerable evidence that they are at the centers of proto clusters. Here we present results of a search for emission line galaxies in the field around the radio source 1138-262 at z=2.2. Based on the success of this project, we are now conducting a large program on the VLT to detect and study clusters around a few selected high redshift radio galaxies, over a redshift range from 2 to 4, which is presented by Kurk et al. in this proceedings.

2. A forming cluster at redshift 2.2

We have initially selected the radio galaxy MRC 1138-262 at z=2.156 as our best field to search for companion galaxies, since there are many observations indicating that it is probably a forming massive galaxy at the center of a dense region. These include: its extremely clumpy optical and near-IR morphology, similar to that predicted by simulations of forming massive galaxies (Pentericci et al 1998); the extremely distorted radio source, with evidence of interaction with the dense surrounding gas; the extreme Faraday rotation of 6600 rad m^{-2}, the largest amongst all high redshift radio galaxies known (Carilli et al. 1997),

indicating the presence of dense magnetized gas on large scales; and the luminous Lyα halo, extending for more than 200 kpc (Kurk et al. 2001).
Initial VLT narrow band imaging centered on the redshifted Lyα line revealed the presence of $\sim$ 50 candidate Lyα emitters, with restframe equivalent width larger than 20 Å (Kurk et al. 2000). Follow-up multi object spectroscopy has revealed 14 galaxies and 1 faint QSO, at approximately the same distance as the radio galaxy, and at least 4 other objects with spectral energy distribution consistent with them being at $z \sim 2.2$ (Pentericci et al. 2000). All the Lyα emitting galaxies, with the exception of the faint QSO, have redshift in the range 2.16$\pm$0.02 and are within a projected physical distance of less than 0.75 h_{100}^{-1} Mpc from the radio source ($q_0 = 0.5$). There is no is no clear enhancement of Lyα emitters toward the radio galaxy, although the distribution of most of the emitters seems to be elongated in the direction of the radio source (Figure 1). The velocity distribution suggests that there are two galaxy subgroups, having velocity dispersions of 280 and 500 km s^{-1} respectively and a relative velocity of 1800 km s^{-1} (see Figure 1 in Kurk et al. this proceedings)
Assuming an isothermal sphere model, the mass contained inside a 0.75h^{-1} Mpc radius, is respectively $0.3 \times 10^{14} h_{100}^{-1}$ and 0.9×10^{14} h_{100}^{-1} $M_\odot$, which implies a total mass in excess of 10^{14} $M_\odot$ for the whole system.

2.1. Hα emitters

Prompted by the success of our optical narrow band search, we also obtained near infrared deep images of the 1138-262 field, with ISAAC, using a narrow band filter centered around 2.07μm which encompasses Hα emission from galaxies at a redshift around 2.15. Since Hα emission is not as sensitive to dust extinction as Lyα these observations are mainly sensitive to dusty galaxies.
Two ISAAC fields were observed, both inside the FORS field: one centered around the radio galaxy and one North East of this, covering 6 confirmed Lyα emitters. The observations were also complemented by broad band near infrared images, in K band for the eastern field and in J-H-K band for the central field
Candidate line emitting objects were then selected on the basis of their excess narrow versus broad band flux. We have selected objects with restframe equivalent width $EW_0 > 50$ Å and $\Sigma > 3$, where Σ is the excess signal in the narrow band compared with the noise in both narrow and broad band, as defined in Bunker et al. (1995). As a result we detect 17 candidates Hα emitters in the central field and only 6 in the NE field, with one in common between the two fields (Figure 1). We will assume that the line detected is Hα. Other possibilities that cannot be excluded, although less likely, include fainter lines from higher redshift objects, such as [OII](3727 Å) from z=4.6 galaxies or [OIII](5007 Å) from z=3.2 galaxies, and Pα (at 1.88μm) from z=0.12 galaxies. Confirmation of the real redshift is one of the aims of the planned follow-up spectroscopy that will be carried out next March with ISAAC.
The spatial distribution of the Hα emitters is clearly not homogeneous over the two fields but rather confined to the central parts where there are 3 times as many candidates as in the outer region. Moreover in the central field the emitters seem to be oriented in a similar direction as the radio source and the ionized gas halo that surrounds the galaxy. This is clearly different from the Lyα emitters that show no clustering on this scale.

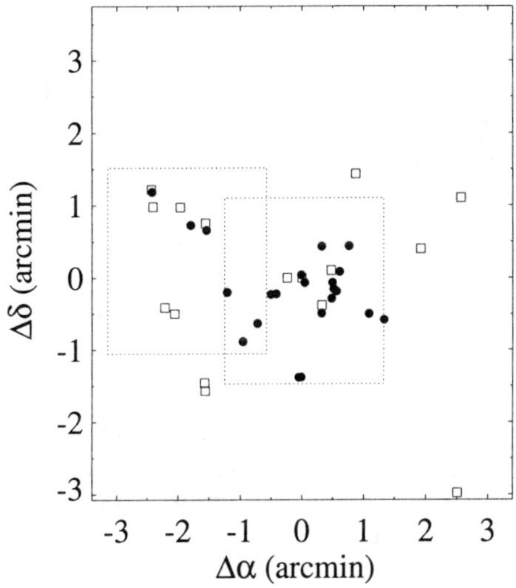

Figure 1. Positions of the Lyα emitters (empty squares) and Hα emitters (circles) around the radio galaxy 1138-262. The total field is the FORS1 field of view, while the two dotted areas represent the two ISAAC fields imaged in the infrared.

3. The galaxy population of the forming cluster

The population of Hα emitters could just be a dusty version of the Lyα emitters. Both Lyα and Hα are produced by recombination of ionized hydrogen and are indicators of star formation rate. However, since Lyα is a resonant line, Lyα photons will be scattered by neutral hydrogen: if there is also dust present in such a region, most Lyα photons will be absorbed before they can escape. Therefore, Lyα emission can only be observed from galaxies with a low or very patchy dust content. This is not the case for Hα radiation which is less susceptible to dust. Hence, with an Hα survey we also target those galaxies with high star formation accompanied by larger amounts of dust.

An indication that this is indeed the case comes from the ratio of SFR derived from Hα emission, and from the UV continuum. In Figure 2 we present such quantities for all the Hα emitters: the SFRs have been derived according to the prescription of Kennicutt (1998), and in particular for the line emission we used SFR($M_\odot$/yr) $= 7.9\times10^{-42}$ L(Hα)(erg s^{-1}), which is appropriate for continuous star formation, and for the UV restframe continuum we used SFR($M_\odot$/yr) $= 1.4\times10^{-28}$ L(1500-2800Å)(erg s^{-1} Hz^{-1}). The ratio of the two quantities is around one or larger in almost all cases, with the exception of the brighter 2-3 objects which might as well be lower redshift interlopers. However if the dust content was the only difference between the Lyα and the Hα emitters, we would not expect a different distribution between the two populations. Possible explanations could then be a difference in age and/or metal content. We are still investigating this issue: preliminary results from the broad band photometry seem to indicate no differences between the two populations, both having spectral energy distributions consistent with young starburst galaxies.

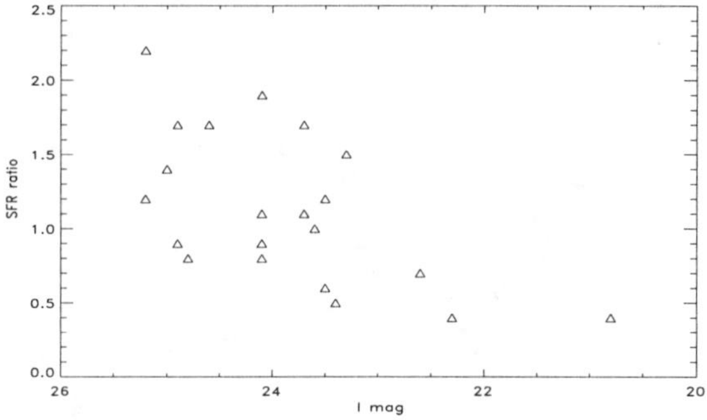

Figure 2. The ratio of SFR derived from Hα emission to the SFR derived from the continuum emission at 2000 Å restframe, versus the total I-band magnitude of the galaxies. For most objects the ratio is well above 1 indicating the presence of dust.

3.1. An enhanced AGN content?

The mean rest frame EW_0 for the Lyα emitters is 60 Å and the distribution is nearly uniform from 15 to 150 Å. There are several objects with EW in excess of 100 Å. This is significantly different from what is obtained by Steidel et al. (2000) for a large sample of galaxies at redshift around 3: they find no galaxies with EW_0 larger than 100 Å. The same considerations can be made for the Hα emitters which have very high restframe EW_0. While our results are most probably due to our observational strategy, that emphasizes large EW_0 objects, it is also possible that some of our cluster galaxies, besides the faint QSO, contain an AGN contribution, since photoionization from hot stars is unlikely to produce such large EW_0 as observed (e.g. Charlot & Fall 1993). An enhanced AGN content for the cluster, is also indicated by the detection of several other faint X-ray emitters in the field (from Chandra observations, Carilli et al 2001), of which one coincides with the QSO, and two others with candidate line emitters.

References

Bunker A.J., Warren, S. J., Hewett, P.C, Clements, D.L., 1995 MNRAS, 273 513

Carilli C.L., Röttgering H.J.A., van Ojik R. et al. 1997, ApJS, 109, 1

Charlot S. and Fall S.M. 1993 ApJ, 415, 580

Kennicutt, R.C. 1998, ARA&A, 36, 189

Kurk J.D., Röttgering H.J.A, Pentericci L. et al. 2000 A&A, 358,L1

Kurk J.D., Röttgering H.J.A, Miley G.K.and Pentericci L. 2001 in REvMexAA astro-ph/0102337

Pentericci L., Röttgering H.J.A, Miley G.K., et al. 1998 ApJ504, 139

Pentericci L., Kurk J.D., Röttgering H.J.A. et al. 2000 A&A, 361, L25

Steidel C.C., Adelberger K.L., Shapley A. et al. 2000 ApJ, 532, 170

Tracing Cosmic Evolution with Galaxy Clusters
ASP Conference Series, Vol. 268, 2002
S. Borgani, M. Mezzetti, and R. Valdarnini, eds.

Evolution of the cluster population with the ROSAT Deep Cluster Survey

Stefano Borgani

INFN - c/o Dipartimento di Astronomia dell'Università, Trieste, Italy

Piero Rosati

ESO - European Southern Observatory, Garching, Germany

1. Introduction

The number density of galaxy clusters has been recognized since more than ten years as a powerful cosmological test. Their abundance in the local Universe and the corresponding redshift evolution provide strong constraints on the amplitude of cosmic density perturbations on scales of about $10h^{-1}$ Mpc (h is the Hubble constant in units of 100 km $s^{-1}Mpc^{-1}$) and on the matter density parameter Ω_m. Over the last five years, remarkable observational progress has been made in constructing large samples of local and distant galaxy clusters with the aim of quantifying the evolution of their space density and providing the basis for follow-up studies of their physical properties. The *ROSAT* satellite is largely responsible for this progress, both with All-Sky Survey data and pointed observations, which have been a gold mine for serendipitous discoveries (see Rosati, Borgani & Norman 2001, for a review).

In this contribution we report on the most recent results from the ROSAT Deep Cluster Survey (RDCS) which has allowed these studies to be pushed beyond $z = 1$ for the first time. After describing results about the evolution of the X-ray luminosity function, we show its cosmological implications on the determination of σ_8, the r.m.s. density fluctuation on 8 h^{-1} Mpc scale, and Ω_m. A more detailed description of our analysis is presented in two separate papers (Borgani et al. 2001, Rosati et al. in preparation).

2. The Evolution of the Cluster X-ray Luminosity Function

The basic quantity to characterize the evolution of the cluster space density is represented by the X-ray luminosity function (XLF). In the left panel of Fig. 1 we show the XLF from the RDCS at different redshift intervals, reaching out to $z \simeq 1$, compared to other measurements from both local and distant surveys. Based on these data, several groups have argued that no significant evolution is observed in the space density of distant clusters with $L_X[0.5-2] \lesssim 3 \times 10^{44}$ erg s^{-1} (see Rosati, Borgani & Norman 2001). As demonstrated by the RDCS, this trend persists out to $z \approx 1$. Measurements of the distant XLF at $L_X \gtrsim L_0^* \simeq 5 \times 10^{44}$ erg s^{-1} are difficult with current samples, due to low number statistics. As a result, the evolution of the high end of the XLF has

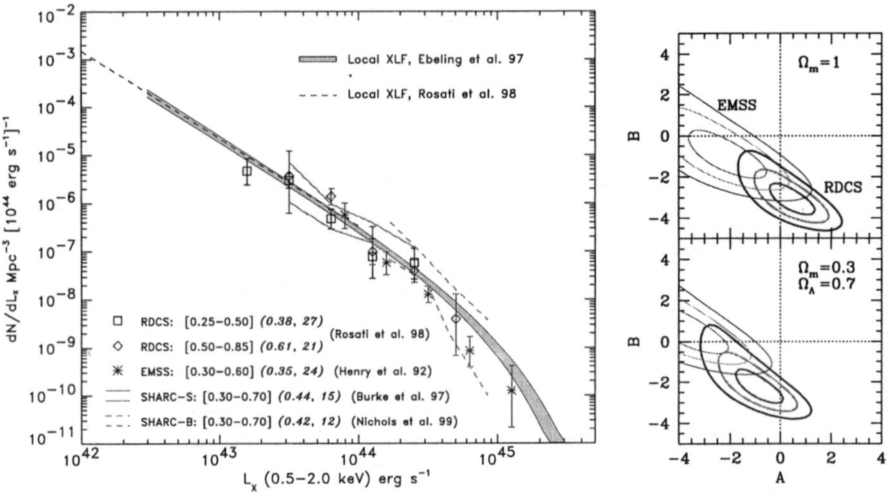

Figure 1. Left panel: distant and local cluster XLFs from the literature. For each sample we indicate the median redshift and the number of clusters of various samples within each redshift bin. Right panel: Maximum–likelihood contours (1, 2 and 3 σ confidence level) for the parameters A and B defining the XLF evolution for the RDCS and EMSS samples, for two different cosmologies.

remained a hotly debated issue, ever since it was first reported in the EMSS (Gioia et al. 1990).

The binned representation of the XLF in Fig. 1 does not provide a full picture of the space density evolution observed in a given sample. In order to fully exploit the amount of information contained in the RDCS, we analyze the unbinned (L_X, z) distribution with a maximum-likelihood (ML) approach, which compares the observed cluster distribution on the (L_X, z) plane with that expected from a given XLF model. We characterize the cluster XLF as an evolving Schechter function, $\phi(L) = \phi_0(1+z)^A L^{-\alpha} \exp(-L/L^*)$, with $L^* = L_0^*(1+z)^B$; where A and B are two evolutionary parameters. Different surveys find consistent values for the faint end slope α, which is not observed to vary as a function of redshift (Fig. 1). For the local XLF, we use here the measurement of the BCS sample by Ebeling et al. (1997), i.e. $\alpha = 1.85$, $L_0^* = 5.7 \times 10^{44}$ erg s^{-1}, $\phi_0 = 3.32\,(10^{-7}\mathrm{Mpc}^{-3}\mathrm{L}_{44}^{\alpha-1})$.

For this analysis, we use the flux–limited RDCS sample with $F_{lim} = 3 \times 10^{-14}$ erg cm^{-2} s^{-1}, which contains of 103 spectroscopically confirmed clusters out to $z_{\max} = 0.85$. Observed flux errors are included in the likelihood computation. The resulting 1σ, 2σ and 3σ c.l. contours in the A-B plane are shown in the right panel of Fig. 1, for two different cosmologies. Best fit values for the $\Omega_m = 1$ case are $A = 0.6^{+1.2}_{-1.7}$ and $B = -3.2^{+1.5}_{-1.1}$ where errors correspond to the 2σ c.l. for two significant fitting parameters. The no evolution model $(A = B = 0)$ is excluded at more than a 3σ confidence level, even when the uncertainties of the local XLF are taken into account. The departure of our best fit model from the no-evolution scenario is due to the small number of observed clusters in the RDCS at $z > 0.5$ with luminosities $L_X \gtrsim L_0^*$ compared to the no-evolution prediction. A redshift dependent inspection of the likelihood also shows that little can be said from RDCS on the evolution of the high end of the XLF at $z \lesssim 0.7$. These findings lead to a consistent picture in which the comoving space density of the bulk of the cluster population is approximately constant out to $z \sim 1$, but the most luminous ($L_X \gtrsim L_0^*$), presumably most massive clusters were indeed rarer at high redshifts.

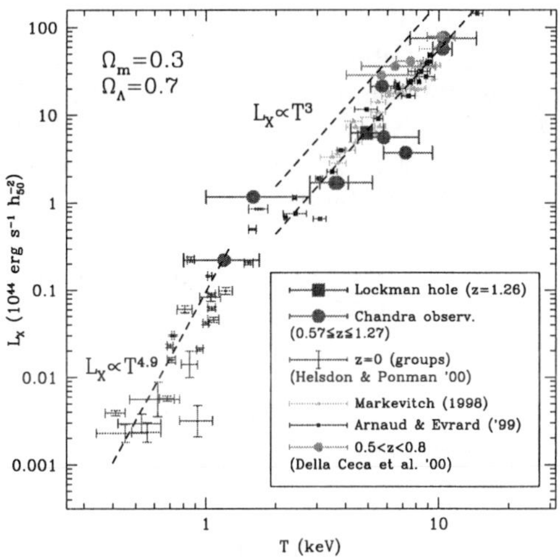

Figure 2. The luminosity–temperature relation for nearby and distant clusters and groups compiled from several sources. The two dashed lines at $T > 2$ keV indicate the slope $\alpha = 3$, with normalization corresponding to the local L_X–T relation (lower line) and to the same relation computed at $z = 1$ for $A = 1$. The dashed line at $T < 1$ keV shows the relation found for groups by Helsdon & Ponman (2000).

3. Constraints on cosmological models

Having quantified the degree of evolution of the cluster XLF, it is worth asking which are its implications on the determination of cosmological parameters. To this purpose, we use the model mass function by Sheth & Tormen (1999). The conversion from masses to X-ray luminosities, which is required to convert a mass function into a XLF, is implemented as follows: **(a)** convert mass into temperature by assuming virialization, hydrostatic equilibrium and isothermal gas distribution; **(b)** convert temperature into bolometric luminosity according to $L_{bol} \propto T^{\alpha}(1+z)^A$. The critical step is represented by the choice for the L_{bol}–T relation. Low redshift data for $T \gtrsim 3$ keV indicates that $\alpha \simeq 2.5$–3, depending on the sample and the data analysis technique, with a reduction of the scatter after account for the effect of cooling flows in central cluster regions (e.g. Markevitch

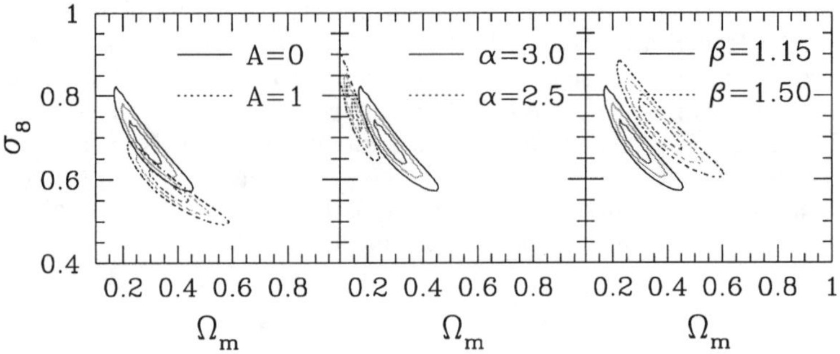

Figure 3. Probability contours in the σ_8–Ω_m plane from the evolution of the L_X distribution of RDCS clusters (from Borgani et al. 2001). The shape of the power spectrum is fixed to $\Gamma = 0.25$. Different panels refer to different ways of changing the M–L_X relation, within theoretical and observational uncertainties.

1998, Arnaud & Evrard 1999). At lower temperatures, evidence has been found for a steepening of the L_{bol}–T_X relation below 1 keV (Helsdon & Ponman 2000). As for the evolution of the L_{bol}–T, Figure 2 shows that no evidence of evolution is observed out to $z \simeq 1.3$. In Figure 3 we show the resulting constraints on the σ_8–Ω_m, in the case of flat geometry, by changing in different ways the M–L_X relation (we assume here $\Gamma = 0.25$ for the shape parameter of the power spectrum). For the reference analysis, we take $\alpha = 3$ and $A = 0$ for the L_{bol}–T relation, and $\beta = 1.15$ for the ratio between the kinetic energy of DM particles and the specific gas thermal energy. A critical density model is always ruled out, with $\Omega_m < 0.6$ at 3σ c.l., for conservative assumptions on the uncertainties in the M–L_X relation. For our reference model, we find $\Omega_m = 0.35^{+0.13}_{-0.10}$ and $\sigma_8 = 0.66^{+0.06}_{-0.05}$ at 1σ c.l. Based on these results, we point out that deep flux–limited X–ray cluster samples, like RDCS, which cover a large redshift baseline ($0.1 \lesssim z \lesssim 1.2$) and include a fairly large number of clusters ($\gtrsim 100$) do indeed place significant constraints on cosmological models.

References

Arnaud, M., & Evrard, A.E. 1999, MNRAS, 305, 631

Borgani, S., et al. 2001, ApJ, 561, 13

Della Ceca, R., et al. 2000, A&A, 353, 498

Ebeling, H., et al. 1997, ApJ, 479, L101

Gioia, I.M., et al. 1990, ApJ, 356, L35

Helsdon, S.F., & Ponman, T.J. 2000, MNRAS, 315, 356

Markevitch, M. 1998, ApJ, 504, 27

Rosati, P., Borgani, S., & Norman, C. 2002, ARAA, submitted

Rosati, P., et al. 1998, ApJ, 492, L21

Sheth, R.K., & Tormen, G. 1999, MNRAS, 308, 119

Tracing Cosmic Evolution with Galaxy Clusters
ASP Conference Series, Vol. 268, 2002
S. Borgani, M. Mezzetti, and R. Valdarnini, eds.

X-ray temperature and morphology of z>0.8 clusters of galaxies

B. Holden

University of California, Davis & Insitute for Geophysics and Planetary Physics, L-413, P.O. 808, Livermore, CA 94551-0808

S. A. Stanford

University of California, Davis & Insitute for Geophysics and Planetary Physics, L-413, P.O. 808, Livermore, CA 94551-0808

P. Rosati

European Southern Observatory, Karl-Schwarzschild-Str. 2, D-85748 Garching bei München

P. Tozzi

Osservatorio Astronomico di Trieste, Via G. B. Tiepolo, 11, 34131 Trieste

G. Squires

California Institute of Technology, MS 105-24, 1201 East California Blvd, Pasadena, CA, 91125

S. Borgani

INFN, Sezione di Trieste, Via A. Pascoli, I-06100 Perugia

P. Eisenhardt

Jet Propulsion Laboratory, California Institute of Technology, MS 169-327, 4800 Oak Grove Drive, Pasadena, CA, 91109

Abstract.
We discuss our current progress in studying a sample of $z > 0.8$ clusters of galaxies from the ROSAT Distant Cluster Survey. To date, we have Chandra observations for four of the ten clusters. We find that the morphology of two of these four are quite regular, with deviations from circular of less than 5%, while two are strikingly elliptical. When the temperatures and luminosities of our sample are grouped with six other high-redshift measurements, there is no measured evolution in the luminosity-temperature relation. We identify a number of X-ray emitting point sources that are potential cluster members. These could be sources of intracluster medium heating, adding the entropy necessary to explain the cluster luminosity-temperature relation.

1. Introduction

Our sample of $z > 0.8$ clusters of galaxies is part of the ROSAT Distant Cluster Survey (RDCS; Rosati *et al.* 1998). The RDCS contains 137 clusters of galaxies covering 50 deg^2 to an X-ray flux limit of 1×10^{-14}erg s^{-1} cm^{-2} in the 0.5-2.0 keV band. This includes the highest redshift, X-ray selected clusters galaxies known to date, with ten clusters at $0.8 \leq z \leq 1.3$. The luminosity and redshift distribution of our high redshift sample is plotted in Figure 1, along with that distribution for the Bright SHARC sample (Romer *et al.* 2000) and the Einstein Medium Sensitivity Sample of clusters of galaxies (Henry *et al.* 1992; Gioia & Luppino 1995) for comparison.

We have targeted this sample for follow-up with the two premier X-ray observatories, Chandra and XMM-Newton. To date, we have Chandra observations for four clusters in that sample, with the details listed in Table 1. The X-ray data for two of the clusters discussed here, RX J0848+4453 and RX J0849+4452, are presented in Stanford *et al.* 2001. For the rest of this work we will assume $\Omega_m = 0.3$, $\Omega_\Lambda = 0.7$ and $H_\circ = 65$ km s^{-1} Mpc^{-1}.

Table 1. Summary of Chandra Observations

Name	α (J2000)	δ (J2000)	z	Lum. (10^{44} erg s^{-1})	Temp. (keV)
RX J0848+4453	08 48 35.8	+44 53 45.5	1.26	$0.64^{+0.25}_{-0.16}$	$1.6^{+0.8}_{-0.6}$
RX J0849+4452	08 48 58.7	+44 51 53.3	1.27	$3.3^{+0.7}_{-0.5}$	$5.8^{+2.6}_{-1.7}$
RX J0910+5429	09 10 44.9	+54 22 07.7	1.10	$2.0^{+.3}_{-0.2}$	$7.2^{+2.2}_{-1.4}$
RX J1317+2911	13 17 21.7	+29 11 18.1	0.80	$1.8^{+0.7}_{-0.4}$	$3.7^{+1.5}_{-0.8}$

2. Morphology

Quantitative morphology for clusters of galaxies usually means fitting a β model. We fit two-dimensional models to the 0.5-2.0 keV photon distributions for the four clusters of galaxies in our sample, after excluding point sources. The model included a constant term for the background in addition to the cluster model. These results are shown in Figure 2 and in Table 2.

Name	β	r_c (kpc)
RX J1317+2911	0.3	1
RX J0910+5429	$0.89^{+0.24}_{-0.23}$	171^{+49}_{-57}
RX J0849+4452	$0.73^{+0.31}_{-0.15}$	81^{+23}_{-23}
RX J0848+4452	0.8	200

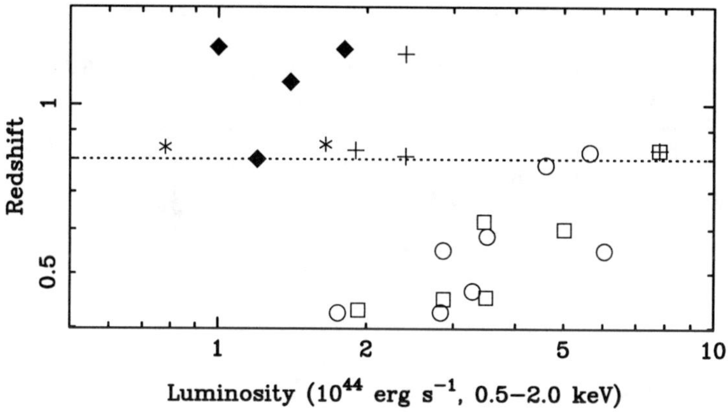

Figure 1. The luminosity and redshift distribution of our sample compared with the Bright SHARC (open squares) and the EMSS (open circles). The four diamonds are discuss in this work, while the remaining are either pending observations (plus symbols) or not observed (asterisks). The dotted line shows our redshift limit. The values of the luminosities are the catalog values as measured by the RDCS, not values from the new Chandra data

Two things are striking. First, two of the models are quite round. The deviations from a circular model are at less than 5% for both. This is in contrast with, say, MS 1054-0321 or RX J0152.7-1357, which show greatly disturbed, asymmetric morphologies. Second, for these round clusters, the best fitting core radii and values of β are normal for low redshift clusters of galaxies. These two clusters are also have the largest number of net counts. For RX J0910+5429, in the 0.5-2.0 keV band we measured a total of 500 counts, including the background, within two core radii. We expected 137.4 counts from the background in the same bandpass. With RX J0849+4452, we found 469 counts in the 0.5-2.0 keV band within two core radii while expecting 142.8 from the background. In contrast, RX J0848+4453 has only 259 counts with 104.6 background counts expected and RX J1317+2911 has 217 events with 54.7 expected. Therefore, we may only be resolving the cores of these systems, and missing a smooth outer region because of the lack of events. Nonetheless, both have strange morphologies not well described with a β model. Because of the poorer statistics and the strange morphologies for these clusters, we quote no error bars on our measured parameters for the β model. RX J0848+4453 is also strongly contaminated by point sources, see Figure 2. Two of these point sources are potential cluster members, based on photometric redshifts, and are likely to be active galactic nuclei based on their hardness ratios. After excluding these sources, the flux we measured for RX J0848+4453 was significantly lower than what we measured in the original RDCS.

For the two clusters with well constrained morphologies, RX J0849+4452 and RX J0910+5429, we used an aperture of twice the core radius to measure

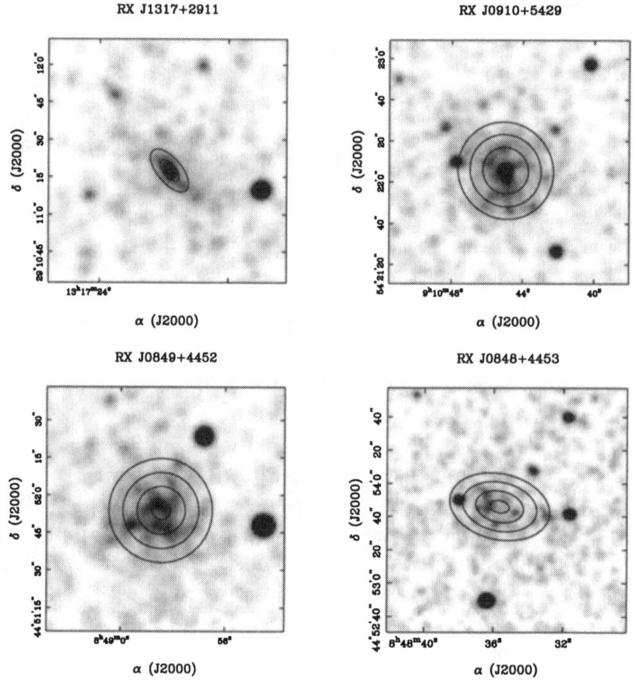

Figure 2. Images of the four clusters in our sample. We plot the smoothed photon distribution with the best fitting β model overlaid as contours. Each model has four contour levels corresponding to 90% of the peak value, 50% of the peak value or the core radius, 25% of the peak value and 12.5% of the peak value. The exception is RX J1317+2911 which has contours of 12.5%, 3%, 2% and 1% of the peak amplitude.

the temperature and flux. We than use the best fitting β model to compute the total flux. In Table 1, we quote the best fitting temperatures and luminosities. For the two remaining clusters, we used a curve of growth analysis to pick a total flux aperture. We used that aperture to measure the temperature as well. These results are also included in Table 1.

3. Luminosity-Temperature Relation

We plot, in Figure 3, the luminosities and temperatures for our data and a number of other high redshift clusters. For comparison, we also plot the relation for low redshift clusters of galaxies as measured by Markevitch (1998) and the low redshift group data of Helsdon & Ponman (2000). When our data are combined with the other high redshift clusters of galaxies, it appears that there is minimal or no evolution in the luminosity-temperature relation.

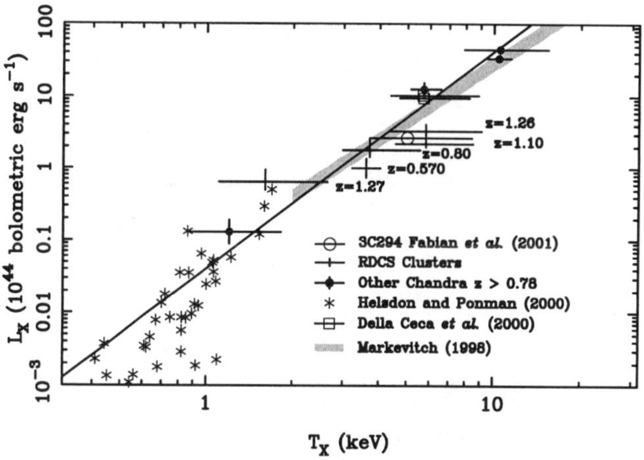

Figure 3. The Luminosity-Temperature relation for high redshift clusters of galaxies. The z > 0.78 clusters come from Borgani *et al.* (2001). The measurements from della Ceca *et al.* (2000) are only those clusters with z > 0.8, not the entire sample in that paper. The points marked by redshifts are various RDCS clusters, including the cluster at z=0.570 from Holden *et al.* (2001). The solid line represents are best fit to the relation.

As a rough test for evolution, we fit the relation $L_{Bol} \propto T^{\alpha}$ to the high redshift data, including the four clusters in our sample, the two $z > 0.8$ clusters (RX J1716.6+6708, and RX J0152.7-1357) from della Ceca et al. (2000), three clusters (MS1137.5+6625, 1WGAJ1226.9, & CDFS-CL1) summarized in Borgani et al. (2001) and the results for MS1054.4-0321 from Jeltema *et al.* (2001) (see also these proceedings). We fit the relation using the method of Akritas and Bershady (1996) which accounts for errors in both the temperature and luminosity. For our sample of ten clusters and groups, we found the best fit slope to be $\alpha = 3.0 \pm 0.5$ (90% confidence limits) at our median redshift of $z = 0.83$. Our measured slope differs by slightly more than one standard deviation from $\alpha = 2.63 \pm 0.27$ (90% confidence limits), the relation of Markevitch (1998). Our result is in good agreement with the slope of $\alpha = 3.1 \pm 0.6$ from Allen & Fabian (1998). We note here that we fit only isothermal models to all of the clusters in our sample. Therefore, we compare our results with the Model A from Allen & Fabian (1998), which used similar assumptions.

4. A Potential Source of Intracluster Heating?

In Figure 4, we plot a color magnitude diagram for the 4 clusters in our sample. We have shifted the colors and magnitudes to the median redshift for our sample

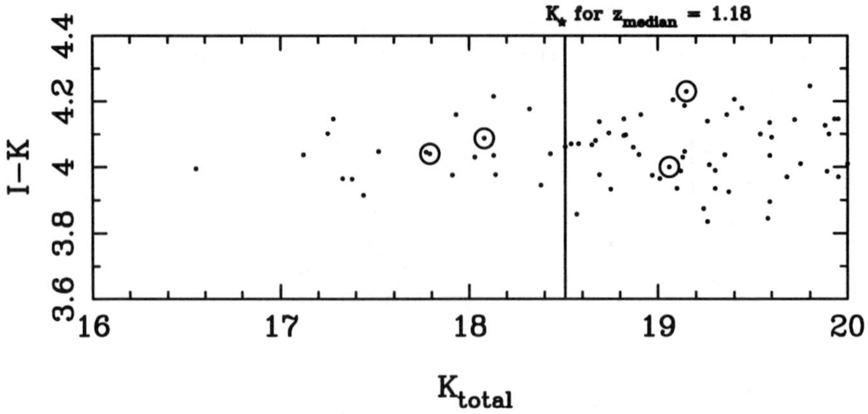

Figure 4. The color-magnitude diagram for all four clusters in our sample. The colors and magnitude for each cluster galaxy shifted to the median redshift of our sample, $z = 1.18$. Circled objects are galaxies within one standard deviation of the color-magnitude relation and are X-ray point sources less than $2'$ from the cluster center.

of 1.18. We circle in X-ray point sources that are within one standard deviation of the mean color-magnitude relation and within $2'$ of the cluster center. So, each of these galaxies are possible cluster members and they appear to have X-ray emission. The number of these objects in entirely consistent with the results of Barger et al. (2001) who find 4% of $L_\star$ or bright galaxies are X-ray emitters and 7% of their entire sample. As the potential sources of the X-ray emission are active galactic nuclei or a very hot interstellar medium, these objects could be important sources of entropy for the intracluster medium (ICM), e.g., Bower (1997) or Ponman, Cannon & Navarro (1999). In Figure 5, we show a hard X-ray image of RX J0910+5429 along with the same soft image shown in Figure 2. There is an obvious hard excesses to the south of the core of the cluster. The hard excess is statistically significant at the 99% confidence limit when compared with both the central region of the cluster (within an equal area centered slightly to the north of the centroid) and when compared with the background. One explanation is we are observing a merger event and the resulting shock. Another is the circled X-ray source in the southeast of the cluster, one of the candidates in Figure 4, is heating the ICM.

Therefore, if we understand these hard X-ray sources and, most importantly, learn how much energy and over what range redshifts they are adding to the ICM, we can directly test the idea of AGN heating of the ICM (see Valageas & Silk 1999; Wu, Fabian & Nulsen 2000) and answer the question of the origin of the ICM pre-heating.

5. Summary

We have observed with the Chandra telescope four clusters of galaxies in our sample of ten at $z > 0.8$. Two of these four are relaxed looking clusters with

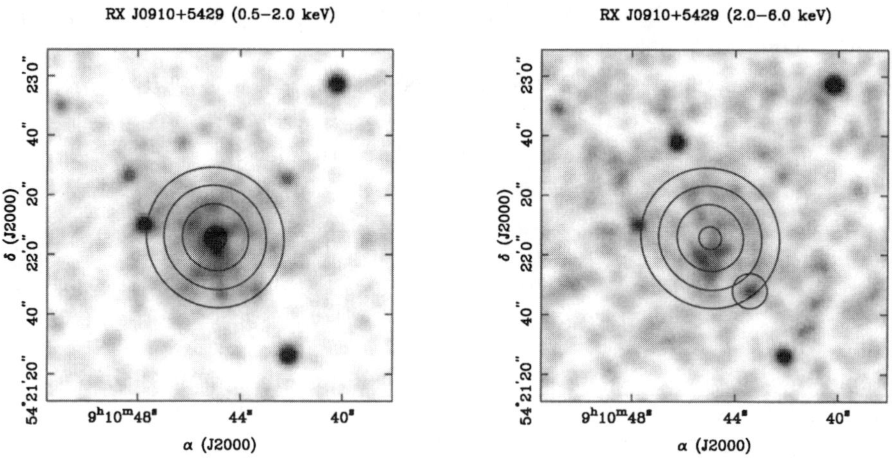

Figure 5. Images of RX J0910+5429 in the ROSAT band and in a hard band from 2.0 - 6.0 keV. We plot the smoothed photon distribution with the best fitting β model overlaid as contours. Each model has four contour levels corresponding to 90% of the peak value, 50% of the peak value or the core radius, 25% of the peak value and 12.5% of the peak value. The circle, in the right panel, is around a potential hard X-ray emitting cluster member.

values for β and the core radius entirely consistent with low redshift clusters. The other two, however, have elongated morphologies and, in one case, an extreme value for β and the core radius.

Despite the wide range of morphologies, all of our clusters agree with the low redshift luminosity temperature relation. Including other results, we can see that the L-T relation has little evolution over almost one order of magnitude in temperature out to a $z_{median} = 0.83$.

Finally, we identify a potential source of cluster members for the additional entropy needed to explain the L-T relation. Further investigation into these objects could shed light on the origin and evolution of the intracluster medium.

References

Akritas, M. G. & Bershady, M. A. 1996, ApJ, 470, 706

Allen, S. W. & Fabian, A. 1998, MNRAS, 297, L57

Barger, A., Cowie, L. L., Mushotzky, R. F., & Richards, E. A. 2001, AJ, 121, 662

Borgani, S., Rosati, P., Tozzi, P., Stanford, S. A., Eisenhardt, P., Lidman, C., Holden, B., della Ceca, R., Norman, C., & Squires, G. 2001, ApJ, submitted, astro-ph/0106428

Bower, R. G. 1997, MNRAS, 288, 355

della Ceca, R., Scaramella, R., Gioia I.M., Rosati, P., Fiore, F. & Squires, G. 2000, A&A, 353, 498

Fabian, A., Crawford, C. S., Ettori, S. & Sanders, J. S. 2001, MNRAS, 322, L11

Gioia, I. & Luppino, G. 1994, ApJS, 94, 583

Helsdon, S. F. & Ponman, T. J. 2000, MNRAS, 315, 356

Henry, J. P., Gioia, I. M., Maccacaro, T., Morris, S. L., Stocke, J. T. & Wolter, A. 1992, ApJ, 386, 408

Holden, B., Stanford, S. A., Rosati, P., Squires, G., Tozzi, P., Fosbury, R. A. E., Papovich, C. Eisenhardt, P., Elston, R., & Spinrad, H. 2001, AJ, 122, 629

Markevitch, M. 1998, ApJ, 504, 27

Ponman, T. J., Cannon, D. B., & Navarro, J. F. Nature, 397, 135

Romer, A.K., Nichol R.C., Holden, B. P., Ulmer, M.P., Pildis, R.A., Adami, C., Burke, D.J., Collins, C.A., Merrelli, A.J., & Metevier, A.M. 2000, ApJS, 126, 209

Rosati, P., Della Ceca, R., Norman, C., & Giacconi, R. 1998, ApJ, 492, L21

Stanford, S. A., Holden, B. P., Rosati, P., Tozzi, P., Borgani, S., Eisenhardt, P., & Spinrad, H. 2001, ApJ, 552, 504

Valageas, P. & Silk, J. 1999 A&A, 347, 1

Wu, K. K. S., Fabian, A. C., & Nulsen, P. E. J. 2000, MNRAS, 318, 889

Tracing Cosmic Evolution with Galaxy Clusters
ASP Conference Series, Vol. 268, 2002
S. Borgani, M. Mezzetti, and R. Valdarnini, eds.

Tracing Cosmic Evolution with an *XMM* Serendipitous Cluster Survey

A. Kathy Romer

Physics Department, Carnegie Mellon University, 5000 Forbes Avenue, Pittsburgh, PA 15213, USA

Pedro T. P. Viana

Centro de Astrofísica da Universidade do Porto, Rua das Estrelas s/n, 4150 Porto, Portugal

Christopher A. Collins

Astrophysics Research Institute, Liverpool John Moores University, U.K.

Andrew R. Liddle

Astronomy Centre, University of Sussex, Brighton, UK.

Robert G. Mann

Institute for Astronomy, University of Edinburgh, Royal Observatory, Blackford Hill, Edinburgh, UK.

Robert C. Nichol

Carnegie Mellon University, Physics Department, 5000 Forbes Avenue, Pittsburgh, PA 15213, USA.

Abstract.
This paper describes updated predictions, as a function of the underlying cosmological model, for a serendipitous galaxy cluster survey that we plan to conduct with the *XMM-Newton* X-ray Satellite. We have included the effects of the higher than anticipated internal background count rates and have expanded our predictions to include clusters detected at $> 3\sigma$. Even with the enhanced background levels, we expect the XCS to detect sufficient clusters at $z > 1$ to differentiate between open and flat cosmological models. We have compared the XCS cluster redshift distribution to those expected from the *XMM* Slew Survey and the *ROSAT* Massive Cluster Survey (MACS) and find them to be complementary. We conclude that the future existence of the XCS should not deter the launch of a dedicated X-ray survey satellite.

$T > 4$ keV	$\Omega_0 = 0.3, \Omega_\Lambda = 0.7$			$\Omega_0 = 1.0, \Omega_\Lambda = 0.0$			$\Omega_0 = 0.3, \Omega_\Lambda = 0.0$		
	$> 8\sigma$	$> 8\sigma$	$> 3\sigma$	$> 8\sigma$	$> 8\sigma$	$> 3\sigma$	$> 8\sigma$	$> 8\sigma$	$> 3\sigma$
	i	*ii*	*iii*	*i*	*ii*	*iii*	*i*	*ii*	*iii*
$z > 0$	750	720	890	80	80	80	1100	1010	1530
$z > 0.3$	700	660	830	50	50	50	1060*	970	1490
$z > 1$	170	150	300	1	1	2	480	390	900

Table 1. The expected number of $T > 4$ keV clusters detected by the XCS as a function of cosmology. Column *i*, the original XCS predictions for $> 8\sigma$ detections from R01. Columns *ii* and *iii*, updated XCS predictions based on the measured in flight internal background count rate. All the results assume a final survey area of 800 sq. degrees. *There was a typo in the the printed version of R01, this is the correct value.

1. Introduction

In an earlier paper (Romer et al. 2001, R01 hereafter) we described the expected catalog properties and scientific applications of an *XMM-Newton* Cluster Survey (XCS) based on serendipitous detections in pointed observations. We revisit these predictions below, in light of improved knowledge of the in flight performance of *XMM*. We also compare the XCS to other on-going X-ray surveys.

2. The Effect of an Enhanced Internal Background Count Rate on the XCS

Recently it has come to light that the quiescent internal background count rates in the EPIC detectors are roughly ten times higher than was anticipated prior to the *XMM-Newton* launch (see Lumb 2001). We have, therefore, recomputed the expected properties of the XCS cluster catalog using a total (external plus internal) background level that is roughly double that used in R01. In Table 1, we compare the new predictions (column *ii*) with those published in Table 4 of R01 (column *i*). For example, for an open model ($\Omega_0 = 0.3$, $\Omega_\Lambda = 0$), we now predict that the XCS will detect 390 $T > 4$ keV, $z > 1$ clusters at $> 8\sigma$, compared to 480 in R01. Changes at that level should not severely limit the ability of XCS to differentiate between open and flat cosmological models.

We have also included in Table 1 predictions for the number of XCS clusters that would be detected at $> 3\sigma$. We did not include low signal to noise detections in R01 because they will be hard to identify using an extent criterion alone. However, when combined with Planck and/or SDSS data, it may be possible to identify these objects in a timely and quantifiable manner, so we include them here for completeness.

The redshift distribution of the $> 3\sigma$ and $> 8\sigma$ XCS detections can be seen in Figure 1.

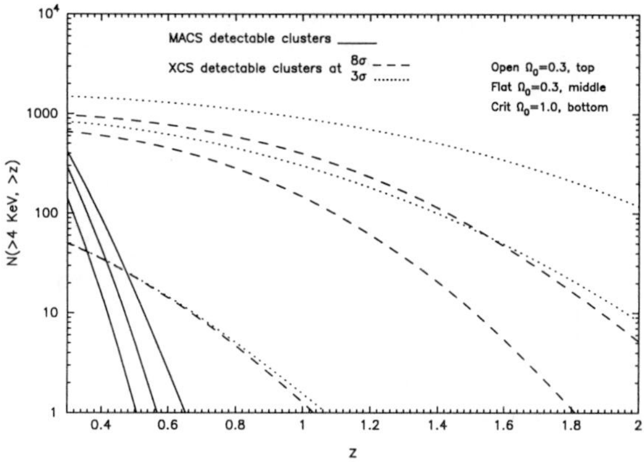

Figure 1. Predicted redshift distribution ($z > 0.3$) of MACS clusters (solid lines) and of the XCS Survey (dashed and dotted lines).

3. Comparisons with the MACS and *XMM* Slew Surveys

The Massive Cluster Survey (MACS, Ebeling, Edge & Henry 2001) has been very successful at identifying high redshift, high luminosity, clusters detected in 22,735 square degrees of the ROSAT All Sky Survey. The wide areal coverage of the MACS (nearly 30 times that of the XCS!) results in sensitivity to a very interesting class of clusters, those at medium redshift and high luminosity ($L_x > 5 \times 10^{44}$ erg cm^{-2} s^{-1}; 0.1-2.0 keV). MACS clusters are already being used for a variety science programs including studies of cluster evolution and the Sunyaev-Zel'dovich Effect.

Using the R01 methodology and based on the MACS selection function presented in Figure 5 of Ebeling et al. (2001) we have been able to predict the number of clusters that would be detected by MACS as a function of cosmology. We present those results in Figure 1 and Table 1. Overall, we can expect MACS to detect between 100 and 500 clusters at $z > 0.3$, depending on cosmology.

An *XMM* Slew Survey has been proposed (*e.g.* Lumb & Jones 2000). An *XMM* Slew survey would have two important advantages over the XCS; it would be drawn from random parts of the sky (rather than from regions that surround known X-ray targets) and it would cover significantly more area. Lumb & Jones (2000) estimated an annual survey rate of 4000 square degrees to a flux limit of 2×10^{-13} erg cm^{-2} s^{-1} (0.5-2.0 keV). This compares to an estimated annual survey rate for the XCS of $\simeq 80$ square degrees (the XCS catalog will not have a single flux limit, but a typical value will be $\simeq 1.5 \times 10^{-14}$ erg cm^{-2} s^{-1} in the 0.5-2.0 keV band, see R01).

We have attempted to predict the properties of a Slew Survey cluster catalog using the R01 methodology. At the time of writing, the Slew Survey had only just begun and the actual sensitivity level was unknown (Lumb, private communication). We, therefore, present results for two different (0.5-2.0 keV) flux limits; 2×10^{-13} erg cm^{-2} s^{-1} and 3×10^{-12} erg cm^{-2} s^{-1}, see Figure 2. We have assumed that 4000 square degrees would be covered in one year by the

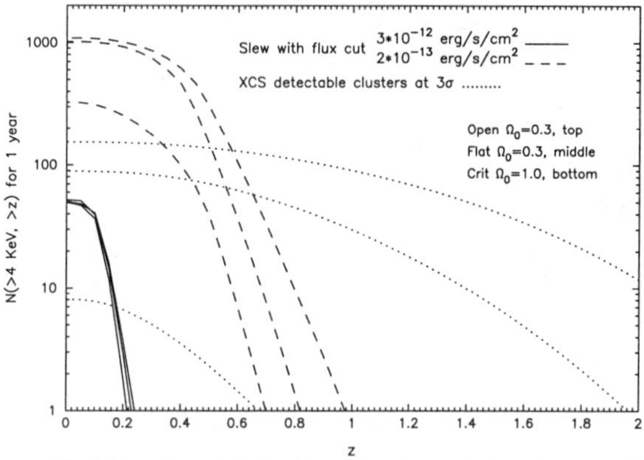

Figure 2. Predicted redshift distribution of the clusters detected during the first year of an *XMM* Slew Survey (solid and dashed lines) and of the XCS Survey (dotted lines).

Slew Survey. For comparison we have shown the expected numbers of clusters that would be detected at $> 3\sigma$ in one year of the XCS (or 80 square degrees).

A Slew Survey with 2×10^{-13} erg cm^{-2} s^{-1} sensitivity would produce a very impressive cluster catalog at low and medium redshifts. Such a catalog would have a variety of applications including constraints on σ_8 and Ω_0 and studies of cluster evolution and Planck foregrounds.

In summary, the MACS and the *XMM* Slew Surveys have much better low redshift sensitivity than XCS by virtue of their large areal coverage. However, neither will be able to differentiate between open and flat low Ω_0 models, because they lack sensitivity to high redshift clusters. These surveys should be seen as complementary too, not in competition with, the XCS.

The XCS faces many challenges, including the complexity of its selection function and our lack of understanding of cluster evolution. The future existence of the XCS should not deter the launch of a dedicated X-ray survey satellite.

References

Ebeling, H., Edge, A. C., & Henry, J. P. 2001, ApJ, 553, 668

Lumb, D. H. & Jones, L. R. 2000, Large Scale Structure in the X-ray Universe, Proceedings of the 20-22 September 1999 Workshop, Santorini, Greece, eds. Plionis, M. & Georgantopoulos, I., Atlantisciences, Paris, France, p.313

Lumb, D. H., 2001, XMM-Newton CCF Release Note, *EPIC Background Files*, http://xmm.vilspa.esa.es/docs/documents/CAL-TN-0016-1-0.ps.gz

Romer, A. K., Viana, P. T. P., Liddle, A. R., & Mann, R. G. 2001, ApJ, 547, 594 (R01)

We would like to acknowledge the assistance of Juliet McKensie, Shane Zabel and Kevin Bandura with the talk preparations.

Tracing Cosmic Evolution with Galaxy Clusters
ASP Conference Series, Vol. 268, 2002
S. Borgani, M. Mezzetti, and R. Valdarnini, eds.

Brightest Cluster Galaxies and the Las Campanas Distant Cluster Survey

Dennis Zaritsky

Steward Observatory, 933 N. Cherry Ave., University of Arizona, Tucson, AZ, 85721, USA

Amy E. Nelson

Department of Astronomy and Astrophysics, Univ. of Calif. at Santa Cruz, Santa Cruz, CA, 95064, USA

Luc Simard

Steward Observatory, 933 N. Cherry Ave., University of Arizona, Tucson, AZ, 85721, USA

Anthony H. Gonzalez

Harvard-Smithsonian Center for Astrophysics, 60 Garden St., Cambridge, MA, 02138, USA

Julianne J. Dalcanton

Department of Astronomy, University of Washington, Box 351580, Seattle, WA, 98195, USA

Abstract. We present our study of the evolution of brightest cluster galaxies based on data from the Las Campanas Distant Cluster Survey (Gonzalez*et al.* 2001) and HST optical and IR imaging. We briefly discuss the technique that enabled us to use short (~ 3 min) exposures and a small (1m) telescope to efficiently survey ~ 130 sq. deg. of sky for high-z galaxy clusters. Follow-up imaging is used to construct a sample of several tens of BCGs at $0.3 \leq z \leq 0.9$ with which to explore their evolution. In particular, we examine the luminosity and color evolution of these galaxies. We confirm the previous results that 1) BCGs in different environments evolve differently (eg. Burke, Collins, & Mann 2000), 2) BCGs, particularly those in low-mass clusters, must be accreting significantly since $z \sim 1$ (Aragon-Salamanca *et al.* 1998). From measurments of the colors vs. redshift, we conclude that this accretion must consist of old stellar populations. Using HST NICMOS and WFPC2 images, we find preliminary evidence for evolution in the sizes of BCGs that is consistent with the accretion model presented in standard hierarchical models (Aragon-Salamanca *et al.* 1998), but puzzling differences between NICMOS and WFPC2 scale lengths precludes a definitive conclusion.

1. Introduction

While definition dictates that every galaxy cluster must have a brightest galaxy, these galaxies are not simply the bright tail of the luminosity function (Oemler 1976; Tremaine & Richstone 1977)). Locally, these galaxies are excellent standard candles (Hoessel 1980) and they appear to remain so to $z \sim 1$ (Aragon-Salamanca et al. 1993; Collins, Burke, and Mann 2000). They appear to be intricately tied to their host cluster because at least some have semimajor axes that extend well beyond 500 kpc (cf. Gonzalez et al. 2000). The evolution of these galaxies is therefore part galaxy evolution, part cluster evolution. The Las Campanas Distant Cluster Survey (LCDCS) provides a new, large sample of clusters with which to construct significant samples of BCGs at $z > 0.5$.

2. The Survey

The basic premise of our cluster detection technique is that we utilize the light from unresolved cluster galaxies. Therefore, we do not need to obtain deep images of the sky to detect a statistically significant number of cluster galaxies, rather we need an image that contains a statistically significant number of cluster photons. The method is described by Dalcanton (1996), Zaritsky *et al.* (1997), and Gonzalez *et al.* (2001).

Our current survey is based on drift scan observations obtained with the Las Campanas 1m Swope telescope and the Great Circle Camera (Zaritsky, Bredthauer, & Shectman 1996). We obtained two scans through each region of the survey area, each with an effective exposure time of $\sim$ 90 s. The power of the technique is evident in the detection of clusters out to $z \sim 1.1$ with a 1m telescope and 3 min exposures. Briefly, the technique involves several flat-fielding passes, masking of bright stars, removal of resolved galaxies and faint stars, and smoothing with a kernel that corresponds roughly to the size of cluster cores at $z \sim 0.6$. We refer the interested reader to Gonzalez *et al.* (2001) for a description of the data reduction techniques and a complete listing of all the candidate clusters.

3. Follow-Up Observations

The most telescope-intensive part of the project has been the follow-up observations aimed at testing our cluster candidates and calibrating methods to estimate the cluster redshift and mass (see below). From imaging in V, I, and K', we construct a sample of BCGs drawn from photometry of 63 confirmed galaxy clusters (not all are imaged in each filter).

With surveys that produce thousands of candidate galaxy cluster identifications, it becomes impractical to obtain redshift and mass measures via spectroscopy for a significant fraction of the candidates. Ideally, these quantities can be estimated from the survey data themselves. In doing so, we will necessarily sacrifice precision on a cluster-by-cluster basis, but as long as the uncertainties are well understood, the sheer number of clusters allows high precision measures of the statistical properties of the sample. There really is no choice in this matter - to obtain a sufficient number of galaxy spectroscopic redshifts for mass

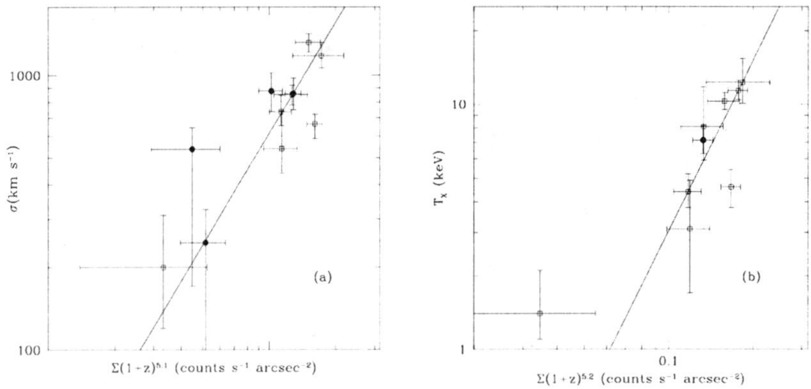

Figure 1. X-ray temperature and galaxy velocity dispersion vs. optical central surface brightness, Σ

measurements of a sample of ONLY 20 clusters requires $\sim$ 35 nights on an 8 to 10m class telescope.

We use cluster spectroscopic redshifts when available, or photometric redshifts based on the color of the red sequence of early-type galaxies in the cluster (Nelson *et al.* 2001a). The calibration of the mass indicator is much more speculative primarily because of the dearth of mass estimates for high redshift clusters. Only one previously known X-ray cluster at $z > 0.35$ lies within our survey area. To augment this, we did small drift scans around 17 other clusters with X-ray luminosity and/or temperature measurements. Unfortunately, the relationships between Σ and other mass indicators (L_x, T_x, or velocity dispersion) are somewhat poorly defined (Figure 1). Two aspects are particularly vexing: 1) we need to determine not only the relation between these quantities but we also need to quantify the scatter to successfully apply this method, and 2) most of the data are for clusters at the low end of the redshift range. Extracting the full potential of this, or any other survey, is predicated on developing a reliable mass estimator and understanding its uncertainties. Nevertheless, these correlations enable us to split the sample into likely low-L_X and high-L_X subsamples to investigate the role of environment on BCG properties.

Selecting the BCG is often non-trivial because of contamination and the definition of a selection radius. Due to space limitations we do not discuss this issue further here (a full discussion is presented by Nelson *et al.* 2001a).

4. Luminosity and Color Evolution and the Nature of the Accretion

We show the I-band BCG Hubble diagram in Fig 2. In the lower panel we bin the data along redshift to reveal the mean trends. At $z < 0.6$ the average BCG appears to follow the no-evolution prediction, while at high redshifts they appear to begin to edge toward the passive evolution model. We find that this peculiar behavior is due to the inclusion of BCGs from a wide range of

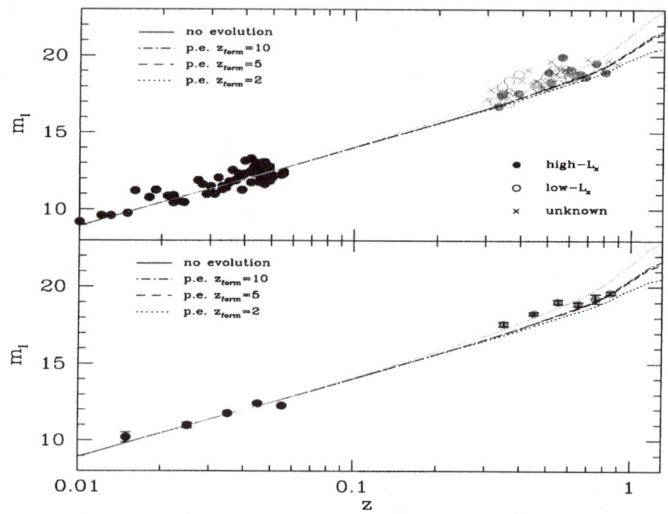

Figure 2. I-band BCG Hubble Diagram. Symbols in the upper panel represent individual BCGs, symbols in the lower panel represent the means in redshift bins. The BCGs at $z < 0.1$ are from Graham *et al.* Clusters with 'unknown' values of L_X at $z > 0.6$ are almost certainly high-L_X clusters because those are the only ones sufficiently bright to be detected. The model lines are described in the upper left of each panel (p.e. = passive evolution).

clusters. In particular, as we discuss next, we find that the BCGs in clusters with $L_X > 2 \times 10^{44}$ ergs s^{-1} lie along the passive evolution models, while the BCGs in fainter clusters lie along the no-evolution models. The different relative fraction of the two types of BCGs at different redshifts leads to the initially perplexing Hubble diagram.

Following the suggestion by Burke, Collins, & Mann (BCM) that 2×10^{44} ergs s^{-1} may be an appropriate division of cluster types at these redshifts, we have divided our sample at the corresponding value of the central surface brightness. Because at $z > 0.6$ our clusters fall primarily in the high L_X sample, the high L_X sample is consistent with passive evolution models. The low L_X sample, which is primarily our lower redshift sample, is consistent with no-evolution models.

Aragon-Salamanca, Baugh and Kauffmann (1998) interpreted consistency with no-evolution models to indicate the presence of accretion (because no-evolution is non-physical). They inferred that these galaxies accreted 2 to 4 times their mass since a redshift of one. BCM found that these results were derived primarily from a sample of BCGs in low-L_X clusters and concluded that much lower (by a factor of 2) rates were upper limits on the accretion for BCGs in the more massive environments. Qualitatively, we confirm both results (accretion is necessary in BCGs in low-L_X environments and less (or no) accretion is necessary in BCGs in high-L_X environments).

Our data in V, I, and K' allow us to investigate the color evolution of BCGs. In contrast to the situation for luminosity evolution, the color evolution

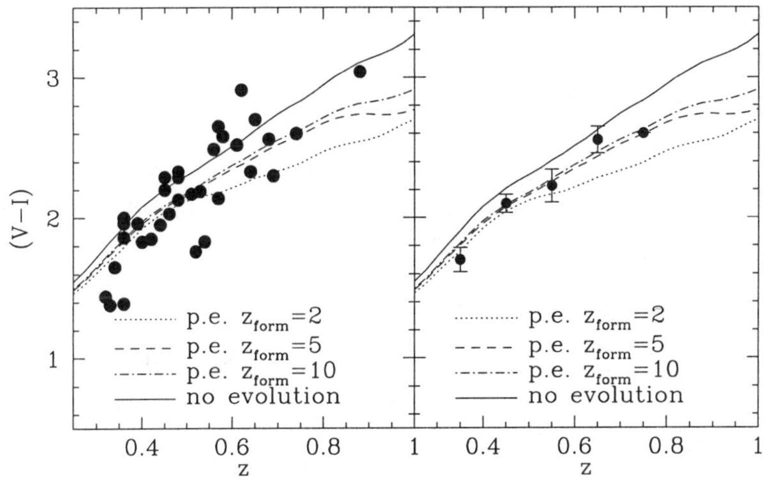

Figure 3. $V - I$ Color Evolution of BCGs. Left panel contains all of our BCGs for which V and I data are available and the right panel contains the same data binned by redshift. The sole galaxies beyond $z \sim 0.85$ is not included in the binned version.

is consistent with passive evolution for both the low and high L_X samples (Figure 3). In all cases the stellar populations of these galaxies appear to be rather old, passively evolving. Therefore, whatever accretion is occurring, it consists of old stellar populations, not the accretion of gas (i.e. a cooling flow) or of gas-rich galaxies, which would presumably lead to a starburst. Interestingly, this result fits in well with the recent discovery of red mergers at high redshift (van Dokkum *et al.* 2000 in MS1054-03).

5. The Search for Direct Evidence of Accretion

If significant accretion is occurring, can we see the structural properties of BCGs change with redshift? If BCGs have velocity dispersions that are nearly independent of mass (as seen locally Malumuth & Kirshner 1981, 1985; Oegerle & Hoessel 1991), then changes in their masses must be reflected by changes in their sizes. If BCG surface brightness profiles can be characterized by deVaucouleurs profiles (see Graham *et al.* 1996; Gonzalez *et al.* 2000 for a variety), then a measure of their effective radii will provide a measure of their size. The nature of the profiles at large radius is still a subject of debate (cf. Schombert 1986; Graham *et al.* 1996; Gonzalez *et al.* 2000), so it is possible that BCGs could change size and yet retain a constant scale length at smaller radii (therefore, a null result with regard to scale length changes as a function of redshift may not rule out accretion).

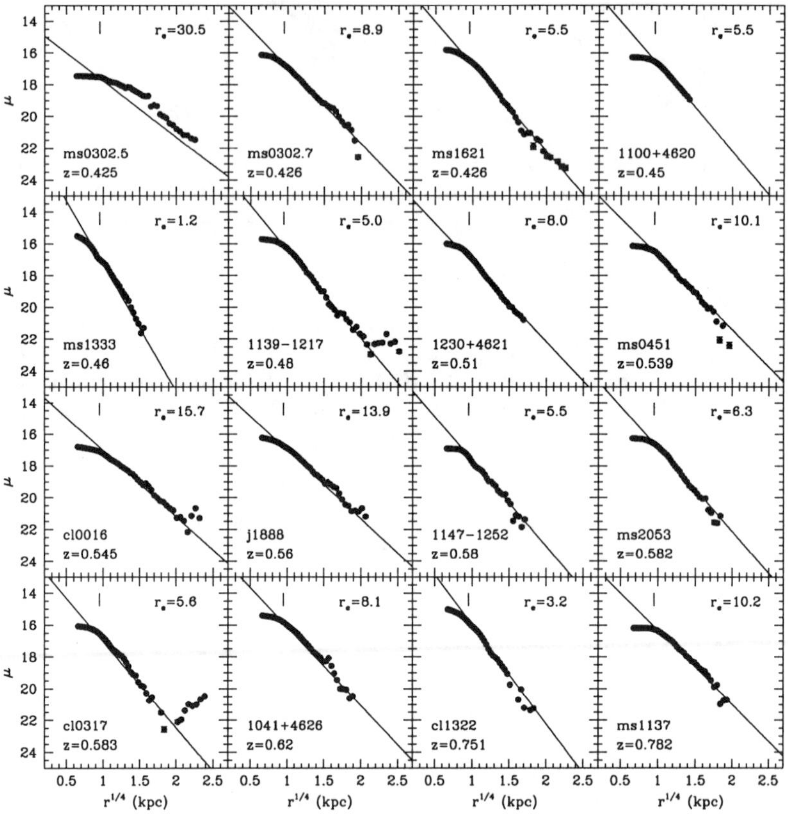

Figure 4. H-band NICMOS Surface Brightness Profiles of High-z BCGs. The vertical tick mark in each panel represents the resolution limit.

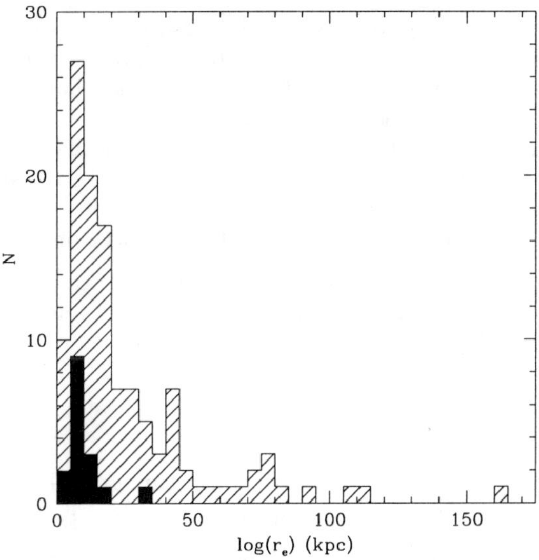

Figure 5. Comparison of Scale Lengths of Local and Distant BCGs. Hashed histogram is the distribution of effective radii for local BCGs and the filled histogram is the result from our NICMOS snapshots (Nelson *et al.* 2001).

Using NICMOS snapshots of 16 BCGs in a heterogeneous sample of literature clusters at a mean redshift of ~ 0.5, to which we have fitted de Vaucouleurs profiles using the Gim2D software (Simard 1998), we measure the distribution of scale lengths and compare to that measured from a local sample by Graham *et al.* For all but one of the galaxies the $r^{1/4}$-law profiles fit well (Figure 4). Comparison to similar measurements on archival WFPC2 images of the BCGs for which some images are available, suggests that non-uniform sky variations in the IR may lead us to underestimate the scale length by as much as 30% systematically. Although the high-redshift sample has a mean scale length that is smaller than that of the local sample by somewhere between 20 and 50% (Figure 5), this result is preliminary because of the nagging systematic problems in resolving differences in scale lengths obtained from NICMOS and WFPC2 data (possibly due in part to physical effects like color gradients). Neither the WFPC2 nor NICMOS images match the rest wavelength of the local sample, and the differences we find between WFPC2 and NICMOS suggest that scale length depends sensitively on color. The magnitude of the accretion necessary to account for the luminosity evolution is within reach of size measurements of BCGs, once the color-dependence of the scale-lengths is understood.

6. Conclusions

Using a completely independent sample of high-redshift BCGs we have confirmed the results of previous investigators: 1) BCG properties vary as a function of environment, 2) BCGs in low-L_X clusters require significant accretion (a factor of 2 or more since $z \sim 1$) to produce the observed Hubble diagram, and 3) BCGs in high-L_X clusters require much less (or possibly no) accretion to produce the observed Hubble diagram. Additionally, we find that the accretion must consist of old stellar populations (because the colors of both BCG populations are consistent with passive evolution models) and that the sizes of high-redshift BCGs appear to be smaller than that of the local counterparts by a factor that is consistent with the accretion model. Further study of BCGs may provide the first truly direct observations of the hierarchical growth of a galaxy population.

Acknowledgments. DZ acknowledges financial support from the David and Lucile Packard Foundation, the Sloan Foundation, and the NSF CAREER program (AST 97-33111).

References

Aragon-Salamanca, A., Baugh, C.M., & Kauffmann, G. 1998, MNRAS, 297, 427

Aragon-Salamanca, A., Ellis, R.S., Couch, W.J., & Carter, D. 1993, MNRAS, 262, 764

Burke, D.J., Collins, C.A., & Mann, R.G. 2000, ApJ, 532L, 105

Dalcanton, J.J., 1996, ApJ, 466, 92

Gonzalez, A.H., Zabludoff, A.I., Zaritsky, D., & Dalcanton, J.J. 2000, ApJ, 536, 561

Gonzalez, A.H., Zaritsky, D., Dalcanton, J.J., & Nelson, A.E. 2001, ApJS, in press

Graham, A., Lauer, T.R., Colless, M., & Postman, M. 1996, ApJ, 465, 534

Hoessel, J.G., 1980, ApJ, 241, 493

Malumuth, E.M., & Kirshner, R.P. 1981, ApJ, 251, 508

Malumuth, E.M., & Kirshner, R.P. 1985, ApJ, 291, 8

Nelson, A.E., Gonzalez, A.H., Zaritsky, D., & Dalcanton, J.J., 2001a, ApJ., submitted

Nelson, A.E., Gonzalez, A.H., Zaritsky, D., & Dalcanton, J.J., 2001b, ApJ., submitted

Nelson, A.E., Simard, L., Zaritsky, D., Dalcanton, J.J., & Gonzalez, A.H. 2001c, ApJ., submitted

Oegerle, W.R., & Hoessel, J.G. 1991, ApJ, 375, 15

Oemler, A. 1976, 209, 693

Schombert, J.M. 1986, ApJS, 60, 603

Simard, L. 1998, in 'Astronomical Data Analysis Software and Systems VII', ASP Conference Series, 145 (R. Albrecht, R. N. Hook and H. A. Bushouse, eds)

Tremaine, S.D., & Richstone, D. 1977, ApJ, 212, 311

Zaritsky, D., Nelson, A.E., Dalcanton, J.J., & Gonzalez, A.H., 1997, ApJL, 480, L91

Zaritsky, D., Shectman, S.A., & Bredthauer, G. 1996, PASP, 108, 104

LARGE-SCALE STRUCTURE WITH CLUSTERS

Tracing Cosmic Evolution with Galaxy Clusters
ASP Conference Series, Vol. 268, 2002
S. Borgani, M. Mezzetti, and R. Valdarnini, eds.

Large–Scale Structure Studies with Clusters of Galaxies

R. C. Nichol
Dept. of Physics, Carnegie Mellon University, 5000 Forbes Ave., Pittsburgh, PA-15232

Abstract. I present here a review of Large-Scale Structure (LSS) studies using clusters of galaxies. First, I re-evaluate the 'pros' and 'cons' of using clusters for such studies, especially in this era of large galaxy redshift surveys. Secondly, I provide an historical review of the Cluster Correlation Function and show that the latest measurements of ξ_{cc} from Abell and X–ray catalogs are in excellent agreement. Thirdly, I review the latest measurements of the power spectrum of clusters which provide strong constraints on the cosmological parameters (*e.g.* Ω_m) and models of structure formation. Moreover, I highlight the recent discovery of "Baryon Wiggles" in the local cluster p(k) which is in perfect agreement with the recent CMB data. Lastly, I examine recent advances in the measurement of the X–ray Cluster Luminosity Function and emphasize the importance of accurately determining the selection function of future cluster surveys.

1. Introduction

I present here an incomplete and personal review of Large-Scale Structure (LSS) studies using clusters of galaxies. For a more complete overview of clusters as cosmological probes, the reader is referred to several excellent recent reviews by Biviano (2000), Schindler (2001a), Borgani & Guzzo (2001), Guzzo (2001) and Haiman, Mohr & Holder (2001).

1.1. Pros and Cons of Using Clusters for LSS Studies

As cosmologists, we want to study and understand the distribution of matter in the Universe as a function of space and time *i.e.* mass(x,y,z,t). This will allow us to constrain cosmological parameters, models of structure formation as well as understanding the physical processes that control the distribution of baryons. Ultimately, we want to directly compare our observations of the Universe to cosmological simulations *e.g.* the VIRGO-GIF project (Kauffmann et al. 1999). This is hard to do since we need to know how the light we see, traces the underlying matter we simulate.

Historically, clusters of galaxies have been used for this task because: *a)* They survey large volumes of space thus producing a fair sample of the Universe; *b)* Clusters can be seen over a large range in redshift, thus probing cosmological evolution; *c)* They maintain the imprint of the initial conditions so simple analytical relationships can be used to explain their properties (*e.g.* Jenkins et al.

2001); *d)* Clusters are the largest gravitationally bound objects in the Universe, so we can weigh the Cosmos using them; *e)* Clusters are full of hot baryons, so we can study how mass follows light on cluster scales.

However, clusters do have their problems. First, they live in the tail of the underlying mass distribution and thus may be highly biased tracers of this distribution. This has been a major concern for a long time (Kaiser 1986), but recent observational and theoretical work (see Miller et al. 2001a, Narayanan et al. 2000) shows that a simple linear biasing model ($b = 1.5$), between normal galaxies and clusters, does work well over an important range of scales ($0.03 \leq k \leq 0.15h^{-1}\,\mathrm{Mpc}^{-1}$. As an aside, I note that because of this fact, clusters may be an excellent probe of the gaussianity of the Universe since our models and simulations of the Universe assume gaussianity at the beginning (*e.g.* Chiu, Ostriker, & Strauss 1998; Robinson, Gawiser & Silk 1999; Matarrese, Verde & Jimenez 2000; Kerscher et al. 2001). Kerscher et al. (2001) has already found statistical evidence for non-gaussianity based on the REFLEX sample of X–ray clusters, while Verde et al. (2001) highlight that the CMB and high redshift galaxies maybe a better way to studying this issue.

Another concern with clusters is the assumption of hydrostatic equilbrium (for the gas) and the virial theorem (for the galaxies). For example, in the recent hydrodynamical/N-body simulations of Ricker & Sarazin (2001), an off-axis merger of two systems (mass ratio of 3:1) can produce large-scale turbulent motions with eddys up to several 100 kpcs. Even after nearly a Hubble time, these motions persist as subsonic turbulence in the cluster cores, providing 5-10% of the support against gravity. Roettiger, Burns & Loken (1996) and Ritchie & Thomas (2001) also found that major merger events can knock the cluster out of hydrostatic equilibrium for several Gyrs (see Schindler et al. 2001b for a recent review of cluster simulations).

To assess the importance of such mergers on the whole cluster population, I show in Fig. 1 the expect fraction of clusters ($M \geq 10^{14}\,M_\odot$ or $L_{\rm X}(0.5-2.0) > 10^{43}$ erg/s) that have experienced a merger in the last 2.5Gyrs. This figure shows that as we push to higher redshifts the vast majority of clusters should show significant evidence for a recent merger that may severely effect their physical state (see Mathiesen & Evrard 2001). Fig. 1 is in agreement with the recent *Chandra* observations reported in Henry (2001) where 75% of $z > 0.75$ clusters showed significant substructure. Therefore, if we wish to continue to use clusters as cosmological tracers, we must understand such effects in greater detail.

In light of these concerns, I challenged the audience at the conference to discuss the following question over the coffee break: "In this era of galaxy redshift, lensing and photometric surveys[1], is there still a need for cluster surveys of the Universe?". For example, in Fig. 2, I show initial results from the Sloan Digital Sky Survey (SDSS) Luminous Red Galaxy (LRG) survey which is designed to create a pseudo–volume limited sample of 100,000 giant ellipticals out to $z \sim 0.4$ (although the sample does reach out to $z \sim 0.6$). This survey will sample the LSS on Gigaparsec scales – comparable with the largest cluster surveys – well beyond the turn–over in the matter power spectrum (see Section 3).

[1]Surveys like the SDSS, 2dF, LSST, SNAP, POI, DEEP & VIRMOS

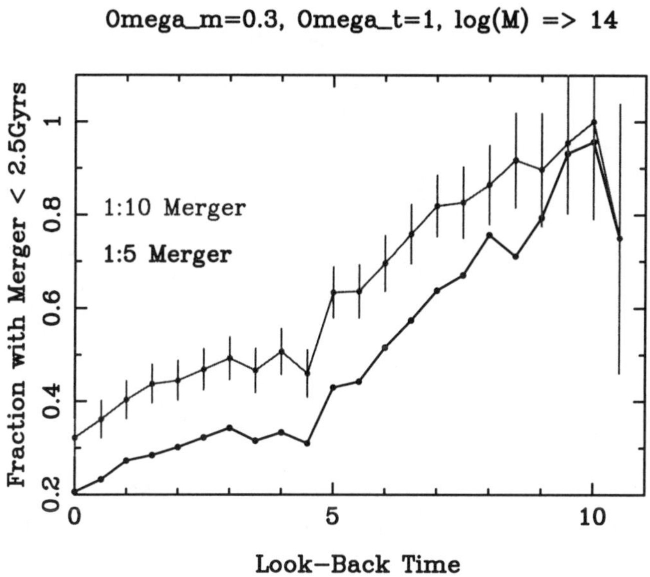

Figure 1. The fraction of $M \geq 10^{14}\, M_{\odot}$ clusters that have experienced a major merger in the last 2.5Gyrs versus look-back time. The red points are for a 10:1 merger, while the black points are for a 5:1. These data were kindly provided by Joanne Cohn based on simulations similar to the work presented in Cohn, Bagla & White (2000). The simulations use $\Omega_m = 0.3$, $\Omega_\Lambda = 0.7$ and $\sigma_8 = 0.9$. The error bars are $\sqrt{N}$ of clusters detected in the simulations.

The LRGs are still biased tracers but there is compelling evidence that we can fully understand their selection function (see Eisenstein et al. 2001 for details).

Returning to my question, I will emphasize that clusters are still the best way to: *a)* Constrain σ_8 (see Viana & Liddle 1999); *b)* Determine Ω_m (see Sections 3 & 4), and; *c)* Study galaxy evolution in dense regions.

2. Cluster Correlation Function

The Cluster Correlation Function ($\xi_{cc} = (\frac{r}{r_o})^{-\gamma}$ where r_o is the correlation length defined as $\xi_{cc}(r_o) = 1$) has had a long and controversial history. In the 1980's, it was one of the main constraints on CDM models of structure formation (see White et al. 1987). The main controversy surrounding measurements of ξ_{cc} has been the effects of optical projection effects on the Abell catalog *i.e.* the contamination of the 2D richness of clusters because of the field or a nearby clusters (Sutherland et al. 1988) and/or the super-position of two groups along the line–of–sight to create a "Phantom Cluster". The hypothesis has been that these projection effects have artificially boosted the amplitude of Abell measurements of ξ_{cc} thus making it inconsistent with CDM models (Sutherland et al. 1988; Dekel et al. 1989; Efstathiou et al. 1992). There are many advocates of

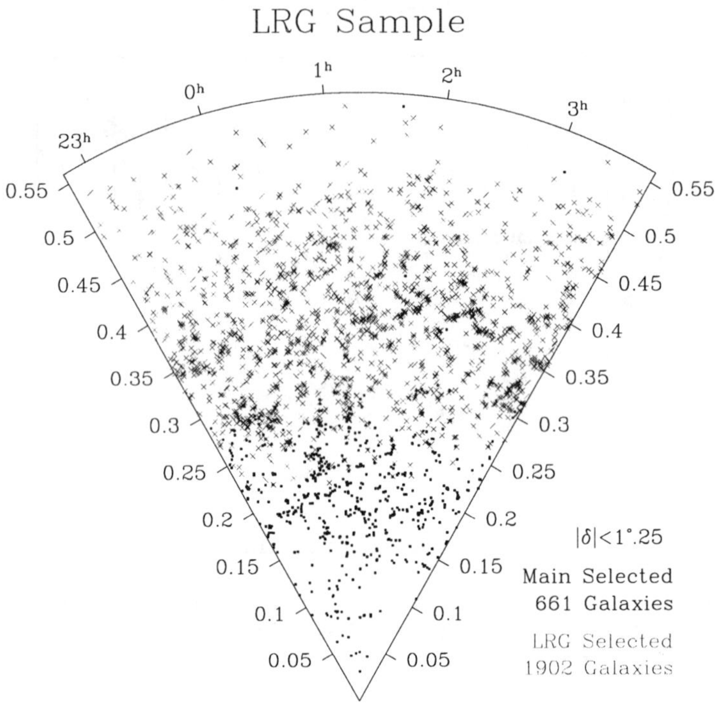

Figure 2. Wedge plot of a small fraction of SDSS spectroscopic data collected to date. This illustrates the power of using the SDSS–LRG sample for studying the LSS. Figure kindly provided by Daniel Eisenstein.

this hypothesis (*e.g.* Dalton et al. 1992; Nichol et al. 1992 & 1994; Romer et al. 1994), but there are also many defenders of the Abell catalog (Postman et al. 1992; Miller et al. 1999). Over the last decade, such concerns have driven the creation of new optical and X-ray cluster surveys *e.g.* the EDCC (Lumdsen et al. 1992), APM (Dalton et al. 1992), DPOSS (Gal et al. 2000) and REFLEX surveys (Collins et al. 2000).

In Fig. 3, I show an historical overview of ξ_{cc} measurements over the last 20 years. There are two interesting points to make about this figure: First, the correlation length of Abell ξ_{cc}'s have decreased with time. Secondly, X-ray measurements of r_o have increased with time. These two trends can be understood via the relative space density of clusters in these different surveys. In the case of the Abell survey, the volume surveyed by the different samples of Abell clusters has not increased much since the first measurements, but the number of systems used has steadily increased from ~ 100 to over 600. Thus the

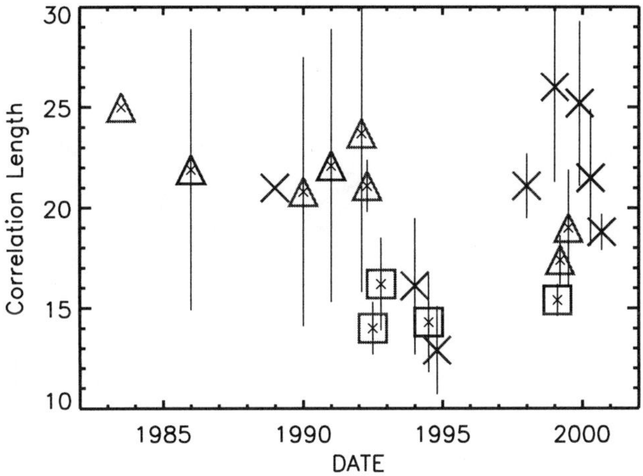

Figure 3. The Correlation Length (r_o in Mpc) of ξ_{cc}, measured by various authors, as a function of time. Triangles are Abell, squares are optical non-Abell samples and crosses are X–ray samples. Grey points have a slope of the correlation function between $1.7 \leq \gamma \leq 2.0$, while black points are outside this range. Please note that I have not scaled these measurements by the space density of the cluster sample used in the measurement. I took the r_o value and error quoted in the abstract of the papers and, in the case of multiple fits to the data, I took the r_o from the largest sample. Data from Ling et al. 1986; West & van der Bergh (1991); Peacock & West 1992; Miller et al. 1999; Lahav et al. 1989; Romer et al. 1994; Borgani et al. 1999a; Abadi et al. 1998; Lee et al. 1999; Moscardini et al. 2000; Collins et al. 2000; Dalton et al. 1992; Nichol et al. 1992 & 1994; Postman et al. 1992; Huchra et al. 1990; Bahcall & Soneria 1983.

space density of Abell clusters used in determining ξ_{cc} has decreased over time. The opposite is true for the X–ray samples; the number of clusters used has increased, but so has the volume sampled, thus the space density has slightly decreased over time (see Croft et al 1997).

If we focus on recent measurements with similar slopes ($1.7 \leq \gamma \leq 2.0$), then the agreement between present measurements of ξ_{cc} is very good. This is illustrated in Fig. 4 where I show the ξ_{cc} from nearly 1000 clusters (both Abell and X–ray). The great debate over the value of the correlation length of ξ_{cc} is probably over especially given that the pi-sigma diagrams shown in Fig. 5 highlights that the clustering in both the Abell and X–ray surveys is now nearly isotropic. This removes one of the main arguments for projection effects in the Abell catalog and is probably due to the larger samples now available averaging over the effect of LSS and/or the effect of a single supercluster. We look forward

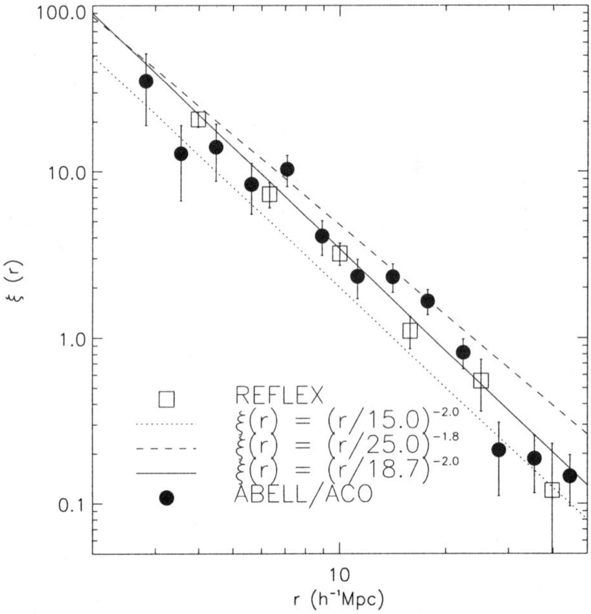

Figure 4. The correlation function for the Abell/ACO sample of Miller et al. (1999) and from the REFLEX sample of Collins et al. (2000). No correction has been made for the space densities of the surveys. Data kindly provided by Chris Collins and Chris Miller.

to new ξ_{cc} measurements from large, homogeneous samples of clusters taken from the SDSS, REFLEX and 2dF surveys.

3. Cluster Power Spectrum

In the last ten years, many authors have switched from measuring correlation functions to measurements of the cluster power spectrum since it allows for a more direct comparison with the theoretical models. For example, increasing $\Omega_m h^2$ pushes the epoch of matter-radiation equality back in time and moves the peak in the p(k) to low k values (Tegmark 1999 [2]). Therefore, the detection of the turn–over in the local power spectrum (p(k)) of matter provides a powerful constrain on both the matter density of the Universe and models of structure formation (*e.g.* CDM).

As yet, there has been no definitive detection of a turn–over (or peak) in the local matter p(k), but we are getting close (Szalay et al. 2001; Miller et

[2]See the CMB movies on Max Tegmark's homepage which illustrate this effect *i.e.* http://www.hep.upenn.edu/~ max/

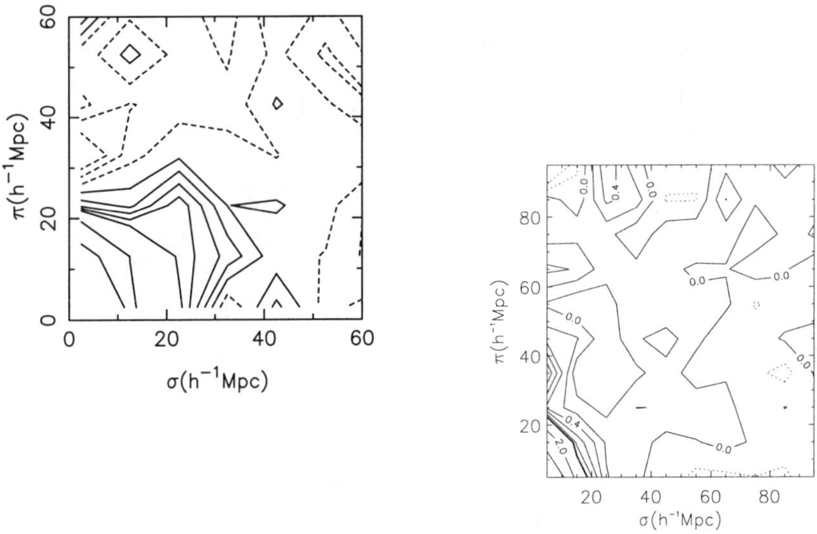

Figure 5. The π–σ diagrams for the two surveys presented in Fig 4. This shows the radial (σ) and transverse (π) components of the correlation function. The left–hand plot is for the REFLEX sample, while the right–hand side is from the Abell/ACO sample of Miller et al. (1999). These figures were kindly provided by Chris Collins and Chris Miller.

al. 2001a; Percival et al. 2001; Huterer et al. 2001) as new surveys of the local Universe are approaching volumes than demand a turn–over to satisfy the COBE measurements at very low k. This lack of a detection naturally drives estimates of Ω_m from these p(k)'s towards low values *e.g.* Miller et al. (2001a) recently estimated $\Omega_m h^2 = 0.12^{+0.03}_{-0.02}$ using a large sample of clusters to probe Gigaparsec scales. Similarly, Schuecker et al. (2001) presented the p(k) from the REFLEX sample and finds it favors low–density CDM models as the peak in the p(k) is yet undetected.

In addition to the hunt for the turn–over in the p(k), several authors have looked for "bumps" and "wiggles" in the local matter p(k). For example, Einasto et al. (1997) reported a "bump" in the cluster p(k) which could be do to excess power in the initial p(k) coming out of Inflation (a physical mechanism for such a feature however, remains unclear; see Atrio–Barandela et al. 2000). Alternatively, one may expect fluctuations in the observed p(k) because of the acoustic oscillation of baryons in the early Universe; sometimes called the "Baryon Wiggles". Eisenstein et al. (1998) recently looked for this signature in the p(k) measurements from galaxy and cluster surveys but was unable to find a satisfactory fit between the cosmological models (with acoustic oscillations) and the data available at that time. This year, Miller et al. (2001a,b) and Percival et al. (2001) have revisited this problem and have found compelling evidence

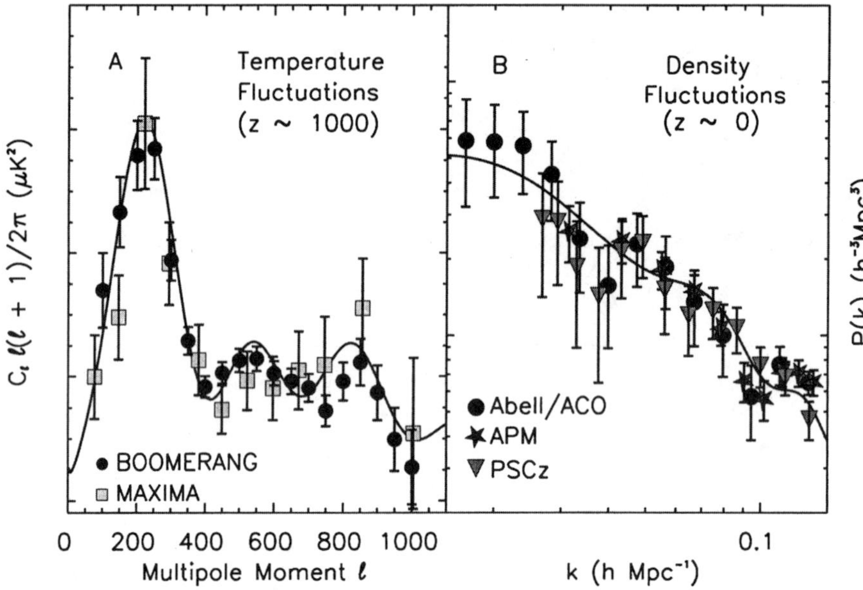

Figure 6. Top: On the right we show the p(k) data from clusters and galaxies as discussed by Miller et al. (2001a). The best fit model (with acoustic oscillations) is shown. On the left is the 2001 CMB balloon data. Here we plot the same model as shown on the right; *this is not a fit to the CMB data.* As discussed in Miller et al. (2001b), there is good agreement between the two epochs.

for the detection of the "Baryon Wiggles" in the local matter p(k) that is fully consistent with the detection of the same "Baryon Wiggles" in the recent CMB data (MAXIMA. DASI & BOOMERANG). This concordance is shown in detail in Fig. 6. In summary. in 2001, we have witnessed the simultaneous discovery of the "Baryon Wiggles" in the local ($z \sim 0.1$) and distant ($z \simeq 1000$) Universe providing yet another major success for the Hot Big Bang cosmological model.

4. The X–ray Cluster Luminosity Function

In addition to studying the clustering of clusters, it is important to understand the distribution of cluster properties (*i.e.* the luminosity, temperature and mass functions) and how these scale with each other and as a function of redshift. This will allow us to determine if clusters are a self–similar population (*i.e.* their properties can be directly predicted from using just the virial theorem and hydrostatic equilbrium, see Kaiser 1986) or are there extra contributions to the energy which breaks the simple scaling laws predicted by self-similarity. In recent years, there have been significant advances in understanding these scaling relationships (Horner et al. 1999; Sheldon et al. 2001; Mathiesen 2001;

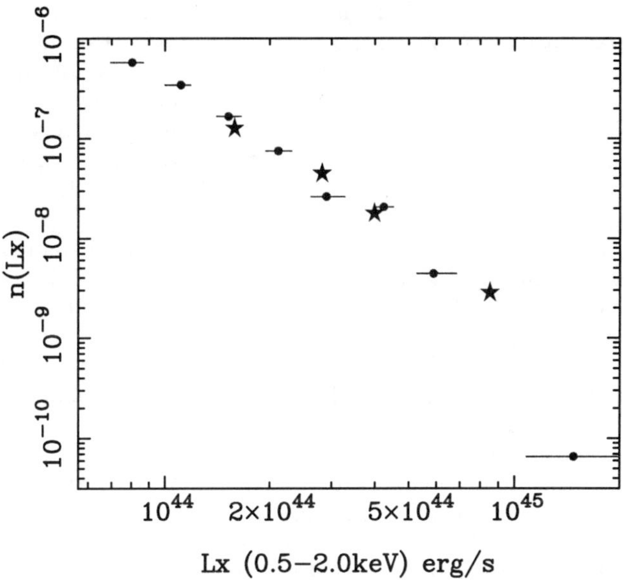

Figure 7. The X–ray Cluster Luminosity Function for the REFLEX (red; Boehringer et al. 2001) and Bright SHARC (blue; Adami et al. 2000) cluster surveys. The REFLEX probes to $z = 0.3$, while the Bright SHARC surveys the range $0.3 < z \sim 0.8$. The error bars on the REFLEX sample are discussed in Boehringer et al. (2001), while a detailed analysis of the errors on the Bright SHARC are given in Adami et al. (2000) and Romer et al. (2000). The units on $n(L_x)$ are $h^{-3}\mathrm{Mpc}^{-3}\,(L_x(44))^{-1}$. Data kindly provided by Hans Bohringer and Christophe Adami.

Neumann & Arnaud 2001) but we can expect continued advancements over the next decade because of new X–ray (XMM & *Chandra*) and lensing (Joffre et al. 2000) surveys of clusters as well as more detailed Hydro/N–body cluster simulations (see Ricker & Sarazin 2001). One example is the XMM-Newton Ω_m Project (Bartlett et al. 2001) which is designed to obtain accurate temperature measurements ($\simeq 10\%$ error) for all of the high redshift SHARC clusters (Romer et al. 2000; Burke et al. 1997) thus obtaining an estimate of the high redshift L_x–T_x relation (over a decade in L_x) as well as constraining Ω_m.

In the absence of a large, homogeneous sample of cluster temperatures, masses and velocity dispersions, authors have focused in recent years on the X–ray Cluster Luminosity Function (XCLF). The reader is referred to Gioia et al. (2001) and Boehringer et al. (2001) for a good recent overview of ROSAT & *Einstein* measurements of the XCLF as a function of redshift. The key debate with the XCLF is whether the bright–end (brighter than $L^\star \simeq 4 \times 10^{44}$ erg/s) of the luminosity function has evolved since $z \sim 0.8$, thus producing a deficit of high redshift, luminous systems. As discussed by many authors, the evolution of such massive clusters is a strong constraint on Ω_m and thus it is vital to quantify how the luminous end of the XCLF changes with look–back time (*e.g.* Oukbir & Blanchard 1997; Viana & Liddle 1999; Reichart et al. 1999; Borgani et al. 1999b; Henry 2001). If $\Omega_m \sim 0.2$ (*e.g.* from p(k) of clusters above), then the amount of evolution out to $z \simeq 0.8$ should be small and thus the signal we are looking for will be weak. In this case, we must be very careful about systematic uncertainties in our distant cluster surveys. Alternatively, if $\Omega_m = 1$, we expect approximately an order of magnitude decrease in the space density of $L_{\rm x} = 10^{45}$erg/s systems to $z = 0.8$.

In Fig. 7, I show the bright–end of the XCLF derived from the REFLEX cluster survey (Boehringer et al. 2001) and the Bright SHARC (Romer et al. 2000; Adami et al. 2000). These are two of the most well–understood surveys of X–ray clusters available to date (in terms of their selection functions *etc.*). The agreement between the two is remarkable but the number of luminous clusters ($> L^\star$) in both surveys is still relatively small because of the tiny space densities of such clusters. Furthermore, to obtain a "high precision" measurement of the XCLF, we need to carefully model the selection functions of these, and future, surveys. This was recently emphasized by Adami et al. (2000) who performed extensive simulations of the Bright SHARC selection function and found that the shape and morphology of the clusters had a dramatic effect on the efficiency of detection. When this is combined with the effect seen in Fig. 1 – *i.e.* that a large fraction of high redshift clusters are likely to have signs of recent interactions (see also Henry 2001) – it becomes essential to model such effects in the selection function of any distant cluster survey. Therefore, the selection functions of future surveys will need to include cosmological (*e.g.* Ω_Λ), morphological (*e.g.* profile of clusters) and survey design (*e.g.* off-axis PSF) effects.

5. Future Cluster Surveys

As emphasized above, future cluster surveys need to improve in both quality and quantity. I highlight here four on–going efforts that will hopefully address both of these points (this is by no means a complete list). First, the MACS survey (Ebeling et al. 2001) is mining the ROSAT All–Sky Survey (RASS) for previously undiscovered high redshift ($0.3 < z < 0.7$) luminous clusters. The survey has been highly successful and has nearly 100 new massive clusters to date. Secondly, Gladders & Yee (2000) are embarked on a large–area (100deg^2), two–passband, optical survey in search of high redshift clusters. Their technique makes use of the fact that cluster ellipticals are strongly clustered in color and are the reddest (rest–frame) galaxies in existence. They plan to expand their technique to $\simeq$ 1000deg^2.

As discussed in Romer et al. (2001), the XMM data archive will be a gold–mine for searching for distant X–ray clusters. Over the life–time of the satellite, XMM is expected to serendipitously survey $\simeq$ 800deg^2 which should yield thousands of cluster detections. Most importantly, the XMM Cluster Survey (XCS; Romer et al. 2001) is predicted to find zero ($\Omega_m = 1$), twelve ($\Omega_m = 0.3$) or fifty ($\Omega_m = 0.3 + \Omega_\Lambda = 0.7$) massive clusters ($T_{\rm x} > 6\,$keV) at redshifts greater than one (see Romer et al. 2001). This will provide an excellent constraint on both Ω_m and Ω_Λ as well as providing a rich dataset of cluster temperatures and morphologies.

Finally, the Sloan Digital Sky Survey (SDSS) should produce several high quality, large–area, cluster surveys. For example, in Nichol et al. (2000), I outlined a new algorithm called "C4" which uses the 5 passband SDSS photometry and photometric redshifts, to look for clusters in a 7–dimensional space (4 colors, 2 angular, 1 radial). This removes all concerns about projection effects (see Section 2) and, when combined with the RASS data, can find clusters out to $z \simeq 0.7$.

Before I leave this section, I will emphasize again the importance of determining the selection functions of these new surveys. As the data, procedures and algorithms used in these new surveys get more complex, then we can not naively assume the volume of these surveys is just calculated from their flux limit and their area on the sky. We will need to turn to simulations of the Universe to help us *e.g.* using semi–analytical, N–body techniques (see Kauffmann et al. 1999) or fast analytical approximations like PTHALO (Scoccimarro & Sheth 2001).

I thank the organizers of the conference for inviting me. I would like to also thank my collaborators on the SDSS, XMM–Newton Ω_m project, XCS and SHARC surveys. I especially thank Chris Miller, Chris Collins and Kathy Romer for their help, guidance and reading of this manuscript.

6. References

Abadi, M., et al. 1998, ApJ, 507, 526
Adami, C., et al. 2000, ApJS, 131, 391
Atrio-Barandela, F., et al. 2000, ApJL, see astro-ph/0012320
Bahcall, N. A. & Soneira, R. M. 1983, ApJ, 270, 20
Bartlett, J., et al. 2001, see astro-ph/0106098
Biviano, A., 2000, see stro-ph/0010409
Boehringer, H., et al. 2001, see astro-ph/0106243
Borgani, S., Plionis, M., & Kolokotronis, V. 1999a, MNRAS, 305, 866
Borgani, S., et al. 1999b, ApJ, 527, 561
Borgani, S. & Guzzo, L. 2001, Nature, 409, 39
Burke, D. J., et al. 1997, ApJ, 488, L83
Chiu, W. A., Ostriker, J. P., & Strauss, M. A. 1998, ApJ, 494, 479
Cohn, J. D., Bagla, J. S., & White, M. 2001, MNRAS, 325, 1053
Collins, C. A. et al. 2000, MNRAS, 319, 939
Croft, R. A. C., et al. 1997, MNRAS, 291, 305
Dalton, G. B. et al. 1992, ApJ, 390, L1
Dekel, A., Blumenthal, G. R., Primack, J. R., & Olivier, S. 1989, ApJ, 338, L5
Ebeling, H., Edge, A. C., & Henry, J. P. 2001, ApJ, 553, 668
Efstathiou, G. et al. 1992, MNRAS, 257, 125
Einasto, J., et al. 1997, Nature, 385, 139
Eisenstein, D. J., Hu, W., Silk, J., & Szalay, A. S. 1998, ApJ, 494, L1
Eisenstein, D. J., et al. see astro-ph/0108153
Gal, R. R., et al. 2000, AJ, 120, 540
Gioia, I. M., et al. 2001, ApJ, 553, L105
Gladders, M. D. & Yee, H. K. C. 2000, AJ, 120, 2148
Guzzo, L., 2001, see astro-ph/0102062
Haiman, Z., et al. 2001, see astro-ph/0103049

Henry, J. P., 2001, astro-ph/0109498
Horner, D. J., Mushotzky, R. F., & Scharf, C. A. 1999, ApJ, 520, 78
Huchra, J. P., Geller, M. J., Henry, J. P., & Postman, M. 1990, ApJ, 365, 66
Huterer, D., Knox, L., & Nichol, R. C. 2001, ApJ, 555, 547
Jenkins, A., et al. 2001, MNRAS, 321, 372
Joffre, M. et al. 2000, ApJ, 534, L131
Kaiser, N. 1986, MNRAS, 222, 323
Kauffmann, G. et al. 1999, MNRAS, 303, 188
Kerscher, M. et al. 2001, A&A, 377, 1
Lahav, O., Fabian, A. C., Edge, A. C., & Putney, A. 1989, MNRAS, 238, 881
Ling, E., Frenk, C., Barrow, J., 1986, MNRAS, 223, 21
Lumsden, S. L. et al. 1992, MNRAS, 258, 1
Matarrese, S., Verde, L., & Jimenez, R. 2000, ApJ, 541, 10
Mathiesen, B. F. 2001, MNRAS, 326, L1
Mathiesen, B. F. & Evrard, A. E. 2001, ApJ, 546, 100
Miller, C. J., Batuski, D. J., Slinglend, K. A., & Hill, J. M. 1999, ApJ, 523, 492
Miller, C. J., Nichol, R. C., & Batuski, D. J. 2001a, ApJ, 555, 68
Miller, C. J., Nichol, R. C., & Batuski, D. J. 2001b, Science, 292, 2302
Moscardini, L. et al. 2000, MNRAS, 314, 647
Narayanan, V. K., Berlind, A. A., & Weinberg, D. H. 2000, ApJ, 528, 1
Neumann, D. M. & Arnaud, M. 2001, A&A, 373, L33
Nichol, R. C., Briel, O. G., & Henry, J. P. 1994, MNRAS, 267, 771
Nichol, R. C. et al. 1992, MNRAS, 255, 21P
Nichol, R. C., et al. 2000, astro-ph/0011557
Oukbir, J. & Blanchard, A. 1997, A&A, 317, 1
Peacock, J., & West, M., MNRAS, 259, 494
Percival, W., 2001, MNRAS, see astro-ph/0105252
Postman, M., Huchra, J. P., & Geller, M. J. 1992, ApJ, 384, 404
Reichart, D. E., et al. 1999, ApJ, 518, 521
Ricker, P., Sarazin, C., 2001, ApJ, see astro-ph/0107210
Ritchie, B. W., Thomas, P. E., MNRAS, see astro-ph/0107374
Robinson, J., Gawiser, E., & Silk, J. 2000, ApJ, 532, 1
Roettiger, K., Burns, J. O., & Loken, C. 1996, ApJ, 473, 651
Romer, A. K. et al. 1994, Nature
Romer, A. K. et al. 2000, ApJS, 126, 209
Romer, A. K. et al. 2001, ApJ, 547, 594
Schindler, S., 2001a, see astro-ph/0109040
Schindler, S., 2001b, see astro-ph/0107008
Schuecker, P. et al. 2001, A&A, 368, 86
Scoccimarro, R., Sheth, R. K., 2001, see astro-ph/0106120
Sheldon, E. S. et al. 2001, ApJ, 554, 881
Sutherland, W. 1988, MNRAS, 234, 159
Szalay, A. S., 2001, ApJ, see astro-ph/0107419
Tegmark, M. 1999, ApJ, 514, L69
Verde, L. et al. 2001, MNRAS, 325, 412
Viana, P. T. P. & Liddle, A. R. 1999, MNRAS, 303, 535
West, M., & van der Bergh, S., 1991, 373, 1
White, S. D. M., Frenk, C. S., Davis, M., & Efstathiou, G. 1987, ApJ, 313, 505

Tracing Cosmic Evolution with Galaxy Clusters
ASP Conference Series, Vol. 268, 2002
S. Borgani, M. Mezzetti, and R. Valdarnini, eds.

The REFLEX Cluster Survey: Probing the Mass Distribution in the Universe

Luigi Guzzo

Osservatorio Astronomico di Brera, Milano/Merate, Italy

Hans Böhringer

Max-Planck-Institut für extraterr. Physik, Garching, Germany

Chris A. Collins

Liverpool John Moores University, Liverpool, U.K.

Peter Schuecker

Max-Planck-Institut für extraterr. Physik, Garching, Germany

& the REFLEX Team[1]

Abstract. We summarize some of the major results obtained so far from the REFLEX survey of X-ray clusters of galaxies, concentrating on the latest measurements of the cluster X-ray luminosity function and two-point correlation function. The REFLEX luminosity function provides the most homogeneous census of the distribution function of masses in the local Universe, representing a unique zero-redshift reference quantity for evolutionary studies. On the other hand, the observed clustering of REFLEX clusters is very well described by the correlation function of a low-Ω_M CDM model. Also, the bidimensional correlation map $\xi(r_p, \pi)$ shows no stretching along the line of sight, indicating negligible spurious effects in the sample, with at the same time a clear compression of the contours as expected in the presence of coherent motions.

1. The REFLEX Survey

The REFLEX[1] (ROSAT-ESO Flux-Limited X-ray) cluster survey is currently the largest sample of X-ray selected clusters of galaxies from the ROSAT All-Sky Survey with 1) a statistically homogeneous and fairly well-understood selection

[1] The REFLEX Team includes: H. Böhringer (MPE), L. Guzzo (OAB), C.A. Collins (LJMU), P. Schuecker (MPE), G. Chincarini (OAB), R. Cruddace (NRAL), A.C. Edge (Durham), S. De Grandi (OAB), D.M. Neumann (Saclay), T. Reiprich (MPE), S. Schindler (LJMU), P.A. Shaver (ESO), W. Voges (MPE)

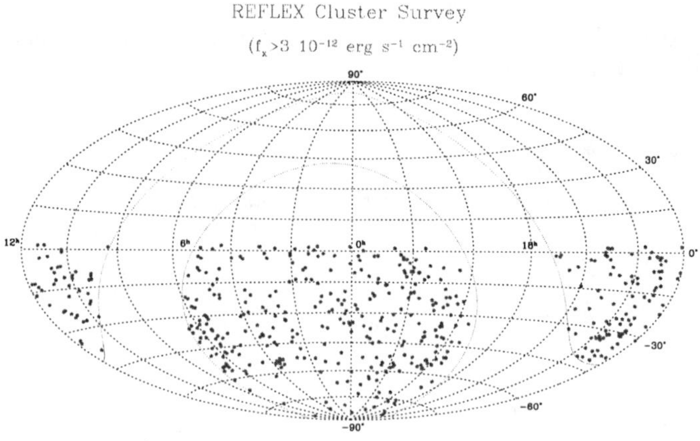

Figure 1. The sky distribution of REFLEX clusters. The dotted band marks the region of the galactic plane ($|b_{II}| > 20°$) which is excluded from the survey.

function; 2) measured redshifts. The survey contains 452 clusters over the southern celestial hemisphere ($\delta < 2.5°$), at galactic latitudes $|b_{II}| > 20°$ and is more than 90% complete to a flux limit of 3×10^{-12} erg s^{-1} cm^{-2} (in the ROSAT band, 0.1–2.4 keV). X-ray fluxes and source extensions were re-measured from the RASS with a dedicated algorithm, thus avoiding the limitations of the standard analysis software in the characterisation of extended sources. The details of the whole identification process and a critical discussion of potential sources of incompleteness are presented in Böhringer et al. (2001a). Redshifts for all but 3 REFLEX clusters have been measured during a long Key Programme (1992-2000) using ESO telescopes (e.g. Guzzo et al. 1999), that collected more than 3500 galaxy redshifts over almost 500 X-ray targets. Fig. 1 shows the distribution on the sky of the 452 clusters in the REFLEX survey, centered on the South Galactic Cap region (the largest contiguous area covered), while Fig. 2 plots their X-ray luminosity against their redshift. Given the survey flux limit (defined by the lower boundary of the point distribution), at large redshifts we are allowed to detect only the very bright, massive clusters. On the other hand, the very large solid angle of the survey allows for an extremely large volume to be explored (4.24 steradians, corresponding to 8.7×10^8 h^{-3} Mpc^3 out to $z = 0.3$ in an $\Omega_M = 0.3$, $\Omega_\Lambda = 0.7$ cosmology), such that these extremely rare objects on the tail of the cluster X-ray luminosity function have a significant probability to be detected. It is not by chance, in fact, that the most luminous cluster known is still that discovered by the REFLEX survey in 1994, i.e. RXCJ1347.4-1144 (Schindler et al. 1995). Another example is shown in Fig. 4. Such a large volume is also ideal for sampling the very large modes of density fluctuations in the Universe (Schuecker et al. 2001). From the distribution of Fig. 2 one can compute the mean density of clusters as a function of distance, which indicates a very good completeness (i.e. constant density) out to at least $z = 0.2$, possibly above. It is also evident how one can use the REFLEX survey to select sub-

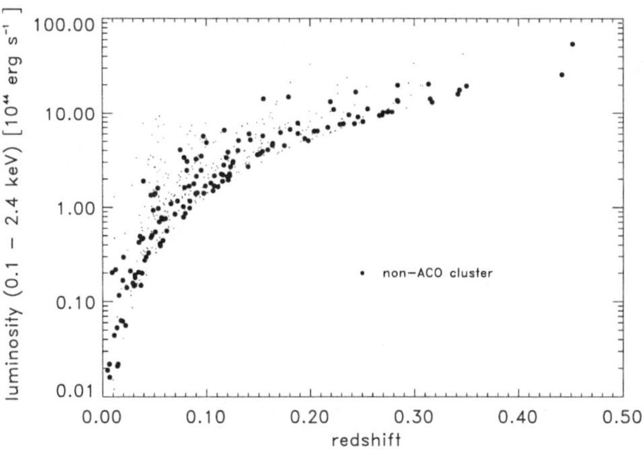

Figure 2. The distribution of X-ray luminosities (ROSAT band) as a function of redshift for all REFLEX clusters, with non-Abell clusters marked by filled circles. The two highest-redshift clusters are RXCJ1206.2-0848 at $z = 0.441$ (whose RGB image is shown in Fig. 4) and RXCJ1347.4-1144, the most X-ray luminous cluster currently known, at $z = 0.452$ (Schindler et al. 1995).

samples of clusters within well-defined ranges in L_X (i.e. mass) and redshift. A number of follow-up studies of this kind are indeed ongoing (see Böhringer et al. 2001b for an overview), and more will certainly follow when the catalogue is released in early 2002. The filled circles in Fig. 2 represent 142 clusters which do not appear in the visually-selected Abell/ACO catalogue (Abell 1958; Abell et al. 1989). It is interesting to see that these clusters are distributed basically at any redshift. This is a demonstration of how the "richness" criterion (the reason why most of these objects did not enter the Abell selection) is a poor indicator of the cluster mass, and how incomplete in mass a purely visually-selected cluster survey could be. We have also performed a direct X-ray analysis of the RASS data at all Abell-ACO cluster positions, measuring their X-ray flux. This gave the rather encouraging result that the REFLEX survey in fact detects *all* Abell-ACO clusters within its flux and area boundaries[2].

2. The Cluster X-ray Luminosity Function

The X-ray luminosity function (XLF) of galaxy clusters is an excellent observable incarnation of the cluster mass function, given that in clusters of galaxies X-ray luminosity is well related to the cluster mass (e.g. Reiprich & Böhringer 2002). For this reason, estimating the XLF was one of the prime targets of the REFLEX

[2]Even more, the REFLEX survey misses only 1 of the so-called Supplementary Abell clusters i.e. the extension to the main ACO catalogue, where those objects that did not meet all original criteria while still looking as *bonafide* clusters were listed.

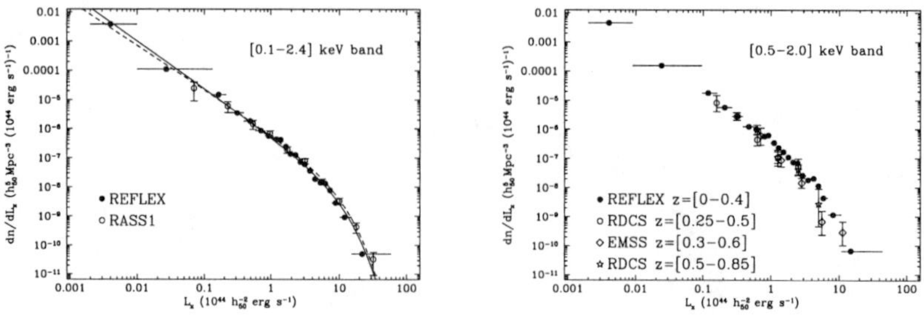

Figure 3. Left: The REFLEX cluster X-ray luminosity function (XLF, Böhringer et al. 2002), compared to the previous estimate obtained from the RASS1 Bright Sample, the brighter forerunner of REFLEX (De Grandi et al. 1999). Right: Comparison of the REFLEX XLF, with that of two representative distant cluster samples, the RDCS (Rosati et al. 1998) and EMSS (Henry et al. 1992), showing the moderate reduction in the number density of very luminous clusters at high redshifts. The REFLEX XLF is here computed in the [0.5–2.0] band, to allow for an homogeneous comparison.

survey. It was also one of the first results obtained by the project along the way, from the higher–flux RASS1 Bright Sample (De Grandi et al. 1999). The recent XLF from the whole survey (Böhringer et al. 2002) is compared to that early result in the left panel of Fig. 3. The improvement in the error bars of the latest REFLEX XLF (smaller than the size of the dots) gives an immediate idea of its quality. Such an accurate local measurement represents a standard reference of crucial importance for evolutionary studies, i.e. for comparison with XLF measured at high redshift. One such example is shown in the right panel of the same figure, where the REFLEX XLF is plotted together with those from the RDCS (Rosati et al. 1998) and EMSS (Henry et al. 1992) deep surveys. With the REFLEX data pivoting the local abundance, the evidence for mild evolution of the XLF bright end at high z can be tested to a high accuracy (e.g. Borgani et al. 2001; Henry 2001).

3. The Clustering of REFLEX Clusters

Clusters have a long history as tracers of large-scale structure (see Nichol, this volume) and, in particular, surveys of X-ray clusters have their own specific advantages (see Lahav et al. 1989 and Romer et al. 1994 for early applications, and Borgani & Guzzo 2001, Guzzo 2001 and Henry 2001 for recent reviews). First of all, once the characteristics of the X-ray instrument used are specified, a clean selection function can be constructed. This is by all means not an obvious task for an optically-selected survey. Second, fairly precise predictions can be made from the models for both the mass function and the clustering of objects above a given mass (see e.g. Borgani, this volume and Moscardini et al. 2000). These can be reliably translated into "observer space" in terms of X-ray

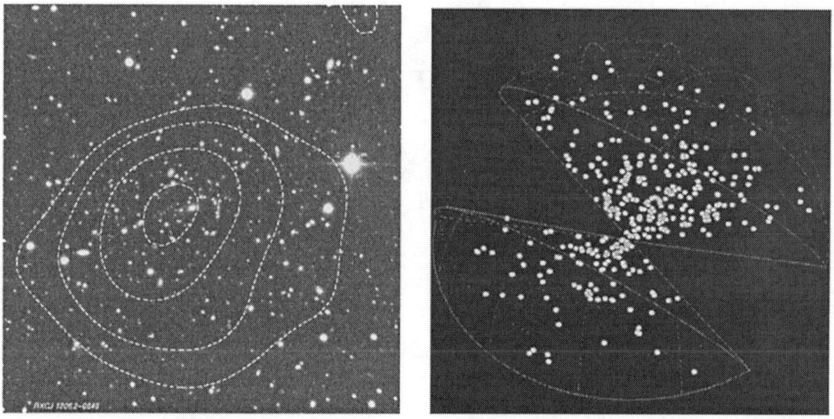

Figure 4. Left: BVR combined mage of the spectacular cluster RXCJ1206.2-0848, the second most distant object in the REFLEX survey, at z=0.441. Right: The spatial distribution of REFLEX clusters, out to 600 h^{-1} Mpc (plot from Borgani & Guzzo 2001). Note the presence of agglomerates and "chains", demonstrating the strong clustering of clusters among themselves.

luminosities and fluxes. These advantages allow one to select samples uniformly by mass over large areas (avoiding also other effects that plague optically-selected samples as line-of-sight projections), and provide the baseline for the REFLEX survey clustering results.

The right panel of Fig. 4 plots the 3D distribution of REFLEX clusters within a radius of 600 h^{-1} Mpc. Despite the fading with distance due to the flux-limited selection function, a number of superstructures with sizes 100 h^{-1} Mpc are evident: clusters are clearly still strongly clustered on such scales. This can be quantified by the two-point correlation function, that we plot for the whole flux-limited survey in Fig. 5 (Collins et al. 2000). The shape of ξ_{cc} is tightly constrained by the REFLEX data, and closely reproduces an amplified version of the galaxy two-point correlation function over almost two decades of scales, a classic prediction of biasing theory (Kaiser 1984). In the same figure, we also plot the predictions of two low-density CDM models from Moscardini et al. (2000), which are able to correctly reproduce both the shape and amplitude of the observed REFLEX ξ_{cc}. We remark that in this case the amplitude is not a free parameter as for galaxy correlation functions, since the bias value for the specific REFLEX selection function can be computed using an appropriate theory for the clustering of massive haloes (e.g. Mo & White 1996 and subsequent refinements), given the fairly straightforward relation between X-ray luminosity and mass (see Borgani & Guzzo 2001 for more details and references). Analogous results have been obtained from the power spectrum, as discussed in the contribution by P. Schuecker to this volume (and in detail by Schuecker et al. 2001). Another prediction which can be precisely made for a survey like REFLEX is the dependence of the correlation length on the survey flux limit. The expectation values for different CDM variants are compared to

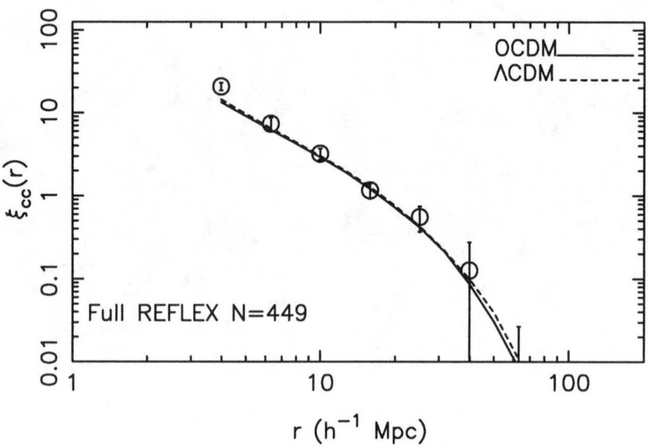

Figure 5. The REFLEX two-point correlation function compared with the predictions for the same survey selection function of two low-density CDM models, respectively open with $\Omega_m = 0.3$, $\Omega_\Lambda = 0.0$ (OCDM) and flat with $\Omega_m = 0.3$, $\Omega_\Lambda = 0.7$ (ΛCDM), as computed by Moscardini et al. (2000). (Here r is the redshift-space separation, usually called s in galaxy surveys.)

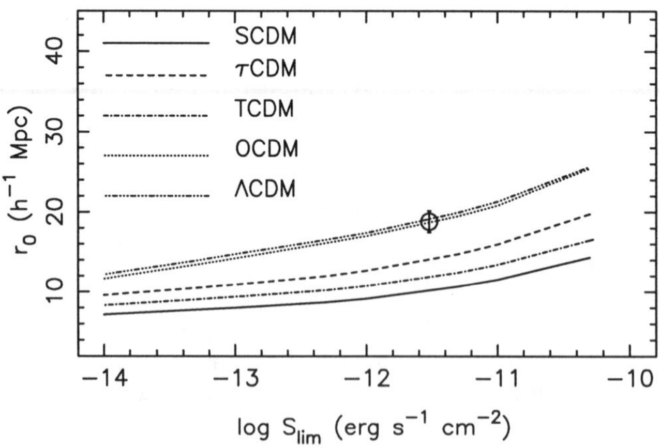

Figure 6. Comparison of the value of correlation length r_0 from the full REFLEX flux-limited survey with the predictions as a function of limiting X-ray flux for a range of CDM-type cosmological models computed by Moscardini et al. (2000).

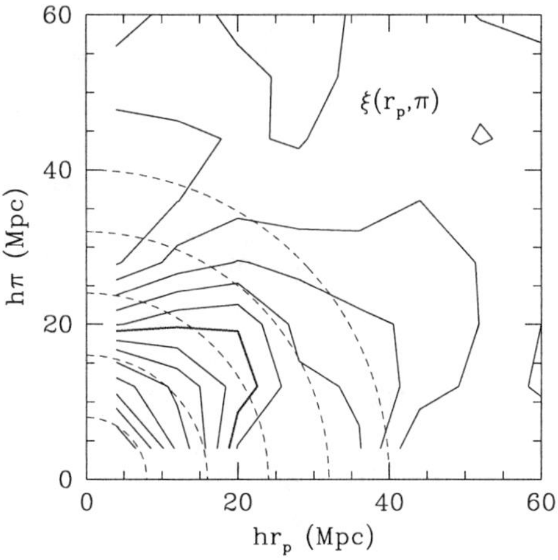

Figure 7. The bi-dimensional correlation function $\xi(r_p, \pi)$, used to evidence redshift-space anisotropies in the clustering of REFLEX clusters. The diagram shows no stretching of the contours along the line of sight, indicating that projection biases and redshift errors are small in the REFLEX survey. On the contrary, the evident large-scale compression is a possible fingerprint of coherent motions. The dashed circles show how a perfectly isotropic distribution would look like.

the observed correlation length from the full REFLEX in Fig. 6, again indicating that low-Ω_M models provide the best consistency with observations (Collins et al. 2000).

Redshift-space anisotropies in the clustering pattern of clusters can be evidenced through the bidimensional correlation function $\xi(r_p, \pi)$ [or analogously $\xi(\sigma, \pi)$]. Classical analyses of Abell samples show strong elongations along the line-of-sight π direction (e.g. Miller et al. 2001), which are typically not seen in automatic cluster catalogues (e.g. Nichol et al. 1992). These arise by a combination of projection effects, redshift errors and true spatial anisotropies in small–size catalogues. The plot of $\xi(r_p, \pi)$ for REFLEX (Fig. 7), in fact shows no evidence for stretching along the line of sight, indicating how these spurious effects are negligible in this survey. On the other hand, the contours are significantly compressed at large separations, a typical signature of streaming motions (proportional to the quantity $\beta = \Omega_M^{0.6}/b$) which is for the first time seen in a cluster sample (Guzzo et al. 2002; see also Padilla & Baugh 2001).

We would like to conclude by mentioning that the statistical quality of the results presented here will significantly improve in the near future, thanks to the extension of REFLEX to fainter fluxes (by a factor ~ 2). The REFLEX-2 sample is in fact being constructed and will bring the total number of clusters over the same area to ~ 800 (see e.g. Böhringer et al. 2001b).

References

Abell, G.O., 1958, ApJS, 3, 211

Abell, G.O., Corwin, H.G. Jr., & Olowin, R.P., 1989, ApJS, 70, 1

Böhringer, H., et al. (the REFLEX Team), 2001a, A&A, 369, 826

Böhringer, H., et al. (the REFLEX Team), 2001b, *The Messenger*, 106, in press

Böhringer et al. (the REFLEX Team), 2002, ApJ, in press (astro-ph/0106243)

Borgani S., & Guzzo L., 2001, *Nature*, 409, 39

Borgani, S., et al., 2001, ApJ, in press (Nov 2001 issue, also astro-ph/0106428)

Collins, C.A., et al. (the REFLEX Team), 2000, MNRAS, 319, 939

De Grandi, S., et al. (the REFLEX Team), 1999, ApJ, 513, L17

Guzzo, L. 2001, in *Where's the Matter? Tracing Dark and Bright Matter with the New Generation of Large Scale Surveys*, proc. of meeting held in Marseille (June 2001), M. Treyer & L. Tresse eds., Frontier Group (astro-ph/0111309)

Guzzo, L., et al. (the REFLEX Team), 1999, *The Messenger*, 95, 27

Guzzo, L., et al. (the REFLEX Team), 2002, in preparation

Henry, J.P., et al., 1992, ApJ, 386, 408

Henry, J.P., 2001, in "AMiBA 2001: High-z Clusters, Missing Baryons, and CMB Polarization", in press (astro-ph/0109498)

Kaiser, N., 1984, ApJ, 284, L9

Lahav, O., et al., 1989, MNRAS, 238, 881

Miller, C.J., Nichol, R.C., & Batuski, D.J., 2001, ApJ, 555, 68

Mo, H.J. & White S.D.M., 1996, MNRAS, 282, 347

Moscardini, L., et al., 2000, MNRAS, 314, 647

Nichol, R.C., et al., 1992, MNRAS, 255, 21p

Padilla, N.D., Baugh, C.M., 2001, MNRAS, submitted (astro-ph/0104313)

Reiprich, T.H., & Böhringer, H., 2002, ApJ, in press (astro-ph/0111285)

Romer, A.K., et al., 1994, *Nature*, 372, 75

Rosati, P., et al., 1998, ApJ, 492, L21

Schindler, S.A., et al. (the REFLEX Team), 1995, A&A, 299, L9

Schuecker et al. (the REFLEX Team), 2001, A&A, 368, 86

Tracing Cosmic Evolution with Galaxy Clusters
ASP Conference Series, Vol. 268, 2002
S. Borgani, M. Mezzetti, and R. Valdarnini, eds.

Clustering and Substructuring of REFLEX Clusters of Galaxies

Peter Schuecker and Hans Böhringer

Max-Planck Institut für extraterrestrische Physik, Garching, Germany

Luigi Guzzo

Osservatorio Astronomico di Brera, Merate, Italy

Chris A. Collins

Liverpool John Moores University, Liverpool, UK

Abstract. Results of a power spectrum analysis of the REFLEX cluster number density fluctuations using Fourier modes and Kahunen-Loewe eigenvectors are presented and their cosmological implications discussed. First results of a systematic analysis of the substructure of a representative sample of 470 X-ray cluster images obtained from the ROSAT All-Sky Survey are presented.

1. Clustering of REFLEX Clustering of Galaxies

Whereas the spatial distribution of *galaxies* is still not fully understood due to the incomplete knowledge of the physical processes connected to galaxy formation and evolution, the spatial distribution of *clusters* of galaxies is mainly driven by gravitational processes and thus provides a more direct view of the large-scale distribution of matter in the Universe. This is especially the case when clusters are selected in X-rays. The structures are characterized and compared with predictions from structure formation models using lower-order statistics like the the power spectrum, i.e., the Fourier transform of the classical two-point spatial correlation function. However, on scales up to $1\,h^{-1}$ Gpc the anisotropy of basically all recent cluster survey volume becomes apparent leading to leakage effects. Therefore, the plane waves used in the standard power spectrum analysis are no longer orthogonal on very large scales and should be replaced by another set of basis functions fulfilling this fundamental criterion. The conditions of orthogonality and statistical orthogonality immediately lead to a set of basis functions, the eigenvectors of the sample correlation matrix called the Karhunen-Loeve (KL) eigenvectors. Consequently, the KL eigenvectors provide by construction the optimal set of basis functions which are ideally suited for the analyses of the cluster number density fluctuations on the largest scales. In the following the power spectrum results obtained with the ROSAT ESO Flux-Limited (REFLEX) X-ray cluster survey using both the standard Fourier method and the KL eigenvector method are described and their cosmological implications dis-

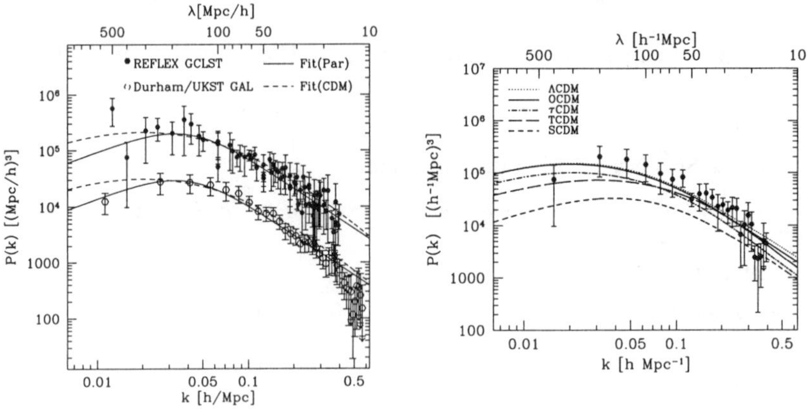

Figure 1. Left: REFLEX cluster and Durham/UKST galaxy power spectra. Right: comparison of one REFLEX power spectrum with predictions from different structure formation models.

cussed. The REFLEX sample (Böhringer et al. 2001) used for the power spectrum analyses includes 452 clusters (449 with redshifts), detected in the RASS database south of the Declination +2.5 deg and down to the nominal X-ray flux limit of $3.0 \times 10^{-12}\,\mathrm{erg\,s^{-1}\,cm^{-2}}$ in the ROSAT energy band (0.1-2.4 keV). REFLEX covers 4.24 sr excluding the area ±20 deg around the galactic plane and 0.0987 sr around the Magellanic Clouds and is basically restricted to redshifts $z \leq 0.3$ (a few clusters reach $z = 0.45$). Several tests are performed showing no significant incompletenesses of the REFLEX clusters with X-ray luminosities brighter than $10^{43}\,\mathrm{erg\,s^{-1}}$ up to scales of about $1\,h^{-1}$ Gpc. Power spectra, $P(k)$, of comoving cluster number densities are estimated for flux- and volume-limited subsamples. The most important result is the detection of a broad maximum within the comoving wavenumber range $0.022 \leq k \leq 0.030\,h\,\mathrm{Mpc}^{-1}$ (Fig. 1, left). The data suggest an increase of the power spectral amplitude with X-ray luminosity. Compared to optically selected cluster samples the REFLEX $P(k)$ is flatter for wavenumbers $k \leq 0.05\,h\,\mathrm{Mpc}^{-1}$ thus shifting the maximum of $P(k)$ to larger scales. The smooth maximum is not consistent with the narrow peak detected at $k = 0.05\,h\,\mathrm{Mpc}^{-1}$ using the Abell/ACO richness ≥ 0 data. In the range $0.02 \leq k \leq 0.4\,h\,\mathrm{Mpc}^{-1}$ general agreement is found between the slope of the REFLEX $P(k)$ and those obtained with optically selected galaxies. A semi-analytic description of the biased nonlinear power spectrum in redshift space gives the best agreement for low-density Cold Dark Matter models with or without a cosmological constant (Fig. 1, right). For more details see Schuecker et al. (2001a). The results are also in good agreement with the REFLEX correlation function measured by Collins et al. (2000). Using the Karhunen-Loeve eigenvectors of the REFLEX survey, fluctuations of the number densities of 452 REFLEX clusters of galaxies are measured on scales up to $1\,h^{-1}$ Gpc. A likelihood estimate of the shape parameter of the linear theory CDM power spectrum using the eigenvectors gives $\Gamma = 0.14^{+0.13}_{-0.07}$ for the shape parameter of the linear

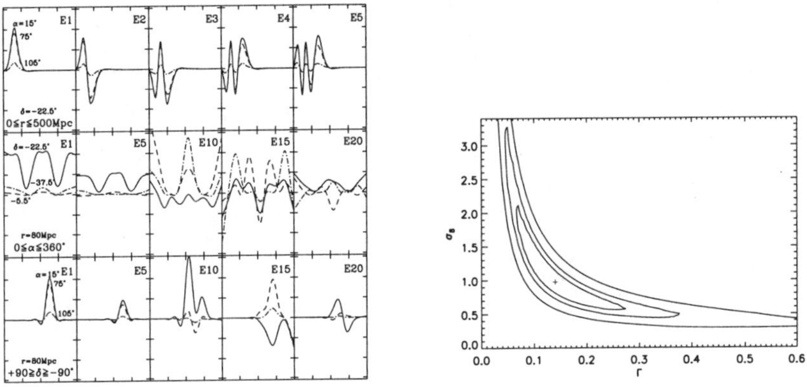

Figure 2. Left: examples of REFLEX eigenvectors used for the KL analysis. Right: significant contours (1σ (one degree of freedom), 1σ (two degrees of freedom), 2σ (two degrees of freedom) for the shape parameter of the linear theory transfer function and its normalization.

theory transfer function, in good agreement with the results obtained with the Fourier analysis (Schuecker et al., to be submitted).

2. Substructures in REFLEX and BCS clusters of galaxies

Current structure formation scenarios suggest an hierarchical growth of cosmic objects where the merging of subclumps turns out to be a fundamental process. The frequencies of subclumps (substructures) and elongations are thus useful statistical quantities with a direct relation to cosmology. A more detailed study would ideally use deep pointed X-ray observations to get high signal-to-noise temperature and surface brightness maps. Unfortunately, X-ray data from public archives generally allow precise measurements on interesting but not necessarily representative clusters of galaxies. For the majority of the clusters needed for a statistically representative sample, detailed studies are at present and in the near future not achievable. What is needed for cosmological investigations is a systematic and comprehensive study of substructure characteristics for a large set of clusters compiled in an homogeneous way.

One possibility towards a systematic study of substructure is offered by the large area ROSAT All-Sky Survey (RASS). However, both the energy resolution of 40 percent FWHM at 1 keV and the limited energy range $0.1 \leq E \leq 2.4\,\mathrm{keV}$ of the ROSAT position-sensitive proportional counter (PSPC) limit temperature measurements of the intracluster gas, especially for $kT \geq 4\,\mathrm{keV}$. This and the fact that the small number of RASS photons counted in the direction of a typical galaxy cluster in our samples do neither allow the determination of useful temperature maps nor the proper quantification of the amount of substructure, e.g., in the form of mass estimates of the individual subclumps for the majority of a fair cluster sample. The analysis described below thus gives more weight to a representative study of a large sample of clusters, mainly restricted to sub-

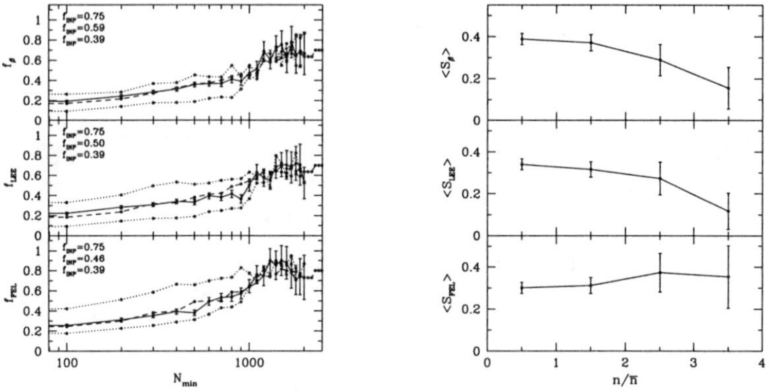

Figure 3. Left: significance curves for different substructure statistics as a function of the minimum number of RASS photons (continous lines) compared with predictions of an empirical model. Right: substructure-density relation: the significance for regular, unimodal clusters decreases with increasing local cluster number density.

structure detection and significance determination in terms of significance for the degree of substructure for a whole sample or for specific subsamples. For the analyses in addition to the REFLEX sample the BCS sample (Ebeling et al. 1998) is used which includes 201 clusters (statistical sample), detected north of the Declination $\delta = 0\,\mathrm{deg}$, excluding the area $\pm 20\,\mathrm{deg}$ around the galactic plane using the RASS data. The formal flux limit of the resulting sample is $4.4 \times 10^{-12}\,\mathrm{erg\,s^{-1}\,cm^{-2}}$ (0.1-2.4 keV). The combined sample referred to as REFLEX+BCS contains 470 clusters (excluding the BCS clusters in common with REFLEX) with a formal flux limit of $4.4 \times 10^{-12}\,\mathrm{erg\,s^{-1}\,cm^{-2}}$. After correction for several systematic effects using an empirical model, 52 ± 7 percent of the REFLEX+BCS clusters are found to be substructured in metric apertures of 1 Mpc radius ($H_0 = 50\,\mathrm{km\,s^{-1}\,Mpc^{-1}}$) applying the β, LEE, and Fourier Elongation statistics (Fig. 2, left). More details are given in Schuecker et al. (2001b). Future simulations will show statistically which mass spectrum of major and minor mergers contributes to this number. Another important result is the discovery of a substructure-density relation, analogous to the morphology-density relation for galaxies. Here, clusters with asymmetric or multi-modal X-ray surface brightness distributions are located preferentially in regions with higher cluster number densities (Fig. 2, right).

References

Böhringer, H., Schuecker, P., Guzzo, L., et al., 2001, A&A, 369, 826

Collins, C.A., Guzzo, L., Böhringer, H., et al., 2000, MNRAS, 319, 939

Schuecker, P., Böhringer, H., Guzzo, L., et al., 2001a, A&A, 368, 86

Schuecker, P., Böhringer, H., Reiprich, T.H., Feretti, L., 2001b, A&A (in press, see astro-ph/0109030)

Tracing Cosmic Evolution with Galaxy Clusters
ASP Conference Series, Vol. 268, 2002
S. Borgani, M. Mezzetti, and R. Valdarnini, eds.

The Large-Scale Environment of Dynamical Young Clusters of Galaxies

Manolis Plionis

Institute of Astronomy & Astrophysics, National Observatory of Athens, I.Metaxa & B.Pavlou, P.Penteli, 152 36 Athens, Greece

Spyros Basilakos

Astrophysics Group, Blackett Laboratory, Imperial College, London, UK

Abstract. Using the APM cluster data (Dalton et al. 1997) we investigate whether the dynamical status of clusters is related to the large-scale structure of the Universe. The methods used have been calibrated using a limited number of APM/Abell clusters for which ROSAT pointed observations were available. We find that cluster substructure is strongly correlated with the tendency of clusters to be aligned with their nearest neighbour and in general with the nearby clusters that belong to the same supercluster. Furthermore, dynamically young clusters are more clustered than the overall cluster population. These are strong indications that clusters develop in a hierarchical fashion by merging along the large-scale filamentary structures within which they are embedded.

1. Introduction

In the popular hierarchical clustering scenario, clusters are supposed to form by merging and accretion of galaxies along filamentary protostructures. This structure formation scenario is seen to work in cosmological N-body CDM simulations, while observational studies have provided indications that our Universe follows such a scenario (cf. West 1994; West, Jones & Forman 1995)

If we can unambiguously identify the dynamical status of clusters (eg. presence of significant substructures, X-ray temperature gradients, compression of X-ray emitting gas along the merging direction, cf. Sarazin 2000; Schindler 1999) then we can search for indications of cluster formation along the lines of the hierarchical scenario.

Mohr et al. (1995), Rizza et al. (1998) and Kolokotronis et al. (2001) have investigated the frequency of cluster substructure using in a complementary fashion optical and X–ray data. The advantage of using X–ray data is that the X–ray emission is proportional to the square of the gas density (rather than just density in the optical) and therefore it is centrally concentrated, a fact which minimises projection effects. The advantage of using optical data is the shear size of the available cluster catalogues and thus the statistical significance of the emanating results. Kolokotronis et al. (2001) calibrated various substructure measures using APM and ROSAT data of 22 Abell clusters and found that in most cases using X–ray or optical data one can identify substructure unambiguously. Only in $\sim 20\%$ of the clusters they studied did they find projection effects in the optical that altered the X-ray definition of substructure. Their conclusion was that solely optical cluster imaging data can be used in order to identify the clusters that have significant substructure.

An interesting observable, that was thought initially to provide strong constraints on theories of galaxy formation, is the tendency of clusters to be aligned with their nearest neighbour as well as with other clusters that reside in the same supercluster (cf. Binggeli 1982; Plionis 1994). Analytical (Bond 1986) and numerical (cf. Splinter et al. 1997; Onuora & Thomas 2000) work have shown that such alignments, expected naturally to occur in "top-down" scenarios, are also found in hierarchical clustering models of structure formation like the CDM.

2. Substructure & Alignment Measures

Evrard et al. (1993) and Mohr et al. (1995) have suggested as an indicator of cluster substructure the shift of the center-of-mass position as a function of density threshold above which it is estimated. The *centroid-shift* (sc) is defined as the distance between the cluster center-of-mass, which is a function of density threshold, and the highest cluster density-peak. The significance, σ, of such centroid variations to the presence of background contamination and random density fluctuations, are estimated using Monte Carlo cluster simulations in which, by construction, there is no substructure (see details in Plionis 2001). According to Kolokotronis et al. (2001), a large and significant value of sc is a clear indication of substructure in the optical APM cluster data.

Furthermore, in order to investigate the alignment between cluster orientations, we use the relative position angle between cluster pairs, $\phi_{i,j}$. In an isotropic distribution we will have $\langle \phi_{i,j} \rangle \simeq 45°$. A significant deviation from this would be an indication of an anisotropic distribution which can be quantified by $\delta = \sum \phi_{i,j}/N - 45$ (Struble & Peebles 1995). In an isotropic distribution we have $\langle \delta \rangle \simeq 0$, while the standard deviation is given by $\sigma = 90/\sqrt{12N}$. A significantly negative value of δ would indicate alignment and a positive misalignment.

3. Results: Alignments vs Substructure:

We have correlated the alignment signal with the substructure significance indication, σ, in order to see whether there is any relation between the large-scale environment, in which the cluster distribution is embedded, and the internal cluster dynamics. In the left panel of figure 1 we present the alignment signal, $\langle \delta \rangle$, between cluster nearest-neighbours (filled dots) and between all pairs (open dots) with pair separations $< 20\ h^{-1}$ Mpc. Evidently, there is a strong correlation between the strength of the alignment signal and the substructure significance level.

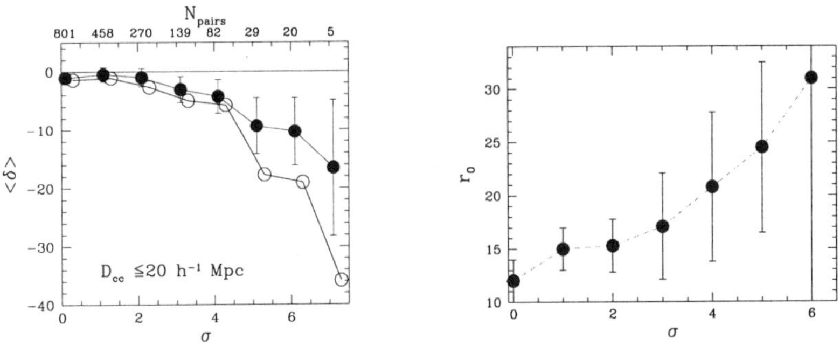

Figure 1. Left: Alignment signal as a function of substructure significance level. Filled circles correspond to nearest-neighbours and open circles to all neighbours within 20 h^{-1} Mpc. Right: The APM cluster correlation length as a function of substructure significance level.

4. Results: Local Density vs Substructure

Where do clusters with significant substructure reside? To answer this question we measure the spatial 2-point cluster correlation function for different substructure significance levels. In the right panel of figure 1 we plot the correlation length, r_0, as a function of cluster substructure significance level, which is clearly an increasing function of σ. The conclusion of this analysis is that clusters showing evidence of dynamical activity reside in high-density environments. Our results are in agreement with a similar environmental dependence found in the REFLEX and BCS X-ray cluster sample (Schüecker et al., this volume) and for cooling flow clusters with high mass accretion rates (Loken, Melott & Miller 1999).

As an individual example we present in figure 3 the distribution on the sky of a dense supercluster, found with a percolation radius of 5 h^{-1} Mpc. Embedded in the plot are the galaxy density distributions of the clusters that constitute the supercluster. It is interesting that the position angles of the clusters (between 16° and 48°) are aligned also with the orientation of the filamentary supercluster, which has a position angle of $\sim 30°$.

References

Bingelli B., 1982, A&A, 250, 432

Bond, J.R., 1986, in *Galaxy Distances and Deviations from the Hubble Flow*, eds. Madore, B.F., Tully, R.B., (Dordrecht: Reidel), p.255

Dalton G. B., Maddox S. J., Sutherland W. J., Efstathiou G., 1997, MNRAS, 289, 263

Evrard A.E., Mohr J.J., Fabricant D.G., Geller M.J.,1993, ApJ, 419, L9

Kolokotronis, V., Basilakos, S., Plionis, M., Georgantopoulos, I., 2001, MNRAS, 320, 49

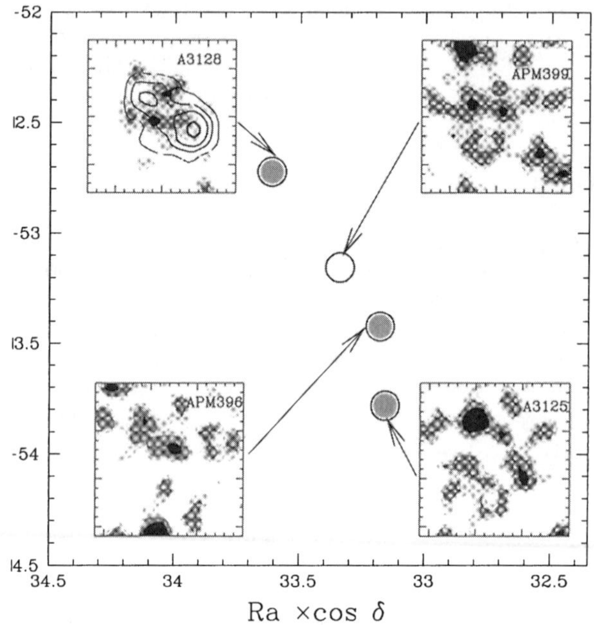

Figure 2. An example of a dense APM supercluster. Filled dots signify clusters with significant substructure. For A3128 we overlay the ROSAT PSPC X-ray contours.

Loken, C., Melott, A.L., Miller, C.J., 1999, ApJ, 520, L5

Mohr, J.J., Evrard, A.E., Fabricant, D.G., Geller, M.J., 1995, ApJ, 447, 8

Onuora, L.I., Thomas, P.A, 2000, MNRAS, 319, 614

Plionis M., 1994, ApJS, 95, 401

Plionis M., 2001, in Clusters and the High-Redshift Universe observed in X-rays, XXIth Moriond Astrophysics Meeting, eds. Neumann et al., *in press*

Richstone, D., Loeb, A., Turner, E.L., 1992, ApJ, 393, 477

Rizza E., Burns J. O., Ledlow M. J., Owen F. N., Voges, W., Blito M., 1998, MNRAS, 301, 328

Sarazin, C.L., 2001, in Merging Processes in clusters of Galaxies, eds. Feretti, L., Gioia, M., Giovannini, G., (Dordrecht: Kluwer).

Schindler S., 1999, Vulcano Workshop 1999, Multifrequency Behaviour of High Energy Cosmic Sources, eds. Giovanelli F., Sabau-Graziati L.

Splinter, R.J., Melott, A.L., Linn, A.M., Buck, C., Tinker, J., 1997, ApJ, 479, 632

Struble, M.F., Peebles, P.J.E., 1985, AJ, 90, 582

West, M. J., 1994, MNRAS, 268, 79

West, M. J., Jones C., Forman W., 1995, ApJ, 451, L5

Tracing Cosmic Evolution with Galaxy Clusters
ASP Conference Series, Vol. 268, 2002
S. Borgani, M. Mezzetti, and R. Valdarnini, eds.

Clustering properties of galaxy clusters in present and future surveys

Lauro Moscardini and Sabino Matarrese

Università degli Studi di Padova, I-35100 Padova, Italy

Paola Andreani

Osservatorio Astronomico di Padova, I-35122 Padova, Italy

Matthias Bartelmann and Houjun Mo

Max-Planck-Institut für Astrophysik, D-85748 Garching, Germany

Abstract. We use a theoretical model to predict the clustering properties of galaxy clusters. We consider a set of present catalogues, defined both in the optical and X-ray bands, to constrain the cosmological parameters. Finally, we use our model to make predictions on the correlation length of galaxy clusters expected in future surveys, including catalogues of objects detected through their thermal Sunyaev-Zel'dovich emission.

1. Modelling the clustering of galaxy clusters

Clusters of galaxies are the largest collapsed objects in the Universe. Their properties are largely determined by the gravitational collapse in the cosmological density field, and so their distribution in the Universe depends both on cosmology and on the initial density perturbations. Therefore, the study of the clustering of clusters may provide important diagnostics for models of structure formation.

An algorithm to describe the clustering in the past light-cone has been developed in Matarrese et al. (1997). The final expression for the observed spatial correlation function $\xi_{\rm obs}$ in a given redshift interval $\mathcal{Z}$ is

$$\xi_{\rm obs}(r) = \frac{\int_{\mathcal{Z}} dz_1 dz_2 \overline{\mathcal{N}}(z_1)\overline{\mathcal{N}}(z_2) b_{\rm eff}(z_1) b_{\rm eff}(z_2) \xi_{\rm m}(r, z_{\rm ave})}{[\int_{\mathcal{Z}} dz_1 \overline{\mathcal{N}}(z_1)]^2} , \qquad (1)$$

where $\overline{\mathcal{N}}(z) \equiv \mathcal{N}(z)/r(z)$, $\mathcal{N}(z)$ is the actual redshift distribution of the catalogue, $\xi_{\rm m}$ is the dark matter (DM) covariance function and $z_{\rm ave}$ is a suitably defined intermediate redshift between z_1 and z_2. An important ingredient entering the previous equation is the effective bias $b_{\rm eff}$, which can be expressed as a weighted average of the 'monochromatic' bias factor $b(M, z)$ of objects of some given intrinsic property M: $b_{\rm eff}(z) \equiv \mathcal{N}(z)^{-1} \int_{\mathcal{M}} d\ln M'\, b(M', z)\, \mathcal{N}(z, M')$, where $\mathcal{N}(z, M)$ is the number of objects actually present in the catalogue with redshift in the range z, $z + dz$ and M in the range M, $M + dM$, whose integral

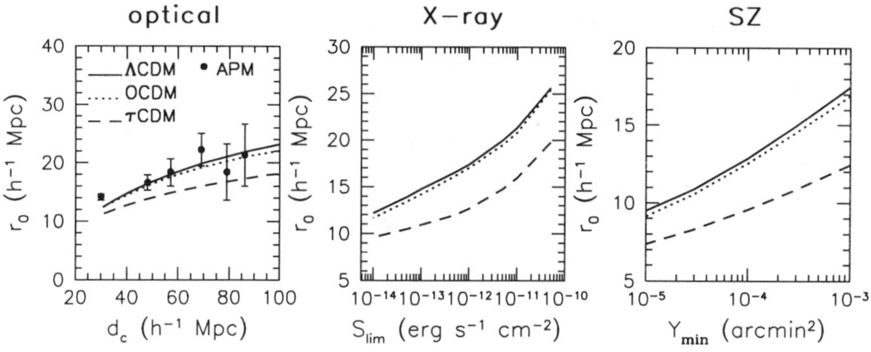

Figure 1. The correlation length r_0 as a function of the intracluster distance d_c, the X-ray limiting flux $S_{\rm lim}$ and the SZ limiting sensitivity $Y_{\rm min}$. In the left panel we show also the observational estimates from APM.

over $\ln M$ is $\mathcal{N}(z)$. Galaxy clusters are expected to form by the hierarchical merging of smaller mass units. Consequently it is possible to fully characterize their properties by the mass M of their hosting DM haloes at each redshift z. One can then estimate their comoving mass function and the monochromatic bias by using the relations in Sheth & Tormen (1997). The model is completed by computing the redshift evolution of the DM covariance function $\xi_{\rm m}$ using the fitting formula in Peacock & Dodds (1996).

In order to predict the abundance and clustering of galaxy clusters in different catalogues we need to relate the sample characteristics to a corresponding halo mass at each redshift. For the optical catalogues, the various samples are characterized by a different cluster number density. In this case we fix the minimum mass of the hosting DM haloes (which in our model represents the lower limit of the integral of bias) in such a way that the comoving cumulative mass function reproduces the number density of clusters in the range of redshift sampled by the original catalogues. For X-ray selected clusters, it is easy to relate the limiting flux to the minimum halo mass (see e.g. Moscardini et al. 2000a,b). In fact the flux S in a given band corresponds to an X-ray luminosity $L_X = 4\pi d_L^2 S$ in the same band, where d_L is the luminosity distance. The quantity L_X can be converted into the total luminosity $L_{\rm bol}$ by performing band and bolometric corrections. Local observations (but also recent data out to $z \approx 0.4$) are consistent with a simple relation between the cluster bolometric luminosity and the temperature: $L_{\rm bol} \propto T^3$. Finally, to convert the temperature into the mass of the hosting DM halo, we assume virial isothermal gas distribution and spherical collapse: $T \propto M^{2/3}$. Following Bartelmann (2001), for SZ clusters we assume that a cluster is detectable if its integrated Compton y-parameter is larger than a given sensitivity limit $Y_{\rm min}$: $Y \equiv \frac{k\,\sigma_T\,T}{m_e\,c} \int d^2\vec{\theta}' \int dl\; n_e(\vec{\theta}', l) \geq Y_{\rm min}$, where n_e is the electron density. Consequently, the quantity Y depends upon the cluster mass through the previous mass-temperature relation.

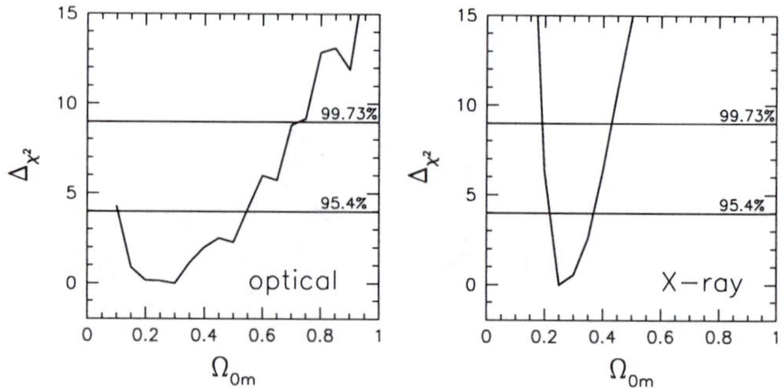

Figure 2. The variation of Δ_{χ^2} around the best-fitting value of Ω_{0m}.

In Figure 1 we show the dependence of the correlation length r_0 on the limits used to define the surveys in the different bands: the intracluster distance d_c (optical band), the limiting flux $S_{\rm lim}$ (X-ray band) and the limiting sensitivity $Y_{\rm min}$ (SZ emission). We find that r_0 is an increasing function of the limits defining a cluster survey: objects with larger intracluster distance or limiting X-ray flux or minimum SZ sensitivity limit are expected to be more clustered, independently of the cosmological model.

2. Application to present and future data

We start by discussing some of the constraints on the cosmological parameters which can be set by performing a maximum-likelihood analysis (see an extended discussion in Moscardini, Matarrese & Mo 2001). We compare the theoretical predictions to a set of existing observational data, both in the optical (APM and EDCC) and X-ray bands (RASS1, BCS, XBACs and REFLEX). Note that to date no surveys of SZ clusters are available to be included in this analysis. In Figure 2 we show the results of our analysis when we consider as a free parameter Ω_{0m} only. We fix a flat geometry, a shape parameter $\Gamma = 0.2$ and σ_8 to reproduce the cluster abundance. The left panel refer to the analysis of the optical catalogues. In this case the minimum of χ^2 is in the range $\Omega_{0m} = 0.2{-}0.3$ and we find $\Omega_{0m} \lesssim 0.5$ at 95.4 per cent level. The results for the X-ray catalogues are shown in the right panel. Also in this case the most likely value for Ω_{0m} is $0.2 - 0.3$ but the allowed region is much smaller: $0.2 \lesssim \Omega_{0m} \lesssim 0.35$ (95.4 per cent level).

We use now our model to predict the correlation length expected in future surveys. In the optical band, a large improvement of our knowledge on the properties of the large-scale structure will be obtained when the Sloan Digital Sky Survey will be completed. This survey is expected to be nearly as deep as an existing catalogue, the ESO Imaging Survey, which however is much smaller. We choose to compute our predictions for a future optical catalogue by fixing the minimum mass needed to reproduce the EIS cluster density for $z \leq 0.3$ and $z > 0.3$. In the X-ray band it has been proposed to use the XMM/Newton satellite to build a very deep large-scale structure survey covering a region of 64 square degrees. The expected limiting flux will be approximately $S_{\rm lim} = 5{\times}10^{-15}$

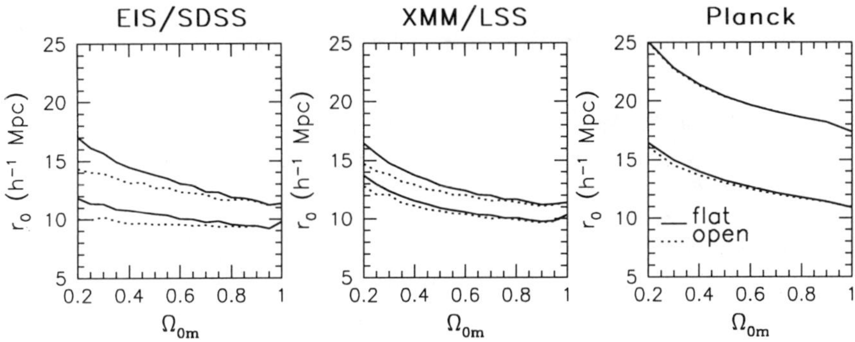

Figure 3. The predicted value of the correlation length for different future surveys as a function of Ω_{0m}. Lower and upper lines refer to clusters with $z \leq 0.3$ and $z > 0.3$, respectively.

erg cm^{-2} s^{-1} in the 0.5–2 keV band. Finally, the Planck satellite will produce the first all-sky survey of galaxy clusters detected through their thermal SZ emission, reaching a sensitivity limit of $Y_{min} = 3 \times 10^{-4}$ arcmin^2.

In Figure 3 we show the predicted value of the correlation length for future surveys as a function of Ω_{0m}. Here we fix $\Gamma = 0.2$ and σ_8 to reproduce the cluster abundances. As a general result, we find that the presence of a cosmological constant increases the correlation length by a factor always smaller than 15 per cent. Moreover we find that varying Ω_{0m} changes r_0 only by a factor of at most 30 per cent. More precisely, we find a slight decrease of the correlation length with increasing Ω_{0m}. Finally all the catalogues display a positive redshift evolution of the clustering, i.e. r_0 is larger for clusters at high redshifts. This result is due to the large increase of the bias factor with redshift which overcompensates the corresponding decrease of the DM correlation function. In fact, considering high redshifts, galaxy clusters become rarer objects, connected to higher peaks of the density fluctuations.

References

Bartelmann, M. 2001, A&A, 370, 754

Matarrese, S., Coles, P., Lucchin, F., & Moscardini, L. 1997, MNRAS 286, 115

Moscardini, L., Matarrese, S., De Grandi, S., & Lucchin, F. 2000a, MNRAS 314, 647

Moscardini, L., Matarrese, S., Lucchin, F., & Rosati, P. 2000b, MNRAS 316, 283

Moscardini, L., Matarrese, S., & Mo, H.J. 2001, MNRAS in press

Peacock, J.A., & Dodds, S.J. 1996, MNRAS 280, L19

Sheth, R.K., & Tormen, G. 1999, MNRAS 308, 119

Tracing Cosmic Evolution with Galaxy Clusters
ASP Conference Series, Vol. 268, 2002
S. Borgani, M. Mezzetti, and R. Valdarnini, eds.

The $\sigma_8(\Omega_M)$ constraint from X-ray clusters revised

Elena Pierpaoli, Douglas Scott

Department of Physics and Astronomy, University of British Columbia, BC V6T1Z1 Canada

Martin White

Harvard-Smithsonian Center for Astrophysics, Cambridge, MA 02138

Abstract. We revisited the constraint on the $\sigma_8 - \Omega_M$ relation as derived from the local abundance of galaxy clusters in the light of several recent developments: (1) the availability of well-defined samples of local clusters with relatively accurate X-ray temperatures; (2) new theoretical mass functions for dark matter haloes which provide a good fit to large numerical simulations; (3) more accurate mass-temperature relations from larger catalogues of hydrodynamical simulations; and (4) the requirement to consider closed as well as open and flat cosmologies to obtain full multi-parameter likelihood constraints for CMB and SNe studies.

1. Introduction

The strongest LSS constraint on parameter estimation comes from the estimate of the mass variance on scales of $8\,h^{-1}$ Mpc (σ_8), as determied by the local abundance of galaxy clusters.

There have been several developments in terms of both theory and observation which suggest it would be useful to revisit this constraint. Firstly the addition of *ASCA* temperatures means that there is now a well defined local temperature function for clusters, with relatively small errors in temperature. Secondly a number of large N-body simulations have accurately determined the mass function of virialized haloes, finding non-negligible deviations from the old Press-Schechter theory. For example the extremely large N-body simulations of the Virgo consortium have highlighted systematic departures from the PS predicted mass functions which alter the constraints on the power spectrum normalization coming from the cluster abundance. Thirdly, more ambitious hydrodynamical simulations of cluster formation have resulted in improvements in the relationship between mass and temperature and a better estimate of its scatter. Finally, the increased sophistication of multi-parameter cosmological studies (in particular driven by recent CMB anisotropy measurements) requires that cosmological models with general curvature be considered.

We constructed a new sample of clusters drawn from the literature and used this sample to obtain improved results on σ_8, as a function of the matter density and cosmological constant in a universe with general curvature.

2. The method

The abundance of rich clusters is tied to the normalization of the power spectrum (extrapolated to the present using linear theory) through the Press-Schechter theory and its extensions (see Sheth & Tormen 1999). Our constraint will be on the variance of the density field, smoothed on some (comoving) scale R corresponding to a mass $M = (4\pi/3)\bar{\rho}R^3$ where $\bar{\rho}$ is the background density. The point of comparison between theory and observation will be the temperature function of rich clusters, i.e. the number density of clusters within a temperature range dT about T.

We shall make use of the fact that rich clusters are large virialized structures dominated by gravity, and thus we can factorize the problem and proceed in two steps. First we shall use the abundance of dark matter haloes of a given mass drawn from extremely large N-body simulations. Next we shall use a mass-temperature relation, calibrated from hydrodynamical simulations, to convert from the (unobservable) virial mass to the IGM temperature, effectively using the larger volume of the N-body simulations to improve the statistics of the hydro simulations.

In order to use the $M-T$ realtion we evaluated the value of the density contrast at collapse, Δ_c, for a generic cosmology. Fitting formulas for Δ_c can be found in Pierpaoli, Scott, & White (2001). The values of the normalization parameterof the $M-T$ relation, β spread from near 1 up to 1.6 when determined from simulations, with no obvious peak of preferred values. We adopt the mean opinion on this issue, allowing for $\beta = 1.3$, with a 10% systematic error (i.e. variation between the simulations) in the mass, plus a 10% statistical error.

For the cluster sample, we have constructed it in a similar way to that of Markevitch (1998), although we use temperatures from White (2000) when available, supplemented by other temperatures from the literature. Importantly our sample is effectively *temperature*-selected at each redshift. Because of the large completeness correction required at low temperature, we make a cut of all clusters with $T_X < 3.5$ keV. This also corresponds approximately to a restriction to clusters with temperatures which are dominated by gravitational physics.

Our final sample contains 38 clusters and s shown in Fig. 1a. However, since we will be considering the temperature errors in our Monte Carlo error analysis, it is important to include clusters which could scatter *into* the sample. We therefore include additional clusters with $T_X + 2\sigma$ being above our 3.5 keV cut-off, or above the temperature-derived flux limit at their redshift.

Rather than fitting to a measure of $n(> kT_X)$ at some fiducial T_X, we try to fit to all the data in a way which accounts for the Poisson distribution of clusters at each temperature using a likelihood method which takes full account of the errors. Given a set of data with which to compare, we compute the likelihood function for any given theory as follows. We break the range of temperatures under consideration into a large number of narrow bins. Then for a given Monte-Carlo realization of the temperatures of the set of clusters we place the clusters into the appropriate temperature bins, giving an occupation number $\eta_i = 0$ or 1 for each bin. For a predicted number density $\mathcal{N}_i = n(T)dT$, depending on our cosmological parameters, the mean occupation number of each bin is $\mu_i = \mathcal{N}_i V_i \ll 1$. Here V_i is the volume of space to which clusters in bin i can be seen. Then the likelihood of observing this combination of clusters is simply

$$\ln \mathcal{L} = \sum_i \left[(\eta_i - 1)\mu_i + \eta_i \ln(1 - \exp(-\mu_i))\right] , \tag{1}$$

where by assumption only one of the two terms is non-zero for each bin i.

3. Results and conclusions

Our main result is the power spectrum normalization as a function of cosmological model. We quote the normalization for the variance on a scale $R_{\rm Cl}$ that corresponds to a cluster of 6.5 keV forming now. This is advantageous because the value of $\sigma(R_{\rm Cl})$ is almost independent of the power-spectrum shape Γ, while at fixed $\sigma(R_{\rm Cl})$ the value of σ_8 can vary by as much as 15% as Γ spans the currently allowed range, 0.19–0.37. Fitting formulae accurate at the 1% level for $R_{\rm Cl}$ and $\sigma(R_{\rm Cl})$ in the range $0.2 \leq \Omega_{\rm M} \leq 0.8$ and $0.3 \leq \Omega_\Lambda \leq 1$ are given by:

$$R_{\rm Cl} = p_0\, \Omega_{\rm M}^{-(p_1 + p_2\,\Omega_{\rm M} + p_3\,\Omega_\Lambda)}\, h^{-1}{\rm Mpc}\,; \tag{2}$$

$$\sigma_{\rm Cl} \equiv \sigma(R_{\rm Cl}) = p_0\, \Omega_{\rm M}^{-(p_1 + p_2\,\Omega_{\rm M} + p_3\,\Omega_\Lambda)}\,. \tag{3}$$

Here $\vec{p} = (9.086, 0.354, 0.058, 0.049)$ gives the coefficients in Eq. (2), while those for Eq. (3) can be found in Table 1, along with fits to the $\pm 1\sigma$ limits on $\sigma(R_{\rm Cl})$.

For the concordance model ($\Omega_{\rm M} = 0.3$) we find $R_{\rm Cl} \simeq 14.8\, h^{-1}$Mpc. A less precise, but simpler fit for flat models is given by

$$R_{\rm Cl} \simeq 9\, \Omega_{\rm M}^{-0.41} h^{-1}{\rm Mpc}. \tag{4}$$

This fits to better than 1% over $0.2 \leq \Omega_{\rm M} \leq 1$.

	p_0	p_1	p_2	p_3
mean	0.505	0.150	−0.233	0.048
-1σ	0.473	0.124	−0.234	0.053
$+1\sigma$	0.532	0.171	−0.222	0.044

Table 1. Coefficients of the fitting formula, Eq. (3), for the mean and errors of $\sigma(R_{\rm Cl}) = p_0 \Omega_{\rm M}^{-(p_1 + p_2 \Omega_{\rm M} + p_3 \Omega_\Lambda)}$.

For explicit constraints on σ_8 we verified that the following equation is a good fit with a maximum error of 2%, in the range $0.19 \leq \Gamma \leq 0.37$, $0.2 \leq \Omega_{\rm M} \leq 0.8$ and $0.3 \leq \Omega_\Lambda \leq 1$:

$$\sigma_8 = \sigma_{\rm Cl}\, q_0\, \Omega_\Lambda^{q_1}\, \Omega_{\rm M}^{-(q_2 + q_3 \Omega_{\rm M})}\, \Gamma^{(q_4 + q_5 \Omega_{\rm M})}. \tag{5}$$

Here the vector of coefficients is given by $\vec{q} = (1.049, 0.029, 0.546, 0.350, 0.278, -0.312)$. For flat models (i.e. $\Omega_\Lambda = 1 - \Omega_{\rm M}$) this corresponds approximately to

$$\sigma_8 \simeq (0.495^{+0.034}_{-0.037})\, \Omega_{\rm M}^{-0.60} \tag{6}$$

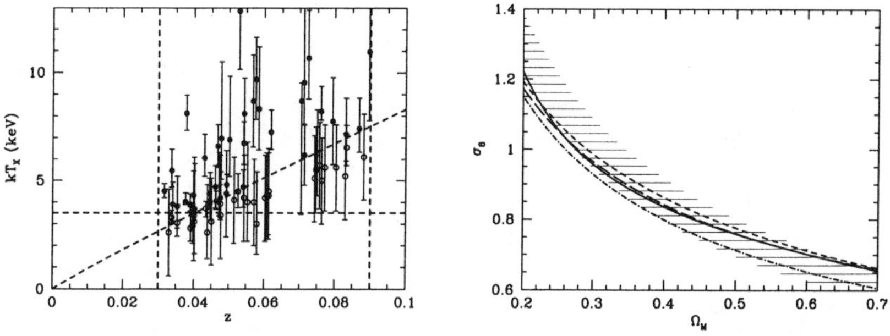

Figure 1. **a** (Left) The sample of X-ray clusters used in our analysis, with dashed lines being the boundaries, and open symbols marking those which can scatter inside in the Monte Carlos. **b** (Right) The normalization σ_8 for flat cosmologies with $\Gamma = 0.23$ as derived here with 1–σ limits (Shaded area) compared with some previous results.

for $\Gamma = 0.23$ explicitly, and for general Γ:

$$\sigma_{\rm Cl} = 0.462^{+0.027}_{-0.029}\Omega_{\rm M}^{-0.194\{^{-0.179}_{-0.218}\}}, \tag{7}$$

where the exponents in brackets refer to the best power-laws for the $+1\sigma$ and -1σ limits. This result is compared with those of Viana & Liddle (1998), Wang & Steinhardt (1998), Eke et al. (1996), Borgani et al. (1999) in Fig. 1b.

We find that our uncertainty in σ_8 is dominated by systematics in the $M{-}T$ relation. The statistical scatter in the $M{-}T$ relation has an almost negligible effect on the central value and the error. The temperature errors themselves give a skewed distribution to σ_8, but this effect is sub-dominant. We performed several checks on both our best-fits and their uncertainties. The reader is addressed to Pierpaoli et al. (2001) for further details on these issues.

4. References

References

Borgani, S., Rosati, P., Tozzi, P., Norman, C. 1999, ApJ, 517, 40

Eke, V., Cole, S., Frenk, C.S. 1996, MNRAS, 282, 263

Markevitch M. 1998, ApJ, 504, 27

Pierpaoli, E., Scott, D., White, M. 2001 MNRAS, 325, 77

Sheth, R.K., Tormen, G., 1999, MNRAS, 308, 119

Viana, P.T.P., Liddle, A. 1999, MNRAS, 303, 535

Wang, L., Steinhardt, P.J. 1998, ApJ, 508, 483

White D.A. 2000, MNRAS, 312, 663

Tracing Cosmic Evolution with Galaxy Clusters
ASP Conference Series, Vol. 268, 2002
S. Borgani, M. Mezzetti, and R. Valdarnini, eds.

Using supercluster geometry as a cosmological probe

S. Basilakos

Astrophysics Group, Imperial College London, Blackett Laboratory, Prince Consort Road, London SW7 2BW, UK

V. Kolokotronis

Institute of Astronomy & Astrophysics, National Observatory of Athens, I.Metaxa & B.Pavlou, Palaia Penteli, 15236 Athens, Greece

M. Plionis

Institute of Astronomy & Astrophysics, National Observatory of Athens, I.Metaxa & B.Pavlou, Palaia Penteli, 15326 Athens, Greece

Abstract. We study the properties of superclusters detected in the Abell/ACO cluster catalogue. We identify the superclusters utilizing the 'friend-of-friend' procedure, and then determine supercluster shapes by using the differential geometry approach of Sahni et al. (1998). We find that the dominant supercluster morphological feature is filamentariness. We compare our Abell/ACO supercluster results with the corresponding ones generated from two different CDM cosmological models in order to investigate statistically which of the latter models best reproduces the observational results.

1. Introduction

The classical pattern of the distribution of matter on large scales supports the idea that the galaxy clusters are not randomly distributed but tend to aggregate in larger systems, the so called superclusters (Bahcall 1988 and references therein). Individual galaxy superclusters and their properties (shape, size etc) have been investigated by different authors (West 1989; Jaaniste et al. 1997 and references therein). It has been found that the vast majority of the superclusters are flattened triaxial objects, while Plionis, Valdarnini, & Jing (1992) found a preference for prolate (filament-like) superclusters.

In order to study in an objective manner the distribution of superclusters and their physical properties, it is necessary to develop objective algorithms and to apply them onto well controlled data. Indeed different methods like minimal spanning trees, shape statistics (Sahni & Coles 1995 references therein), genus-percolation statistics (Gout, Melt & Dickinson 1986) and Minkowski functionals (Mecke et al. 1994) have been used in order to describe the global geometrical and topological properties of the matter distribution utilizing angular and redshift surveys of galaxies as well as N-body simulations.

To this end, in this work, we utilize the Abell/ACO cluster catalogue in order: (i) to investigate whether we can reliably identify superclusters and measure their shapes in flux-limited galaxy samples, (ii) to measure the shape and size distribution of the Abell/ACO superclusters and (iii) to investigate whether these distributions can be used as a cosmological probe.

2. Data and Supercluster Detection

In our analysis we use the volume-limited sample of the Abell/ACO cluster catalogue, with $| b | \geq 30°$ and limited within $315h^{-1}$Mpc or $z \sim 0.11$ (see Einasto et al. 1997). As a result our sample contains ~ 926 clusters. In order to find the Abell/ACO superclusters, we use a very common procedure based on a 'friend-of-friend' algorithm. In particular, all mutually linked pairs (within a critical radius) are joined together to form groups (for those having common boundaries) and groups with more than two clusters are identified as candidate superclusters. To this end, the critical radius can be defined directly from the following equation (Peebles 2001):

$$R_{\rm cr} \simeq \left(\frac{3-\gamma}{\Omega_s \langle n \rangle r_\circ^\gamma} \right)^{\frac{1}{3-\gamma}} \tag{1}$$

where $\Omega_s \simeq 6.28$ steradians (solid angle), $\langle n \rangle \simeq 1.42 \pm 0.34 \times 10^{-5} h^3 \text{Mpc}^{-3}$ (the Abell/ACO mean number density) and $\gamma = 1.8$ with $r_\circ \simeq 20 \pm 2h^{-1}$Mpc (the correlation properties). Taking the latter parameters into account, we compute $R_{\rm cr} \simeq 27 \pm 4h^{-1}$Mpc (in agreement with Einasto et al. 1997). Finally, using higher or much lower values of $R_{\rm cr}$ we tend to connect superclusters and percolate the whole volume.

2.1. Shape Statistics

Shapes are estimated for those "superclustrers" that consist of 8 or more clusters, utilizing the moments of inertia (I_{ij}) method to fit the best triaxial ellipsoid to the data (Carter & Metcalfe 1980). We diagonalize the inertia tensor: $\det(I_{ij} - \lambda^2 M_3) = 0$ (where M_3 is 3×3 unit matrix), obtaining the eigenvalues α_1, α_2, α_3 (where α_1 is the semi-major axis) from which we define the shape of the configuration since, the eigenvalues are directly related to the three principal axes of the fitted ellipsoid. The volume of each supercluster is then $V = \frac{4\pi}{3}\alpha_1\alpha_2\alpha_3$.

The shape statistic procedure, that we use, is based on a differential geometry approach, introduced by Sahni et al. (1998) [for application to the PSCz data see Basilakos, Plionis & Rowan-Robinson 2001]. Here we review only some basic concepts. A set of three shapefinders are defined having dimensions of length; $\mathcal{H}_1 = VS^{-1}$, $\mathcal{H}_2 = SC^{-1}$ and $\mathcal{H}_3 = C$, with S the surface area and C the integrated mean curvature. Then, it is possible to define a set of two dimensionless shapefinders K_1 and K_2, as: $K_1 = \frac{\mathcal{H}_2 - \mathcal{H}_1}{\mathcal{H}_2 + \mathcal{H}_1}$ and $K_2 = \frac{\mathcal{H}_3 - \mathcal{H}_2}{\mathcal{H}_3 + \mathcal{H}_2}$, normalized to give $\mathcal{H}_i = R$ ($K_{1,2} = 0$) for a sphere of radius R. Therefore, based on these shapefinders we can characterize the morphology of cosmic structures (underdense or overdense regions) according to the following categories: (i) pancakes for $K_1/K_2 > 1$; (ii) filaments for $K_1/K_2 < 1$; (iii) triaxial for $K_1/K_2 \simeq 1$ and

(iv) spheres for $\alpha_1 \simeq \alpha_2 \simeq \alpha_3$ and thus $(K_1, K_2) \simeq (0,0)$. For the quasi-spherical objects the ratio K_1/K_2 measures the deviation from pure sphericity.

3. The Geometrical Properties

In Figure 1 (left panel) we present the "shape spectrum" (broken line) and the multiplicity function (open symbols) of the identified superclusters. From the shape spectrum plot, it is obvious that the dominant feature of the Abell/ACO superclusters is filamentariness; ie., $K_1/K_2 < 1$ (in agreement with previous studies). Regarding extreme shaped superclusters, we have found, 4 very filamentary superclusters with $K_1/K_2 < 0.45$ (among which the Near Shapley and Hercules superclusters), 1 triaxial and 1 extreme pancake-like structures with $K_1/K_2 > 3$ (one of which is the Perseus-Pegasus supercluster).

To complement this, we utilize the completed IRAS flux-limited 60-μm redshift survey (PSCz) which is described in Saunders et al. (2000). We identify, in the smooth galaxy density field of the PSCz catalogue, high density regions (superclusters) and estimate their shapes (for details see Basilakos et al. 2001). In Figure 1 (right panel) we present a direct comparison of the PSCz and Abell/ACO geometrical properties, out to 240 h^{-1} Mpc. The two shape-profiles are in quite good agreement. This is a further indication that the two density fields are consistent with each other out to this distance.

3.1. Comparison with Cosmological Models

We use mock Abell/ACO catalogues (having similar to the observed Abell/ACO characteristics) generated from two large cosmological N-body simulations of Colberg et al. (2000), in order to investigate whether supercluster properties can discriminate between models. In particular, we consider two different cold dark matter models covering Hubble volumes, which are: (1) a flat low-density CDM model with $\Omega_\Lambda = 0.7$ and shape parameter $\Gamma = 0.17$, $h = 0.7$ and $L_{\rm box} = 3000h^{-1}$Mpc and (2) a critical density universe with $\Gamma = 0.21$ (τCDM), $h = 0.5$ and $L_{\rm box} = 2000h^{-1}$Mpc. The CDM models are normalized by the observed cluster abundance at zero redshift; (Eke, Cole, & Frenk 1996). We average results over 64 ΛCDM and 27 τCDM independent mock Abell/ACO catalogues extending out to a radius of 315 h^{-1}Mpc. We analyse the mock Abell/ACO supercluster properties in the same fashion as that of the observed catalogue and we compare the outcome of this procedure in Figure 1 (left panel), where we plot the detected supercluster shape-spectrum and multiplicity function for two models and Abell/ACO data.

In order to quantify the differences between models and data we perform a standard χ^2 test, assuming Gaussianly distributed errors. This statistical test proves that the model which is excluded by the data, by the multiplicity function comparison, at a relatively high significance level is the τCDM model ($\mathcal{P}_{>\chi^2} = 3 \times 10^{-3}$), while the ΛCDM model reproduces the observed supercluster shape-spectrum and multiplicity function ($\mathcal{P}_{>\chi^2} = 0.80$). To validate our analysis we test the robustness of our method by comparing the models among themselves. We find that the shape-spectrum is insensitive to the different cosmologies, probably because supercluster shapes reflect the Gaussian nature of

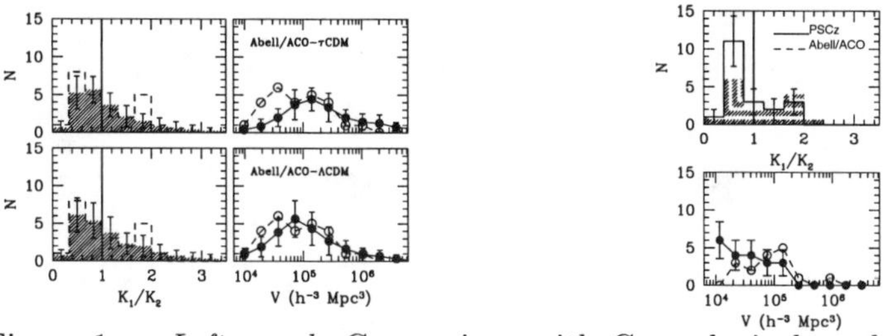

Figure 1. *Left panel*: Comparison with Cosmological models: The Abell/ACO results are represented by open symbols and broken lines. *Right panel*: Comparison between the Abell/ACO and the PSCz geometrical properties.

the initial conditions which are common to all models. However, the supercluster multiplicity function is a strong discriminant between the models.

4. Conclusions

We have studied the properties of superclusters detected in the Abell/ACO cluster catalogue. To determine supercluster shapes we use a differential geometry approach and find that the dominant supercluster morphological feature is filamentariness. Finally, we have compared our supercluster results with the corresponding ones generated from the analysis of two cosmological models (τCDM and ΛCDM) and we find that the model that best reproduce the observational results is the ΛCDM model ($\Omega_\Lambda = 0.7$).

References

Bahcall, N. A., 1988, ARA&A, 26, 631

Basilakos, S., Plionis, M., Rowan-Robinson, M., 2001, MNRAS, 223, 47

Carter, D. & Metcalfe, J., 1980, MNRAS, 191, 325

Colberg, J. M., et al., 2000, MNRAS, 319, 209

Einasto, M., Tago, E., Jaaniste, J., Einasto, J., Andernach, H. 1997, A&AS, 123, 119

Eke, V., Cole, S., Frenk, C. S., 1996, MNRAS, 282, 263

Gott, J. R., Dickinson, M., Melott, A. L., 1986, ApJ, 306, 341

Jaaniste, J., Einasto, M., Einasto, J., Andernach, H., Muller, V., 1997, A&A, 329, 1

Mecke, K. R., Buchert, T., Wagner, H., 1994, A&A, 288, 697

Peebles, P. J. E., 2001, astro-ph/0101127

Plionis, M., Valdarnini, R., Jing, Y. P., 1992, ApJ, 398, 12

Sahni, V. & Coles, 1995, Phys. REp., 262, 1

Sahni, V., Sathyaprokash, B. S., Shandarin, S., 1998a, ApJ, 495, L5

Saunders, W., et al., 2000, MNRAS, 317, 55

West, J. M., 1989, ApJ, 347, 610

CLUSTERS WITH SZ

Tracing Cosmic Evolution with Galaxy Clusters
ASP Conference Series, Vol. 268, 2002
S. Borgani, M. Mezzetti, and R. Valdarnini, eds.

Searching and studying clusters with the SZ effect

James G. Bartlett

Observatoire Midi–Pyrénées, Toulouse, France
New institution: *APC – Université Paris 7, Paris, France*

Abstract. I discuss galaxy cluster surveys based on the Sunyaev–Zel'dovich effect and their relevance for cosmological studies. The unique aspects of cluster selection by this method are emphasized and certain issues of surveying are addressed. Finally, I briefly present prospects for upcoming surveys.

1. Introduction

Galaxy cluster catalogs are useful for cosmological studies for a number of reasons: cluster abundance, distribution and evolution reflect the structure formation process, which depends sensitively on certain cosmological parameters, such as the matter density $\Omega_{\rm M}$ (Oukbir & Blanchard 1992, 1997; Bahcall 1998; Bartlett 1997; Borgani et al. 1999 & these proceedings; Eke et al. 1998; Henry 1997; Viana & Liddle 1999); and cluster internal properties provide clues to the nature of dark matter and on certain mechanisms associated with galaxy formation, such as heating and feedback.

The construction of well–defined catalogs appropriate for statistical studies demands good understanding of the cluster selection criteria (see Postman, these proceedings). These criteria in turn depend on the type of observations – optical, X–ray, radio – as well as on the data analysis proceedures. Each type of observation has its own advantages and disadvantages that we may refer to as its intrinsic biases. For example, X–ray cluster selection eliminates the problem of projection effects suffered by optical observations, but on the other hand is significantly influenced by the rather poorly understood physics taking place in the core region.

The question that arises is, what kind of selection is best? The answer clearly depends on the utlimate use of the catalog, and for cosmological studies we may argue that the most ideal selection would be based on cluster mass. This is because cluster mass and redshift are the two primary theoretical cluster descriptors. From this line of argument, understanding cluster selection criteria is reduced to relating the observational selection criteria to cluster mass and redshift. This is a useful viewpoint, for it provides a framework for quantifying the biases associated with each type of observation for the purposes of cosmological studies. We do not argue that one particular observational method is the best; rather, that the important point is to use several methods with *different* biases.

The Sunyaev–Zel'dovich (SZ) effect (Sunyaev & Zel'dovich 1972) will soon open the way to a new method of surveying for clusters. Several ground–based instruments optimized for SZ surveying are currently under construction, and the Planck satellite will supply an almost full–sky catalog of several 10^4 clusters at the turn of the decade. In anticipation and in the spirit just outlined, we shall now examine in more detail this new kind of survey (the 'SZ band'), and in particular the nature of its cluster section.

This potential of surveying for clusters with the SZ effect has been appreciated for some time. To my knowledge, the first to give source count predictions were Korolyov, Sunyaev, & Yakubstev (1986), and many authors have since considered the issue. An indicative, useful, though by no means exhaustive, list of references focused on SZ surveying (i.e, counts and redshift distributions of SZ detected clusters) would include: Bond & Meyers (1991), Bartlett & Silk (1994), Markevitch et al. (1994), Barbosa et al. (1996), Eke, Cole, & Frenk (1996), Aghanim et al. (1997), Colafrancesco et al. (1997), da Silva et al. (2000), Holder et al. (2000), Lo et al. (2000), Benson, Reichardt, & Kamionkowski (2001), Fan & Chuieh (2001), Kay, Liddle, & Thomas (2001), Kneissl et al. (2001), Springel, White, & Hernquist (2001), Xue & Wu (2001)[1].

2. SZ Cluster Selection

In the non–relativistic limit (which applies in practice to all but the hottest clusters), the SZ effect may be written as

$$\begin{aligned}\delta i_\nu(\hat{n}) &= y(\hat{n}) j_\nu \\ y &= \int_{los} \frac{kT}{mc^2} n\sigma_T dl\end{aligned}$$

where δi_ν is the brightness change induced in the cosmic microwave background (CMB) by compton scattering off electrons (mass m) in the hot intracluster medium (ICM) at temperature T and electron density n. The Compton y–parameter measures the amplitude of the energy transfer to the photons (σ_T is the Thompson cross section); and the function j_ν describes the spectral form of the distortion. It is convenient to consider the effect integrated over the entire cluster face:

$$Y = \int d\Omega\, y(\hat{n}) \propto f_{\rm gas} \frac{M\langle T\rangle}{D_{\rm ang}^2} \tag{1}$$

where now M is the cluster (virial) mass, $f_{\rm gas}$ is the ICM gas mass fraction, $\langle T\rangle$ is the (true) mean electron temperature (i.e., particle weighted), and $D_{\rm ang}(z)$ is the angular–size distance. Notice that Y is measured in, for instance, square arcmins.

It is from these equations that we begin to understand the nature of SZ cluster selection (or detection). Let us emphasize four aspects:

1. The spectral signature j_ν is unique – negative at low frequency, positive at high frequency, with a fixed zero point at $\lambda \sim 1.4$ mm – and the same

---

[1] The reader will appreciate the precarious position of the author in giving such a list. In offering my apologies to those that I have unjustly, but without malice, forgotten, I ask no more than to be corrected by a kind email in reminder.

for all clusters (j_ν depends only on the observation frequency). There is therefore no k–correction;

2. The integrated effect $\propto D_{\rm ang}^{-2}$, which is the statement that the surface brightness is independent of redshift z. This is in contrast to other emission mechanisms, that vary as the luminosity distance $D_{\rm lum} = (1+z)^2 D_{\rm ang}$, and which suffer from 'cosmic dimming' (the redshift);
3. The integrated signal is directly proportional to the total ICM thermal energy $f_{\rm gas} M\langle T\rangle$ (this is essentially the definition of the mean temperature $\langle T\rangle$). We may therefore expect a tight correlation between the measurable, Y and the cluster mass, e.g., $Y \propto f_{\rm gas} M^{5/3}$, especially in the more massive systems where non–graviational effects appear less important.
4. Cluster detection and subsequent modeling may be based on the same measurable quantity Y.

A particularily useful quantity for characterizing the selection properties of a survey is the minimum detectable mass as a function of redshift, $M_{\rm det}(z)$. For example, it determines the source counts as

$$\frac{dN}{d\Omega} = \int_0^\infty dz \, \frac{dV}{dz d\Omega} \int_{M_{\rm det}(z)}^\infty dM \, \frac{dn}{dM}(M,z) \tag{2}$$

where dn/dM is the mass function. A comparison of $M_{\rm det}$ from all–sky X–ray and SZ (the Planck curve) surveys is made in Figure 1. The X–ray survey flux limit corresponds to that of the RASS. An interferometer–like similar to those proposed (see below) and capable of surveying several square degrees from the ground is also shown. To make the figure, a non–evolving $L_x - T$ relation of the form given by Arnaud & Evrard (1999) and a $T - M$ relation from Evrard, Metzler, & Navarro (1996) have been adopted. The cosmological model is an open, low–density model, and Eq. (1) was used with a constant gas mass fraction $f_{\rm gas} = 0.06h^{-1.5}$ (Evrard 1997)

The figure is illustrative, especially because we have no idea of the evolution of the $L_x - T$ relation to such redshifts. It is nevertheless instructive to note the different behavior of the X–ray and SZ curves, and in particular the almost flat asymptote of the SZ curves; in fact, they demonstrate a slow *decrease* towards high redshift (this is independent, of course, of the $L_x - T$ relation). In other words, SZ selection corresponds essentially to a mass selection. The conclusion depends in detail on the constancy of $f_{\rm gas}$ with redshift, but the potential consequences are interesting: for example, this implies that there is no $V_{\rm max}$ for a SZ detected cluster. Even more significant is that SZ selection therefore finds the *same kinds of objects* (mass–wise) over a large range of redshifts. For evolutionary studies of clusters, this is an important advantage over other selection techniques that force one to compare ever more massive objects detected at high redshift to less massive local samples.

3. Aspects of Surveying

The discussion so far as assumed that clusters are unresolved – that the integrated SZ flux of Eq. (1) is directly measured. This situation applies in practice

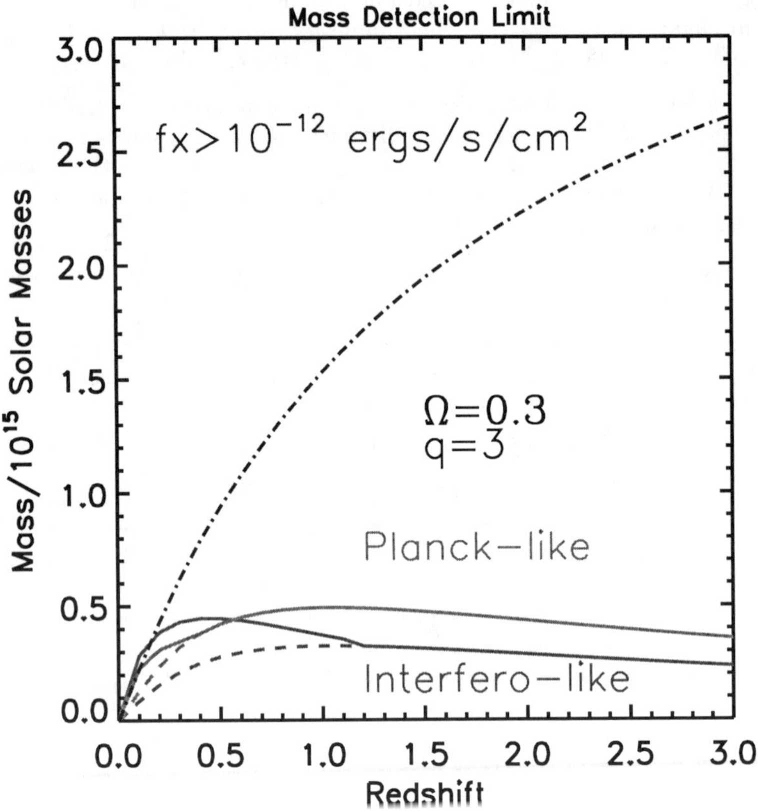

Figure 1. Minimal detectable mass as a function of redshift for a Planck–like all–sky survey, an interferometer–like (limited sky, ground–based) survey and for a flux–limited X–ray survey similar to the RASS. For the SZ curves, the dashed lines account for cluster extension, while the solid lines assume that clusters are point sources. The Planck–like survey is based on the Planck 2.1 mm channel performance *only*: $Y = 10^{-4}$ arcmin^2/beam and an angular resolution of $\theta_{\rm fwhm} = 5$ arcmins (this is most surely an optimistic estimate as the Planck SZ maps will be weighted by channels with differing resolution and sensitivity, and foreground removal will further degrade the sensitivity). The interferometer–like curve assumes $\theta_{\rm fwhm} \sim 2$ arcmin and a sensitivity of $Y = 5 \times 10^{-5}$ arcmin^2.

to the Planck survey, with angular resolution of at best 5 armins. Ground–based instruments, on the other hand, will operate at higher resolution and clusters will appear extended, complicating the selection criteria: some clusters may be *resolved out*, relative to a hypothetical point–source approximation. Detection depends now not only on Y, but also on the detailed SZ cluster profile; and photometry becomes an issue, as with any extended source.

Many ground–based instruments will operate, at least at first, at a single frequency. In this case, primary CMB fluctuations (in the background) become an important source of noise, and one must extract extended sources over a fluctuating background. Another potentially important effect is confusion by point sources, either randomly distributed on the sky, or preferentially in cluster cores. Without spectral information, their removal requires higher resolution imaging; for this reason, some ground–based interferometer projects plan to control this confusion with additional telescopes placed on long baselines.

There are certainly other issues, for example other foregrounds, that we could list as problematic. However, I would instead like to address a specific question that often arises when preparing a survey: In a given amount of observing time, should one go (integrate) 'deep', or instead 'wide', covering more sky to less depth? Actually, the real approach is to try to do several surveys, i.e., in some sense to do both. All the same, it is useful to try to answer the question in terms of maximizing the number of detected objects. This depends on the slope of the source counts *as a function of survey sensititivity*[2].

3.1. Deep or Wide?

We calculate the counts using Eq. (2). For unresolved sources, the detection mass is only a function of sensitivity, $Y_{\rm noise}$, and redshift – $M_{\rm det}(Y_{\rm noise}, z)$. Additional factors are important in the case of resolved source detection, such as angular resolution ($\Omega_{\rm pix}$, the solid angle of a 'pixel') and source profile: $M_{\rm det}(Y_{\rm noise}, z, \Omega_{\rm pix}, profile)$. A simple analytic model of extended source detection in a map may be employed to examine cluster selection (Bartlett 2000). It supposes that one can identify and then integrate over all the pixels covering a cluster (out to the virial radius) to access the total cluster signal Y for detection; it is therefore the best one can hope to do.

Using this approach, the effects of resolution on $M_{\rm det}$ are presented in Figure 1, where the dashed lines (for the SZ curves) give $M_{\rm det}(z)$ once cluster extension has been taken into account. Clearly, clusters at low/intermediate redshifts are *resolved out* relative to the point–source calculation, because they are covered by several noisy pixels; this is most evident for the higher resolution interferometer–like curve. The difference for a low resolution survey, such as the Planck survey, is relatively minor, as can been seen.

This point is important when it comes to predicting the number of objects to be expected. Clearly the predicted number will be lower when correctly accounting for resolution effects. Even more relevant to the question at hand (deep or wide), is the fact that, as seen in Figure 2 (and in Figure 4 below), the actual counts as a function of sensitivity are steeper than the point–source

[2]For resolved sources, these counts are not necessarily the same as the counts as a function of integrated cluster signal Y.

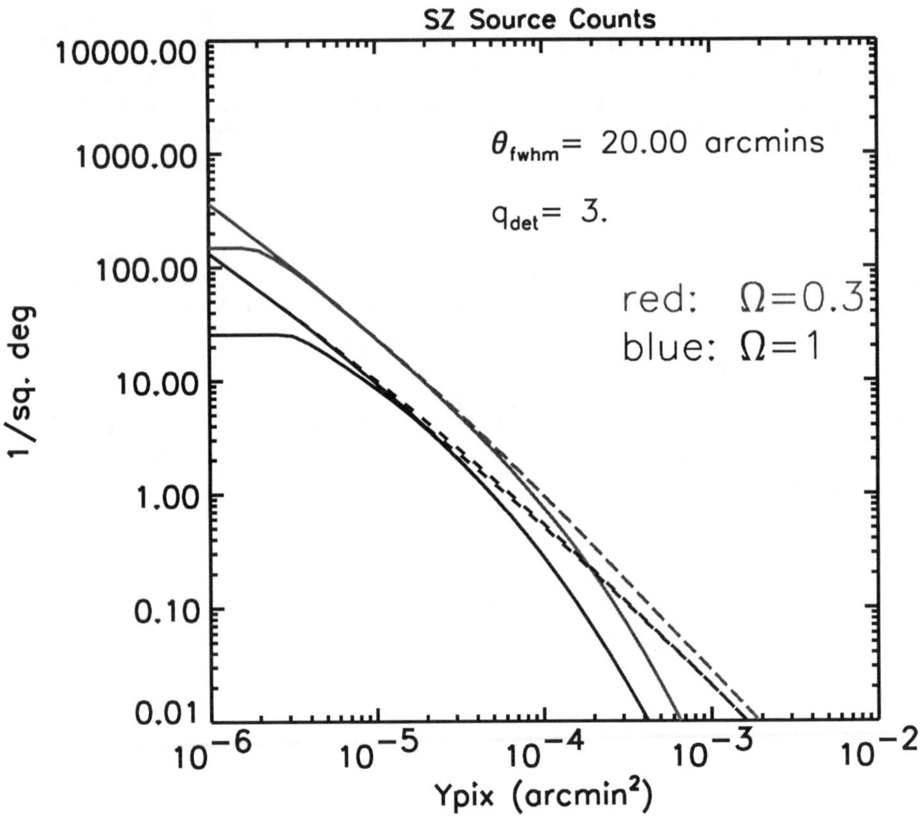

Figure 2. Integrated SZ counts as a function of *pixel noise* expressed in terms of Y (arcmin^2). Two cosmological models are used to illustrate the effects of source extension. The upper (red) curves are for an open model and the lower (blue) curves are for a critical model. The solid curves take into account finite source extension with a simple analytical model (Bartlett 2000), while the dashed lines give the results of the point–source approximation. In all cases, two different low–mass cutoffs – 10^{13} & 10^{14} solar masses – are applied in order to highlight the contribution of low–mass objects. The effect of the latter cutoff is clearly seen as the point where the counts flatten off at low Y.

calculation would suggest. In fact, at the bright end they are steeper than $Y_{\rm noise}^{-2}$, the critical slope for answering our question. Counts steeper than -2 imply that more objects will be found by integrating deep on a small patch of sky (as long as $Y_{\rm noise} \sim t^{-1/2}$, which usually applies). At the faint end, the counts flatten out and rejoin the point–source calculation, which is always flatter than $Y_{\rm noise}^{-2}$. Two important conclusions follow:

1. the point–source calculation gives the wrong answer in suggesting that one should go wide;

2. the correct answer is found by accounting for resolution effects and indicates that, for each experimental set–up, there is an optimal sensitivity corresponding to the point where the counts reach a slope of -2.

4. Expectations

A number of dedicated instruments optimized for SZ surveying have recently been proposed. At the time of writing, many of these projects are funded and under construction, including:

- Interferometer arrays
  - Arcminute MicroKelvin Imager (AMI) (`http://www.mrao.cam.ac.uk/telescopes/ami/index.html`)
  - Array for Microwave Background Anisotropy (AMiBA) (`http://www.asiaa.sinica.edu.tw/amiba/`)
  - SZ Effect Imaging Array (`http://www.astro.uiuc.edu/ jmohr/SZE/index.html`)
- Bolometer cameras
  - ACBAR (`http://cfpa.berkeley.edu/ swlh/research/acbar.html`)
  - BOLOCAM (`http://www.astro.caltech.edu/ lgg/bolocam/bolocam.html`)
- Planck satellite (`http://astro.estec.esa.nl/Planck/`)

Several authors have recently made predictions for these various instruments (Aghanim et al. 1997; Holder et al. 2000; Lo et al. 2000; Benson et al. 2001; Kay et al. 2001; Kneissl et al. 2001). In Figures 3 & 4 I show some expectations for the Planck Survey and for the AMI interferometer, both for two extreme cosmologies: a critical model and a purely open model with $\Omega_{\rm M} = 0.3$. A flat low–density model with non–zero cosmological constant will fall between these two cases, although somewhat closer to the purely open model.

In the following, it should be kept in mind that the predicted total number of detected clusters is subject to large modeling uncertainties and can vary by as much as a factor of ~ 5. A particularly important source of uncertainty comes from the uncertainty in the amplitude of the density perturbations σ_8. To be conservative, the predictions are given for 'low σ_8': ($\sigma_8 = 0.8, \Omega_M = 0.3$) and ($\sigma_8 = 0.5, \Omega_M = 1$) (e.g., Eke et al. 1998). Use of higher values quoted in the literature (e.g., Blanchard et al. 2000; Pierpaoli, Scott, & White 2001) would substantially increase the predicted total number of clusters. Another source of uncertainty lies in the potential evolution of the cluster gas mass fraction f_{gas}. The lack of evidence for strong evolution of f_{gas} with redshift or for an important dependence on cluster mass may simply reflect a lack of sufficient data. In this light, it is important to note that the SZ signal for unresolved clusters (e.g., Planck detections) depends *only* on the global quantity f_{gas}, and not on the details of the gas distribution; X-ray observations, as a contrast, are heavily influenced by the physics of the core, which is very poorly understood and difficult to model.

The Planck predictions are given in terms of redshift distributions: shown as a function of redshift z are the total number of clusters with $Y > 10^{-4}$ armin^2 at redshifts *larger* than z[3]. The numbers at $z = 0$ therefore give the total number of detectable clusters, and we see that on the order of 1 cluster/sq.deg. is expected for the survey, depending on the cosmology. At $z \sim 1$ the two models differ wildly. There are essentially no clusters expected in the critical model. On the other hand, about 10% of clusters are expected to lie at $z > 1$ in the open scenario. These conclusions assume that f_{gas} remains relatively constant with redshift and mass. The presence of an important population of clusters at high redshift would be a strong indication of a low-density universe; but the lack of such a population could be accommodated in the open model by decreasing f_{gas} with z. The result in any case would be an important piece of cosmological information.

The predictions for an AMI-like instrument (resolution of a few arcminutes or better) are given in terms of integrated source counts as a function *beam sensitivity*. Both cosmological models are once again shown, each one for two different low-mass cutoffs (10^{13} and 10^{14} solar masses) in order to separate the contribution of low and high mass objects. The dashed curves show the equivalent counts in a point source approximation, and we observe that the latter fails rather badly because most clusters are resolved at this resolution. The dotted lines in the background follow the critical slope of Y^{-2}. This kind of instrument reaches its optimal sensitivity at $\sim$ *few* $\times 10^{-5}$ arcmin^2/beam, where the predicted counts attain a slope of -2. At this sensitivity, one may expect, as a conservative round number and within the modeling uncertainties mentioned, on the order of several clusters/sq.deg.

5. Conclusion

It will shortly be possible to perform purely SZ-based surveys, opening up what may be refered to as the 'SZ band' for cluster studies. A number of ground-based instruments will realize such surveys within a few years over patches of

[3]i.e., the integrated counts from $z = \infty$ down to z, as a function of z and at the fixed sensitivity of 10^{-4} arcmin^2.

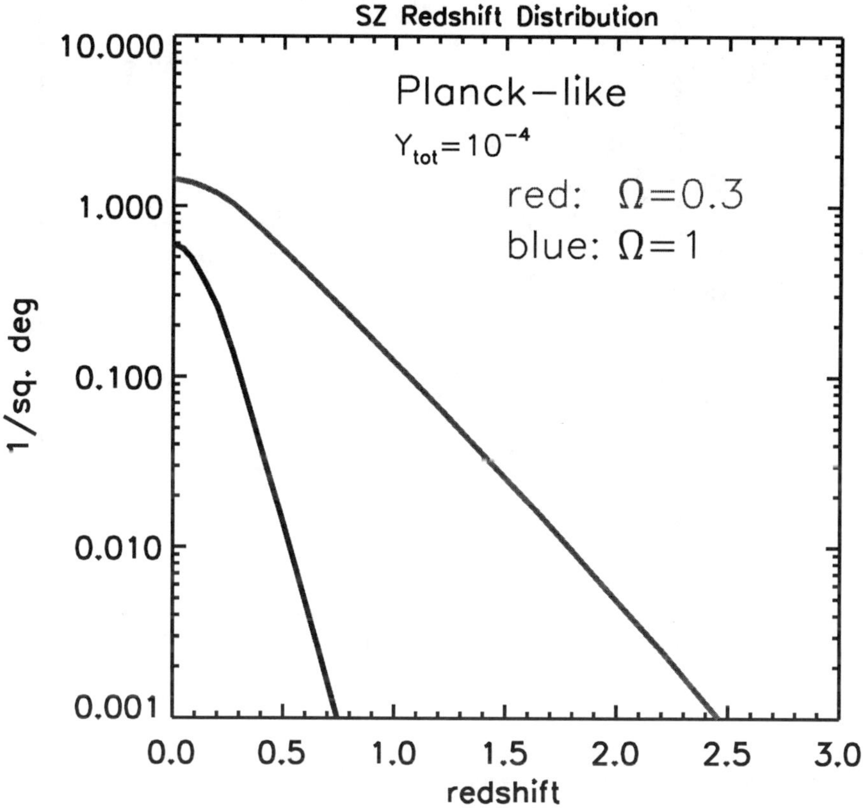

Figure 3. Planck–like cumulative redshift distributions for two extreme cosmologies. The y–axis gives the counts at $Y > 10^{-4}$ arcmin^2 for objects at redshifts $> z$, plotted as a function of z (x–axis). The upper (red) curve corresponds to the open model, while the lower (blue) curve is for the critical model. In addition to the various modeling uncertainties mentioned in the text, the predicted number of clusters depends on the final sensitivity of the survey, which for the fixed observing time of the mission will depend on various factors, including the weighting given to each of the bands, the effects of foreground removal, etc... This is an area of current study.

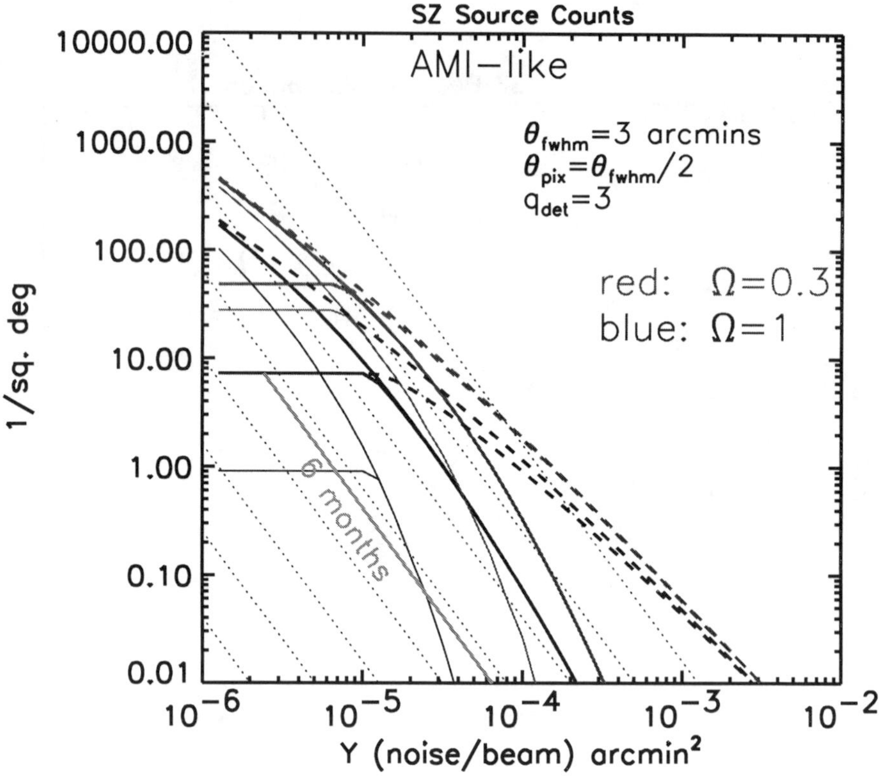

Figure 4. Integrated counts for an AMI–like instrument as a function of survey sensitivity in terms of *noise rms/beam*. The color code is the same as above. The thick solid lines show the total counts, while the thin solid lines show only unresolved sources. Each model is plotted for two different low–mass cutoffs (10^{13} and 10^{14} solar masses); the higher mass cutoff in each case can be seen to flatten off at the low–noise end. The dashed lines show the results of the (invalid) point–source approximation, for both mass cutoffs and for the open model (above) and the critical model (below). Finally, the dotted lines in the background indicate the critical slope of −2, steeper than which deep surveys yield more objects than wide, shallow ones.

sky covering several square degrees; and by the end of the decade, the Planck mission will produce a catalog of several 10^4 SZ clusters over the majority of the sky. Although predictions for the number of clusters expected for these surveys suffer from modeling uncertainties[4], various studies indicate roughly on the order of ~ 1 cluster/sq.deg. for Planck and $\sim$ *few* clusters/sq.deg. for the deeper ground–based efforts under way.

These surveys represent a new method for studying clusters that will find its place alongside X–ray and optical methods. As a population, clusters provide a unique statistical sample for probing cosmology and the large–scale structure formation process. Each observational approach to surveying for clusters has its proper advantages and disadvantages (collectively, its biases). In light of cosmological studies, it is most instructive to quantify observational biases in terms of cluster mass and redshift, the two primary theoretical cluster descriptors. The specific advantages inherent to SZ surveying are listed above under Section 2 and include the fact that the integrated SZ signal is proportional to a physical quantity – the ICM thermal energy.

For a more complete understanding of the cluster population in view of its application as a cosmological probe it is essential to combine a variety of surveying (catalog selection) methods with different biases. SZ surveying will soon add an important new, independent and complementary approach in this endeavor.

References

Aghanim, N., de Luca, A., Bouchet, F.R., Gispert, R., & Puget, J. L. 1997, A&A, 325, 9

Arnaud, M., & Evrard, A.E. 1999, MNRAS, 305, 631

Bahcall, N., & Fan, X. 1998, ApJ, 504, 1 &

Barbosa, D., Bartlett, J.G., Blanchard, A., & Oukbir, J. 1996, A&A, 314, 13

Bartlett, J.G. 1997, in ASP Conf. Ser. Vol. 126, From Quantum Fluctuations to Cosmological Structures, ed. D. Valls–Gabaud et al. (San Francisco: ASP) 365

Bartlett, J.G. 2000, astro–ph/0001267

Bartlett, J.G., & Silk, J. 1994, ApJ, 423, 12

Benson, A.J., Reichardt, C., & Kamionkowski, M. 2001, astro–ph/0110299

Blanchard, A., Sadat, R., Bartlett, J.G., & Le Dour, M. 2000, A&A, 362, 809

Bond, J.R. & Meyers, S.T. 1991, in Trends in Astroparticle Physics, ed. D. Cline (Singapore: World Scientific)

Borgani, S., Rosati, P., Tozzi, P., & Colin, N. 1999, ApJ, 517, 40

Colafrancesco, S., Mazzotta, P., Rephaeli, Y., & Vittorio, N. 1997, ApJ, 433, 454

da Silva, A.C., Barbosa, D., Liddle, A.R., & Thomas, P.A. 2000, MNRAS, 317, 37

[4] Which is not surprising: if there were no uncertainties the surveys would be of little value!

Eke, V.R., Cole, S., & Frenk, C.S. 1996, MNRAS, 282, 263
Eke, V.R., Cole, S., Frenk, C.S., & Henry, P.J. 1998, MNRAS298, 1145
Evrard, A.E. 1997, MNRAS, 292, 289
Evrard, A.E., Metzler, C.A., & Navarro, J.F. 1996, ApJ, 469, 494
Fan, Z. & Chiueh, T. 2001, ApJ, 550, 547
Henry, J.P. 1997, ApJ, 489, L1
Holder, G.P., Mohr, J.J., Carlstrom, J.E., Evrard, A.E., & Leitch, E.M. 2000, ApJ, 544, 629
Kay, S.T., Liddle, A.R., & Thomas, P.A. 2001, MNRAS, 325, 835
Kneissl, R., Jones, M.E., Saunders, R., Eke, V.R., Lasenby, A.N., Grainge, K., & Cotter, G. 2001, astro–ph/0103042
Korolyov, V.A., Sunyaev, R.A., & Yakubtsev, L.A. 1986, Sov. Astron. Lett., 12, L141
Lo, K.Y., Chiueh, T.H., Martin, R.N., Kin–Wang Ng, Liang, H., Ue-li Pen, & Chung-Pei Ma 2000, astro–ph/0012282
Markevitch, M., Blumenthal, G.R., Forman, W., Jones, C., & Sunyaev, R.A. 1994, ApJ, 426, 1
Oukbir, J., & Blanchard, A. 1992, A&A, 262, L21
Oukbir, J., & Blanchard, A. 1997, A&A, 317, 1
Pierpaoli, E., Scott, D., & White M. 2001, MNRAS,325, 77
Springel, V., White, M., & Hernquist, L. 2001, ApJ, 549, 681
Sunyaev, R.A. & Zel'dovich, Ya.B. 1972, Comm. Astrophys. Space Phys., 4, 173
Viana, P.T.P., & Liddle, A.R. 1999, MNRAS, 303, 535
Xue, Y.–J., & Wu, X.–P. 2001, ApJ, 552, 452

Tracing Cosmic Evolution with Galaxy Clusters
ASP Conference Series, Vol. 268, 2002
S. Borgani, M. Mezzetti, and R. Valdarnini, eds.

Masses of High-Redshift Clusters via SZ Effect Observations

Laura Grego

Harvard-Smithsonian Center for Astrophysics, 60 Garden St., Cambridge, MA 02138

Marshall Joy

Dept. of Space Science, SD50, NASA Marshall Space Flight Center, Huntsville, AL 35812

John E. Carlstrom

Dept. of Astronomy & Astrophysics, University of Chicago, Chicago, IL 60637

Samuel LaRoque

Dept. of Astronomy & Astrophysics, University of Chicago, Chicago, IL 60637

Daisuke Nagai

Dept. of Astronomy & Astrophysics, University of Chicago, Chicago, IL 60637

Kyle Dawson

Physics Department, University of California, Berkeley, Berkeley, CA 94720

Harald Ebeling

Institute for Astronomy, 2680 Woodlawn Dr., Honolulu, HI 96822

Erik D. Reese

Dept. of Astronomy & Astrophysics, University of Chicago, Chicago, IL 60637

William L. Holzapfel

Physics Department, University of California, Berkeley, Berkeley, CA 94720

Abstract. We present Sunyaev-Zel'dovich Effect (SZE) observations of distant, highly X-ray luminous clusters of galaxies. We use the SZE data to constrain their total masses, independent of X-ray observations. To do this, we assume the clusters have the same gas mass fraction as that derived from SZE measurements of a sample of known massive clusters, and then infer each cluster's mass from its SZE data. In the systems with published X-ray temperatures, we find good agreement between our SZE-derived temperatures and those inferred from X-ray spectroscopy; in the system without X-ray derived temperatures, the SZE data provide the first confirmation that it is indeed a massive system.

The abundance of clusters at high redshift is critically sensitive to the values of the cosmological parameters and so the demonstrated ability to determine cluster temperatures and masses from SZE observations independent of X-ray data illustrates the power of using deep SZE surveys to probe the distant universe.

1. Introduction

The existence of galaxy clusters at high redshift can place powerful constraints on the physical and cosmological parameters of structure formation models (Bahcall & Cen 1992; Luppino & Gioia 1995; Oukbir & Blanchard 1997; Donahue et al. 1998; Eke et al. 1998; Haiman, Mohr, & Holder 2001). The greatest leverage is provided by the most massive and distant clusters (e.g., Holder, Haiman & Mohr 2001); in fact to constrain Ω_Λ, one must use clusters with redshifts greater than 0.5. Ongoing and planned surveys will generate large numbers of massive cluster candidate objects. As the science proceeds into the realm of surveys, one outstanding technical problem will be how to best determine the masses of the high-redshfit candidate objects.

Most methods for cluster mass measurement become resource-intensive at large distances. For example, a cluster's mass can be estimated from a measurement of the temperature of its intracluster medium (ICM), which relates to its mass through an observationally-determined (c.f. Finoguenov, Reiprich, & Böhringer 2001; Horner, Mushotzky, & Scharf 1999) and theoretically-supported (c.f. Evrard, Metzler, & Navarro 1996) mass-temperature relation. To give a fiducial point, the X-ray spectrum from a 120 kilosecond observation of a massive ($kT_e \sim 7.5$ keV), distant (redshift z =0.783) cluster with the *Chandra* ACIS instrument yields a measurement of the ICM temperature to about 10% accuracy at 90% confidence. A temperature measurement of similar accuracy is expected to require an observation 5-10 times shorter with the *XMM* facility.

Weak gravitational lensing observations also permit a measurement of a cluster's total mass. (Mass densities which generate strong lensing events are confined to the cores of galaxy clusters; we require measurements of the cluster masses within radii as near as possible to the virial radius.) Deep imaging over large angular scales is required for weak lensing mass estimates, and the imaging efficiency decreases with cluster distance, as the surface brightness of the lensed galaxies decreases as well as the number of galaxies available to be lensed.

Here, we present a method for determining the mass of a galaxy cluster via a measurement of the cluster's Sunyaev-Zel'dovich Effect (SZE). This technique was originally presented in Joy et al. (2001). It is particularly valuable for distant clusters, as the surface brightness of the SZE is independent of distance. It is also independent of the X-ray spectral and gravitational lensing mass measurements;

together, these three mass determinations can provide checks on the acccuracy of and systematic uncertainties in each method.

The Sunyaev-Zeldovich Effect provides a measure of the ICM pressure integrated along the line of sight (Sunyaev & Zeldovich 1972; Birkinshaw 1999). The SZE is manifested as a change in the observed brightness temperature of the Cosmic Microwave Background (CMB) radiation that results from passage through the cluster's thermally ionized gas:

$$\frac{\Delta T_{thermal}}{T_{CMB}} = f(\nu, T_e)\frac{k_B \sigma_T}{m_e c^2}\int n_e T_e dl, \tag{1}$$

where T_{CMB} is the microwave background temperature; σ_T is the Thomson scattering cross section; and m_e, n_e, and T_e are the electron mass, density, and temperature. The frequency dependence of the SZE is contained in the factor $f(\nu, T_e)$, and includes relativistic effects; in the Rayleigh-Jeans regime of the CMB spectrum, $f(\nu) \sim -2$.

The integrated SZE flux is proportional to the gas mass in the cluster, weighted by the gas temperature:

$$\int d\Omega \Delta T \sim M_{gas}\langle T_e\rangle D_A^{-2}, \tag{2}$$

where D_A is the angular diameter distance to the cluster. The total gravitational mass of the cluster can be determined from $n_e(r)$, which is constrained by spatially-resolved SZE measurements, and the electron temperature. Therefore the gas mass fraction, f_{gas}, of the observed cluster can be determined directly from the measured SZE and the cluster's temperature. Under the assumption that the cluster gas mass fraction approaches a universal value, we can turn this relation around and instead estimate thc cluster's ICM temperature (and thence the mass) as that which reproduces the universal f_{gas}.

We use this method to measure the masses of two distant clusters recently discovered in X-ray surveys, and a cluster at similar redshift of known ICM temperature and mass, to provide a test of the method.

2. Method

The gas mass fraction is calculated from an SZE observation as a function of the cluster's ICM temperature. If we parametrize the electron density distribution as an isothermal beta-model (Cavaliere & Fusco-Femiano 1976, 1981) with a core radius r_c

$$n_e(r) = n_{e0}\left(1 + (r/r_c)^2\right)^{-3\beta/2}, \tag{3}$$

the spatial distribution of the gas density can be recovered from the projected SZE effect:

$$\Delta T(\theta) = \Delta T(0)\left(1 + \frac{\theta^2}{\theta_c^2}\right)^{\frac{1}{2}-\frac{3\beta}{2}}, \tag{4}$$

where $\theta = r/D_A$, $\theta_c = r_c/d_A$, and $\Delta T(0)$ is proportional to $n_{e0} \times T_e$. The central electron density can therefore be recovered from this relation:

$$n_{eo} = \frac{-\Delta T(0)}{T_{CMB}} \frac{m_e c^2}{2k\sigma_T} \frac{1}{T_e} \left(D_A \int_{-\infty}^{+\infty} \left(1 + \left(\frac{\theta}{\theta_c}\right)^2\right)^{-3\beta/2} d\theta \right)^{-1} \tag{5}$$

where the integral, *dl*, is along the line of sight. The gas mass is simply the electron number density integrated through the cluster volume, multiplied by the mean atomic weight of the ionized species:

$$M_{gas} = \mu m_p \int n_e(r) dV. \tag{6}$$

If the ICM is in hydrostatic equilibrium in the cluster potential, the cluster's total mass can be determined from the gas density distribution:

$$M_{tot}(< r) = -\frac{k_B T_e r}{\mu m_p G} \left(\frac{\mathrm{dln} n_e(r)}{\mathrm{dlnr}} \right). \tag{7}$$

To calculate μ, we assume the gas has solar metallicity as measured by Anders & Grevesse (1989). We also assume that μ is constant throughout the ICM. Therefore, the gas mass fraction as measured from a cluster is inversely proportional to the square of the gas temperature:

$$f_{gas} = \frac{M_{gas}}{M_{total}} \propto T_e^{-2} \tag{8}$$

We use the results from Grego et al. (2000) as the universal f_{gas}. This gas mass fraction was derived from SZE measurements and the published X-ray temperatures of 18 massive clusters, using the relations above. The clusters have kT_e >~5 keV, and range in redshift from $0.14 \leq z \leq 0.83$. The SZE data are interferometric, made at centimeter wavelengths (26-36 GHz) at the Owens Valley Radio Observatory (OVRO) and Berkeley-Illinois-Maryland Association (BIMA) millimeter-wave observatories. The mean f_g within r_{500} for these 18 clusters, is $0.081^{+0.009}_{-0.011} h_{100}^{-1}$ (statistical uncertainty at 68% confidence level, assuming Ω_M=0.3, Ω_Λ=0.7).

We use this mean value as the universal f_{gas}. We estimate the ICM temperature from new SZE observations of two distant and highly X-ray luminous clusters, recently discovered in deep ROSAT X-ray images. Cl J1226.9+3332, a cluster at redshift $z = 0.89$, was discovered in the WARPS survey (Ebeling et al. 2001; Scharf et al. 1997). Cl J0152.7−1357, at redshift $z = 0.83$, was detected in the RDCS, SHARC, and WARPS surveys (Della Ceca et al. 2000; Romer et al. 2000; Ebeling et al. 2000). Based on their X-ray luminosities ($L_X[0.5 - 2\mathrm{keV}] \gtrsim 2 \times 10^{44} \mathrm{h}_{100}^{-2}$ erg s^{-1}), these clusters are thought to be highly massive.

To provide a standard with which to compare our results on these clusters, we also determine the temperature of a well-studied cluster at a similar redshift (z = 0.83) and luminosity: MS 1054.4−0321, the most distant cluster discovered in the EMSS survey (Gioia & Luppino 1994; Donahue et al. 1998; Hoekstra et

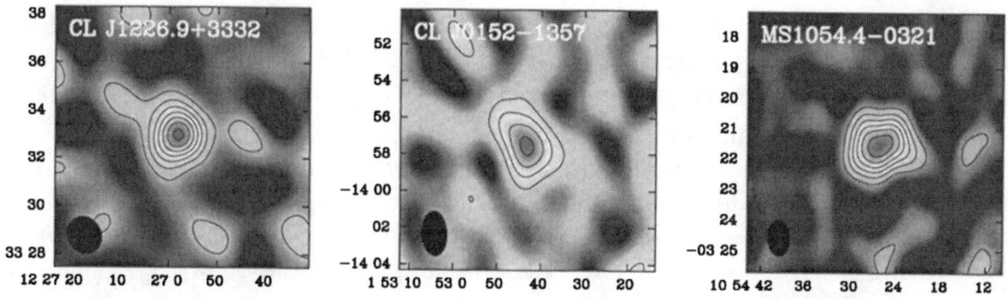

Figure 1. Synthesized images of the SZE decrement in Cl J1226.9+3332, Cl J0152.7−1357, and MS 1054.4−0321. Left panel: Synthesized image of Cl J1226.9+3332, obtained by applying a Gaussian taper with a half-power radius of 1 kλ to the interferometer $(u-v)$ data, yielding a resolution of 99.3″ × 87.4″ at position angle (p.a.) 32°. Contours are multiples of 290 μJ (1.5σ), and the rms is 190 μJ. Center panel: Synthesized image of Cl J0152.7−1357, obtained by applying a Gaussian taper with a half-power radius of 1 kλ to the $u-v$ data, yielding a resolution of 151″ × 87″.9 at p.a. 4°. Contours are multiples of 480 μJ (1.5σ), and the rms is 320 μJ. Right panel: Synthesized image of MS 1054.4−0321, obtained by applying a Gaussian taper with a half-power radius of 2 kλ to the $u-v$ data, yielding a resolution of 73″.1 × 45″.5 at p.a. 2°. Contours are multiples of 120 μJ (1.5σ), and the rms is 80 μJ.

al. 2000). The temperature of MS 1054.4−0321 was estimated by Donahue et al. from *ASCA* data to be $12.3^{+3.1}_{-2.2}$ keV.

The SZE images of these three clusters are shown in Figure 1. The derived temperatures and the total mass within 65″ for each cluster (via Equation 5) are reported in Table 1. The range in temperature and mass reflect the values which give a gas mass fraction within the sample mean of the Grego et al. (2000) f_{gas} value. We make a number of assumptions in these calculations: that the ICM is isothermal; is spherically symmetric; and that the gas is in hydrostatic equilibrium, supported only by thermal pressure. However, the gas mass fractions for Cl J1226.9+3332 and Cl J0152.7−1357 are calculated in exactly the same way as those used to determine the universal f_{gas}. We have used the universal value of f_{gas} in a comparative rather than an absolute way, and

Table 1. Cluster Properties derived from SZE measurements

Cluster Name	T_e^{Xray} (keV)	T_e^{SZ} (keV)	$M_{total}[<65'']$ ($10^{14}h^{-1}M_\odot$)
Cl J1226.9+3332	–	$9.8^{+4.7}_{-1.9}$	2.7 ± 0.5
Cl J0152.7−1357	$6.5^{+1.7}_{-1.2}$	$8.7^{+4.1}_{-1.8}$	2.1 ± 0.7
MS 1054.4−0321	$12.3^{+3.1}_{-2.2}$	$10.4^{+5.0}_{-2.0}$	2.3 ± 0.3

Recent work confirms the accuracy of the SZE mass measurement method. Jeltema et al. (2001) analyzed a 91 ks *Chandra* ACIS-S observation of MS 1054.4 −0321 and find the emission-weighted temperature to be $10.4^{+1.7}_{-1.5}$ keV at 90% confidence, lower than but consistent with the temperature of $12.3^{+3.1}_{-2.2}$ keV inferred from ASCA data in Donahue et al. (1998). We also compare the total mass calculated for this cluster via the SZE method with the mass determined by weak lensing. Hoekstra et al. (2000) infer a total mass of $5.4 \pm 0.6 \times 10^{14} h_{100}^{-1}$ $M_\odot$ within a 94 $''$ aperture. The SZE data imply a value of $M_{total}^{SZE}(< 94'') = 4.6 \pm 0.8 \times 10^{14} h_{100}^{-1}$ $M_\odot$, consistent with the lensing measurements. Thus we conclude that estimating cluster masses from SZE data and a measurement of the cluster redshift is a viable means of determining cluster temperatures and masses without X-ray or lensing data.

3. SZE Survey

A number of SZE-sensitive sky surveys are planned for the near future, with various depths and covered areas. The sensitivity of a SZE survey can be expressed as a redshift independent mass limit (Bartlett & Silk 1994; Barbosa et al. 1996; Holder et al. 2000). This follows from Equation 2 and the assumption of a universal f_{gas}.

SZE surveys can be grouped into two types of surveys: those which cover large areas to relatively shallow depths, similar to what can be expected with the Planck Surveyor Satellite; or smaller, relatively deep surveys, such as those expected with upcoming ground-based efforts. The left panel of Figure 2, from Carlstrom et al. (2000), shows the approximate expected limiting mass for the two types of surveys. The instrumental sensitivity used to represent the Planck-like survey is a limit on the integrated flux density of 15 mJy at 30 GHz. For the deep ground-based surveys, the flux density limit is 0.5 mJy. The SZE survey mass limits are also compared to an approximate mass limit for an XMM serendipitous survey, assuming a simple X-ray flux limit. A deep SZE survey will be able to probe somewhat lower masses at redshifts past $z \sim 1$, which promises to be a very interesting regime.

The right panels of Figure 2 show the expected cluster counts per square degree for a ΛCDM model ($\Omega_M = 0.3, \Omega_\Lambda = 0.7, h = 0.65, \sigma_8 = 1$). A deep SZE survey can expect to find roughly one cluster per square degree with $z > 1.5$. On the other hand, even though a shallow survey has a relatively high mass limit, a survey that covers half the sky would find ∼10000 clusters.

The SZA (Sunyaev-Zel'dovich Effect Array), a project in development, will add eight 3.5 m telescopes to the existing array of six 10.4 meter telescopes comprising the OVRO Millimeter Array. The full array will have two receivers, to operate in the centimeter wavelength (26-36 GHz) and millimeter wavelength (75-115 GHz) regimes. The SZA will use a wideband digital correlator, which can correlate 8 GHz bandwith with about 20 MHz resolution. The 3.5 meter telescopes are specified to be built with rms surface accuracy of 30 μm, so that the array has the capability to operate at higher frequencies at an improved site. The SZA survey is expected to have sensitivity similiar to the curves labeled "0.5 mJy" in Figure 2.

Such a survey will be able to probe the universe to high redshift and clusters of a cosmologically interesting range of masses. As the cluster masses can be determined from the SZE data itself, the SZA survey can provide constraints on cosmological parameters independent of other determinations.

Acknowledgments. We thank Cheryl Alexander, Rick Forster, Steve Padin, Dick Plambeck, Steve Scott, David Woody, and the staff of the BIMA and OVRO observatories for their support, and Laurence Jones, Eric Perlman, and Caleb Scharf for providing data on Cl J1226.9+3332 prior to publication. ER gratefully acknowledges support from NASA GSRP Fellowship NGT5-50173. This work is supported by NASA LTSA grants NAG5-7986 (JC,MJ,WH) and NAG5-8253 (HE). Radio astronomy at the BIMA millimeter array is supported by NSF grant AST 96-13998. The OVRO millimeter array is supported by NSF grant AST96-13717. The funds for the additional hardware for the SZE experiment were from a NASA CDDF grant, a NSF-YI Award, and the David and Lucile Packard Foundation.

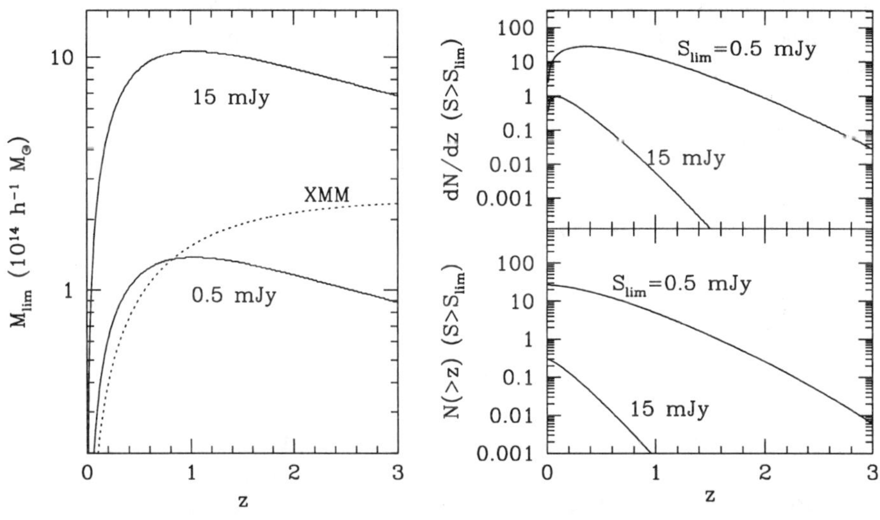

Figure 2. Illustration of the projected yield of SZE cluster surveys. Left: Mass limits as a function of redshift for a typical wide-field type of survey (sensitivity limit of ~ 15 mJy at 30 GHz) and for a typical deep survey (~ 0.5 mJy). The approximate XMM serendipitous survey limit is also shown. Right: Differential (top) and cumulative (bottom) counts per square degree as a function of redshift for the two SZE surveys shown at left, assuming a ΛCDM cosmology (Holder et al. 2000).

References

Anders, E. & Grevesse, N. 1989, Geochim. Cosmochim. Acta, 53, 197

Bahcall, N. A., & Cen, R. 1992, ApJ, 398, L81

Barbosa, D., Bartlett, J. G., Blanchard, A., & Oikbir, J. 1996, A&A, 314, 13

Bartlett, J. G., & Silk, J. 1994, ApJ, 423, 12

Birkinshaw, M. 1999, Phys. Rep., 310, 97

Carlstrom, J. E., Joy, M., & Grego, L. 1996, ApJ, 456, L75

Carlstrom, J. E., Joy, M., Grego, L., Holder, G., Holzapfel, W. L., LaRoque, S., Mohr, J. J., Reese, E. D. 2000, in "Constructing the Universe with Universe with Clusters of Galaxies, IAP 2000 meeting, Paris, France, Durret & Gerbal Eds.

Cavaliere, A., & Fusco-Femiano, R. 1976, A&A, 49, 137

—. 1978, A&A, 70, 677

Donahue, M., Voit, G. M., Gioia, I., Lupino, G., Hughes, J. P., & Eke, V. R., Cole, S., Frenk, C. S., & Henry, J. P. 1998, MNRAS, 298, 1145

Evrard, A. E. 1997, MNRAS, 292, 289

Evrard, A. E., Metlzer, C. A., & Navarro, J. F. 1996, ApJ, 469, 494

Finoguenov, A., Reiprich, T. H., & Böhringer, H. 2001, 368, 749

Gioia, I. M. & Luppino, G. A. 1994, ApJS, 94,583

Grego, L. 1999, PhD thesis, Caltech

Grego, L. Carlstrom, J. E., Joy, M. K., Reese, E. D., Holder, G. P., Patel, S., Cooray, A. T., & Holzapfel, W. L. 2000, ApJ, 539, 39

Haiman, Z., Mohr, J. J., & Holder, G. P. 2001, ApJ, 553,545

Hoekstra, H., Franx, M., & Kuijken, K. 2000, ApJ, 532, 88

Holder, G. P., Haiman, Z., & Mohr, J. J. 2001, astro-ph/0105396, ApJL in press

Holder, G. P., Mohr, J. J., Carlstrom, J. E., Evrard, A. E., & Leitch, E. M. 2000, ApJ, 544, 629

Horner, D. J., Mushotzky, R. F., & Scharf, C. A. 1999, ApJ, 520, 78

Jeltema, T. E., Canizares, C. R., Bautz, M. W., Malm, M. R., Donahue, M., and Garmire, G. P., ApJ (accepted), astro-ph/0107314

Joy, M.K., LaRoque, S., Grego, L., Carlstrom, J. E., Dawson, K., Ebeling, H., Holzapfel, W. L., Nagai, D., & Reese, Erik D. 2001, ApJ, 551, L1

Stocke, J. T. 1998, ApJ, 502, 550

Luppino, G. A., & Gioia, I. M. 1995, ApJ, 445, L77

Oukbir, J., & Blanchard, A. 1997, A&A, 317, 1

Reese, E. D., Mohr, J. J., Carlstrom, J. E., Joy, M., Grego, L.,

Holder, G. P., Holzapfel, W. L., Hughes, J. P., Patel, S. K., &

Donahue, M. 2000, ApJ, 533, 38

Sunyaev, R. A., & Zel'dovich, Y. B. 1972, Comments Astrophys. Space Phys., 4, 173

Viana, P. T. P., & Liddle, A. R. 1996, MNRAS, 281, 323

Tracing Cosmic Evolution with Galaxy Clusters
ASP Conference Series, Vol. 268, 2002
S. Borgani, M. Mezzetti, and R. Valdarnini, eds.

SZ surveys with the Arcminute MicroKelvin Imager

Rüdiger Kneissl and Michael E. Jones

Astrophysics Group, Cavendish Laboratory,
University of Cambridge, Madingley Road, Cambridge CB3 0HE, UK

Abstract. The Arcminute MicroKelvin Imager (AMI) is an instrument currently under construction in Cambridge designed to produce a survey of galaxy clusters via the Sunyaev-Zel'dovich effect. It consists of two interferometric arrays, both operating at 12–18 GHz; one of ten 3.7-m antennas to provide good temperature sensitivity to arcminute-scale structures, and one of eight 13-m antennas (the present Ryle Telescope). Based on hydrodynamical simulations in combination with the Press-Schechter expression we predict that AMI will discover at least several tens of clusters every year with $M_{\rm tot} \geq 10^{14} M_{\odot}$ under pessimistic assumptions, possibly hundreds of clusters; the numbers depend on factors such as the mean matter density, the density fluctuation power spectrum and cluster gas evolution. The effective mass limit of the survey is almost redshift independent.

1. Introduction

A wealth of astrophysical and cosmological information can be obtained from the study of galaxy clusters. In order to fully exploit the potential of clusters for learning about the universe, however, it is necessary to be able to select them according to uniform intrinsic properties over a large range in redshift. The promise of SZ surveys is that they can select clusters over a very large redshift range, in particular to the high redshifts where the first clusters form, with well-understood and physically meaningful selection criteria (e.g. Barbosa et al. 1996, Holder et al. 2000, Kneissl et al. 2001).

Interferometers offer several significant advantages over total power measurements: they are not susceptible to scan-synchronous systematics; they can use the rotation of the earth to modulate the sky signal in a way that is distinguishable from other signals such as crosstalk or groundspill; they are insensitive to most atmospheric emission; and they can simultaneously measure, and separate, the SZ effect and point source contamination.

To observe the SZ effect efficiently, however, the interferometer must have baselines which are well matched to the angular scale of the cluster. Each pair of antennas in an interferometer responds only to a narrow range of fourier components of the sky brightness, centred at λ/D where D is the antenna separation, and of width $\sim \lambda/d$ where d is the antenna size. Clusters at moderate to high redshift have angular sizes of a few arcminutes, and hence most of their power in the fourier plane is on scales of a few hundred to a thousand wavelengths.

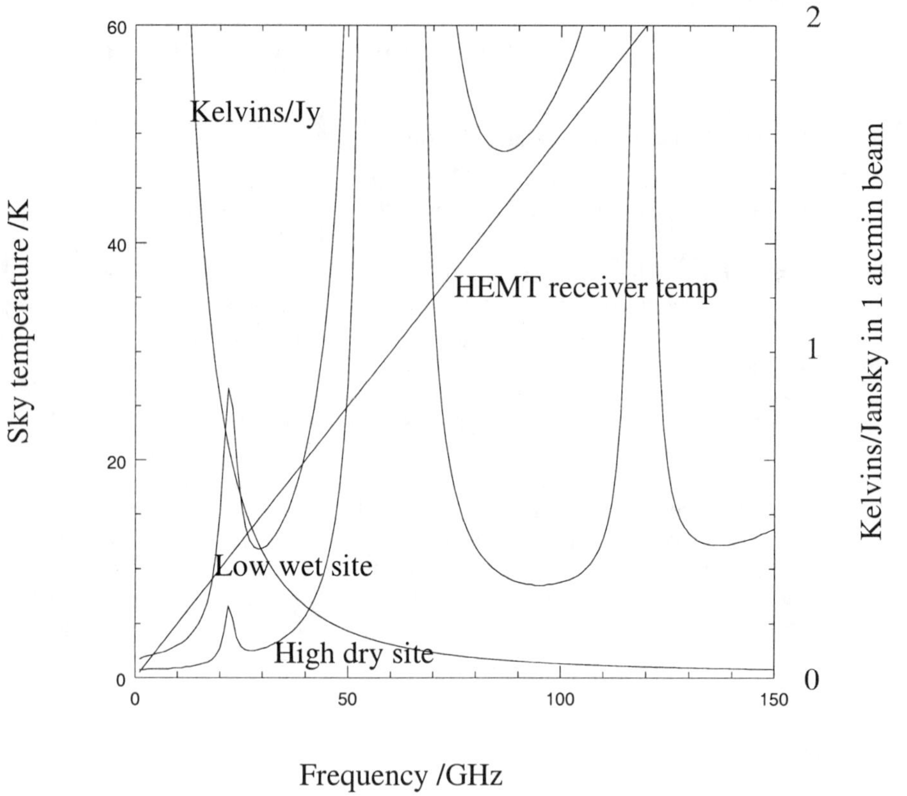

Figure 1. Contributions to system temperature as a function of frequency; description in the text below

2. Design parameters of AMI

The design requirements for an SZ survey instrument are: maximum sensitivity, hence low system temperature and high bandwidth; shortest baselines of $\sim 200\lambda$ (confusion from primary CMB anisotropies becomes a problem at shorter baselines); and ability to cope with foregrounds. The first question is the choice of observing frequency—this controls all subsequent aspects of the design.

Figure 1 shows the contributions to the system temperature of a telescope due to the atmosphere, at a sea level site such as Cambridge and a high dry site such as Mauna Kea, and from typical HEMT receivers. Adding these two contributions together (total system noise is the sum of the amplifier and sky noise; plus other contributions from losses, spillover, the CMB etc.), it can be seen that there is a clear advantage of operating at low frequency, even from a relatively poor site. However, this is offset by the other feature shown on the plot, the temperature contribution of a source of given flux density (scale on the right), which falls as approximately ν^{-2}. Since extragalactic point sources are the main contaminant for SZ observations, this is an important consideration. Several strategies are possible. The approach we adopt with AMI is to exploit the low system temperature available at low frequency, but to provide sufficient flux sensitivity at higher resolution to detect and then remove the point sources, since we have the large antennas of the Ryle Telescope available.

Operating at 15 GHz from Cambridge it is possible to achieve total system temperatures below 25 K. With the wavelength fixed at $\lambda \simeq 2$ cm, the remaining design parameters are fairly constrained. The available bandwidth is the K_a waveguide band of 12–18 GHz; the minimum required baseline of $\sim 200\lambda$ fixes the antenna size at ~ 4 m. The number of these antennas is fixed by the flux sensitivity available for source subtraction from the Ryle Telescope and the 15 GHz source counts (Taylor et al. 2001) to about ten, i.e. a collecting area equal to about one RT antenna.

3. Sensitivity and expected survey results

The flux sensitivities are 2 mJy $s^{-1/2}$ and 20 mJy $s^{-1/2}$ for the large and small arrays, in 6 and 20 arcmin fields of view respectively. The temperature sensitivity depends on the exact array configuration used; a representative number would be about 15 μK in a week, in a 1.5 arcminute beam and 20 arcmin field of view.

To determine how many galaxy clusters AMI can detect, it is useful to create realistic SZ sky maps simulated for different cosmologies. This approach is particularly helpful for interferometric measurements, because the limited coverage in Fourier space can lead to ringing in the map plane affecting the detection efficiency and to a reduction in the observed flux of a single cluster depending on it's size and shape. To produce a map, the Press-Schechter expression is used in combination with individual cluster templates from ten ΛCDM hydrodynamical cluster simulations (Eke, Navarro and Frenk 1998). Each cluster is observed at eight different redshifts out to $z \approx 1.1$, and templates are produced from three orthogonal directions to maximise the variety in the apparent cluster shapes.

We simulate the cluster detection process for a blank field AMI observation in two different ways: a detailed, but time consuming, simulation of the response of the interferometric array to the structure on the sky, and a simplified simulation, which only operates in the image plane and uses a compensated beam profile that is constructed to match the synthesised beam resulting from the more detailed simulations. We find that after one year of observation AMI will have completed a cluster survey down to roughly 1×10^{14} $M_\odot$ with 20 ($\Omega_0 = 1$) to 70 ($\Omega_0 = 0.3$) clusters for our pessimistic assumptions about σ_8 and f_g, or several hundred clusters for more realistic assumptions. The details of these results can be found in Kneissl et al. (2001). The SZ cluster catalogue and the differential cluster counts at different flux levels will be prime observational results from the AMI blank field survey. Figure 2 shows the mass–redshift distribution we expect for the catalogued clusters in a $\Omega_0 = 0.3$ cosmology.

4. Conclusions

The cluster survey will permit optical and X-ray follow-up to measure individual cosmological parameters, such as the mean matter density, the power spectrum

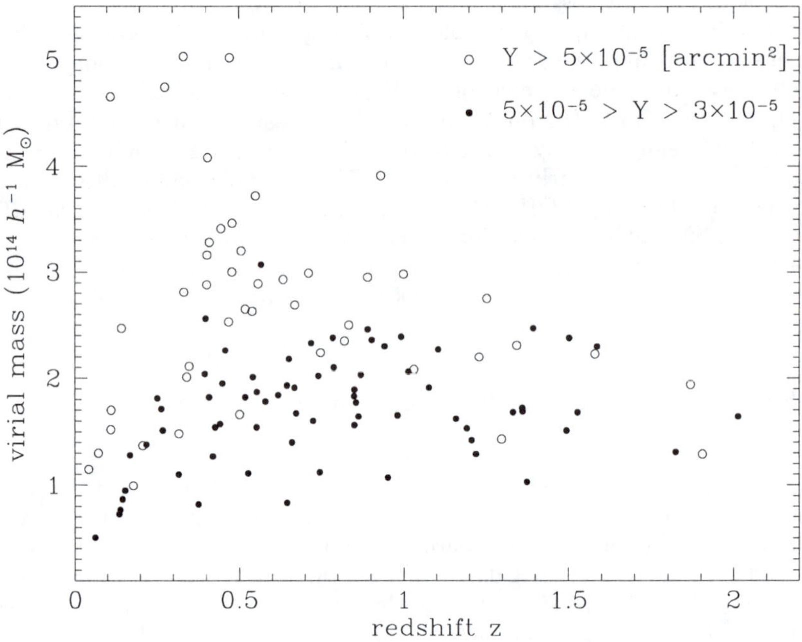

Figure 2. Masses and redshifts of the clusters detectable in the AMI survey for the $\Omega_0 = 0.3$ cosmology. The sample is nearly mass-limited and more high redshift clusters can be found in a deeper survey.

amplitude, and gas density and temperature structures and their evolution, to good approximation. Those clusters at high redshift will be key targets for multi-waveband studies of cluster evolution. First results from AMI are expected by mid-2003.

References

Barbosa, D., Bartlett, J. G., Blanchard, A., Oukbir, J. 1996, A&A, 314, 13

Eke V.R., Navarro J.F., Frenk C.S. 1998, ApJ, 503, 569

Holder, G. P., Mohr. J. J., Carlstrom, J. E., Evrard, A. E., Leitch, E. M. 2000, ApJ, 544, 629

Kneissl, R., Jones, M. E., Saunders, R., Eke, V.R., Lasenby, A.N., Grainge, K., Cotter, G. 2001, MNRAS, in press

Taylor, A. C., Grainge, K., Jones, M. E., Pooley, G. G., Saunders, R., Waldram E. M. 2001, MNRAS, 327, L1

INTERNAL STRUCTURE AND DYNAMICS OF CLUSTERS

Tracing Cosmic Evolution with Galaxy Clusters
ASP Conference Series, Vol. 268, 2002
S. Borgani, M. Mezzetti, and R. Valdarnini, eds.

Tracing the Cluster Internal Dynamics with Member Galaxies

Andrea Biviano

INAF – Osservatorio Astronomico di Trieste, via G.B. Tiepolo 11, I-34131 Trieste, Italy – biviano@ts.astro.it

Abstract. The analysis of the spatial distribution and kinematics of galaxies in clusters allows one to determine the cluster internal dynamics. In this paper, I review the state of the art of this topic. In particular, I summarize what we have learned so far about galaxy orbits in clusters, and about the cluster mass distribution. I then compare four methods that have recently been used in the literature, by applying them to the same data-set. The results stress the importance of reducing *systematic* besides *random* errors, for a reliable determination of cluster mass profiles.

1. Introduction

Already in 1931, Hubble & Humason (1931) provided rough estimates of the velocity dispersions of three galaxy clusters, ranging from 500 to 1200 km s^{-1}. This is more or less the range of values found for rich clusters nowadays. A few years later, Zwicky (1933, 1937) and Smith (1936), determined the total masses of the Coma and the Virgo cluster, respectively, by application of the virial theorem to the cluster galaxies. The cluster masses they determined[1], $> 3.3 \times 10^{14} M_{\odot}$ (Coma), and $7.7 \times 10^{14} M_{\odot}$ (Virgo), are in reasonable good agreement with nowadays estimates. Of course, these early estimates used galaxies as unbiased tracers of the underlying mass distribution, but this was just a convenient assumption. Relaxing this assumption can have a large effect on the cluster mass estimate (see, e.g., Merritt 1987). In fact, galaxies are *unlikely* to be distributed like the mass, simply because different cluster galaxy populations have different distributions (e.g. emission-line galaxies, ELG hereafter, and non-ELG, see Biviano et al. 1997). That is why we need to determine the cluster mass distribution, if we want to determine its total mass. And if we use the cluster galaxies for this purpose, we must at the same time determine their orbital anisotropies.

In this contribution, I review our current understanding of the internal dynamics of clusters, as determined from the spatial and velocity distributions of cluster member galaxies. Due to space limitation, I focus my review on the determination of the cluster mass, and mass-to-light profiles, and on the orbital anisotropy of cluster galaxies. Readers interested in cluster substructures are

[1] Throughout this paper I use a Hubble constant H_0=70 km s^{-1} Mpc^{-1}. Quoted values are scaled accordingly.

referred to the recent review of Girardi & Biviano (2001), and those interested in a complete historical review of this topic are referred to Biviano (2000).

2. The Jeans approach

Here I briefly describe the traditional Jeans approach (see Binney & Tremaine 1987). For the sake of simplicity, I assume to deal with a non-rotating spherically symmetric system, in a steady-state. The Abel equations relate the projected density profile, $I(R)$, and projected velocity dispersion profile, $\sigma_p(R)$ (the observables), to the de-projected density profile, $\nu(r)$ and, respectively, the radial and tangential components of the velocity dispersion profile, $\overline{v_r^2}$, and $\overline{v_t^2}$. R and r are the projected and 3D clustercentric radial distances, respectively. When $\nu(r)$, $\overline{v_r^2}$ and $\beta(r)$ are determined, the mass profile follows from the Jeans equation.

In principle, one should determine the de-projected profiles from the observed ones, but this is not simple, unless the anisotropy term, $\beta(r) \equiv 1 - \overline{v_t^2}/\overline{v_r^2}$ is zero. The general case requires solving an integro-diffential equation (eq.4-62 in Binney & Tremaine 1987; see Binney & Mamon 1982; Solanes & Salvador-Solé 1990). This "direct" approach is complicated when the data are noisy. Therefore, the simpler "inverse" route is usually chosen. One adopts suitable functional forms for the 3D profiles – $\nu(r)$, $\overline{v_r^2}$, $\beta(r)$, or the mass profile, $\rho(r)$ –, projects them in 2D, and determines their best-fit parameters by comparison with the observables (see, e.g., Carlberg et al. 1997a, and Sect. 3.2).

The Jeans approach has several potential problems. First of all, one generally assumes the system to be spherically symmetric, but clusters are not. However, the number of available galaxies in each single cluster is limited, and it is often necessary to build a "composite" cluster by joining together the data for several clusters, after proper normalization (see, e.g., Biviano et al. 1992, 1997; Carlberg et al. 1997a). This composite cluster is spherically symmetric by construction.

Evidence for cluster rotation has been claimed for a couple of clusters (see, e.g., Biviano et al. 1996; Dupke & Bregman 2001) but in general we can safely assume that rotational motions are not important.

The assumption of steady-state is not always valid. E.g. galaxy groups are probably still in the phase of collapse (Giuricin et al. 1988). On the other hand, in a low-density Universe, clusters evolve quite rapidly, and their formation process should be over by now. Following Ellingson et al. (2001), we estimate that the infall rate of field galaxies in nearby clusters is $\lesssim 1\%$ of the total cluster population per Gyr. Time-dependent terms in the Jeans equation could be needed in more distant clusters, or when galaxies are subject to tidal disruption (e.g. spirals in cluster centres, see Katgert et al. 2002).

The Jeans equation is derived from the collisionless Boltzmann equation, and its applicability is limited to collisionless fluids. Clusters of galaxies are not collisionless in their inner regions (e.g. Biviano et al. 1992; den Hartog & Katgert 1996), and dynamical friction and mergers could be important for very massive custer galaxies. It is then wise to exclude the brightest cluster members from the sample (Katgert et al. 2002).

Interlopers are not a serious problem, when one deals with a composite sample of several hundred galaxies. Different methods of interloper rejection give similar results (Fadda et al. 1996). Subclustering is a more serious issue, since galaxies in subclusters are distributed in a different way from the smooth cluster population (Biviano et al. 2001; see also Katgert's contribution in these proceedings). It is better to identify galaxies belonging to subclusters, and to exclude them from the sample (Katgert et al. 2002).

Finally, the basic problem of the "Jeans" approach is that the solution is intrinsically degenerate; we can either assume $\beta(r)$ and determine $M(< r)$, or *viceversa,* but we cannot simultaneously constrain $M(< r)$ and $\beta(r)$ from the observables (see, e.g., King 1972; Bailey 1982). A possible solution consists in modelling the whole velocity distribution of cluster galaxies, and not only its second moment (Merritt 1987; Merritt & Saha 1993). Another possibility is to break this degeneracy by the use of several different tracers of the same gravitational potential (Katgert et al. 2002; see Sect. 3.6).

3. Mass profiles and orbital anisotropies: results so far

Seven decades since its first application to Coma (Zwicky 1933) the virial theorem has not found simple alternatives yet (see Perea et al. 1990 and references therein). The main improvement over the classical virial method has been the introduction of the "surface-term" (The & White 1986; see also Sect. 3.4). The solutions to the direct Jeans approach provided by Binney & Mamon (1982) and Solanes & Salvador-Solé (1990), have never been applied to galaxy cluster data-sets so far, because these data are noisy. Most analyses therefore model the profiles instead of deriving them directly from the data. If the models are not general enough, this procedure can lead to biased results. That is why non-parametric approaches have been developed, mostly by Merritt and collaborators (Merritt & Saha 1993; Merritt & Tremblay 1994; Merritt & Gebhardt 1994; Gebhardt & Fisher 1995).

Despite these technical improvements, the question of whether or not the cluster mass is distributed like the galaxies has not received a definite answer yet. In clusters, as well as in groups, recent works provide evidence both in favour (Carlberg et al. 1997a; Mahdavi et al. 1999; van der Marel et al. 2000), and against (Koranyi et al. 1998; Rines et al. 2000; Carlberg et al. 2001) the hypothesis of a constant mass-to-light ratio with radius. Cluster mass density profiles are found to be in agreement with the 'universal' profile of Navarro, Frenk & White (1997; NFW hereafter), but other models are acceptable as well (Carlberg et al. 1997b; van der Marel et al. 2000; Geller, Diaferio & Kurtz 1999). From a theoretical point of view, the situation is not much clearer. Both decreasing and increasing mass-to-light profiles are predicted, resulting from the processes of tidal stripping (Mamon 1999), or, respectively, of dynamical friction and merging (Fusco-Femiano & Menci 1998).

Numerical simulations show that dark matter particles are characterized by a moderate radial anisotropy, increasing out to $\sim 2\,r_{200}$ (e.g. Diaferio 1999). However, the idea of shells of collapsing material around clusters has been around since the work of Gunn & Gott (1972). This work stimulated several investigations, and the evidence for the infall of spirals into clusters has been accumulating over the years. Indirect evidence comes from the larger velocity dispersion of spirals as compared to that of early-type galaxies (Moss & Dickens 1977; Sodré et al. 1989). More direct evidence is provided by the analyses of the Tully-Fisher distance-velocity diagram in nearby clusters (Tully & Shaya 1984; Gavazzi et al. 1991).

Additional constraints on the orbital anisotropy of cluster galaxies come from the analysis of their velocity distributions. Most studies conclude that the orbits of early-type galaxies are quasi-isotropic, and those of late-type galaxies moderately radial (Mohr et al. 1996; Biviano et al. 1997; Adami et al. 1998; van der Marel et al. 2000, see also Sect. 3.3), with the exception of Ramírez & de Souza (1998, see Sect. 3.1). Similar results are also found by the application of dynamical models (e.g. Natarajan & Kneib 1996; Carlberg et al. 1997c; Mahdavi et al. 1999).

Other methods for constraining the galaxy orbits include Pryor & Geller's (1984) and Merrifield's (1998). Pryor & Geller try to constrain the orbits of HI-deficient galaxies by noting that cluster core-crossing is a necessary condition for gas stripping. Merrifield suggests using the radio or X-ray trails of galaxies in order to constrain their orbits.

3.1. Ramírez & de Souza

Ramírez & de Souza (1998) propose to use $\mid u \mid \equiv <\mid v_j \mid> / \sigma_p$ as a robust shape estimator of the velocity distribution of cluster members. The value of the anisotropy of galaxy orbits is obtained by comparing the observed values of $\mid u \mid$ with those derived from a kinematical model. The model is rather over-simplified, and not very realistic: the velocity distribution is Gaussian at any radial distance, and both $\overline{v_r^2}$ and $\overline{v_t^2}$ are constant.

Van der Marel et al. (2000) point out that the assumption of Gaussianity is not acceptable for anisotropic models. Another problem is that Ramírez & de Souza (1998) do not compute σ_p and $<\mid v_j \mid>$ on the same sample. E.g., when they consider the subsample of ellipticals, σ_p is still computed on all cluster galaxies. This effectively makes the $\mid u \mid$ parameter a *scale* rather than a *shape* estimator. They find that ellipticals have a lower $\mid u \mid$ than spirals, but this is just a restatement of the finding of Moss & Dickens (1977; see also Adami et al. 1998), that ellipticals have a lower velocity dispersion than spirals, and do not support Ramírez & de Souza's (1998) claim of radial anisotropy.

3.2. Carlberg et al.

Carlberg et al. (1997a,b) analyse the internal dynamics of galaxy clusters from CNOC, by the "inverse" Jeans approach (see Sect. 2). They adopt *ad hoc* functional forms for $\nu(r)$, $\overline{v_r^2}$, and $\beta(r)$, project them in 2D, and constrain their parameters via a χ^2 comparison with the observables. The advantage of their approach is that the chosen functions are well behaved, and easy to invert, but the solution for $M(<r)$ is to some extent imposed *a priori* (Merritt & Gebhardt 1994).

Their results are based on $\sim$ 1000 galaxy members in a composite of 14 "regular" clusters. They find that galaxies trace the mass, to within $\pm 30\%$, and the anisotropy is null or at most mildly radial, $0 \lesssim \beta \lesssim 0.5$. The mass profile is found to be consistent with a NFW profile, but other profiles (Hernquist 1990) provide equally acceptable, if not better, fits.

Carlberg et al. (2001) have recently applied the same technique on a composite sample of $\sim$ 700 galaxies in $\sim$ 200 groups from the CNOC2. They find a strongly increasing mass-to-light ratio with radius, at variance with the findings of Mahdavi et al. (1999). The construction of a composite group sample is anyway problematic, due to uncertainties in the mean velocity and velocity dispersions of individual groups. Moreover, removing interlopers in small galaxy groups is a difficult task. Finally, groups may be far from virialization (Giuricin et al. 1988). Therefore, I feel that Carlberg et al.'s result needs confirmation by independent analyses.

3.3. van der Marel et al.

The CNOC sample has recently been re-analysed by van der Marel et al. (2000) using the method developed by van der Marel (1994). They assume a three-parameter family of models for the mass density profile, $\rho(r)$, and a set of values for the anisotropy β, taken to be a constant over radii, and determine the model parameters from the best-fit to $\sigma_p(R)$. The galaxy density profile is derived via Abel-inversion from a parametric representation of $I(R)$. Their approach improves over Carlberg et al.'s, since both the adopted model for $\rho(r)$, and the function representing $I(R)$, are quite general. On the other hand, a constant β is unlikely to be a realistic model.

In order to break the intrinsic degeneracy between $\rho(r)$ and β, van der Marel et al. take into account the shape of the whole velocity distribution of cluster members (following the idea of Merritt & Saha 1993). They are thus able to conclude that mass follows light to within $\pm 30\%$, β ranges from -0.8 to 0.1, and a NFW profile is an acceptable fit to $\rho(r)$ – but not unique.

3.4. Girardi et al.

Girardi et al. (1998) work out the masses of 170 clusters, the largest sample analysed so far, using the virial theorem with the surface-term correction. In order to work out this correction, they first determine $\nu(r)$ from the Abel inversion of a (simple) parametric representation of $I(R)$. Assuming $\rho(r) \propto \nu(r)$, they then determine which of three $\beta(r)$-models provides the best-fit to $\sigma_p(R)$. From $\beta(r)$ and $\rho(r)$, $\overline{v_r^2}(r)$ and the surface-term correction are derived. In principle, it is then possible to compute the corrected virial mass at any radius, hence the mass profile, but the solution is certainly biased by the rather arbitrary assumptions.

They find that the surface-term correction is important, of order 20% at the virial radius (in agreement with The & White 1986, and Carlberg et al. 1997b). They find good agreement between their optically-determined cluster masses and those derived from the X-ray data. Using the same algorithm, Girardi & Mezzetti (2001) claim no evidence for the evolution of the internal cluster dynamics up to $z \simeq 0.3$.

3.5. Diaferio & Geller

Based on the results of numerical simulations of structure formation, Diaferio & Geller (1997) and Diaferio (1999) have devised a new method for determining

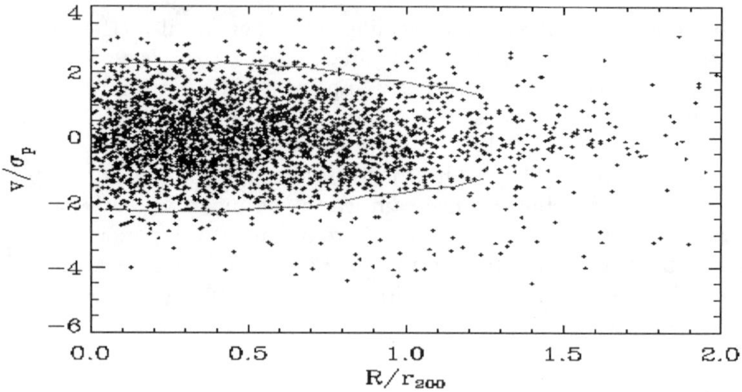

Figure 1. The (R, v) plane of 3365 galaxies in the composite cluster sample drawn from ENACS. Solid lines show the location of the caustics, as determined with the method of Diaferio (1999).

cluster mass profiles. They show that the velocity field of cluster surroundings is determined by the cluster mass distribution, via $GM(<r) = \int_0^r \mathcal{A}^2(x)\,\mathcal{F}_\beta(x)\,dx$ where $\mathcal{A}$ is the amplitude of the caustics in the (R, v) space, and $\mathcal{F}_\beta$ is a function of the gravitational potential and of the anisotropy profile, assumed to be a constant. The caustics are determined by an adaptive kernel method.

The "caustic method" does not rely upon the hypothesis of virial equilibrium, and thus it allows estimating $M(<r)$ to large radii, where other methods fail. On the other hand, the choice of the caustic in the (R, v) plane is rather subjective, and the approximation $\mathcal{F}_\beta \simeq$ constant is valid only beyond 1–2 r_{200}.

Using this method Geller et al. (1999) estimate $M(<r)$ of the Coma cluster out to 14 Mpc, and find it to be well fitted by a NFW profile. Rines et al. (2000) determine the mass-to-light profile of Abell 576 out to 6 Mpc, and find it to be decreasing with radius. A similar indication for little mass outside the virialized cluster region is found by Reisenegger et al. (2000) in the Shapley supercluster. However, the method can constrain $M(<r)$ only to within a factor two (Diaferio 1999), and a constant mass-to-light profile is probably still consistent with the data.

3.6. Katgert & Biviano

Peter Katgert and myself, we are currently analysing the internal dynamics of a composite data-set of $\sim$ 3000 galaxies in clusters from ENACS. We build $M(<r)$ directly from $I(R)$ and $\sigma_p(R)$, instead of modelling $\rho(r)$ and $\beta(r)$ with *ad hoc* chosen functions. In fact, *"using ad hoc parametrized models severely bias the answer regardless of the quality of the data"* (Merritt & Gebhardt 1994). Our approach is fully non-parametric; we use *LOWESS* (Gebhardt & Fisher 1995) to provide a smoothed representation of the observed profiles, which we invert to provide the mass profile in the isotropic case. Using several tracers of the gravitational potential (i.e. ellipticals and S0's, early spirals, late spirals and ELG) we expect to break the intrinsic degeneracy of the problem, to constrain

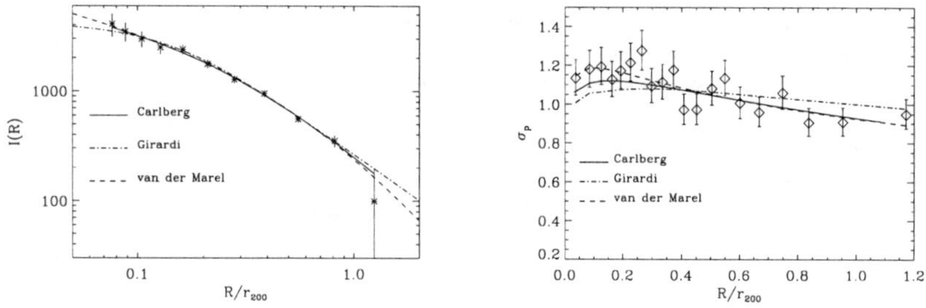

Figure 2. The projected number density profile $I(R)$ (left panel) and the projected velocity dispersion profile $\sigma_p(R)$ (right panel) of the 1557 non-ELG members of the composite ENACS cluster. The labelled lines indicate model fits of these profiles.

the orbits of cluster galaxies, and to test whether different galaxy populations are in equilibrium within the same gravitational potential.

The main problem of this method is numerical instability. This can be solved by smoothing the data, but excessive smoothing must be avoided, not to erase useful information in the observed profiles.

Preliminary results (Biviano et al. 1999; Mazure et al. 2000) indicate that different cluster galaxy populations are characterized by different anisotropy profiles.

4. Four new methods and one data-set

In this Section I compare the results obtained by four different methods on the *same* data-set. I consider 37 ENACS (Katgert et al. 1996, 1998) clusters with at least 45 galaxy redshifts, totalling 3365 galaxies. The observational data are shown in Fig. 1, where I also display the location of the caustics, determined with the method of Diaferio (1999). From these clusters I remove interlopers (following den Hartog & Katgert 1996), and galaxies with a significant probability of being in subclusters (following Biviano et al. 2001). I then build a composite cluster, in which galaxy velocities are scaled with the cluster σ_p's, and galaxy clustercentric distances are scaled with the cluster r_{200}'s. There are 1557 non-ELG and 217 ELG in this "cleaned" composite cluster. I apply the methods of Carlberg et al., Girardi et al., and van der Marel et al. to the sample of 1557 non-ELG. The last method is also applied to the sample of 217 ELG. Diaferio & Geller's metod is instead applied on the uncleaned sample of 3365 galaxies, since it does not require preliminary removal of interlopers, or subclusters. Results of the method of Katgert & Biviano will be presented elsewhere (Katgert et al. 2002).

The advantage of using a composite cluster is that one reduces asymmetries in the galaxy distributions, thus effectively imposing the circular symmetry to the problem. But of course, the composite cluster is an artificial construction, and as such it may bear no resemblance to any real cluster. On the other hand,

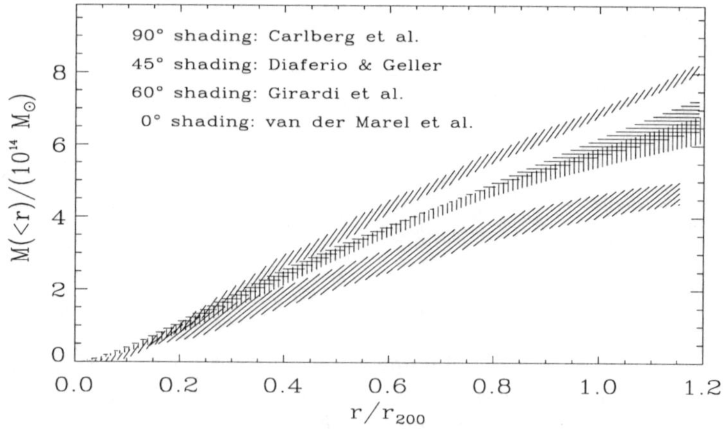

Figure 3. Four different estimates of the mass profile of the non-ELG composite cluster, under the isotropic assumption. Shadings represent (approximate) 1-σ uncertainties.

we may reasonably expect that results obtained for this composite cluster are representative of the average cluster population. It is important to use a homogeneous sample (like ENACS, or CNOC) for properly dealing with observational selection functions. This is particulary important for the determination of $I(R)$ (for further details, see Biviano et al. 2001).

In Fig. 2a,b I plot the $I(R)$ and $\sigma_p(R)$ for the 1557 non-ELG. Also shown are the three fits to these profiles, determined using the three different methods labelled in the Figure, and assuming $\beta(r) \equiv 0$. The most remarkable differences occur where the data are less constraining, i.e. at small and large radii. They are caused by different choices of the fitting functions.

The mass profiles determined with the four different methods (assuming $\beta(r) \equiv 0$ where needed) are shown in Fig. 3. The mass scale is arbitrary and the different values of $M(< r_{200})$ indicate that different effective values of r_{200} are implied by the different methods. Approximate 1-σ confidence intervals are determined by splitting the sample in two, and independently computing $M(< r)$ on each subsample. The four different estimates of the mass profile are not in agreement. the mass at any given radius is uncertain by a factor two, and systematic errors dominate over random errors.

In Fig.s 4a,b I show the integrated mass-to-number and mass-to-luminosity ratios as a function of radius, computed with the four different methods. On average, light seems to trace mass to within ±30 %, at radii $> 0.4 r_{200}$. In the central region, the mass seems to be more concentrated than the galaxies, but less than their light, an indication of luminosity segregation (see, e.g., Biviano et al. 1992).

The determination of the mass for the non-ELG is based on the assumption of isotropy. This assumption is supported by the analysis of the non-ELG velocity distribution, shown in Fig. 5a, alongwith its Gauss-Hermite representation. The Gauss-Hermite moments of this distribution clearly indicate that isotropy (or, at most, mild tangential anisotropy, $\overline{v_r^2}/\overline{v_t^2} \simeq 0.8$) is a valid assumption.

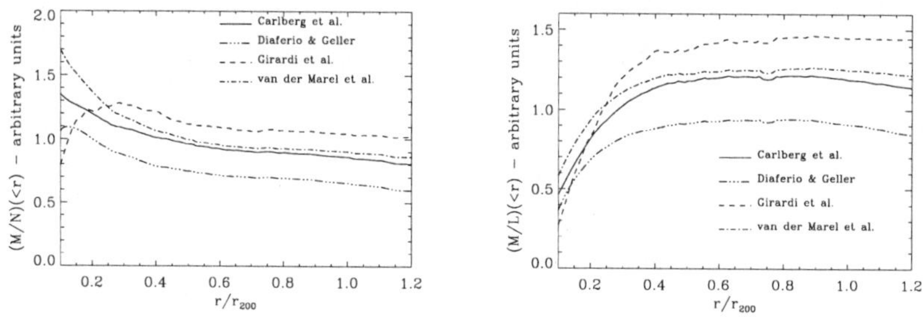

Figure 4. Left panel: the integrated mass-to-number profile determined with four different methods on the the 1557 non-ELG members of the composite ENACS cluster, assuming isotropy. Right panel: the integrated mass-to-light profile determined with four different methods on the the 1557 non-ELG members of the composite ENACS cluster, assuming isotropy.

I then consider the sample of 217 ELG. I assume that the non-ELG mass profile, in the isotropic assumption, is the true cluster mass profile, and solve for $\beta(r)$ of the ELG, using the method of van der Marel et al. I show in Fig. 5b the best-fit χ^2 solution for a constant $\overline{v_r^2}/\overline{v_t^2}$. For consistency, the solutions for non-ELG are also shown, where deviations from the expected $\overline{v_r^2}/\overline{v_t^2} \equiv 1$ solution are indicative of the accuracy of the method. The data indicate a moderate degree of radial anisotropy for the ELG, $1.3 \leq \overline{v_r^2}/\overline{v_t^2} \leq 1.6$, at 68% confidence level. Zero anisotropy models for the ELG are rejected at $> 99.99\%$ probability.

5. Conclusions

The use of member galaxies to determine the internal cluster dynamics is now reaching an unprecedented level of accuracy, thanks to the increasing amount of data (ENACS, Katgert et al. 1996, 1998; CNOC, Yee, Ellingson, & Carlberg 1996; Coma, Geller et al. 1999; Abell 576, Rines et al. 2000), and the development of new algorithms. The size of the data-sets must still be increased if we want to separately consider different cluster galaxy populations (e.g. early-types and late-types).

Whatever the sample size, it is difficult to discriminate among different models for the cluster mass density profile. This is because models either differ at large radii, where the cluster fades into the field, or near the centre, where one samples the cD rather than the cluster potential. The determination of cluster mass profiles appears to suffer from the systematics of different methods. Even on samples of ~ 1000 galaxies, the total mass of a cluster can be uncertain by a factor two. Such an uncertainty also affects our current knowledge of the relative distributions of mass and light. Taking the data at face value, there are now some indications that mass *is* indeed *roughly* traced by light, at least outside the very central region (where there is evidence for luminosity segregation). Finally, independent analyses suggest that different cluster galaxy populations

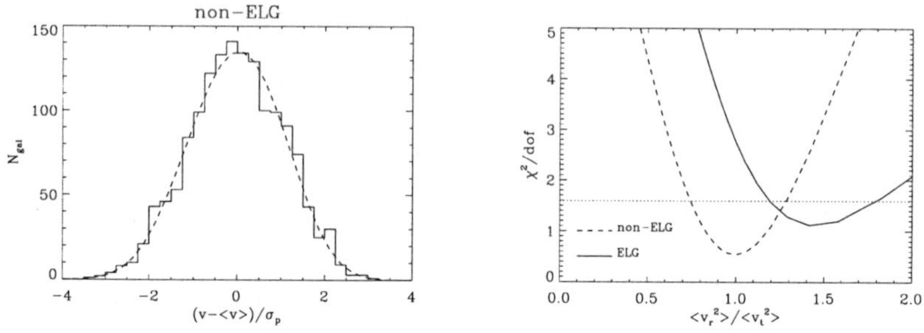

Figure 5. Left panel: The velocity distribution of the non-ELG and its decomposition into Gauss-Hermite moments. Right panel: The χ^2 solutions to the constant anisotropy model for ELG in the gravitational potential of non-ELG (van der Marel et al.'s method). The dotted line represents the cutoff value above which solutions are excluded at > 99% confidence level. The solutions for non-ELG in the same potential are shown as a check for consistency.

are characterized by different orbits. The currently best-buy model has (spectral or morphological) early-type galaxies on fairly isotropic orbits, and late-type galaxies on moderately radial orbits.

Ever since Zwicky (1937) we have been aware of the need for estimating cluster masses by different means. Different methods are affected by different biases, which are sometimes difficult to properly correct for. The perspective for improving upon the current situation is therefore clear. We need to constrain the cluster internal dynamics by the complementary analyses of the distribution of cluster members, as well as of X-ray and gravitational lensing data. For this, coordinated multi-wavelength surveys of galaxy clusters, using several instruments, are needed. Most promising, in this respect, is the upcoming XMM Large Scale Structure Survey[2], but a much more intensive spectroscopic follow-up than currently planned would be required to properly address the issue of internal cluster dynamics.

Acknowledgments. I wish to thank Stefano Borgani, Marino Mezzetti and Riccardo Valdarnini for organizing such an interesting and entertaining workshop. I acknowledge useful discussions with Peter Katgert, Marisa Girardi and Alain Mazure.

This paper is dedicated to the memory of my parents, Angelo and Anita.

References

Adami, C., Biviano, A., & Mazure, A. 1998, A&A, 331, 439

Bailey, M.E. 1982, MNRAS, 201, 271

Binney, J., & Mamon, G. 1982, MNRAS, 200, 361

---

[2]http://vela.astro.ulg.ac.be/themes/spatial/xmm/LSS/index_e.html

Binney, J., & Tremaine, S. 1987, "Galactic Dynamics" (Princeton: Princeton Univ. Press)

Biviano, A. 2000, in "Constructing the Universe with Clusters of Galaxies", ed. F. Durret & D. Gerbal, http://www.iap.fr/Conferences/Colloque/coll2000/contributions/

Biviano, A., Durret, F., Gerbal, D. et al. 1996, A&A, 311, 95

Biviano, A., Girardi, M., Giuricin, G., Mardirossian, F., & Mezzetti, M. 1992, ApJ, 396, 35

Biviano, A., Katgert, P., Mazure, A. et al. 1997, A&A, 321, 84

Biviano, A., Katgert, P., Thomas, T., Mazure, A., & Adami, C. 1999, 1st workshop of the italian network "Formazione ed evoluzione delle galassie", eds.: C. Chiosi, L. Portinari, & R. Tantalo, p.45, http://www.brera.mi.astro.it/docB/galaxy/news.html

Biviano, A., Katgert, P., Thomas, T., Mazure, A., & Adami, C. 2001, A&A, submitted

Carlberg, R.G., Yee, H.K.C., Ellingson, E. 1997b, ApJ, 478, 462

Carlberg, R.G., Yee, H.K.C., Ellingson, E. et al. 1997c, ApJ, 476, L7

Carlberg, R.G., Yee, H.K.C., Ellingson, E. et al. 1997a, ApJ, 485, L13

Carlberg, R.G., Yee, H.K.C., Morris, S.L. et al. 2001, ApJ, 552, 427

den Hartog, R. & Katgert, P. 1996, MNRAS, 279, 349

Diaferio, A. 1999, MNRAS, 309, 610

Diaferio, A. & Geller, M.J. 1997, ApJ, 481, 633

Dupke, R.A., & Bregman, J.N. 2001, ApJ, 547, 705

Ellingson, E., Lin, H., Yee, H.K.C. & Carlberg, R.G. 2001, ApJ, 547, 609

Fadda, D., Girardi, M., Giuricin, G., Mardirossian, F., & Mezzetti, M. 1996, ApJ, 473, 670

Fusco-Femiano, R. & Menci, N. 1998, ApJ, 498, 95

Gavazzi, G., Scodeggio, M., Boselli, A. & Trinchieri, G. 1991, ApJ, 382, 19

Gebhardt, K. & Fischer, P. 1995, AJ, 109, 209

Geller, M.J., Diaferio, A., & Kurtz, M.J. 1999, ApJ, 517, L23

Girardi, M. & Biviano, A. 2002, to appear in "Merging Processes in Clusters of Galaxies", ed. L. Feretti, I.M. Gioia, & G. Giovannini (Dordrecht: Kluwer)

Girardi, M., Giuricin, G., Mardirossian, F., Mezzetti, M., & Boschin, W. 1998, ApJ, 505, 74

Girardi, M. & Mezzetti, M. 2001, ApJ, 548, 79

Giuricin, G., Gondolo, P., Mardirossian, F., Mezzetti, M., & Ramella, M. 1988, A&A, 199, 85

Gunn, J.E. & Gott, J.R. III 1972, ApJ, 176, 1

Hernquist, L. 1990, ApJ, 356, 359

Hubble, E., & Humason, M.L. 1931, ApJ, 74, 43

Katgert, P., Biviano, A. et al. 2002, in preparation

Katgert, P., Mazure, A., den Hartog, R., Adami, C., Biviano, A., & Perea, J. 1998, A&AS, 129, 399

Katgert, P., Mazure, A., Perea, J. et al. 1996, A&A, 310, 8

King, I.R. 1972, ApJ, 174, L123

Koranyi, D.M., Geller, M.J., Mohr, J.J., & Wegner, G. 1998, AJ, 116, 2108

Mahdavi, A., Geller, M.J., Böhringer, H., Kurtz, M.J. & Ramella, M. 1999, ApJ, 518, 69

Mamon, G. A. 1999, in "Dynamics of Galaxies: from the Early Universe to the Present", ASP Conf. Series 197, ed. F. Combes, G.A. Mamon, & V. Charmandaris, p.377

Mazure, A., Biviano, A., Katgert, P., Thomas, T. & Adami, C. 2000, in "Galaxy Dynamics: from the Early Universe to the Present", ASP Conf. Series 197, ed. F. Combes, G.A. Mamon, & V. Charmandaris, p.373

Merrifield, M.R. 1998, MNRAS, 294, 347

Merritt, D. 1987, ApJ, 313, 121

Merritt, D. & Gebhardt, K. 1994, in "Clusters of Galaxies", ed. F. Durret, A. Mazure & J. Tran Thanh Van (Gif-sur-Yvette: Editions Frontières)

Merritt, D. & Saha, P. 1993, ApJ, 409, 75

Merritt, D. & Tremblay, B. 1993, AJ, 106, 2229

Mohr, J.J., Geller, M.J., Fabricant, D.G., Wegner, G., & Thorstensen, J. 1996, ApJ, 470, 724

Moss, C. & Dickens, R.J. 1977, MNRAS, 178, 701

Natarajan, P. & Kneib, J.-P. 1996, MNRAS, 283, 1031

Navarro, J.F., Frenk, C.S. & White, S.D.M. 1997, ApJ, 490, 493 (NFW)

Perea, J., del Olmo, A., & Moles, M. 1990, A&A, 237, 319

Pryor, C., & Geller, M.J. 1984, ApJ, 278, 457

Ramírez, A.C. & de Souza, R.E. 1998, ApJ, 496, 693

Reisenegger, A., Quintana, H., Carrasco, E.R., & Maze, J. 2000, AJ, 120, 523

Rines, K., Geller, M.J., Diaferio, A., Mohr, J.J., & Wegner, G.A. 2000, AJ, 120, 2338

Smith, S. 1936, ApJ, 83, 23

Sodré, L.J., Capelato, H.V., Steiner, J.E. & Mazure, A. 1989, AJ, 97, 1279

Solanes, J.M. & Salvador-Solé 1990, A&A, 234, 93

The, L.S. & White, S.D.M. 1986, AJ, 92, 1248

Tully, R. B. & Shaya, E. J. 1984, ApJ, 281, 31

van der Marel, R. 1994, MNRAS, 270, 271

van der Marel, R., Magorrian, J., Carlberg, R.G., Yee, H.K.C., & Ellingson, E. 2000, AJ, 119, 2038

Yee, H.K.C., Ellingson, E. & Carlberg, R.G. 1996, ApJS, 102, 269

Zwicky, F. 1933, Helv. Phys. Acta, 6, 110

Zwicky, F. 1937, ApJ, 86, 217

Tracing Cosmic Evolution with Galaxy Clusters
ASP Conference Series, Vol. 268, 2002
S. Borgani, M. Mezzetti, and R. Valdarnini, eds.

Segregation of cluster galaxies, and cluster substructure

Peter Katgert

Leiden Observatory, PO Box 9513, 2300 RA Leiden, The Netherlands

Andrea Biviano

INAF, OAT, via G.B. Tiepolo 11, I-34131 Trieste, Italy

Tom Thomas

Leiden Observatory, PO Box 9513, 2300 RA Leiden, The Netherlands

Abstract. We have studied the segregation properties of the galaxies in a sample of 59 rich, nearby clusters which were observed in the ESO Nearby Abell Cluster Survey. In this contribution we focus on the kinematics and spatial distribution of the different types of galaxies, and on the rôle of substructure on the properties and evolution of the galaxies.

1. Introduction

In the ESO Nearby Abell Cluster Survey (Katgert et al. 1996, Katgert et al. 1998; ENACS for short) redshifts and magnitudes were determined for more than 5600 galaxies in about 100 rich ($R_{ACO} \geq 1$) and nearby ($z < 0.1$) Abell cluster candidates. About 90% of the latter were confirmed as real clusters, as they correspond to compact structures in redshift space. Subsequently, galaxy types were obtained for a large fraction of the galaxies with redshifts, in two different ways. First, de Theije & Katgert (1999) used the ENACS spectra to derive a galaxy type from a Principal Component Analysis, in combination with an Artificial Neural Network. They calibrated their spectral galaxy types with several hundred galaxy morphologies from Dressler (1980), and they estimated that the success rate of their classification was about 75%.

Recently, Thomas & Katgert (2002) have determined the morphological types of about 2300 galaxies in the ENACS clusters, from CCD images obtained mostly with the 0.9-m Dutch telescope at La Silla. These morphological types were used to improve the calibration of the galaxy types derived from the spectra, and they were complemented with some morphological types from the literature. The morphological and spectral galaxy types could be successfully integrated into one system, even though the spectra do not allow the identification of ellipticals (the earliest spectral type is E/S0), and of early spirals (Sa–Sb). However, late spirals and emission-line galaxies (ELG) can be classified from the spectrum alone. As is described in detail in Thomas & Katgert (ibid.) this has yielded a sample of 4884 ENACS galaxies for which the galaxy type was determined with a coherent classification scheme.

2. Construction of ensemble clusters, and detection of substructure

For a study of the segregation properties of cluster galaxies of various types, the statistics in the individual clusters is not sufficient. Therefore, we have constructed ensemble clusters of ellipticals, S0's, early spirals (Sa–Sb), late spirals (Sbc–Irr) and emission-line galaxies by combining data for many clusters. Galaxies were included in the ensemble clusters only if $\geq 80\%$ of the galaxies in their parent cluster had been classified from imaging and/or spectrum. This condition defines a sample of 59 clusters with 3056 galaxies of which 2948 (=94%) have a type. In combining data from many clusters, we expressed all projected distances, R, in units of r_{200} (approximately the radius of the virialized core), and we scaled all relative line-of-sight velocities ($v_{rad,gal} - v_{rad,cluster}$), with the global velocity dispersion of all galaxies in the cluster.

The superior statistical weight of the ensemble clusters has its price: the ensemble clusters are much smoother than any of the individual clusters. Yet, substructure may be an important factor in the evolution of clusters and of the galaxies within them. Therefore, we analyzed the surroundings of each galaxy in its parent cluster for evidence of substructure. We quantified substructure by a slightly modified version of the parameter devised by Dressler and Schectmann (1988). This measures the difference of the average velocity of the nearest neighbours of a galaxy and the average velocity of the cluster, and how much the velocity dispersion of the neighbours is smaller than that of the whole cluster. By applying a threshold to the substructure parameter we select galaxies in groups that are either moving in and/or colder than their parent cluster. The size of the substructure groups that we select in this way increases with distance from the center, because the number of neighbours was not varied with position. However, the velocity dispersion of the selected groups appears to be independent of position.

3. Results of segregation analysis

We analysed the segregation properties of E, S0, S_{early}, S_{late}, all spirals and ELG, by comparing the 2-D distributions of R/r_{200} and $\Delta V/\sigma_V$ (Biviano et al. 2002). For all types we considered galaxies *in* and *outside* substructure separately. The comparison was done with a 2-D version of the Kolmogorov-Smirnov test (Peacock 1983; Fasano & Franceschini 1987). We judged two (R, v)-distributions to be different when the probability for them to be drawn from the same parent distribution was less than 5%.

We first analysed each of the ensemble clusters for evidence of luminosity segregation. Only for the ellipticals do we see such evidence: the (R, v)-distribution of the bright E's outside substructure, with $M_R < -22.0$ is significantly different from that of the fainter E's with $M_R \geq -22.0$. None of the other galaxy classes shows luminosity segregation; yet, as a precaution we ignored galaxies of all types with $M_R < -22.0$ in the other segregation tests.

We confirm the well-known morphological segregation of early- and late-type galaxies. In addition, for galaxies of *all* types (E, S0, Sp and ELG), those in and those outside substructure have different (R, v)-distributions, as do the early and late spirals outside substructure. The 2-D KS comparisons of galaxy classes

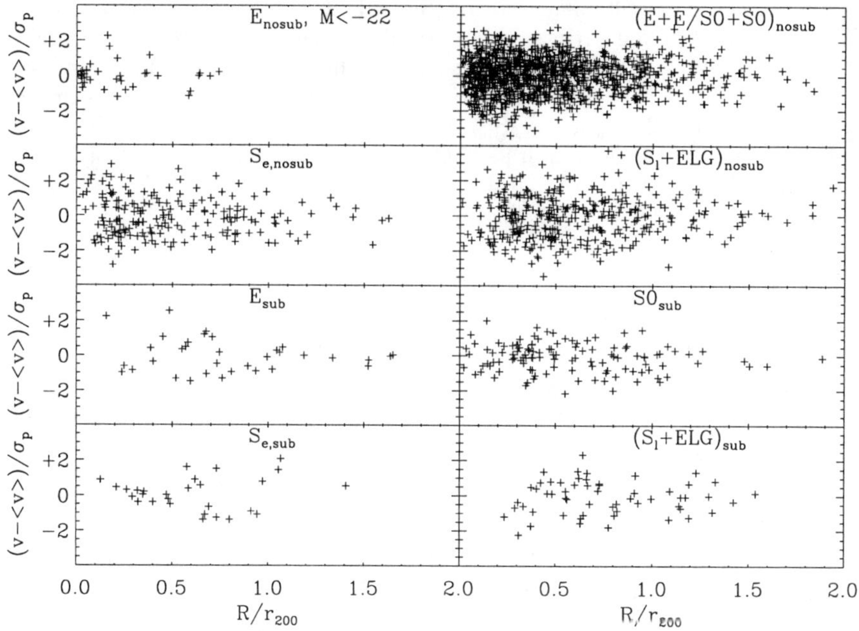

Figure 1. Distributions of normalized projected distance from the cluster centre, R/r_{200}, and normalized relative velocity, $\Delta V/\sigma_V$. All distributions of galaxies outside substructure (upper 4 panels), are different at more than 95% confidence level. For the galaxies in substructure (the lower 4 panels), this is true only for the S0's and [S_{late}+ELG].

outside substructure show that, in addition to the bright E's, we can and must distinguish three galaxy ensembles: viz. [E+S0], S_{early}, and [S_{late}+ELG]. The (R, v)-distributions of [E+S0] and [S_{late}+ELG] are different, and that of the S_{early}'s is different from both. Among the galaxies **in substructure**, we must distinguish at least 2 ensembles, because the (R, v)-distribution of the S0's is different from those of the S_{late}'s and ELG, both separately as well as combined. The data allow the other galaxies to be combined in several ways with these two classes that must be distinguished. In Fig. 1 we show the (R, v)-distributions for the 4 classes outside and in substructure.

The E's and S0's do not have different distributions, nor do S_{late}'s and ELG, and this is true both in and outside substructure. The early spirals outside substructure are a distinct class, the reason for which is not immediately clear.

4. The nature of the morphological segregations

In Fig. 2, we illustrate the character of the differences in the (R, v)-distributions of the three ensembles that must be distinguished outside substructure (except the brightest ellipticals. The [E+S0]'s appear to have the most centrally concentrated projected distribution, and the [S_{late}+ELG] the widest. For $R/r_{200} \lesssim 0.5$ the distribution of the early spirals is similar to that of the [E+S0]'s, while for $R/r_{200} \gtrsim 0.5$ they are distributed more like the [S_{late}+ELG]. In the outer regions (say, beyond 1.0 r_{200}) the velocity dispersions of the three classes are quite similar. Within 0.25 r_{200}, the [S_{late}+ELG] are hottest, but they are not very much hotter than the early spirals. Between 0.25 and 0.75 r_{200}, the early spirals and [E+S0]'s have very comparable velocity dispersions, and the [S_{late}+ELG] are hotter than the other two classes.

References

Biviano, A., Katgert, P., Thomas, T., et al. 2002, A&A(submitted)
de Theije, P., Katgert, P. 1999, A&A, 341, 371
Dressler, A, Schectman, S.A. 1988, AJ, 95, 985
Dressler, A. 1980, ApJS, 42, 565
Fasano, G., Franceschini, A. 1987, MNRAS, 225, 155
Katgert, P., Mazure, A., Perea, J., et al. 1996,A&A, 310, 8
Katgert, P., Mazure, A., den Hartog, R., et al. 1998 A&AS, 129, 399
Peacock, J. 1983, MNRAS, 202, 615
Thomas, T., Katgert, P. 2002, A&A(submitted)

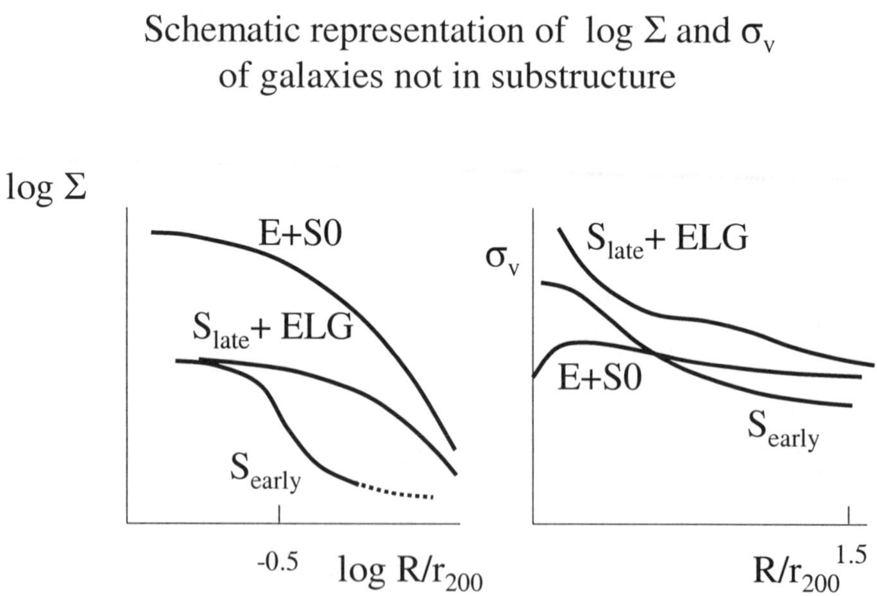

Figure 2. The nature of the morphological segregations. On the left: surface density, and on the right: normalized velocity dispersion.

Tracing Cosmic Evolution with Galaxy Clusters
ASP Conference Series, Vol. 268, 2002
S. Borgani, M. Mezzetti, and R. Valdarnini, eds.

Merging clusters in the Shapley Concentration

S.Bardelli[1], E.Zucca[1], T.Venturi[2], S.DeGrandi[3], S.Ettori[4], S.Molendi[5]

[1]*Osservatorio Astronomico di Bologna, via Ranzani 1, I-40127 Bologna*

[2]*Istituto di Radioastronomia del CNR, via Gobetti 101, I-40129 Bologna*

[3]*Osservatorio Astronomico di Brera, Via Bianchi 46, I-23807 Merate*

[4]*Institute of Astronomy, Madingley Road, CB3 0HA Cambridge*

[5]*Istituto di Fisica Cosmica "Occhialini", via Bassini 15, I-20133 Milano*

Abstract. We present the analysis of multiwavength data on merging clusters in the central part of the Shapley Concentration and we discuss a possible "evolutive" scenario for the cluster complexes.

Rich superclusters are the ideal environment for the detection of cluster mergings, because the peculiar velocities induced by the enhanced local density of the large-scale structure favour the cluster-cluster collisions, in the same way as seen in the simulations. The Shapley Concentration supercluster represents a unique laboratory where it is possible to follow cluster mergings at different stages and to test related astrophysical consequences, as the formation of radio halos, relics and wide angle tail radiosources and the presence of galaxies with enhanced star formation. In particular, its central part is fragmented in three main structures ("cluster complexes"): one is dominated by the cluster A3558 and could be considered the core of the supercluster (see Fig.1); a second is formed by A3528, A3530 and A3532. The third complex is dominated by A3571, and includes A3572 and A3575.
We present the results of an extensive, multiwavelength survey of the central part of the Shapley Concentration: in this contribution we focus on the analysis of radio and X-ray data; results on large-scale distribution and spectral properties of galaxies are presented in the contribution by Zucca, Bardelli & Baldi (these proceedings).

We performed a radio survey at 22cm and 13cm with the ATCA telescope to $\sim$ 1 mJy in the region of the A3558, A3528 and A3571 cluster complexes, obtaining a total of $\sim$ 570 sources (Venturi et al. 2000, 2001a, 2001b). Among these sources, $\sim$ 140 have optical counterpart and $\sim$ 50 have redshifts at the distance of the Shapley Concentration. As an example, the position of the radiosources in the A3558 complex is reported in Fig.2a, superimposed on the wedge diagram of the optical sample.
The bivariate radio-optical luminosity function has been calculated for all cluster complexes and compared with the luminosity function of normal clusters (Ledlow & Owen 1996): in the A3558 complex a significant lack of radiogalaxies is present (see Fig.2b), while in the A3528 and the A3571 complexes we found consistency with Ledlow & Owen.

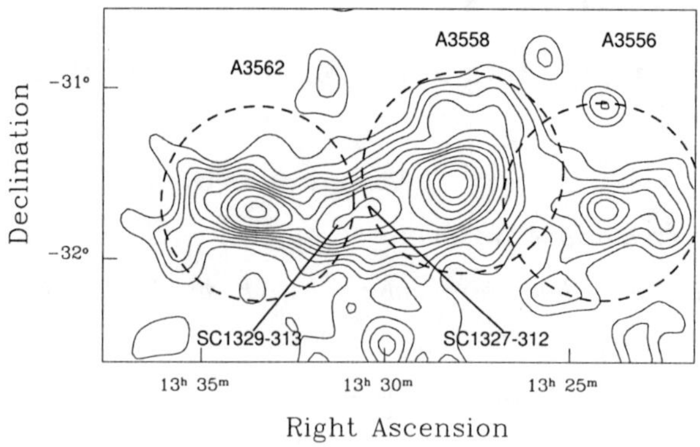

Figure 1. Isodensity contours of galaxies in the A3558 complex. For the three ACO clusters circles of one Abell radius have been drawn.

In the region of interacting clusters, a number of extended radiosources has been detected: the study of these sources is particularly important for deriving physical constraints on the interaction of electrons with the intracluster medium. In particular, Venturi et al. (1998), studying a relic radiosource at the periphery of the A3558 complex, individuated a possible shock front between the A3556 cluster and a smaller group.

Fundamental data on the physics of the intracluster medium during a merging come also from the X-ray band.
Bardelli et al. (1996) and Kull & Böhringer (1999) showed that the three clusters and the two poor groups of the A3558 complex are embedded in a hot gas filament. This fact points toward a major merging event.
A careful analysis of the distribution and the temperature of the intracluster medium in this region has been performed with the use of ROSAT and BeppoSAX data (Ettori et al. 2000, Bardelli et al. 2001b), studying in particular the two poor clusters SC 1327–312 and SC 1329–313 between A3558 and A3562. We do not find evidence of regions where the gas is shocked or significantly heated: the image of SC 1327–312 seems rather symmetric, while the gas profile of SC 1329–313 shows disturbed, "comet-like" shaped isophotes, with the tail pointing toward A3562 and a compression toward SC 1327–312 (see Fig.3).
The lack of heating supports the hypothesis that the merging is at a late stage, after the first core–core encounter, when the main shock front had the time to travel to the external regions of the main clusters.

Discussion

The wealth of information available from radio to optical (both photometry and spectroscopy), up to X-ray energies, suggest the following scenario:
• The A3528 cluster complex is at the very beginning of a merger event, where the two merging entities have just started "to feel each other" (Bardelli et al. 2001a). The gradients in the temperature distribution of the intracluster gas delineate the merging front (Schindler 1996). The radio luminosity function of elliptical galaxies is in good agreement with that of ellipticals not located in

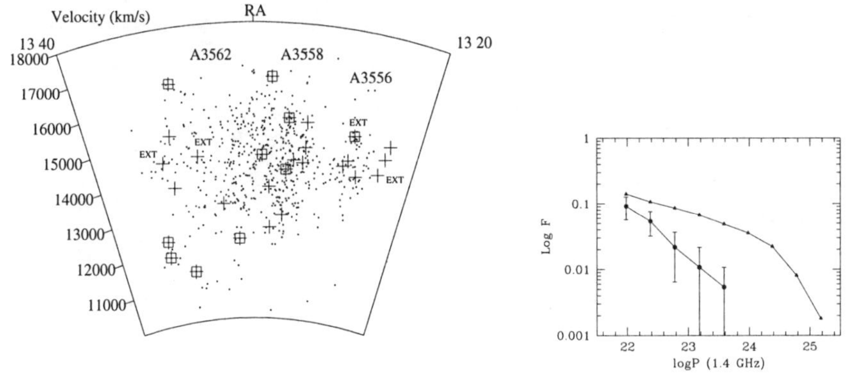

Figure 2. Left: Wedge diagram of $\sim$ 700 galaxies in the A3558 complex. Crosses indicate the position of radiosources with optical identification. Radiogalaxies with steep spectral index have been marked with a square and the extended sources have been highlighted. Right: Bivariate radio-optical luminosity function for the A3558 complex (circles with error bars): note the lack of radiosources with respect to normal clusters (triangles, from Ledlow & Owen).

merging clusters, and no sign of starburst emission, possibly induced by merger shocks, is detected (Venturi et al. 2001a). We suggest that the pre-merging stage had not yet had time to affect the radio emission properties of the cluster galaxies in the complex.

• The A3558 complex is thought to be an advanced merger, where two clusters of similar mass have already undergone the first core-core encounter. The amazing similarity between the galaxy distribution (Bardelli et al. 1998) and the gas density distribution (Ettori et al. 1997, Kull & Böhringer 1999) provides further evidence of the ongoing process. In the radio band it was found (Venturi et al. 2000) that this complex shows a significant deficit of radiogalaxies, suggesting that the major encounter switched off the nuclear radio emission and temporarily inhibited its formation. No radio excess of starburst origin was detected.

• The A3571 complex represents the final stage of a merger event (Venturi et al. 2001b), where A3571 itself is the resulting cluster after virialization of the merger. The distribution of the gas is already relaxed, as well as the galaxy distribution in A3571, while the outer edge of the galaxy distribution is still unrelaxed. The location of the radiogalaxies in A3571 suggests that they had time to develop a nuclear source after the active merging ceased.

The distribution of the galaxies and of the hot gas in these complexes is remarkably similar to the various stages of the cluster-cluster collision modelled by Ricker & Sarazin (2001). Their simulations for a frontal merger of two clusters with equal mass show an A3528-like situation for $t = 0$, which evolves into an A3558 scenario at $t \sim 5$ Gyr, to end up in the A3571 case after ~ 9 Gyr.
Updates about this project:
http://boas5.bo.astro.it/~bardelli/shapley/shapley_new.html

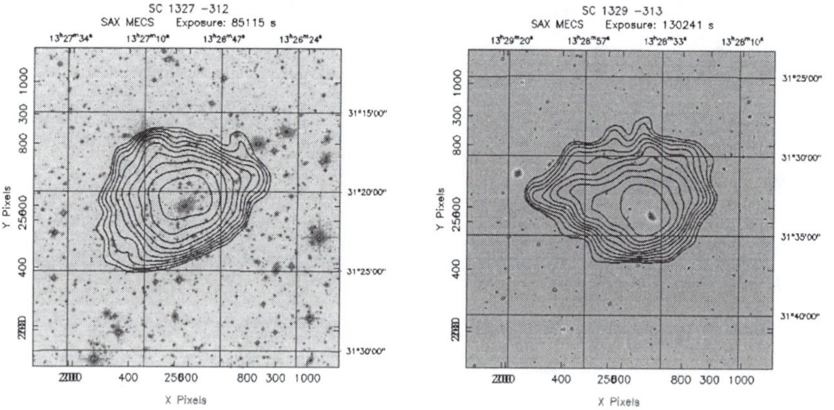

Figure 3. BeppoSAX MECS image of SC 1327–312 (left) and SC 1329–313 (right) in the [2–10] keV range. We show the central region of 8 arcmin radius, superimposed to the Digital Sky Survey.

References

Bardelli S., Zucca E., Malizia A., et al., 1996, A&A 305, 435

Bardelli S., Zucca E., Zamorani G., et al., 1998, MNRAS 296, 599

Bardelli S., Zucca E., Baldi A., 2001a, MNRAS 320, 387

Bardelli S., DeGrandi S., Ettori S., et al., 2001b, A&A submitted

Ettori S., Fabian A.C., White D.A., 1997, MNRAS 289, 787

Ettori S., Bardelli S., DeGrandi S., et al., 2000, MNRAS 318, 239

Kull A., Böhringer H., 1999, A&A 341, 23

Ledlow M.J., Owen F.N., 1996, AJ 112, 9

Venturi T., Bardelli S., Morganti R., Hunstead R.W., 1998, MNRAS 298, 1113

Venturi T., Bardelli S., Morganti R., Hunstead R.W., 2000, MNRAS 314, 594

Venturi T., Bardelli S., Zambelli G., et al., 2001a, MNRAS 324, 1131

Venturi T., Bardelli S., Zagaria M., et al., 2001b, A&A submitted

Ricker P.M., Sarazin C.L., 2001, ApJ in press (astro-ph/0107210)

Schindler S., 1996 MNRAS 280, 309

Tracing Cosmic Evolution with Galaxy Clusters
ASP Conference Series, Vol. 268, 2002
S. Borgani, M. Mezzetti, and R. Valdarnini, eds.

Mass–to–Light Ratio of Galaxy Systems

Marisa Girardi, Patrizia Manzato, Marino Mezzetti

Dip. di Astronomia, Univ. di Trieste, Via Tiepolo 11, Trieste, Italy;

Giuliano Giuricin

Deceased; formerly at Dip. di Astronomia, Univ. di Trieste;

Füsun Limboz

Instanbul Univ., Science Faculty, Astronomy and Space Science Dept.

Abstract.
We study the mass–to–light ratio of galaxy systems from poor groups to rich clusters. We extend a previous work based on 89 clusters obtaining blue luminosity and virial mass for other 52 clusters, 43 poor systems, and ~ 500 groups of two catalogs. A large effort is devoted to the homogeneity of the resulting values. Analyzing a combined sample of 294 systems we find that mass increases faster than the luminosity does: the linear fit gives $M \propto L_B^{1.34\pm0.03}$, and we find a tendency for a steeper increase in the low–mass range. We compare our results to the behavior of M/L_B vs. halo mass as predicted by cosmological numerical simulations coupled with semianalytic modeling of galaxy formation.

1. Introduction

The relative contribution of the dark matter component is usually specified in terms of the mass–to–light ratio, M/L, the total amount of mass relative to the total light within a given scale. Pioneering analyses showed that M/L increases from the bright luminous parts of galaxies to their faint halos to cluster scales, possibly saturating at the level of galaxy systems.

Analyzing 89 clusters with homogeneous mass and luminosity estimates, Girardi et al. 2000 (hereafter G00) found that mass has a slight, but significant tendency to increase faster than the luminosity does, $M \propto L_{B_j}^{1.2-1.3}$. Although this result agrees with those indirectly recovered by fundamental plane analyses, there is a general absence of direct evidence for correlation between M/L and cluster properties (cf. references in G00). Very recent support to G00 results comes from CNOC2 groups (Carlberg et al. 2001; Hoekstra et al. 2001). However, the question is still open: e.g., Hradecky et al. (2000) found that M/L is roughly independent of system mass.

To draw more definitive conclusions about this topic, we extend the work of G00. A Hubble constant of 100 h km s^{-1} Mpc^{-1} is used throughout.

2. Data Sample and Analysis

We consider the sample of nearby clusters of Girardi et al. (1998, hereafter G98) having homogeneous virial mass estimate: G00 computed B_j band luminosities using COSMOS catalog (Yentis et al. 1992) for 89 clusters; here we compute luminosities by using the APS Revised Catalog of POSS I (Pennington et al. 1993) for 52 clusters. We also consider 43 poor systems: the poor clusters by Ledlow, Loken, and Burns (1996), and the rich groups by Zabludoff & Mulchaey (1998), having magnitudes available in COSMOS and/or APS catalogs. Following G98 and G00, we compute mass and luminosity for all poor systems. In particular, we have homogenized photometry converting magnitudes from B_{APS} band to COSMOS B_j band.

Both mass and luminosity are computed within the virial radius R_{vir}.

We extend our analysis to the groups identified by Giuricin et al. (2000) in the Nearby Optical Galaxy (NOG) sample by means of both hierarchical and the percolation "friend–of–friend" methods: the final catalogs contain 475 and 513 groups, respectively (hereafter HG and PG). Due to the small number of member galaxies and the uncertain dynamical status we cannot apply the procedure used for clusters to compute virial masses. We rather use the procedure recently applied by Girardi & Giuricin (2000, hereafter GG00) taking into account the dynamical state of groups. Both mass and luminosity estimates refer to the whole sampled region which is gererally well larger than the virialized region (e.g., Zabludoff & Mulchaey 1998; GG00; Carlberg et al. 2001).

As for a reliable comparison with clusters and poor systems, we rescale both group mass and luminosity to regions within R_{vir}, applying a mean correction to all groups, following the procedure outlined by G00. We estimate that R_{vir} is about half of the sampled region, and the mass and luminosity are 65% of those determined in the whole sampled region. Moreover, when comparing results in B band with those in B_j band, we assume that $(L_{Bj}/L_{Bj,\odot}) = 1.1 \times (L_B/L_{B,\odot})$.

3. Results and Conclusions

We computed the values of the mass–to–light ratio for all systems: clusters (CL), poor systems (PS), percolation and hierarchical NOG groups of increasing richness (PG, PG5, PG7 and HG, HG5, HG7), cf. Figure 1. The general feeling is that M/L_B increases with system mass.

Then we analyze the $M - L$ relation. We combine data of clusters, poor systems, and 132 groups identified in both HG and PG catalogs and with at least five members so to avoid spurious groups. We obtain a fiducial combined sample of 294 systems (CL+PS+GROUP) where we fit, by using the unweighted bisecting procedure, $(M/M_\odot) \propto (L_B/L_\odot)^{1.338\pm0.033}$. Although the above straight line approach can be very useful to show that mass increases faster than luminosity does, it could be no longer adequate to describe $M - L$ relation in a so wide dynamical range, from poor groups to very rich clusters. We attempt also a quadratic fit for the two extreme situations, i.e. minimizing scatter on M–axis and minimizing scatter on L–axis, cf. Figure 2: both show a steeper slope in the low–mass range.

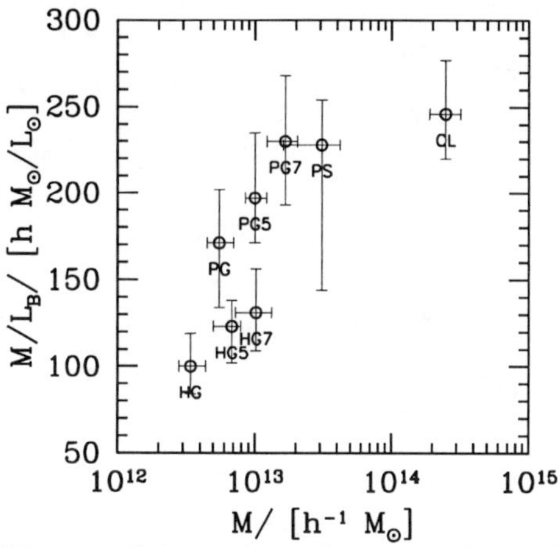

Figure 1. Mass–to–light ratio vs. mass: median values and 90% c.l.

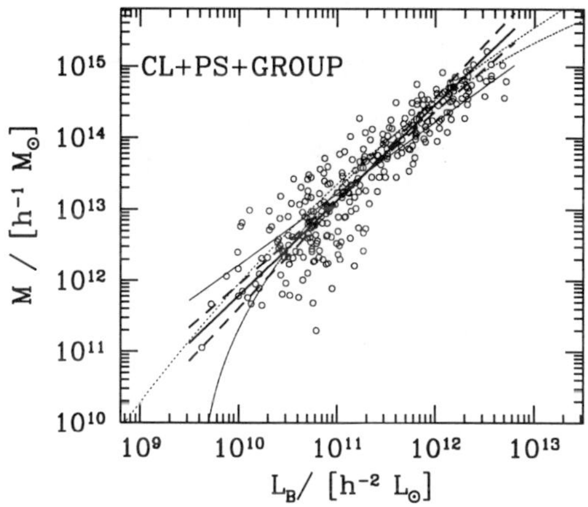

Figure 2. Mass vs. luminosity. Heavy lines represent the linear fits (solid line for the bisecting fit). The faint solid line is the $M \propto L_B$ relation. The two faint dotted lines represent the two quadratic fits.

Figure 3 compares our observational results to the theoretical predictions of Kauffmann et al. (1999) and Benson et al. (2000), obtained by the combination of cosmological N–body simulations and semianalytic modeling of galaxy formation in the framework of cold dark matter (CDM) models for two alternative cosmologies: a low–density model with $\Omega_m = 0.3$ and $\Omega_\Lambda = 0.7$ (ΛCDM), and a high–density model with $\Omega_m = 1$ and shape parameter $\Gamma = 0.21$ (τCDM).

The $\Omega_m = 1$ cosmology is clearly rejected from our observational data. However, the preferred value of Ω_m is not clear since the model with $\Omega_m = 0.3$ by Kauffmann et al. well fits the data, while the results of Benson et al. suggest a smaller value. Kauffmann et al.'s results well resemble the steepness of the increasing of M/L, too.

We suggest that our study will be useful in constraining future models of galaxy formation.

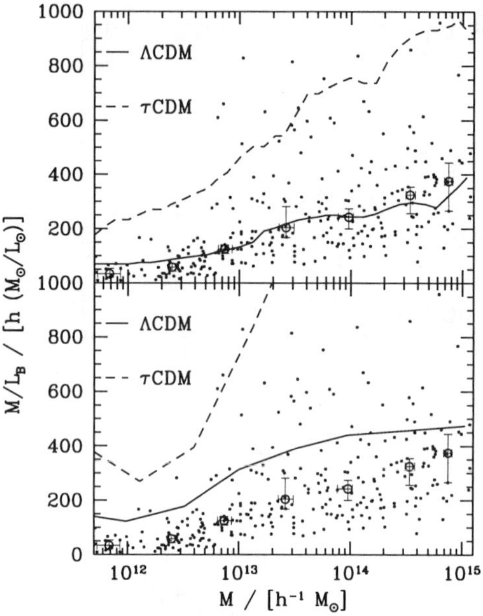

Figure 3. Comparison between our observations and the theoretical predictions for two cosmological models of Kauffmann et al. (with a closure value of $M/L_B[Univ] = 1350h\, M_\odot/L_\odot$; top panel) and Benson et al. (bottom panel). Solid points represent individual data for our combined sample, while open points show median values.

References

Benson, A. J., et al. 2000, MNRAS, 311, 793

Carlberg, R. G., et al. 2001, ApJ, 552, 427

Girardi, M., et al. 2000, ApJ, 530, 62 (G00)

Girardi, M., & Giuricin, G. 2000, ApJ, 540, 45 (GG00)

Girardi, M., et al. , ApJ, 505, 74 (G98)

Giuricin, G., Marinoni, C., Ceriani, L., & Pisani, A. 2000, ApJ, 543, 178

Hoekstra, H., Yee, H. K., & Gladders, M. D. 2001, ApJ, 558, L11

Hradecky, V. et al. 2000, ApJ, 543, 521

Kauffmann, G., et al. 1999, MNRAS, 303, 188

Ledlow, M. J., Loken, C., & Burns, J. O. 1996, AJ, 112, 388

Pennington, R. L., et al. 1993, PASP, 105, 521

Yentis, D., et al. 1992, in Digitised Optical Sky Survey, p. 67, eds. MacGillivray, H. T., & Thomson, E. B., Kluwer Acad. Publ., Dordrecht, Boston, MA.

Zabludoff, A., & Mulchaey, J. S. 1998, ApJ, 496, 39

Tracing Cosmic Evolution with Galaxy Clusters
ASP Conference Series, Vol. 268, 2002
S. Borgani, M. Mezzetti, and R. Valdarnini, eds.

Dark Matter in the center of galaxies and galaxy clusters: ruling out the CDM scenario?

E. D'Onghia[1,2], G. Chincarini[2,3], C. Firmani[2], D. Marchesini[1]

1-Universita' degli Studi di Milano, Via Celoria 16, 20133 Milan, Italy, 2-Osservatorio Astronomico di Brera, Via E. Bianchi 46, 23807 Merate, Italy, 3-Universita' degli Studi di Milano-Bicocca, Piazza dell'Ateneo Nuovo 1, 20126 Milan, Italy.

Abstract. This work is focused on the preliminary results of the observations of Hα rotation curves for some of the late type dwarf and LSB galaxies carried out at the TNG telescope. In light of the observational data and of the N-body simulations we have performed recently, we discuss some of the implications on the nature of the DM particles and the formation of the dark haloes.

1. Introduction

Cold Dark Matter (CDM) models provide a solid framework capable to explain most of the properties of the universe at large scales. In the last few years a very large effort has been spent in the investigation of the detailed structure of the virialized haloes within the CDM scenario, in which dark particles are assumed to be cold and collisionless. Using N-body simulations it was showed that virialized haloes are well described by an universal density profile, the Navarro, Frenk, & White model (1997; hereafter NFW). This profile diverges at the center showing a cuspy core: $\rho \propto 1/r$. Recent high-resolution N-body simulations show that as the numerical resolution is increased, the resulting dark density profile goes as $\rho \propto 1/r^{1.5}$ (Moore et al. 1999). On galactic scales the theory seem to be in conflict with the observations, since the rotation curves of LSB and dwarf galaxies rise too slowly with respect to the predictions of the CDM model. Indeed, the rotation curves of these galaxies call for a finite central density (soft core), in conflict with the cuspy cores predicted by the model. Note that LSB and dwarf galaxies are systems strongly dark matter dominated, so their rotation curves are good tracers of the underlying dark haloes gravitational potential and therefore good candidates to explore the innermost shape of the dark matter distribution. Recently, however, some authors have challenged the existence of soft cores in centers of galaxies measuring Hα rotation curves of late-type dwarf and LSB galaxies. Basically they find two results: HI rotation curves for most of the galaxies are affected by beam-smearing and a good agreement with concentration values predicted by the NFW model (van den Bosch and Swaters 2000).

2. Soft core observational evidence

We have carried out Hα rotation curves at the TNG of dwarf and LSB galaxies. The spatial resolution in the central regions is a factor 10 better than with respect the HI data published in literature for the same galaxies. In Fig.1 we reproduce the circular velocities for the galaxies UGC4325, UGC11861 and F571-8 (empty circles are our Hα observations and filled circles are HI data). From a comparison between optical and radio data it is interesting to note that HI observations for these galaxies are not affected (or are very little) by beam smearing (see Marchesini et al. at this conference, for details). To discriminate between the NFW and King models as representative profile of the primeval dark halo, we built a fiducial galaxy. The dotted lines in the top panels represent a NFW profile while the dotted lines in the bottom panels are due to a King profile. Accounting for the disc formation (dashed lines) we computed next the adiabatic contraction of the DM component due to the cooling of the baryons into the virialized haloes. The final rotation curve is then obtained (solid lines) and compared with the Hα data. The final curve is best fitted for all galaxies assuming that the dark matter in the primeval halo is distribuited as a King profile. On galaxy cluster scale, there are at least two clusters with evidence of soft cores: CL0024+1654 as inferred by strong lensing technique (Tyson et al. 1998) and A1705 as revealed by the Chandra data (Ettori et al. at this conference). The central densities of the galaxies analyzed here or taken from the literature and of CL0024+1654 seem to confirm the evidence that the central density is independent on the halo mass (Firmani et al. 2000). Future observations will increase the sample and put these observational findings on a more robust basis.

3. Implications on the DM nature

The question is: how can we justify the existence of soft cores in galaxies and in galaxy clusters? Warm dark matter (WDM) has been proposed in order to solve the conflict with CDM models. However, even if interesting, this scenario is unable to produce soft cores at galactic scales (see e.g. Avila-Reese et al. 2001). A possible solution is to assume a weakly self-interacting cold dark matter (SIDM) (Spergel & Steinhardt 2000). It is well known that the NFW profile is far from thermal equilibrium, and shows, in fact, a inner positive temperture gradient. Collisions between dark particles would modify the velocity distribution to a Maxwellian distribution generating a constant velocity dispersion. We have developed different approaches to explore the SIDM regime: thermodynamic (Firmani et al. 2000) and dynamic models building a cosmological Boltzmann code (Firmani, D'Onghia, & Chincarini 2000, astro-ph/0010497). Here, we propose a new numerical procedure applied to N-body simulations (D'Onghia, Firmani, & Chincarini 2001, in preparation). Our technique is capable to simulate a single halo of any size and mass in a cosmological framework coupling semi-numerical technique and N-body simulations. The cosmological initial conditions are fixed by the mass aggregation history of the halo using semi-analytical models and only the dynamical evolution of the halo is followed by the N-body code. To achieve this, the HYDRA code has been ad hoc modified and our version can describe simulations both of collisional and collisionless dark haloes. In our simulations we assume a cross section inversely proportional to the particle velocity

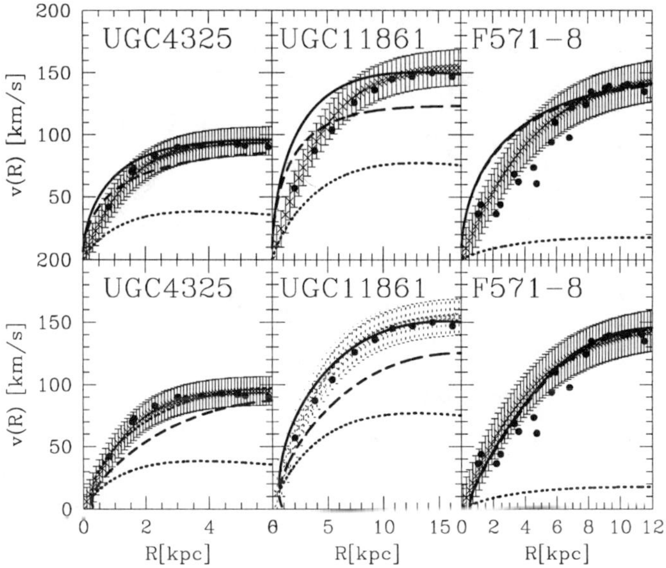

Figure 1. Rotation curves of dwarf galaxies and LSB galaxies. *Top panels*: our Hα observations with error bars (skeletal), HI data (filled points). An initial NFW profile is assumed for the halo (dashed line), the disc (dotted line) and the final rotation curve to be compared with data (solid line).*Bottom panels*: as above, however, an initial King profile is assumed for the dark halo (dashed line).

dispersion: $\sigma/m_x v_{100} = 10^{-24}$ cm^2/Gev with v_{100} the halo velocity dispersion in units of 100 km/s. The top panels of Fig. 2 show the dark density profiles of haloes of $M = 10^{11}\ M_\odot$ (left panel) and $10^{15}\ M_\odot$ (right panel) as obtained in a SIDM model. Each halo have been modeled assuming the cross section value above (in the figure the corresponding density profiles have empty circles) and a cross section three times larger (filled circles). In the same plot solid lines are the NFW profiles. Within a Hubble time a modest self-interaction cross section value creates soft cores in haloes of any size and a three times stronger cross section value produces larger soft cores. Bottom panels of Fig. 2 show the corresponding halo radial dispersion velocity profiles (symbols have the same meanings as above). This is the case in which any trend towards the core collapse is avoided by the competition between a mass accretion determined by the halo merging history and a thermalization process by collisions.

4. Conclusions

Hα rotation curves carried out at TNG of two late-type dwarf galaxies and one LSB galaxy show evidence of soft cores as the mass distribution of CL0024+1654 on galaxy cluster scales. Integrating the information from galaxies to clusters

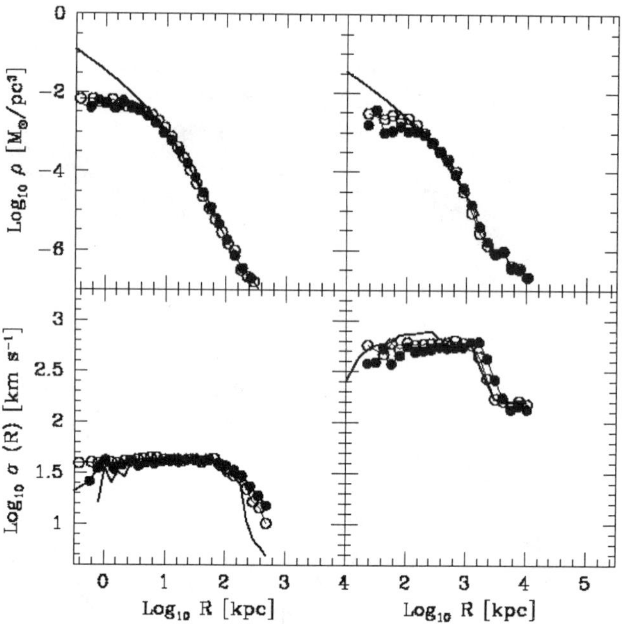

Figure 2. *Top panels.* Dark density profiles in a SIDM scenario as derived by N-body simulations. *Bottom panels.* The corresponding radial dispersion velocity profiles (see the text for details).

of galaxies the halo central density seems to be independent of the halo mass. WDM in unable to produce visible soft cores, but SIDM in a *very weak* self-interacting regime ($\sigma/m_x v_{100} = 10^{-24}$ cm^2/Gev) produces soft cores in agreement with the available observations on different scales.

5. Acknowledgments

E.D. and D.M. are grateful to Richard Bower and Oleg Gnedin for stimulating discussions about this topic during the congress.

References

Avila-Reese V.,Colin P.,Valenzuela O.,D'Onghia E.,Firmani C., 2001, (astro-ph/0010525)

Firmani C.,D'Onghia E.,Avila-Reese V.,Chincarini G., Hernández X., 2000, MNRAS, 315, L29

Moore B., Quinn T., Governato F., Stadel J., Lake G., 1999, MNRAS, 310, 1147

Navarro J., Frenk C.S., White S.D.M., 1997, ApJ, 490, 493

Spergel D., Steinhardt P., 2000, PhRvL, 84, 3760

Tyson J.A., Kochanski G.P., Dell'Antonio I.P., 1998, ApJ, 498, L107

van den Bosch F.C., Swaters R.A., 2001, MNRAS, 325, 1017

ICM - OBSERVATIONS

Tracing Cosmic Evolution with Galaxy Clusters
ASP Conference Series, Vol. 268, 2002
S. Borgani, M. Mezzetti, and R. Valdarnini, eds.

X–ray observations of the ICM Physics

Monique ARNAUD

CEA/DSM/DAPNIA/SAp Saclay, 91191 Gif-sur-Yvette, France

Abstract. X-ray observations of clusters reveal the complexity of the Intra-Cluster Medium physics. The dynamical and thermodynamical state of the ICM is determined by gravitational processes (like shock heating during mergers or matter infall) but also by additional physics, such as non-gravitational heating, radiative cooling and central AGN activity. I summarize our current knowledge, based on the observation of nearby clusters.

1. Introduction

The first X-ray images of clusters of galaxies obtained by the Einstein satellite clearly showed that the space between the galaxies is filled with hot X-ray emitting gas (e.g Jones & Forman, 1984). This Intra-cluster medium (ICM) is the main baryonic cluster component: its mass can reach 5 times the galaxy total mass in rich clusters. The simplest model for the ICM is that of an ideal gas evolving in the gravitational potential of the dark matter (which constitutes $\sim 75\%$ of the cluster mass). Although the ICM evolution is first driven by gravity (as discussed in Section 2), purely gravitational models fails to account for all the observed cluster properties. Additional physics have to be taken into account to fully understand cluster evolution, as we will discuss in Section 3.

2. Gravitational Processes

2.1. Cluster formation history

Because substructures in the ICM are erased on a typical time scale of a few Gyr, the detection of significant substructure is the signature of recent dynamical evolution. The variety of morphologies of clusters, as revealed unambiguously by the Einstein observations (e.g. Jones & Forman, 1992, 1999), and the direct observation of merger events with ROSAT (e.g. Briel et al. 1991) indicate that clusters are still forming at the present epoch, as expected in hierarchical formation models, such as the Cold Dark Matter model.

In such scenario, bound objects arise from the collapse of initial density fluctuations that grow under the influence of gravity, following recurrent merging events. The Universe get more and more structured with time, with voids, filaments and knots constituting the present cosmic 'web'. Clusters are thought to form preferentially at the crossing of filaments and to grow by matter infall/sub cluster mergers along these filaments. The Coma cluster belongs to a

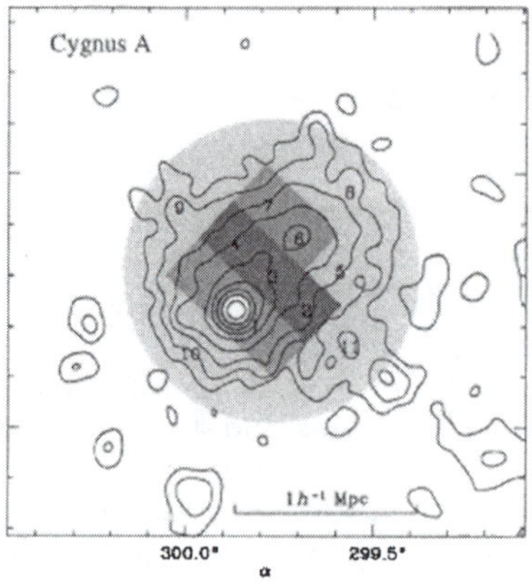

Figure 1. ASCA Temperature map of Cyg-A overlaid on the ROSAT surface brightness contours (from Markevitch et al. 1999).

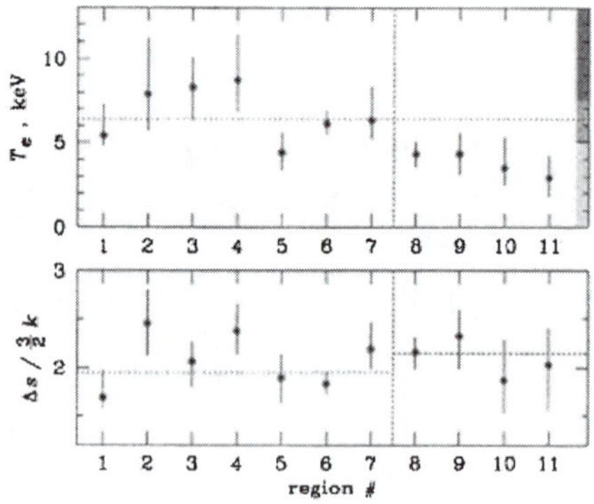

Figure 2. Temperature and specific entropies with 90% errors of the regions indicated in Fig. 1.

large scale filamentary structure that defines the Coma-A1367 supercluster. The analysis of the XMM Mosaic observation (Briel et al. 2001; Arnaud et al. 2001b; Neumann et al. 2001) demonstrates that Coma is currently accreting subclusters along this preferred direction, giving direct support to this standard formation scenario (see Briel et al. 2001b these proceedings).

2.2. Physics of merger events

Merger events are central to theories of cluster formation. When the subclusters collide, shocks are driven in the ICM, the relative motion in mergers being moderately supersonic ($v \sim 2000$ kms/s). Those shocks are thought to be a major source of heating of the ICM and should also accelerate non-thermal relativistic particles (see the detailed reviews of Sarazin 2001 and Buote 2001 on the physics and observation of merger events).

The thermal gas The gas dynamics during merger events has been studied using numerical simulations of cluster collisions (e.g. Evrard 1990; Schindler & Müller 1993; Roettiger et al. 1997; Ricker & Sarazin 2001). Shocks and adiabatic compression result in complex density and temperature maps, the details of which depend in particular on the stage of the merger, the geometry of the collision and the mass ratio of the merging components.

Quantitative analysis of the ICM morphology show that substructures and mergers are common in nearby clusters (e.g. Mohr et al. 1995; Buote & Tsai 1997). A variety of density and temperature features, in qualitative agreement with the prediction of numerical simulation of merger events, have been observed with Rosat and ASCA or BepposSax (see Buote (2001) and references therein). With XMM we are now able to map the temperature structure of the intra-cluster medium with unprecedented accuracy. For instance the observation of A3921 (Sauvageot et al. 2001), a clear case of on going merger, with two subunits clearly separated in the ROSAT image, revealed a filamentary hot structure between the two merging components similar to those observed in numerical

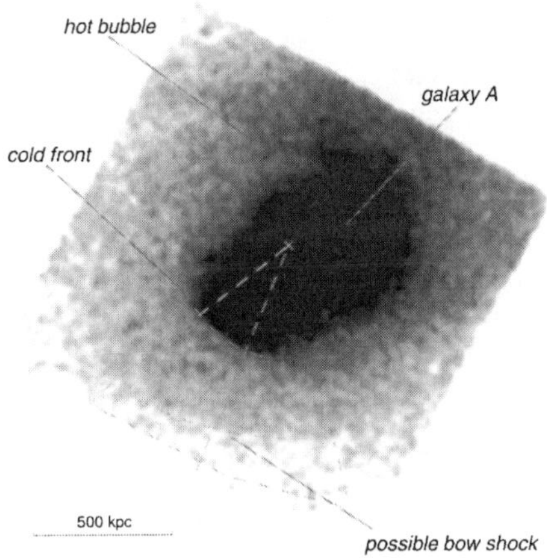

Figure 3. Chandra image of A3667 showing the sharp surface brightness discontinuity ("cold front") in the southwest from Vikhlinin et al. (2001b).

simulations. A direct evidence that shocks induced by mergers do contribute to the ICM heating was provided by the ASCA observation of Cygn-A (Markevitch et al. 1999, Fig 1). From the observed temperature variation in the interaction region between the two colliding subclusters, a collision velocity of ~ 2200 kms/s was derived, close to the free fall velocity. Interestingly the specific entropy map show variations $\sim 3/2$k, indicating a merger energy input similar to the the thermal energy (Fig. 2).

However the physics of mergers events is far from being fully understood. One of the most spectacular discoveries with Chandra is the unexpected presence of a "Cold front" in some merger clusters (Markevitch et al. 2000; Vikhlinin et al. 2001a,b; Mazzotta et al. 2001). For instance the sharp edge in the image of A3667 (Fig. 3) is not a shock as previously suggested, but the boundary of a cool gas cloud traversing the hot ICM. This discovery has important implication for the survival of subclusters during mergers and also indicates that heat conduction is strongly suppressed (Vikhlinin et al. 2001b; Ettori & Fabian 2000).

Statistical studies of the cluster morphology can provide important test of cosmological models of structure formation. In particular, the mass spectrum of subclusters is related to the power spectrum of density fluctuation ($P(k)$) and the present day cluster growth, and thus the dynamical state of clusters, depends strongly on Ω. In practice, attempts to compare the prediction of numerical simulations of structure formation to quantitative statistical studies of nearby cluster morphology have led to contradictory results (Buote, 2001 for a review). This requires a better understanding of the physics of merger events (in particular the ICM relaxation timescale, Nakamura (1995)) and their effect on observable cluster properties.

Particle acceleration and non-thermal emission The ICM is not only filled with hot thermal gas. The presence of large scale magnetic field and relativistic electrons is revealed by radio observations of diffuse synchrotron emission: regular, centrally located, radio halos and/or irregular, peripherally located, radio relics (e.g Clarke & Ensslin 2001, Fig. 4; see Feretti 2000 and Sarazin 2001 for a review). Further evidence is provided by the detection of non thermal hard X–ray emission, interpreted as to be due to inverse compton scattering of the

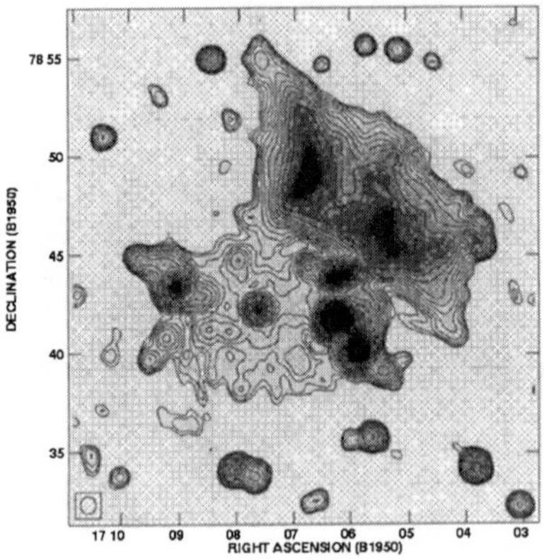

Figure 4. VLA 1368 MHZ radio map of A2256. Both the peripheral radio relics and central radio halo emission are clearly visible (from Clarke & Ensslin, 2001)

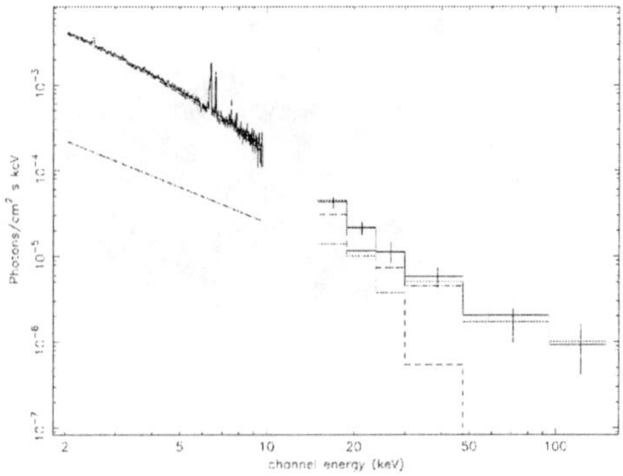

Figure 5. Beppo-SAX/MECS spectrum of A2256. The dashed line is the thermal component. The dotted line is the power-law non thermal component (from Fusco-Femiano et al. 2000).

cosmic microwave background by the relativistic electrons (e.g. Fusco-Femiano et al. 2000, Fig. 5).

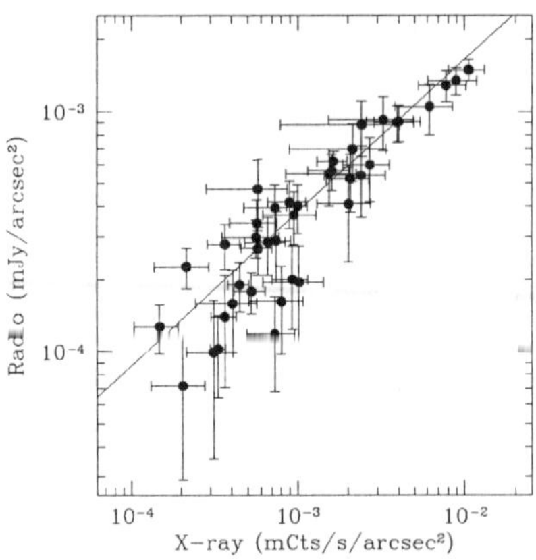

Figure 6. Radio brightness versus X-ray brightness for A2163 in cells of 90" size from Feretti et al. (2001).

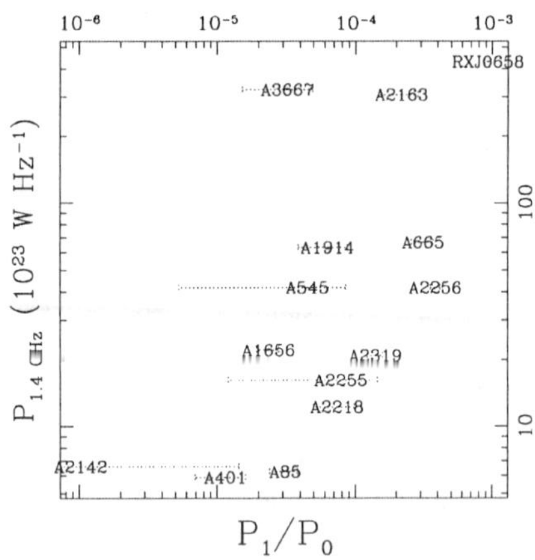

Figure 7. Radio Power (radio halo and radio relics) versus dipole ratio for a sample of clusters from Buote (2001b).

Although the origin of radio halos is not well understood, significant progresses has been made recently. The observed point-to-point correlation between the radio and X-ray surface brightness indicates a direct link between the relativistic and thermal plasma (Govoni et al. 2001; Feretti et al. 2001, Fig. 6). It was known from some time that radio halos are associated with non relaxed clusters, as expected if relativistic electrons are (re)-accelerated by merger shocks. However not all merging clusters present radio emission. Recently Buote (2001b)

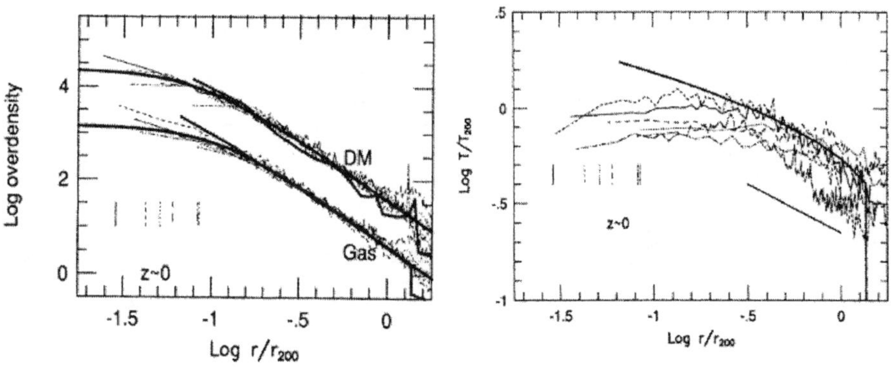

Figure 8. Universal profiles obtained from numerical simulations of the formation and evolution of galaxy clusters in the CDM cosmogony (from Navarro, Frenk & White 1995). The radius is scaled to the virial radius, corresponding to an overdensity of 200 as compared to the mean density of the Universe in which the cluster is embedded. **Left:** Scaled cluster density profiles. The density are normalized to the mean Universe density. **Right:** Scaled temperature profiles. The temperature are normalized to the virial temperature for each cluster.

found a correlation between the dipole power ratio (a measure of the departure from the virialised state) and the power of the radio halo (Fig. 7). Combined with the correlation found previously with the temperature (e.g. Liang et al. 2000) this indicates that radio halos form preferentially in clusters experiencing violent mergers, but that the mass of the system is also a fundamental parameter. This clearly supports the standard picture of merger shock acceleration and can explain the relatively rare occurrence of the radio halos (the energy available, which depends on the mass of the system, must be sufficient to accelerate the particles). An even more direct evidence that acceleration of electrons occurs at the site of shock heating come from the recent Chandra observation of A665 and A2163 (Markevitch & Vikhlinin 2001). A comparison of the radio and temperature maps indicates that radio emission comes preferentially from the hottest regions.

2.3. The Hydrostatic Equilibrium

Between major merger events, the bulk of the ICM is reasonably close to hydrostatic equilibrium (HE) in the gravitational potential. The total mass profiles and mass content of clusters can then be derived from X-ray observations, using the HE equation. This equation relates the gas density and temperature profiles, $\rho_g(r)$ and $T(r)$, to the total mass $M(r)$ enclosed in radius r:

$$\frac{1}{\rho_g(r)}\frac{dP}{dr} = -\frac{GM(r)}{r^2} \quad \text{where} \quad P(r) = \frac{k}{\mu m_p}\rho_g(r)T(r) \text{ is the gas pressure} \qquad (1)$$

In the simplest self-similar model of gravitational collapse (e.g Bertschinger 1985), supported by numerical simulations (e.g. Navarro et al. 1995, 1997; Bryan

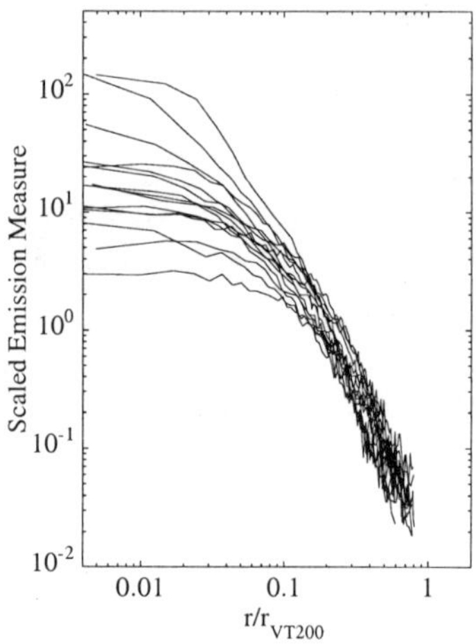

Figure 9. Scaled emission measure profile of a sample of hot clusters (k$T > 4$ keV) from Neumann & Arnaud (1999) using ROSAT data. Beyond $r/r_{200} \sim 0.1$ the profiles look remarkably similar.

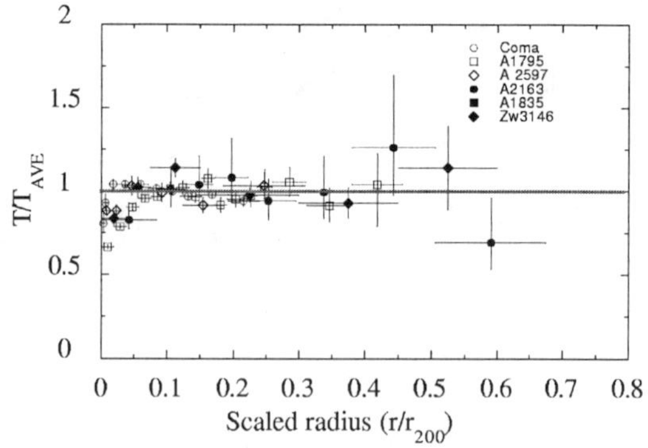

Figure 10. Compilation of scaled kT profiles observed by XMM. The data on Zw3146 and A2597 were kindly provided by R.Mushotzky (private comm.); The data on Coma, A1795, A2163 and A1835 are from Arnaud et al. 2001a, 2001b, Pratt et al. 2001 and Majerowicz & Neumann 2001.

& Norman 1998; Eke et al. 1998) and analytical hierarchical models (e.g Cavaliere et al. 1999) of structure formation, galaxy clusters form a homologous population. The virialised part of a cluster, present at a given redshift, corresponds to a fixed overdensity as compared to the mean density of the Universe at that redshift and the internal shape of clusters of different masses and z are similar. Consequently simple scaling laws relate each physical property to the cluster total mass (or temperature) and redshift. Self similarity applies to both the dark matter component and the hot intra-cluster medium (density and temperature distribution, Fig. 8). How the structural and scaling properties of the observed cluster population compare to these predictions gives us insight into the formation of clusters and the internal physics which governs their evolution.

The gas distribution of clusters has been mapped with unprecedented precision with ROSAT up to large radii. The gas density profiles of *hot* clusters ($T >$ 4keV) do show regularity (Neumann & Arnaud 1999; Vikhlinin, Forman & Jones 1999). The shapes of various clusters, once the radius is scaled to the virial radius, look remarkably similar outside the cooling flow region, supporting the existence of an universal underlying dark matter profile (Fig. 9).

On the other hand, the shape of the temperature profiles in clusters, as measured by ASCA and BeppoSax, was a matter of debate (e.g. Markevitch et al. 1998; White 2000; Irwin & Bregmann 2000; Molendi et al. 2000; Ettori et al. 2000). It was unclear if cluster atmospheres are essentially isothermal or if the temperature decreases with radius (as expected from numerical simulations of cluster formation). As a result total mass estimates are typically uncertain by a factor of two and no real constrains on the shape of the dark matter profiles could be inferred (Neumann & Arnaud 1999; Horner et al. 1999). The

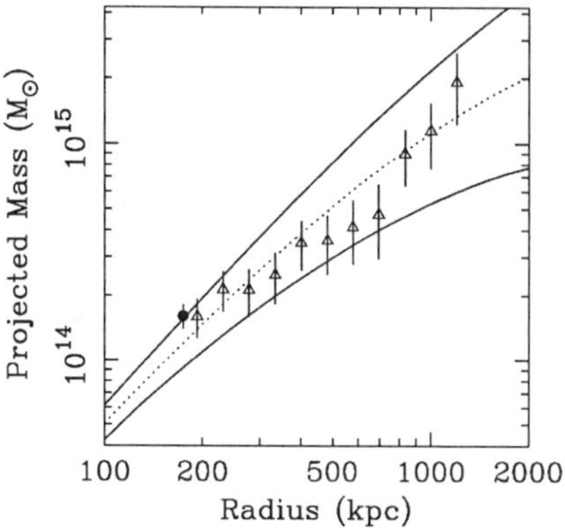

Figure 11. Total mass profile of A2390 (90% confidence range derived from Chandra data defined by the full lines) compared to strong lensing data (filled circle) and weak lensing data (triangles). Fig. from Allen et al. (2001).

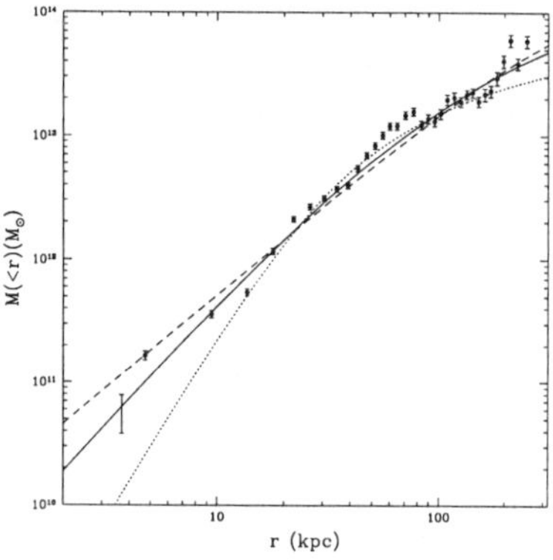

Figure 12. Central mass profile of Hydra-A (data points from Chandra observations) compared to NFW profile (full line) and Moore et al. (1999) profile (dashed line) from David et al. (2001).

large PSF of ASCA (which was energy dependent) and BeppoSAX was a major source of systematic uncertainty, the derivation of the profiles being potentially sensitive to the exact correction for the PSF and the detailed modelling of the non-resolved cooling flow component. XMM-Newton/EPIC has a much better sensitivity and a much smaller PSF than ASCA and SAX, which is essential to decrease statistical and systematical errors. Figure 10 is a compilation of scaled temperature profiles recently observed by XMM (references in the Figure). The profiles are clearly isothermal (within typically 10%) up to half the virial radius. A slight decline of the profile beyond that radius is allowed by the data. However only Zw3146 and A2163 are mapped above $0.5r_{200}$ (in only one annulus). Clearly more data are needed for any definitive statement.

The excellent spatial resolution of Chandra allows a precise estimate of the mass profile of intermediate redshift clusters, which can thus be compared with independent lensing mass estimate. A good agreement was observed for A2390, supporting the reliability of the X-ray method for the mass estimate (Allen, Ettori & Fabian, 2001; Fig. 11). The very central part of nearby clusters can now be resolved and a very peaked central mass distribution was measured in Hydra A (David et al. 2001; Fig. 12) as expected from numerical simulations of cluster formation (e.g. Navarro, Frenk & White 1995), although with a significantly higher concentration parameter.

3. Non–Gravitational Processes

3.1. Breaking of self-similarity

Although we do see regularity on the cluster population - strong correlations between the various physical quantities and similar internal shape of hot clusters - it is now well established that the simplest self-similar model of cluster forma-

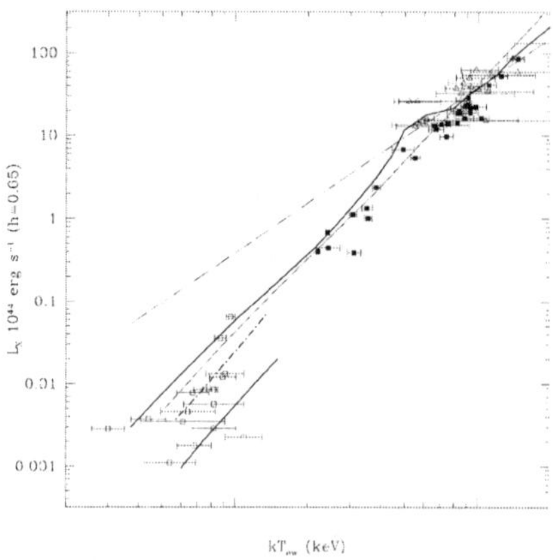

Figure 13. Bolometric luminosity versus temperature (from Tozzi & Norman 2001). Data points from Arnaud & Evrard (1999); Allen & Fabian (1998); Ponman et al. (1996). Dashed line: self -similar model. Full line: model with pre-heating.

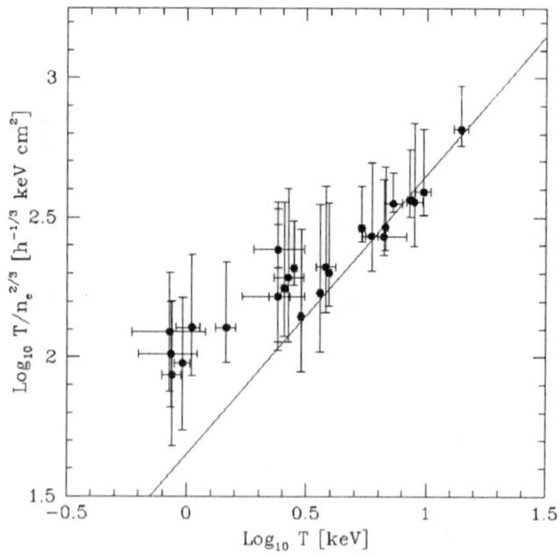

Figure 14. Gas entropy at 0.1 virial radius versus cluster temperature from Ponman et al. (1999) (Rosat and Ginga observations). The solid line is the similarity relation.

tion, purely based on gravitation, fails to explain all the observed properties of the nearby cluster population.

First, nearby clusters deviate from the expected scaling laws. The L_X–T relation is steeper than predicted (Fig. 13), as well as the M_{gas}–T (e.g. Neumann & Arnaud 2001; Mohr et al. 1999) and S_I–T relations (e.g Mohr et al. 2000). Moreover, an entropy excess (the "entropy floor", Fig. 14) was detected in the core of cool systems (Ponman et al. 1999; Lloyd-Davies et al. 2000), a definitive evidence of breaking of self-similarity.

The physical origin of such departures from the simplest model is still a mystery. Non-gravitational pre-heating has been invoked (e.g. Evrard & Henry 1991; Tozzi & Norman 2001; Bialek et al. 2001), although there is no consensus on the astrophysical source responsible for it: early galactic winds (e.g. Loewenstein 2000; Brighenti & Mathews, 2001), AGNs (Valageas & Silk, 1999; Wu et al. 2000, but see also Yamada & Fujita 2001) or possibly both (e.g. Bower et al. 2001). Furthermore, other effects could also play a role, like radiative cooling (Muawong et al. 2000) or variation of the galaxy formation efficiency with system mass (Bryan, 2000) .

The thermodynamical history of the ICM cannot be studied independently from its chemical evolution. The metallicity (Fig. 15) provides independent constraints on galaxy feedback. ICM enrichment through SN driven early winds seems at present the most likely scenario (e.g. Renzini 1997). From the Si abundance, Finoguenov et al. (2001) derived that about 1.6 keV per particle is provided from SN responsible for the ICM pollution (Fig. 16). Early galactic winds are thus definitively a potentially important source of heating of the ICM (see also ref. quoted above). It is unclear, however, how much of this energy goes into the ICM, without being radiated away.

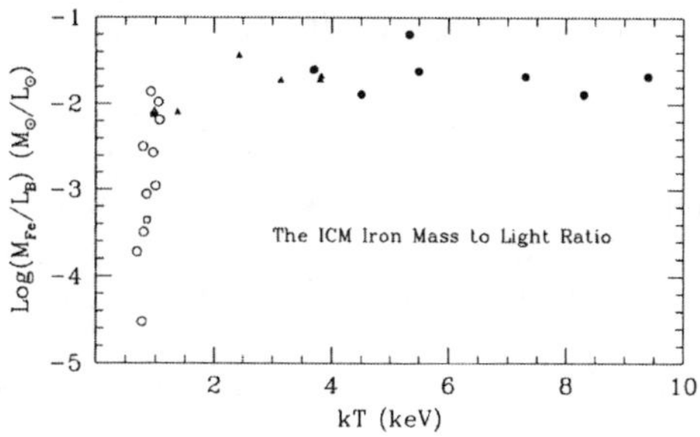

Figure 15. Iron mass over luminosity ratio from Renzini (1997) with measurements from Rosat, Einstein, Exosat and Ginga.

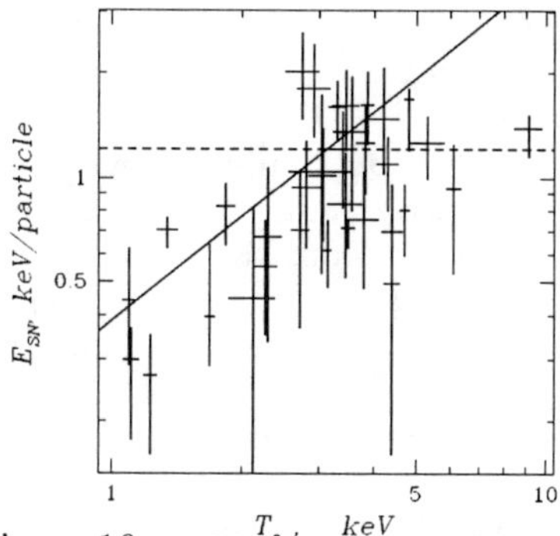

Figure 16. Thermal energy supplied by SNe associated with measured Si Abundances in a sample of nearby clusters observed with ASCA (from Finoguenov et al. 2001).

3.2. The physics in the central core: Cooling and AGN/ICM interaction

The physics in the core of clusters is particularly complex. The cooling time can be less than the Hubble time and radiative cooling has to be taken into account, as well as the impact of the nuclear activity of the central galaxy. This topic is discussed in detail by A.Fabian (these proceedings) and is only briefly reviewed here.

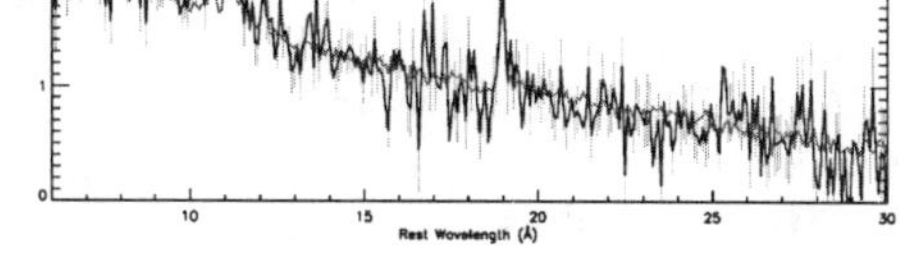

Figure 17. XMM/RGS spectrum of A1835 (Peterson et al. 2001). Red curve: isothermal model with kT = 8.2 keV. Blue line: Model with a hot ambient component plus an isobaric cooling flow component. The predicted strong FeXVII and OVII lines are not observed. Green line: same as blue model without emission below 2.7 keV.

A major discovery with XMM, concerning the physics of gas cooling, is the lack of emission lines from gas below $\sim$ 2 keV in the RGS spectra of A1835 (Peterson et al. 2001; Fig. 17) and A1795 (Tamura et al. 2001). This appears to be a generic problem, further seen in RGS observations of other CF clusters (Peterson and RGS collaboration, private communication) and in the XMM/EPIC observation of M87 (Matsushita et al. 2001). These data, inconsistent with standard cooling flow models, represent a serious theoretical challenge (Fabian et al. 2001a, see also David et al. 2001).

Arcsec spatial imaging with Chandra and high resolution spectroscopy with XMM are providing a new vision of the core of clusters. Not only Chandra (Harris et al. 2000; McNamara et al. 2000; Fabian et al. 2000,2001b; Finoguenov & Jones 2001; Blanton et al. 2001), but also XMM (Böhringer et al. 2001; Belsole et al. 2001) are revealing complex interaction between radio emitting bubbles and the hot ICM.

The interaction of the thermal and radio emitting plasma was studied in detail in the central region of the Virgo cluster, around the giant elliptical galaxy M87 (Belsole et al. 2001). XMM/EPIC establishes unambiguously the thermal nature of the X-ray emission from the eastern and south-western extensions previously seen by ROSAT around M87, which are related to the radio lobes. The extensions have a significantly lower temperature than the surrounding ambient medium. Buoyant radio emitting bubbles powered by the central AGN and dragging cooler material from the center, as proposed by Churazov et al. (2001), seems to qualitatively explain the XMM results but fails to describe the misalignment between X-ray and radio lobes.

4. Conclusion

Galaxy clusters occupy a unique position in the hierarchical scenario of structure formation, as they are still forming today. X-ray observations show that clusters form a relatively regular population, with general properties consistent with the standard hierarchical scenario, but also reveal the complexity of the detailed internal ICM physics. In particular, the physics of merger events, radiative cooling and non-gravitational heating are still poorly understood.

Because the statistical properties of the cluster population (the evolution of the mass function, the frequence of substructure..) depends on the cosmological model, clusters can in principle be used as precise cosmological probes. In practice this requires a good understanding of the internal physics that governs their formation and evolution. On the other hand, independent methods are currently providing concordant, and more and more accurate, constraints on the cosmological parameters. In this context, galaxy clusters should be rather considered as privileged targets to test the physical processes in the structure formation picture, both for the dark matter and the gaseous component. We can expect significant progress in this field. The unprecedented observational capabilities of XMM and Chandra, allow accurate observations of both nearby and distant clusters, and clearly benefit from similar progresses in numerical simulation and theoretical work, as well as in the observations at other wavelengths.

Acknowledgments. I wish to thank the Conference organizers for their invitation to this very interesting meeting.

References

Allen S. W., Ettori S., Fabian A. C. 2001, MNRAS, 324, 877

Allen, S.W., Fabian, A.C., 1998, MNRAS, 297, L57
Arnaud M. & Evrard A.E. 1999, MNRAS, 305, 631
Arnaud M., Aghanim A., Gastaud R. et al. 2001a, A&A, 365, 67
Arnaud M., Neumann D.M., Aghanim A. et al. 2001b, A&A, 365, 80
Belsole E., Sauvageot J.L., Bohringer H. et al. 2001, A&A, 365, L188
Bertschinger E., 1985, ApJS, 58,39
Böhringer H., Belsole E., Kennea J. et al. 2001, A&A, 365, L181
Bialek, J.J., Evrard, A.E., Mohr, J.J. 2001, ApJ, 555, 597
Blanton E.L., Sarazin C.L., McNamara B.R., Wise M.W., 2001, ApJ, 558, L15
Bower R.G., Benson A.J., Lacey C.G et al. 2001, MNRAS, 325, 497
Briel U., Henry J., Schwarz R. et al. 1991, A&A, 246, L10
Briel U., Henry J.P., Lumb D. et al. 2001, A&A, 365, 60
Briel U., Henry J.P., Arnaud M., Neumann D., et al. 2001b, these proceedings
Brighenti, F., Mathews, W.G., 2001, ApJ, 553, 103
Bryan G., Norman M., 1998, ApJ, 495, 80
Bryan G., 2000, ApJ, 544,L1
Buote D.A., Tsai J.C., 1996, ApJ, 458, 27
Buote D. 2001, in "Merging Process in Clusters of Galaxies", edited by L.Feretti, M.Gioia and G.Giovannini (Dordrecht:Kluwer), in press, astro-ph/0106057
Buote D., 2001b, ApJ, 553, L15
Cavallere A., Menci N. Tozzi P., 1999, MNRAS, 308, 599
Churazov E., Brggen M., Kaiser C. et al. , 2001, ApJ, 554, 261
Clarke T.E., Ensslin T.A., 2001, astro-ph/016137
David L., Nulsen, P., McNamara, B. et al. 2001, ApJ, 557, 546
Eke, V.R., Navarro, J.F., Frenk, C.S. 1998, ApJ, 503, 569
Ettori S, Fabian A. 2000, MNRAS, 317, L57
Ettori S., Bardelli S., De Grandi S. et al. 2000, MNRAS, 318, 239
Evrard A. 1990, ApJ, 363, 349
Evrard A. & Henry J.P., 1991, ApJ, 383, 95
Fabian A., Sanders J., Ettori S. et al. 2000, MNRAS, 318, L65
Fabian A., Mushotzky R., Nulsen P., Peterson J. 2001a, MNRAS, 321, L20
Fabian A., Sanders J., Ettori S. et al. 2001b, MNRAS, 321, L33
Feretti, L., 2000, proceedings of the IAU 199 'The Universe at Low Radio Frequencies' in Pune, India, 1999, astro-ph/0006379
Feretti L., Fusco-Femiano R., Giavannini G., Govoni F., 2001, å, 373, 106
Finoguenov A. & Jones C. 2001 ApJ, 547, L107
Finoguenov A. , Arnaud M., David L.P. 2001, ApJ, 555, 191
Fusco-Femaino R. et al. , 2000, ApJ, 534, L7
Govoni F., Ensslin T.A., Feretti L., Giovannini G., 2001, å, 369, 441
Harris D., Nulsen P., Ponman T. et al. , 2000, ApJ, 530, L81
Horner D., Mushotzky R., Scharf C. 1999, ApJ, 520, 78
Irwin J., Bregman J. 2000, ApJ, 538, 543
Jones C., & Forman B. 1984, ApJ, 276, 38
Jones C., & Forman B. 1992 in Clusters and Superclusters of Galaxies (NATO ASI Vol 366) ed. A.C. Fabian, 49
Jones C., & Forman B. 1999, ApJ, 511, 65

Liang et al. , 2000, ApJ, 544, 686
Lloyd-Davies, E.J., Ponman, T.J., Cannon, D.B. 2000, MNRAS, 315, 689
Lowenstein, M., 2000, ApJ, 532, 17
McNamara B., Wise M., Nulsen P. et al. , 200, ApJ, 534, L135
Majerowicz S., Neumann D. 2001, in 'Galaxy Clusters and the High Redshift Universe Observed in X-rays', XXIth Moriond Astrophysics Meeting, in press
Markevitch M., Forman W., Sarazin C., Vikhlinin A., 1998, ApJ, 503, 77
Markevitch M., Sarazin C., Vikhlinin A., 1999, ApJ, 521, 526
Markevitch M., Ponman T., Nulsen P. et al. 2000, ApJ, 541, 542
Markevitch M., Vikhlinin A., 2001, ApJ, in press, astro-ph/0105093
Matsushita K., Belsole E., Finoguenov A., Böhringer H 2001, A&A, submitted
Mazzotta P., Markevitch M., Vikhlinin A. et al. 2001, ApJ, 555, 205
Mohr J.J., Evrard A.E., Fabricant D.G., Geller, M.J., 1995, ApJ, 447, 8
Mohr, J.J., Mathiesen, B., Evrard, A.E., 1999, ApJ, 517, 627
Mohr, J.J., Reese, J.J., Ellingson E. et al. 2000, ApJ, 544, 109
Mohr, J.J., Evrard, A.E. 1997, ApJ, 491, 38
Molendi S., De Grandi S., Fusco-Femiano R., ApJ, 2000, 533, L43
Muanwong, O.,Thomas, P.A., Kay, S.T. et al. , 2001, ApJ, 552, L27
Nakamura F.E., Hattori M., Mineshige S., 1995, å, 302, 649
Navarro J.F., Frenk C.S., White S.D.M. 1995, MNRAS, 274, 720
Navarro J.F., Frenk C.S., White S.D.M., 1997, ApJ, 490, 493
Neuman D.M. & Arnaud M. 1999, A&A, 348, 711
Neumann D.M., Arnaud M., Gastaud R. et al. 2001, A&A, 365, 74
Neumann, D.M., Arnaud, M. 2001, A&A, 373, L33
Peterson J., Paerels F., Kaastra J. et al. , 2001, A&A, 365, L104
Ponman T.J., Bourner P., Ebeleing H., Böhringer H. 1996, MNRAS, 283, 690
Ponman T., Cannon D., Navarro, J. 1999, Nature, 397, 135
Pratt G., Arnaud M., Aghanim, N. 2001, astro-ph/0105431
Renzini, A., 1997, ApJ, 488, 35
Ricker, P.M., Sarazin, C., 2001, ApJ, in press, astro-ph/0107210
Roettiger, K., Loken, C., Burns, J.O., 1997, ApJS, 109, 307
Schindler S., Müller E., 1993, A&A, 272, 137
Sarazin C. 2001, in "Merging Process in Clusters of Galaxies", edited by L.Feretti, M.Gioia and G.Giovannini (Dordrecht:Kluwer), in press, astro-ph/0105418
Sauvageot, J.L., Belsole, E., Arnaud M., Ponman T., 2001, in 'Galaxy Clusters and the High Redshift Universe Observed in X-rays', XXIth Moriond Astrophysics Meeting, in press
Tamura T., Kaastra J. S., Peterson J. R. et al. 2001, A&A, 365, L87
Tozzi P. & Norman C. 2001, ApJ, 546, 63
Valageas P. & Silk J., 1999, å, 347, 1
Vikhlinin A., Forman W., Jones C., 1999, ApJ, 525, 47
Vikhlinin A., Markevitch M., Murray S. 2001a, ApJ, 549, L47
Vikhlinin A., Markevitch M., Murray S. 2001b, ApJ, 551, 160
White D.A., 2000, MNRAS, 312, 663
Wu K.K.S, Fabian, A.C., Nulsen, P.E.J., 2000, MNRAS, 318, 889
Yamada M., Fujita Y., 2001, ApJ, 553, L145

Tracing Cosmic Evolution with Galaxy Clusters
ASP Conference Series, Vol. 268, 2002
S. Borgani, M. Mezzetti, and R. Valdarnini, eds.

X-ray Statistical Properties of the central cool component in clusters of galaxies

Yasushi Ikebe

Max-Planck-Institut für extraterrestrische physik, Postfach 1312, 85741 Garching, Germany

Abstract. Central cool gas component that is often observed from a well-relaxed cluster system has long been interpreted as a consequence of "Cooling Flow" (CF), radiative cooling followed by inflow of Intra-Cluster Medium (ICM). However, recent XMM-Newton spectroscopy has shown no signatures of cooler gas phases below certain temperatures in typical CF clusters (A1795, Tamura et al. 2001; A1835, Peterson et al. 2001). This contradicts the conventional CF model or at least requires a major revision of the model. In order to investigate statistical properties of the central cool component, we performed systematic analysis of *ASCA* data on 85 clusters. We found that 1) temperature of the central cool component strongly depends on the temperature of the main ICM, 2) the cool component is selectively found around a brightest cluster galaxy (BCG) that coincide with the X-ray peak position, and 3) the luminosity-temperature (L–T) relation of the cool component shows nice agreement with the L–T relation of the main ICM. Together with the previous observational fact that, in some of the "CF" clusters, the total gravitating mass is clustering in two distinct spatial scales, a main cluster component and a second small-scale system, we conclude that the central cool component is associated with the second small-scale self-gravitating system that is immersed in the host cluster, and the cool component temperature reflects the gravitational potential depth.

1. Sample and ASCA analysis

For the study we used brightest 85 clusters observed with *ASCA* whose redshifts distribute from 0.004 to 0.201 with a median of 0.046. From each GIS and SIS instrument, a spectrum over a central region of $2'$ radius and one outside $2'$ radius were accumulated. We fitted each set of *ASCA* spectra with the two-temperature model (2T model), in which a hot component is filling entire cluster region, while in the central region a cool component is allowed to coexist with the hot component forming two-phase plasma. We then obtained X-ray luminosity in 0.1-2.4 keV band and temperature of cool and hot component, L_c, T_c, L_h, and T_h. Full results are given in Ikebe et al. (2001). An example of the 2T model fit is shown in Fig. 1 for Abell 1795, which has one of the brightest central cool component.

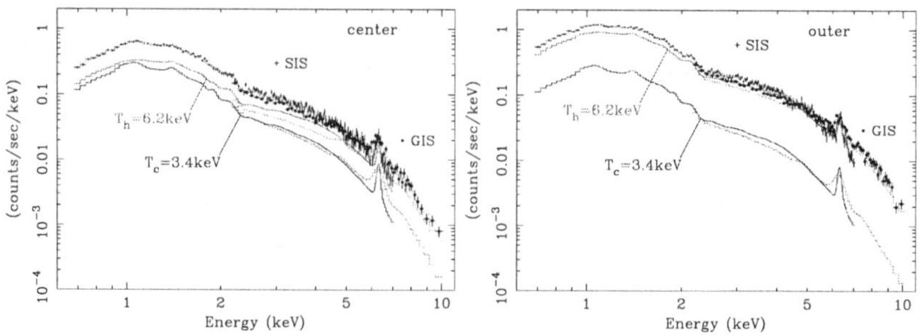

Figure 1. *ASCA* spectra of Abell 1795 extracted from the center (left) and outer (right) regions. Crosses and filled circles show SIS and GIS data, respectively. Flux mixing effects due to the largely extended point-spread functions of X-Ray Telescope have been taken into account and the four spectra are fitted simultaneously with the 2T model. The best-fit model for the SIS and GIS are illustrated with solid and dotted lines respectively.

2. Validity confirmation of the 2T model with XMM-Newton data

The validity of the 2T model was tested with the latest X-ray data of Abell 1795 taken with XMM-Newton. The 11 annular spectra of EPIC-PN were simultaneously fitted with the 2T model, where each annular spectrum is modeled with sum of two isothermal component, and each component is assumed to have common temperature throughout all radii. The fit is good and the best-fit two temperatures are derived to be T_c=3.3 keV and T_h=6.4 keV, which are compared with the temperature profile obtained with the single-phase model (Fig. 2). The T_c and T_h nicely agree with the central coolest temperature and that of the isothermal component in the outer region, respectively. Therefore, the 2T modeling is valid to the most advanced X-ray data and usable to characterize the property of the central cool component and the surrounding ICM.

3. Results and Discussions

Based on the results from the 2T model fitting to the *ASCA* data of the 85 selected clusters, we investigated various correlations. In the study, we classified the cluster sample into three groups, which are Strong Cool Component (SCC), X-ray Dominant (XD) and non-X-ray Dominant (nXD). An SCC cluster is defined as one showing a very strong (statistically very significant) cool component in the *ASCA* spectrum. Among the rest of the clusters, one in which the X-ray peak coincides with the Brightest Cluster Galaxy (BCG) is classified as an XD cluster, while nXD is defined as one without such a galaxy. All the SCC clusters turned out to be XDs, too.

For the SCC clusters, we determined the temperatures and emission measures of each cool and hot component. Figure 3(a) shows the cool component luminosity as a function of the cool component temperature for the SCC clusters. A positive correlation is clearly seen. Figure 4 shows the correlation between

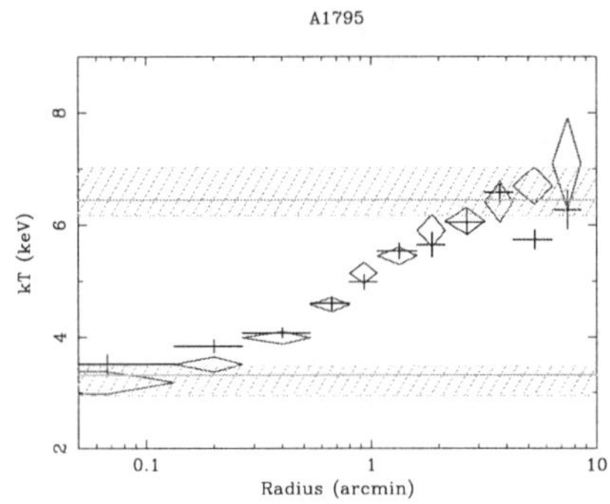

Figure 2. The temperature profile of Abell 1795. The results from the single-phase model given in Tamura et al. (2001), in which each annular spectrum is fitted with an isothermal plasma emission model, are reproduced with crosses (XMM/EPIC-MOS) and diamonds (PN). The cool and hot component temperatures, $T_{\rm h}$ and $T_{\rm c}$, derived with the 2T model are illustrated with crosshatched regions.

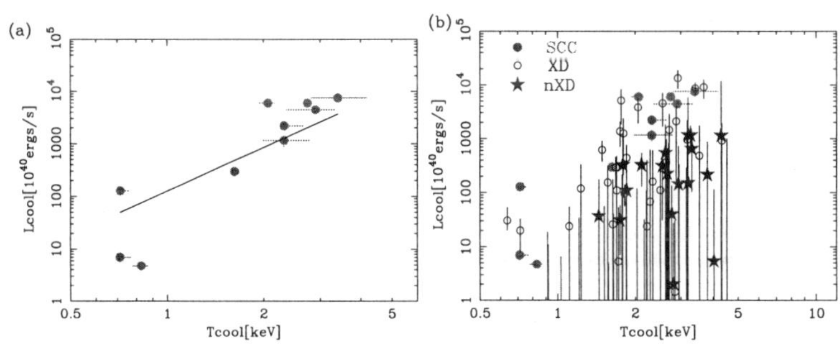

Figure 3. (a) The $L_{\rm c}$–$T_{\rm c}$ relation for the SCC clusters. The line is the best-fit power-law function to the $L_{\rm h}$–$T_{\rm h}$ relation illustrated in Fig. 6. (b) The $L_{\rm c}$–$T_{\rm c}$ relation for all the sample clusters. Filled circles, open circles, and filled stars specify SCC, XD, and nXD clusters, respectively.

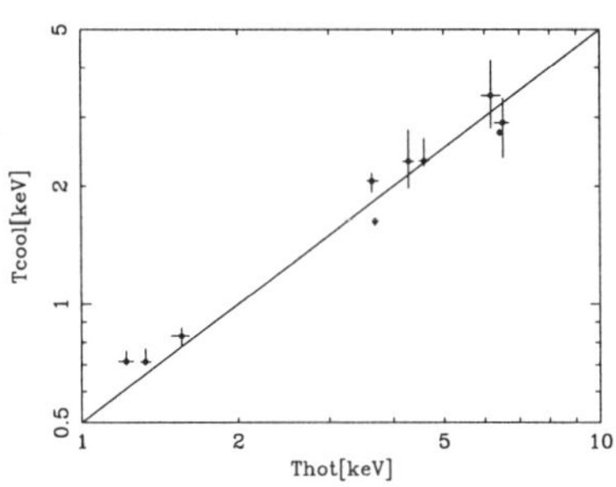

Figure 4. $T_{\rm c}$–$T_{\rm h}$ relation for the SCC clusters. The line represents $T_{\rm c}$=$T_{\rm h}/2$.

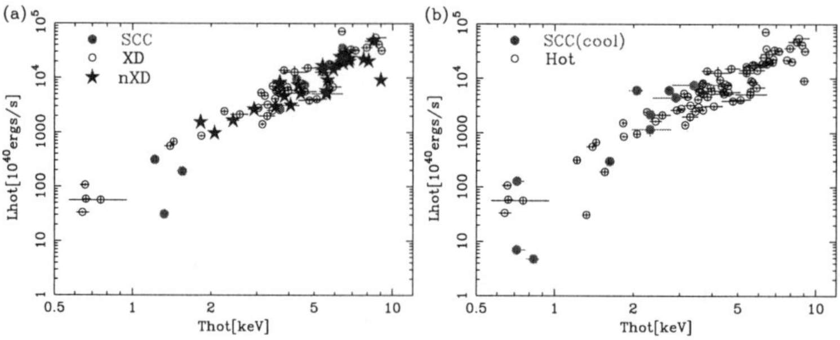

Figure 5. (a) The L_h–T_h relation for all the sample clusters. Filled circles, open circles, and filled stars specify SCC, XD, and nXD clusters, respectively. (b) The L_c–T_c relation for the SCC clusters (filled circles) is overlayed on the L_h–T_h relation (open circles).

the two temperatures of the hot and cool component for the SCC clusters. Surprisingly, the ratios between the two temperatures, T_c/T_h, is virtually constant and the relation, T_c=$T_h/2$ well represents the correlation.

For clusters that do not show very strong cool component, i.e. non-SCC clusters, we fixed T_c at the half value of each mean temperature derived with a single-temperature model fitting, according to the T_c=$T_h/2$ relation. The luminosity of the cool component thus estimated for the non-SCC clusters is illustrated in Fig. 3(b). The results of the SCC clusters are also overlayed. SCC and XD clusters exhibit systematically more luminous cool component than nXD clusters. Actually, many of nXD clusters give only upper limit to L_c.

We obtained the $L - T$ relation of the hot component, which is illustrated in Fig. 5(a). Unlike for the cool component, the three classes do not show any systematic difference. It is then clearly shown that the X-ray characteristics of the XD and nXD clusters are segregated mainly by the central cool component. To compare with the $L - T$ relation of the cool component would be very interesting. In the Fig. 5(b), the L_c–T_c relation for the SCC clusters shown in Fig. 3 is overlayed on the L_h–T_h relation. A surprising agreement is clearly seen. The best fit power-law function to the L_h–T_h relation is compared with the SCC L_c–T_c relation in Fig. 3. The similarity of the two L–T relation suggests that both the cool and hot components are related to individual gravitational bound objects. Therefore, the cool component can be naturally interpreted as ICM filling a self-gravitating system whose size is comparable to a giant elliptical galaxy or a group of galaxies, which is immersed in the host cluster.

References

Ikebe et al. 2001, A&A, submitted
Tamura et al. 2001, A&A, 365, L87
Peterson et al. 2001, A&A, 365, L104

Tracing Cosmic Evolution with Galaxy Clusters
ASP Conference Series, Vol. 268, 2002
S. Borgani, M. Mezzetti, and R. Valdarnini, eds.

A Pilot Study of the Tully-Fisher Relation in Cl0024+1654 at z=0.4

Anne J. Metevier and David C. Koo

Department of Astronomy and Astrophysics, University of California, Santa Cruz, CA 95064 USA

Luc Simard

Steward Observatory, University of Arizona, Tucson, AZ, 85721 USA

Abstract. We present the first determination of the Tully-Fisher relation (TFR) for a distant galaxy cluster. We have measured internal kinematics of seven members of Cl0024+1654 from spatially resolved emission observed in 2D Keck (LRIS) spectra. Measurements of disk structural parameters were made from HST (WFPC2) images. While we do not find evidence for a change in the slope or zeropoint of the cluster TFR, we do find that there was more scatter in the past. We intend to expand this study with kinematic measurements from fifteen more Cl0024 members.

1. Motivation

Since Butcher and Oemler's (1978, hereafter BO78; 1984) pioneering studies of the color evolution of galaxies in clusters, many physical mechanisms have been proposed as causes for this observed evolution. Viable processes can be divided into two groups: those involving interactions between the ICM and cluster galaxies (e.g., ram pressure stripping) and those resulting from the high density of galaxies in the cluster environment (e.g., galaxy harassment). All of these processes involve disks, gas, and star formation. While many recent cluster studies have concentrated on the early-type population of galaxies in clusters (e.g., fundamental plane study by Kelson et al. 2000), we would like to focus on disk systems. To begin our examination of cluster disks, we have undertaken a pilot study of the Tully-Fisher relation (TFR) in Cl0024.

2. Target and Dataset

Cl0024 is an excellent candidate cluster for the study of disk galaxies. As one of the two clusters originally studied by BO78, Cl0024 has a high fraction of blue galaxies. It is one of the richest clusters known, making it a good candidate for a study of intracluster effects. Furthermore, Cl0024 is one of the best-studied clusters. Recent studies of Cl0024 members include a fundamental plane investigation (van Dokkum & Franx 1996), a look at strong emission line galaxies (Koo et al. 1997), and a study of chemical abundances (Kobulnicky & Zaritsky 1999). Morphological classification of Cl0024 galaxies (Smail et al. 1997) has indicated a significant fraction of spirals in the cluster, in agreement with BO78. Furthermore, Cl0024 exhibits a spectacular gravitational lens system with a counter-arc, first discovered by Koo (1987). This system has prompted cluster mass measurements via lensing models (e.g., Broadhurst et al. 2000) which may be compared with X-ray mass estimates, e.g., Soucail et al. (2000).

We have undertaken a Keck (LRIS) spectral survey of Cl0024 (see Metevier et al., in these proceedings, for more details). The wavelength range of this survey covers restframe [OII] through Hα for the cluster, and the spectral and spatial resolution are higher than those of other recent surveys of this cluster (Dressler et al. 1999, Czoske et al. 2001). Furthermore, our slits are aligned with galaxy major axes. These spectra provide us with spatially resolved emission for several cluster members. To complement our spectra, we have obtained WFPC2 images of Cl0024 in two filters (25 ksec in F450W, 20 ksec in F814W) from the HST archive. The high resolution of these images is necessary for deriving disk size scales, inclinations, colors, and structural parameters crucial to Tully-Fisher work.

3. Cl0024 Tully-Fisher Relation

The Tully-Fisher relation (TFR) is a scaling relation between galaxy absolute magnitudes and the terminal velocities of their rotation curves and serves a tool for studying galaxy mass-to-light ratios (M/L). In the field, the TFR has been measured out to z=1 (e.g., Vogt et al. 1996, 1997; Simard & Pritchet 1998). However, the cluster TFR has only been well-studied out to z=0.1 (e.g., Dale et al. 2000). Up to now, a key question has been whether or not emission in distant cluster disk systems is spatially extended enough for rotation curve analysis and Tully-Fisher measurements.

We have measured the rotation curves of a sample of 7 galaxies in the Cl0024 field. Each is a cluster member and lies within the HST images. Furthermore, all 7 galaxies exhibit emission which can be measured out to a radius $\geq 1''$ and are objects for which our spectral slits were aligned with the galaxy major axes. We fit a Gaussian to the following emission lines: Hα, [NII], [OIII], Hβ, and [OII]. Arctan fits (Willick 1999) were then applied to the resulting rotation curves (see Figure 1*a* for an example).

In the future, we will improve these fits by modelling and correcting for the effects of seeing, slit widths, and slit positioning effects. We will also add internal extinction corrections. Furthermore, we are studying cross-correlation routines and 2D modelling as alternatives to our Gaussian emission line fits.

In Figure 1*b*, our Tully-Fisher measurements of Cl0024 members are compared to the z$\simeq$0 cluster TFR derived by Tully & Pierce (2000). While we find no evidence within the presently very small sample for any large changes in the slope or zeropoint of the cluster TFR with redshift, we do see an increase in scatter in the relation at z=0.4. This result is in agreement with the results of Dale et al. (2000), who found a 50% higher scatter in the z=0.08 cluster TFR than in the $0.02<z<0.06$ relation.

We do not have enough data to make stronger statements about the evolution of the cluster TFR. We can, however, explain the higher scatter in the

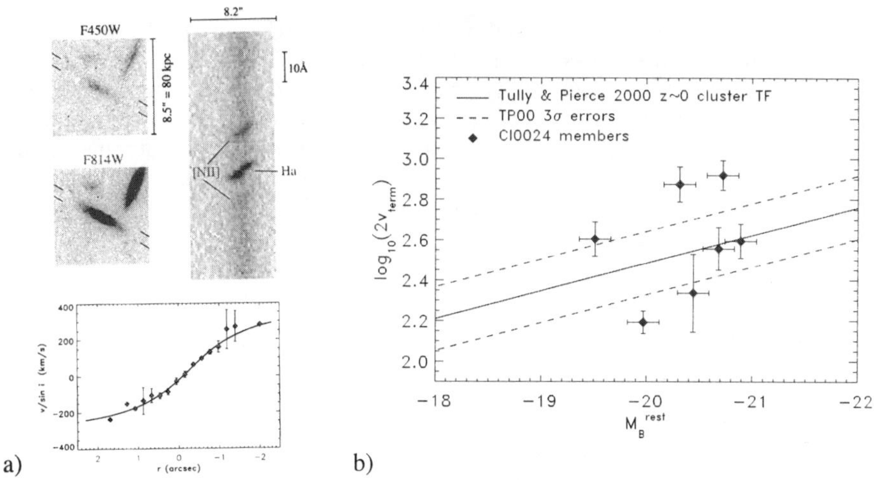

Figure 1. *a*) WFPC2 images, 2D LRIS spectrum, and rotation curve for a Cl0024 member. Rotation curve errors do not include uncertainties in inclination. *b*) Cl0024 Tully-Fisher measurements show more scatter than the local relation derived by Tully & Pierce (2000).

cluster TFR in the past. A galaxy with a high M/L (scattered high on the TFR) in the past may result from the fact that its gas was stripped, e.g., via ram pressure stripping, and hence star formation was quenched. By the present epoch, the gas in such a galaxy may be entirely stripped such that it no longer exhibits emission (i.e., we no longer include it in a Tully-Fisher sample). On the other hand, galaxies scattered low on the TFR (low M/L) in the past may result from enhanced starburst activity. Such galaxies may not be experiencing such dramatic star formation in the present day and therefore their scatter on the diagram is dampened.

4. Results and Future Work

The three key results of our pilot study are the following: **1)** We have shown that some intermediate-z cluster spirals are large enough and have strong enough emission lines to do rotation curve analysis. **2)** Our data already yield 7 rotation curves in the HST-imaged region of Cl0024. We have 15 rotation curve candidates outside this region which will expand our sample. While small, the sample is comparable in size to that studied by, e.g., Vogt et al. (1996). Similar quality data for 5–10 more clusters would tightly constrain the intermediate-redshift cluster TFR. **3)** The scatter in our preliminary measurement of the cluster TFR at z=0.4 is larger than that measured at z=0. Within an admittedly sparse sample, we find no evidence for a change in TF slope or zeropoint.

We will expand our study in the future by examining emission line diagnostics of physical processes affecting Cl0024 disk systems. Our spectral resolution and wavelength range will provide measurements of star formation rates from

Hα, of metallicities via the R23 index (from [OII], [OIII], and Hβ), and of the ionization state of the emission gas via line ratios of Hα, [NII], [SII], [OIII], and Hβ. Evidence for shock ionization might be expected from interactions between cluster galaxies and the ICM. We also add photometric measurements from the archival WFPC2 images of this cluster and structural models of the cluster galaxies. The outcome of these efforts will provide the most complete examination of the processes affecting disk systems in a single distant cluster.

References

Broadhurst, T., et al. 2000, ApJ, 531, L15

Butcher, H. & Oemler, A., Jr. 1978, ApJ, 219, 18

Butcher, H. & Oemler, A., Jr. 1984, ApJ, 285, 426

Czoske, O., et al. 2001, A&A, 372, 391

Dale, D. A. & Uson, J. M. 2000, AJ, 552, 561

Dressler, et al. 1999, ApJS, 515, 79

Kelson, D. D., et al. 2000, ApJ, 531, 184

Kobulnicky, H. A. & Zaritsky, D. 1999, ApJ, 511, 118

Koo, D. C. 1987, in "Large Scale Motions in the Universe", ed. V.C. Rubin & G.V. Coyne

Koo, D. C., Guzman, R., Gallego, J., & Wirth, G. D. 1997, 478, L49

Simard, L., & Pritchet, C. 1998, ApJ, 505, 96

Smail, I., et al. 1997, ApJS, 110, 213

Soucail, G., et al. 2000, A&A, 355, 433

Tully, R. B. & Pierce, M. J. 2000, ApJ, 533, 744

van Dokkum, P. G. & Franx, M. 1996, MNRAS, 281, 985

Vogt, N. P., et al. 1996, ApJ, 465, L15

Vogt, N. P., et al. 1997, ApJ, 479, L121

Willick, J.A. 1999, ApJ, 516, 47

Tracing Cosmic Evolution with Galaxy Clusters
ASP Conference Series, Vol. 268, 2002
S. Borgani, M. Mezzetti, and R. Valdarnini, eds.

Cooling Flows

A.C. Fabian

Institute of Astronomy, Madingley Road, Cambridge CB3 0HA, UK

Abstract. The gas temperature in the cores of many clusters of galaxies drops inward by about a factor of three or more within the central 100 kpc radius. The radiative cooling time drops over the same region from 5 or more Gyr down to below a few 10^8 yr. Although it would seem that cooling flows are taking place, XMM spectra show no evidence for strong mass cooling rates of gas below 1–2 keV. Chandra images show holes coincident with radio lobes and cold fronts indicating that the core regions are complex. The observational situation is reviewed here and ways in which continued cooling may be hidden are discussed, togther with the implications for any heat source which balances radiative cooling.

1. Introduction

The gas density within the central 100 kpc or so of the centre of most clusters of galaxies is high enough that the radiative cooling time of the gas is less than 10^{10} yr. The cooling time drops further at smaller radii, suggesting that in the absence of any balancing heat source much of the gas in the central regions is cooling out of the hot intracluster medium. In order to maintain the pressure required to support the weight of the overlying gas, a slow, subsonic inflow known as a cooling flow develops.

X-ray observations made before Chandra and XMM-Newton were broadly consistent with the cooling flow picture (see Fabian 1994 for a review and Makishima et al 2001 for an opposite view), although several issues remained unresolved. The first issue was the observed X-ray surface brightness profile, which was not as peaked as expected from a homogeneous flow. Instead a multiphase gas was assumed, dropping cold gas over a range of radii. The second was the fate of the cooled gas. At the rates of 100s to more than $1000\,\mathrm{M}_\odot\,\mathrm{yr}^{-1}$ found in some clusters, the central galaxies should be very bright and blue if the cooled gas forms stars with a normal intial-mass-function. In many cases they do have excess blue light indicative of massive star formation (Johnstone, Fabian & Nulsen 1987; Allen 1995; Cardiel et al 1998; Crawford et al 1999), but at rates which are a factor of 10 to 100 times lower than the X-ray deduced mass cooling rate. It has also been argued (e.g. O'Dea et al 1994) that there is no significant sink in terms of cold gas clouds. A third issue involved the shape of the soft X-ray spectrum, which was inconsistent with a simple cooling flow. Absorption intrinsic to the flow was found to be a possible explanation (Allen & Fabian 1997; Allen et al 2001). A final issue was whether the neglect of heating was justified. The effect of gravitational heating as the gas flows was taken into account, but the effects pf any central radio source, which pumps energy into the surrounding gas via jets, together with disturbances due to subclusters plunging into the core every few Gyr were not included due to a lack of quantitative information. Heat flow due to thermal conduction has also been assumed negligible.

The situation with cluster cooling flows has been clarified over the past year, particularly by the high spatial resolution imaging of Chandra and the high spectral resolution of the XMM-Newton Reflection Grating Spectrometer (RGS). Chandra images show much detail in the cores of clusters, with bubbles from radio sources (e.g. Mcnamara et al 2000; Fabian et al 2000) and cold fronts (Markevitch et al 2000; Vikhlinin et al 2001) seen. RGS spectra (Peterson et al 2001; Tamura et al 2001; Kaastra et al 2001) confirm the presence of a range of temperatures in cooling flow clusters but fail to show evidence of for gas cooling below 1–2 keV. Simply put the data are consistent with gas cooling at a high rate to about one third of the mean temperature beyond 100 kpc but then vanishing.

At about the same time, the evidence for both warm (Jaffe & Bremer 1997; Donahue et al 2000; Edge et al 2001) and cold (Edge 2001) molecular gas at the centres of cooling flows clusters has become widespread. In some extreme cases there may be over 10^{11} $M_{\odot}$ of cold gas (Edge 2001).The presence of dust in these regions is also widespread, from evidence of the Balmer decrement in the optical/UV nebulosities commonly seen (e.g. Heckman et al 1989; Crawford et al 1999), dustlanes, and submm and IR detections (Edge et al 1999; Allen et al 2001; Irwin, Bridges & ? 2001). It is therefore possible that some more star formation, and in particular cold gas clouds, may be found in and around central cluster galaxies. There has also been the detection of OVI emission from A2597 with FUSE (Oegerle et al 2001). Lastly, recent numerical simulations of evolving cluster which include radiative cooling of the gas predict cooling flows (e.g. Pearce et al 2000).

We have reached a state where it is tempting for many to assume that some form of heating balances cooling and so dismiss cooling flows altogther. That ignores the how, why and what of the heating, which remains unsolved, although several candidates have been identified (e.g. Tucker & Rosner 1983; Binney & Tabor 1995). Some form of feedback is probably required to prevent all of the gas from being heated up. If feedback does occur we have a good chance to observe how it works, since the region is spatially resolved and optically thin. The process is of wide importance, since it provides the upper mass limit for galaxies (in simulations of the galaxy luminosity function, Kauffmann et al 1994 switch off cooling in massive galaxies).

My own view is to treat it as an intriguing astrophysical puzzle whch can be tackled observationally. Heating from radio sources and infalling subclusters must occur at some level, but whether it can balance radiative cooling over the required spatial scales to better than a factor of a few is not yet clear. Cooling probably does account for the observed star formation and cold gas clouds. A major remaining issue is whether the mass cooling rates are reduced from the earlier X-ray deduced rates by a factor of a few, of ten or a hundred.

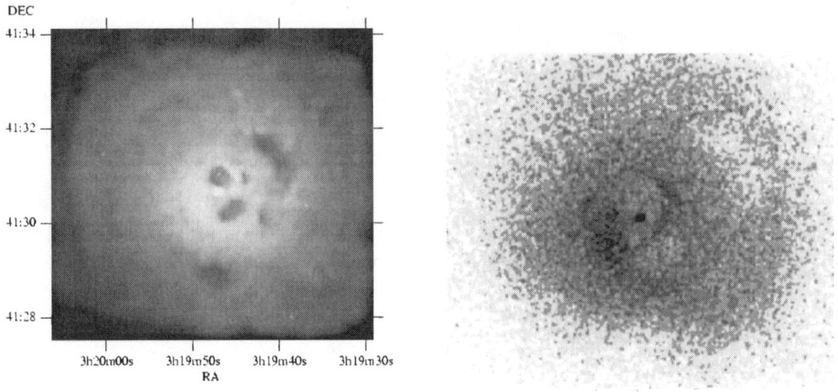

Figure 1. (Left) Adaptively-smoothed 0.5–7 keV ACIS-S X-ray image; (Right) gaussian-smoothed ACIS-I X-ray image of the centre of the Perseus cluster. The ACIS-S image has cosmetic blurring due to a nodal boundary running from just below the nucleus to the W. These effects are absent in the I image. Note that the I image shows better the complex structure of the inner part (compare with Fig. 1). The precise nature of this structure is not clear. The holes may be roughly in the plane of the Sky or arranged in a perpendicular sense with the apparent 'e' shape being part of a helix oriented along our line of sight.

2. Chandra results

Chandra images show structure in cluster cores. The X-ray emission is steeply peaked into the centres of many clusters but there are holes and fronts in the peak. Markevitch et al (2000) found sharply defined cold fronts on A2142, across which the pressure is continuous yet the temperature changes by a factor of about 2. Ettori & Fabian (2000) note that thermal conduction must be heavily suppressed in order that such sharp features can last long enough to be common. The fronts probably indicate that the gas of subclusters does not readily mix with the existing intracluster medium, presumably because they are separate magnetic structures. Also the cores of infalling subclusters may not be strongly shocked in decelerating into the core (see Fabian & Daines 1992; Burns et al priv. comm.).

2.1. The Perseus cluster

Holes in the X-ray surface brightness are seen to coincide with some radio lobes. The best examples are in the Perseus cluster, and were first seen with ROSAT (Böhringer et al 1993). Chandra shows that they have bright rims of X-ray *cool* gas (Fabian et al 2000a). This is contrary to the work of Heinz et al (2000) who predicted that the rims would signify shocks. Other holes coincident with radio lobes are found in Hydra A (McNamara et al 2000; David et al 2001) and many other clusters. The puzzling aspect if radio sources are heating the cooling gas is that in all cases reported the *coolest* gas seen is that closest to the radio lobes.

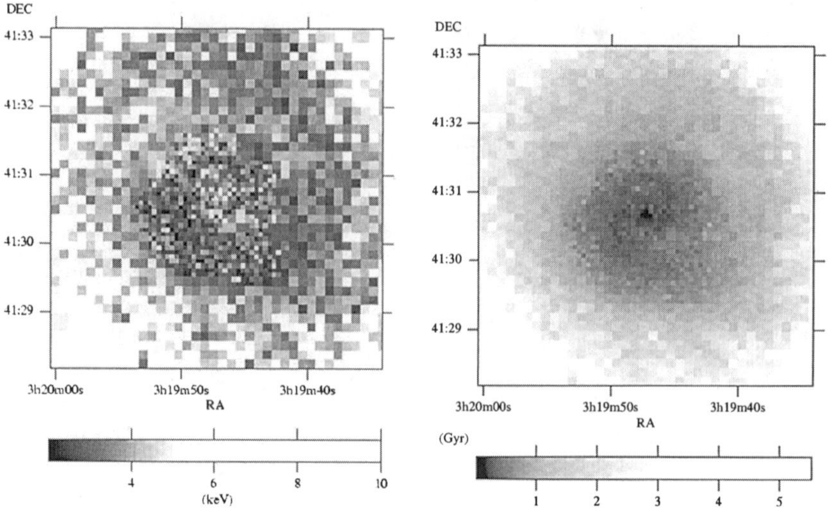

Figure 2. (Left) Temperature and (Right) radiative cooling time maps produced from the X-ray colour ratios (Fabian et al 2000a). Note that the coolest gas ($T \sim 2.5\,\mathrm{keV}$) with the shortest cooling time (~ 0.3 Gyr) lies in the rim around the N lobe and in the E bright blob. Single-phase gas has been assumed for the analysis.

Of course there is much energy going into the lobes, but the energy from the PdV work expended in forming the holes can propagate away as sound waves, and the relativistic energy stored in the bubbles can be lifted away and out of the immediate core by buoyancy.

2.2. A1795 and A2199

The Chandra image of A1795 (Fabian et al 2000b) shows an 80 kpc long soft X-ray feature coincident with an Hα filament found by Cowie et al (1982). It is plausibly a cooling wake trailing behind the central galaxy, which is at the head of the filament. The galaxy is moving around in the core of the cluster at a few hundred $\mathrm{km\,s^{-1}}$. There is no evidence from the temperatures of the gas that this motion has heated the gas significantly. The central galaxy in A2199 may also be oscillating, as deduced from the unusual morphology of its radio source (Burns et al 1983). Again, the coolest gas appears to be close to both the radio source and the central galaxy (Johnstone et al, in preparation).

3. XMM-Newton results

The most striking results have come from the RGS data which show little evidence for gas cooling below 1–2 keV (Peterson et al 2001; Tamura et al 2001; Kaastra et al 2001). Emission lines from FeXX and XVII, for example, should be bright and easily seen if the mass cooling rate is high, but they are absent. EPIC

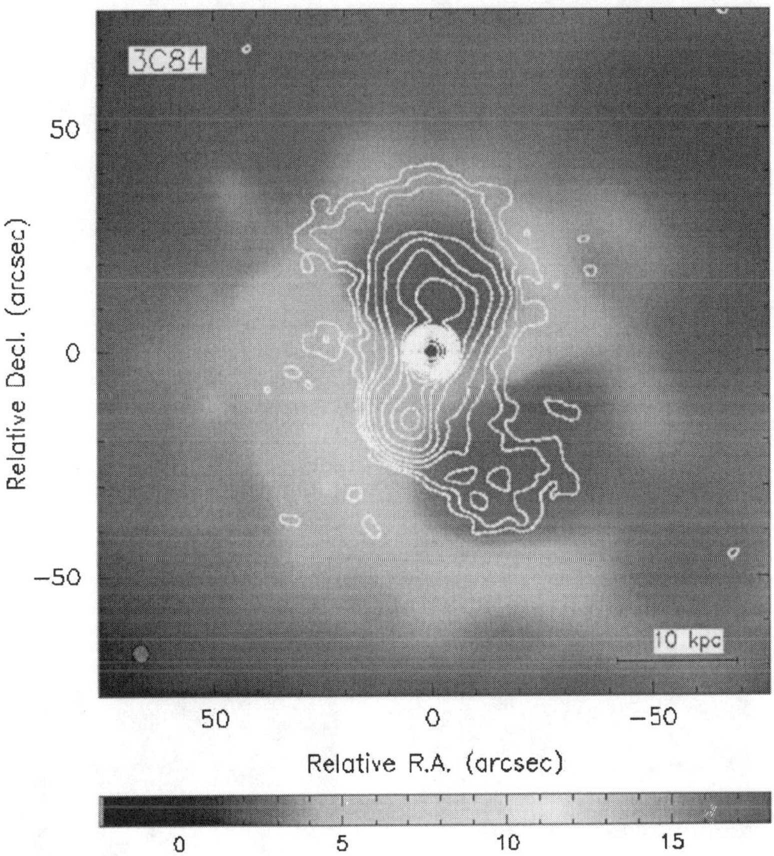

Figure 3. Radio image (1.4 GHz restored with a 5 arcsec beam, produced by G. Taylor; see Fabian et al 2000a) overlaid on an adaptively smoothed 0.5–7 keV X-ray map.

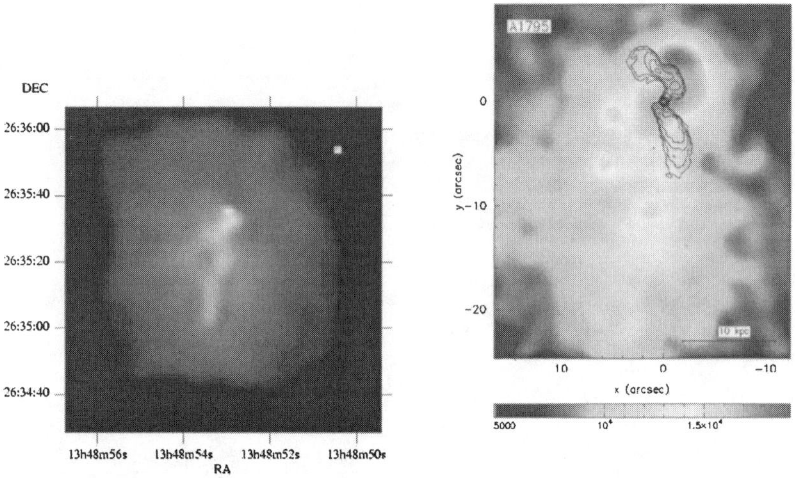

Figure 4. (Left) Adaptively-smoothed X-ray image of the centre of A1795 (Fabian et al 2000b). (Right) Overlay of the 3.6 cm radio emission (Ge & Owen 1993) on the X-ray image.

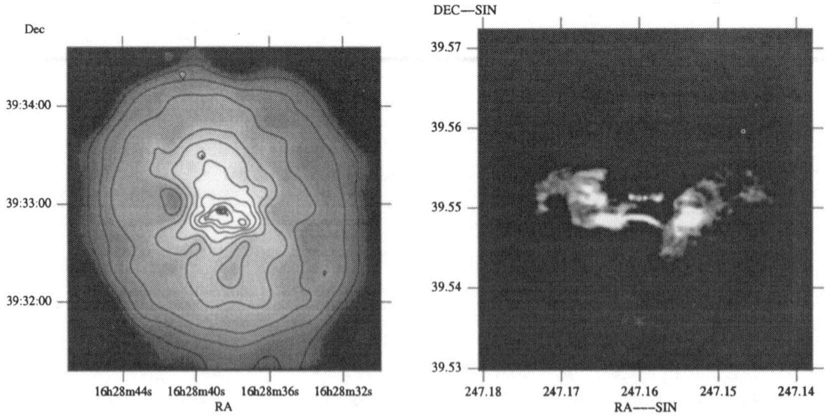

Figure 5. (Left) Adaptively-smoothed 0.5–3 keV X-ray image of the core of A2199 (the contours are logarithmic); (Right) 1.7 GHz radio image (Giovannini et al 1998).

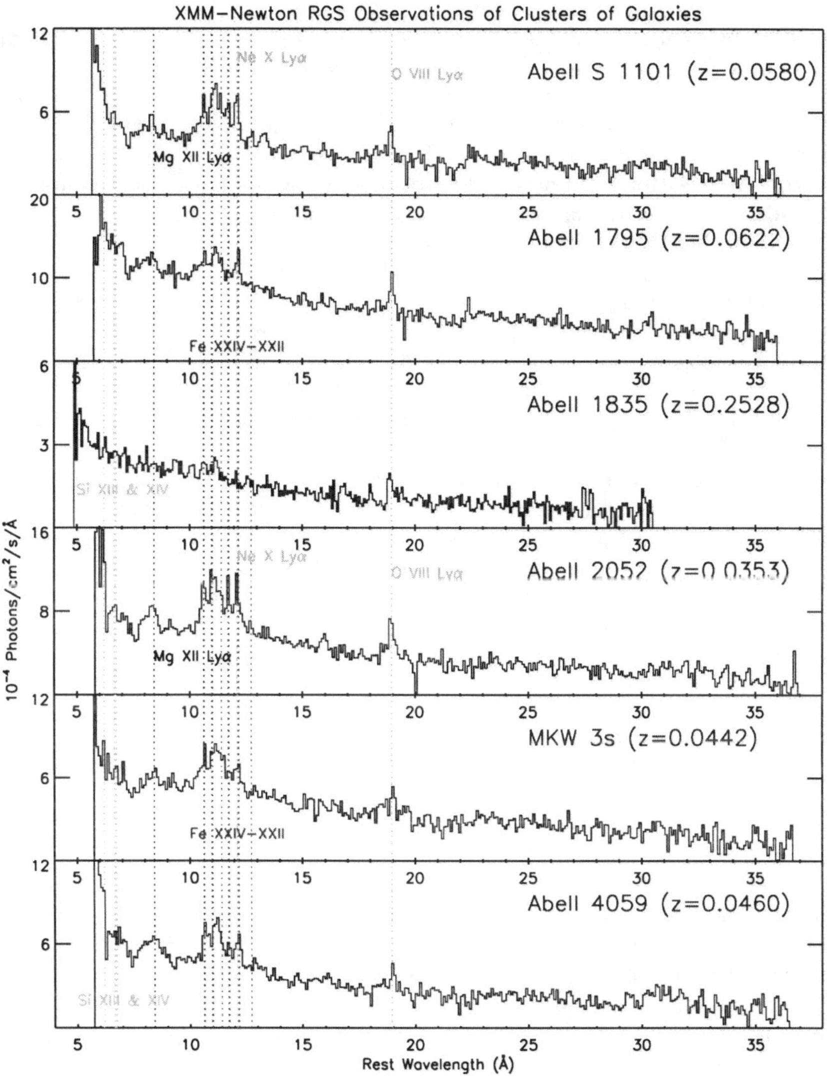

Figure 6. RGS spectra of 6 cooling flow clusters, kindly provided by J. Peterson and J. Kaastra. Emission lines between 10-13 A indicate the presence of cooler gas in these clusters (at 1–3 keV) but the lack of lines between 13 and 18 A shows that gas is not radiatively cooling below 1–2 keV at high rates in any simple unobscured manner.

CCD spectra (e.g. Böhringer et al 2001; Molendi & Pizzolata 2001) confirm this result.

Various explanations have been given (e.g. Peterson et al 2001; Fabian et al 2001). The gas may be cooling and yet appear to vanish when it reaches say 2 keV. Clouds of cold gas may photoelectrically absorb the soft X-rays, or the gas may have become dense enough to separate from the flow and mix in with surrounding hotter gas (Norman & Meiksin 1996). Alternatively it may mix in with colder gas (for example that associated with the optical filaments or the molecular gas – note that this also explains why the filaments are so bright; Fabian et al 2001b).

Another possibility is that the metals in the gas are not uniformly mixed in, but have a bimodal distribution (Fabian et al 2001a). Gas in which ten per cent has a metallicity of 3 times solar and 90 per cent has zero metallicity has the same spectrum as gas at 0.3 solar if cooling is unimportant. When it does cool however, the metal emission lines cool only 10 per cent of the gas and so are much reduced as compared with the situation if they were responsible for cooling all the gas.

If heating is the explanation, then it cannot just be some low level of heat which stops the gas at 1–2 keV, since it would then accumulate at that temperature, contrary to observation. . It has to halt the cooling over the full range of temperatures, and thus radii. How this can happen is a puzzle. If the radio source is responsible, then it may be intermittent. Maybe we do not see the heating phase, which is short lived. The power required to stem the flow during the heating phase then goes up to high values. There may not be any problem in the Virgo cluster around M87 where the energy rquirements are relatively small, but in a massive cluster like A1835 (Peterson et al 2001; Schmidt et al 2001) the necessary power may exceed $10^{46}\,\mathrm{erg\,s^{-1}}$.

4. Summary

The central 100kpc radius region in most clusters has a radiative cooling time shorter than 5 Gyr (Peres et al 1997) and many have cooling times which drop to the centre to a value of only 10^8 yr or so. The gas temperature drops by a factor of 3 or more over this radius range. It seems plausible that the temperature drop and short radiative cooling times are related and that the low temperatures are caused by radiative cooling.

It is then a puzzle as to why gas which has lost two thirds of its thermal energy, and for which the radiative cooling time is very short, is not seen to cool further.

There are two obvious solutions, both of which have some difficulties. The first solution is that the gas does cool but either the soft X-ray emission is absorbed or the cooling is non-radiative and due, say, to mixing. The problem of the fate of cooled gas then remains. The second solution is that some heating balances cooling. The problem here is that the heat has to balance cooling over a wide range of radii and a wide range of timescales. Also observations of radio lobes which are a likely source of heat indicate that they coincide with the coolest gas in cluster cores.

The answer may be more complex, with the major temperature drop being due to a combination of in situ radiative cooling and gas introduced from dense cooling subclusters. Heat from the kinetic energy of infalling subclusters, and turbulence, continues to be dissipated throughout the core, reducing the age of any steady central cooling region to only a few Gyr. An intermittent central radio source powered by accretion from the intracluster medium heats and churns up some of the coolest gas at the centre. Radiatively-cooling clumps (possibly metal rich) fall out of the mean slow inflow once their temperature drops to below one third of the outer temperature and rapidly mix with cooler gas clouds closer to the centre. Massive star formation take place in cooled clumps and spreads dust into the surrounding gas, further enhancing cooling via infrared emission.

References

Allen S. W., 1995, MNRAS, 276, 947

Allen S.W., Fabian A.C., 1997, MNRAS, 286, 583

Allen S. W., Fabian A. C., Johnstone R. M., Arnaud K. A., Nulsen P. E. J., 2001, MNRAS, 322, 589

Binney J., Tabor G., 1995, MNRAS, 276, 663

Bohringer H., Voges W., Fabian A.C., Edge A.C., Neumann D.M., 1993, MNRAS, 264, L25

Böhringer H., Nulsen P.E.J., Braun R., Fabian A.C., 1995, MNRAS, 274, L67

Böhringer H. et al 2000, A&A, 365, L181

Burns JO, Scwendeman O, White RA, 1983, ApJ, 271, 575

Cardiel N., Gorgas J, Arago-Salamanca A., 1998, MNRAS, 298, 977

Cowie L.L., Hu E.M., Jenkins E.B., York D.G., 1983 ApJ, 272, 29

Crawford C.S., Allen S.W., Ebeling H., Edge A.C., Fabian A.C., 1999, MNRAS, 306, 875

David L, Nulsen P.E.J.,McNamara B.R., Forman, W., Jones C., Robertson B., Wise M., 2001, ApJ, 557, 546

Donahue M., Mack J., Voit G. M., Sparks W., Elston R., Maloney P. R., ApJ, 2000, 545, 670

Edge A. C., Ivison R. J., Smail I., Blain A. W., Kneib J.-P., 1999, MNRAS, 306, 599

Edge A. C., 2001, MNRAS, 328, 762

Edge A. C., et al 2001, MNRAS, submitted

Ettori S., Fabian A.C., 2000, MNRAS, 317, L57

Fabian A. C., Daines S. J., 1991, MNRAS, 252, 17

Fabian A.C., 1994, ARAA, 32, 277

Fabian A.C., et al 2000a, MNRAS, 318, L65

Fabian A.C., et al 2000b, MNRAS

Fabian A.C., Mushotzky R.F., Nulsen P.E.J., Peterson J., 2001 MNRAS, 321, L20

Giovannini G., Cotton W.D., Feretti L., Lara L., Venturi T., 1998, ApJ, 493, 632
Ge J.P., Owen F.N., 1993, AJ, 105, 778
Heckman T. M., Baum S. A., van Breugel W. J. M., McCarthy P., 1989, ApJ, 338, 48
Heinz S., Reynolds C.S., Begelman M.C., 1998, ApJ, 501, 126
Irwin
Jaffe W., Bremer M. N., 1997, MNRAS, 284, L1
Johnstone R. M., Fabian A. C., Nulsen P. E. J., 1987, MNRAS, 224, 75
Kaastra J., et al, 2001, A&A 365, L99
Kauffmann G., Guiderdoni B., White S. D. M., 1994, 267, 981
Markevitch M. et al, 2000, ApJ, 541, 542
McNamara B. et al, 2000a, ApJ, 534, L135
Makishima K. et al, 2001, PASJ, 53, 401
Molendi S., Pizzolato F., 2001, ApJ, 560, 194
Norman C., Meiksin A., 1996, ApJ, 468, 97
Pearce FP, Thomas PA Couchman HMP, Edge AC, 2000, MNRAS, 317, 1029
Peres C. B., Fabian A. C., Edge A. C., Allen S. W., Johnstone R. M., White D. A., 1998, MNRAS, 298, 416
Peterson J.A. et al 2001, A&A, 365, L104
Rosner R., Tucker W., 1989, ApJ, 338, 761
Schmidt R.W., Allen S.W., Fabian A.C., MNRAS in press
Tamura T. et al 2001, A&A, 365, L87
Tucker W.H., Rosner R., 1983, 267, 547
Vikhlinin A., Markevitch M., Murray S.S., 2000, ApJ, 549, L47

Tracing Cosmic Evolution with Galaxy Clusters
ASP Conference Series, Vol. 268, 2002
S. Borgani, M. Mezzetti, and R. Valdarnini, eds.

The New Emerging Model for the Structure of Cooling Cores in Clusters of Galaxies

H. Böhringer, K. Matsushita, Y. Ikebe,

Max-Planck-Institut für Extraterrestrische Physik, D-85748 Garching, Germany

E. Churazov

Max-Planck-Institut für Astrophysik, D-85748 Garching, German

Abstract. New X-ray observations with XMM-Newton show a lack of spectral evidence for large amounts of cooling and condensing gas in the centers of galaxy clusters believed to harbour strong cooling flows. Here, we explore these diagnostics of the temperature structure of cooling cores with XMM-spectroscopy. We further find no evidence of intrinsic absorption in the center of the cooling flows of M87 and the Perseus cluster. To explain these findings we consider the heating of the core regions of clusters by jets from a central AGN. We find that the power of the AGN jets as estimated by their interaction effects with the intracluster medium in several examples is more then sufficient to heat the cooling flows. We explore which requirements such a heating model has to fulfill and find a very promising scenario of self-regulated Bondi accretion of the central black hole. In summary it is argued that most observational evidence points towards much lower mass deposition rates than previously inferred for cooling flow clusters.

1. Introduction

X-ray imaging observations have shown that the X-ray emitting, hot gas in a large fraction of all galaxy clusters reaches high enough densities in the cluster centers that the cooling time of the gas falls below the Hubble time, and gas may cool and condense in the absence of a suitable fine-tuned heating source (e.g. Silk 1976, Fabian & Nulsen 1977). ¿From the detailed analysis of surface brightness profiles of X-ray images of clusters obtained with the *Einstein*, *EXOSAT*, and *ROSAT* observatories, the detailed, self-consistent scenario of inhomogeneous, comoving cooling flows emerged (e.g. Fabian et al. 1984, Nulsen 1986, Thomas, Fabian, & Nulsen 1987, Fabian 1994). The main assumptions on which the cooling flow model is based and some important implications are: (i) Each radial zone in the cooling flow region comprises different plasma phases covering a wide range of temperatures. The consequence of this temperature distribution is that gas will cool to low temperature and condense over a wide range of radii. (ii) The gas features an inflow in which all phases with different temperature move with the same flow speed. (iii) There is no energy exchange between the different phases, between material at different radii, and no heating.

Now the first analysis of high resolution X-ray spectra and imaging spectroscopy obtained with *XMM-Newton* has shown to our surprise that the spectra show no signatures of cooler phases of the cooling flow gas below an intermediate temperature which constitutes a problem for the interpretation of the results in the conventional cooling flow picture (e.g. Peterson et al. 2001, Tamura et al. 2001). Another result is that the spectroscopic data are better explained with local isothermality in the cooling flow region (e.g. Böhringer et al. 2001a, Matsushita et al. 2001, Molendi & Pizzolato 2001) also in conflict with the inhomogeneous cooling flow model. Here, we discuss these new spectroscopic results and their implications and point out the way to a new possible model for this phenomenon. The results are mostly based on the detailed observations of the M87 X-ray halo. A detailed description of this study is provided by Böhringer et al. (2001b).

2. Spectroscopic Diagnostics of Cluster Cooling Cores

XMM Reflection Grating Spectrometer (RGS) observations of several cooling core regions show signatures of different temperature phases ranging approximately from the hot virial temperature of the cluster to a lower limiting temperature, T_{low}. Clearly observable spectroscopic features of even lower temperature gas expected for a cooling flow model are not observed. A1835 with a bulk temperature of about 8.3 keV has T_{low} around 2.7 keV (Peterson et al. 2001) and similar results have been derived for A1795 (Tamura et al. 2001). These results are very well confirmed by *XMM* observations with the energy sensitive imaging devices, EPN and EMOS, providing spectral information across the entire cooling core region, yielding the result that (for M87, A1795, and A1835) single temperature models provide a better representation of the data than cooling flow models (Böhringer et al. 2001a, Molendi & Pizzolato 2001) also implying the lack of low temperature components. The very detailed analysis of M87 by Matsushita et al. (2001 and the contribution to this workshop) has shown that the temperature structure is well described locally by a single temperature over most of the cooling core region, except for the regions of the radio lobes and the very center ($r \leq 1$ arcmin, ~ 5 kpc).

Among the spectroscopic signatures which are sensitive to the plasma temperature in the relevant temperature range, the complex of iron L-shell lines is most important. Fig. 1 shows simulated X-ray spectra as predicted for the XMM EPN instrument in the spectral region around the Fe L-shell lines for a single-temperature plasma at various temperatures from 0.4 to 2.0 keV and 0.7 solar metallicity. There is a very obvious shift in the location of the peak making this feature an excellent thermometer.

For a cooling flow with a broad range of temperatures one expects a composite of several of the relatively narrow line blend features, resulting in a quite broad peak. Fig. 2a shows for example the deprojected spectrum of the M87 halo plasma for the radial range 1 - 2 arcmin (outside the inner radio lobes) and a fit of a cooling flow model with a mass deposition rate slightly less than 1 $M_\odot$ yr^{-1} as expected for this radial range from the analysis of the surface brightness

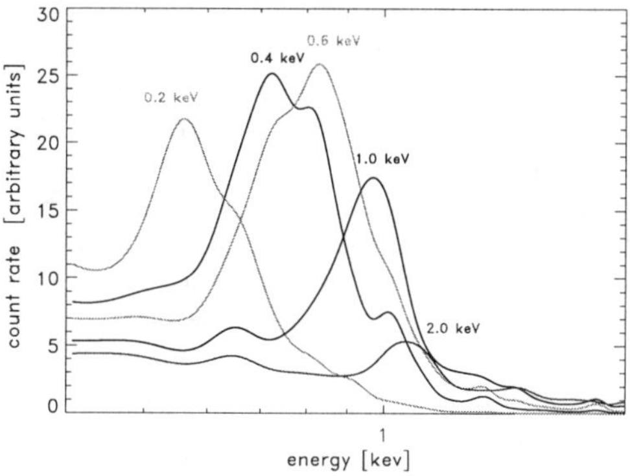

Figure 1. The Fe L-line complex in X-ray spectra as a function of the plasma temperature for a metallicity value of 0.7 solar. The simulations show the appearance of the spectra as seen with the XMM EPN. The emission measure was kept fixed when the temperature was varied.

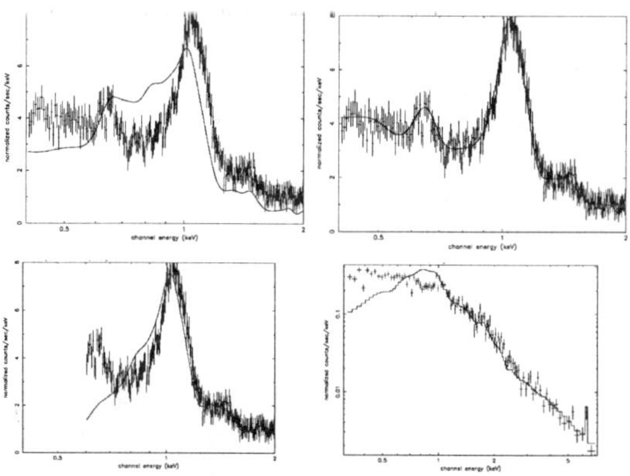

Figure 2. **(a - upper left):** *XMM EPN*-spectrum of the central region of the M87 X-ray halo in the radial range $R = 1 - 2$ arcmin. The spectrum has been fitted with a cooling flow model with a best fitting mass deposition rate of 0.96 $M_\odot$ yr^{-1} and a fixed absorption column density of $1.8 \cdot 10^{20}$ cm^{-2}, the galactic value, and a parameter for T_{low} of 0.01 keV. **(b - upper right):** same spectrum fitted by a cooling flow spectrum artificially constraint to emission from the narrow temperature interval 1.44 - 2.0 keV, where T_{low} was treated as a free fitting parameter. **(c - lower left):** same spectrum fitted with a free parameter for the internal excess absorption. The spectrum was constraint to the energy interval 0.6 to 2.0 keV. **(d - lower right):** *XMM* EPN spectrum of A1795 fitted with a cooling flow model with the galactic value for absorption.

profile (e.g. Stewart et al. 1984, Matsushita et al. 2001). It is evident that the peak in the cooling flow model is much broader than the observed spectral feature. For comparison Fig. 2b shows the same spectrum fitted by a cooling flow model where a temperature of 2 keV was chosen for the maximum temperature and a suitable lower temperature cut-off (1.44 keV) was determined by the fit. The very narrow temperature interval (almost isothermality) is well consistent with the narrow peak. A similar result is obtained for other clusters, e.g. A1795 as shown in Fig. 2d.

Since this diagnostics of the temperature structure is essentially based on the observation of metal lines, an inhomogeneous distribution of the metal abundances in the cluster ICM and a resulting suppression of line emission at low temperatures was suggested as a possible way to reconcile the above findings with the standard cooling flow model by Fabian et al. (2001a and contribution in these proceedings). As shown by Böhringer et al. (2001b) such a scenario will still result in a relatively broad Fe L-line feature and does not solve the problem in this case of M87.

3. Internal absorption

Another possible attempt to obtain consistency is to allow the absorption parameter in the fit to adjust freely. This is demonstrated in Fig. 2c with the same observed spectrum where the best fitting absorption column density is selected in such a way by the fit that the absorption edge limits the extent of the Fe-L line feature towards lower energies. This is actually the general finding with ASCA observations which has shown two possible options for the interpretation of the spectra of cluster core regions: (1) an interpretation of the results in form of an inhomogeneous cooling flow model which than necessarily includes an internal absorption component (e.g. Allen 2000, Allen et al. 2001), or (2) an explanation of the spectra in terms of a two-temperature component model (e.g. Ikebe et al. 1997, 1999, Makishima 2001) where the hot component is roughly equivalent to the hot bulk temperature of the clusters and the cool component corresponds approximately to T_{low}. Thus for the cooling flow interpretation to work and to produce a sharp Fe-L line feature as observed, the absorption edge has to appear at the right energy and therefore values for the absorption column of typically around $3 \cdot 10^{21}$ cm^{-2} are needed (e.g. Allen 2000 and Allen et al. 2001 who find values in the range $1.5 - 5 \cdot 10^{21}$ cm^{-2}).

It is therefore important to perform an independent test on the presence of absorbing material in the cluster cores. Thanks to *CHANDRA* and *XMM-Newton* we can now use central cluster AGN as independent light sources for probing. Using the nucleus and jet of M87 (with *XMM*, see Fig. 3) and the nucleus of NGC1275 (with *CHANDRA*) we find no signature of internal absorption. Thus at least for these two cases it is difficult to argue for internal absorption to obtain consistency of the observations with the cooling flow model.

4. Heating Model

In view of these difficulties of interpreting the observations with the standard cooling flow model, we may consider the possibility that the cooling and mass

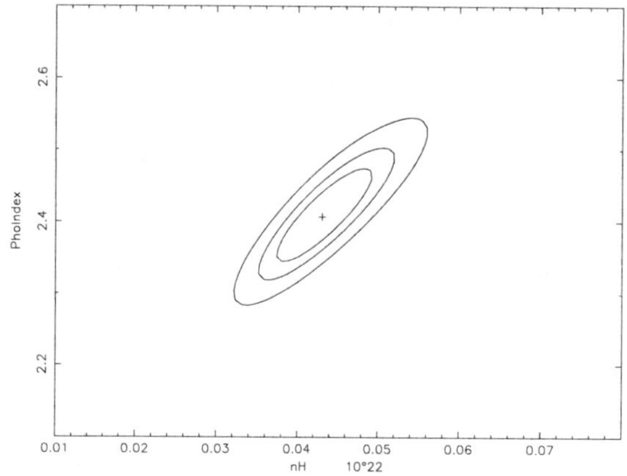

Figure 3. Constraints on the shape of the *XMM EPN*-spectrum of the nucleus of M87. The lines show the 1, 2, and 3σ confidence intervals for the combined fit of the slope (photon index) of the power law spectrum and the value for the absorbing column density, n_H, in units of 10^{22} cm^{-2}.

deposition rates are much smaller than previously thought, that is reduced by at least one order of magnitude. To decrease the mass condensation under energy conservation some form of heating is clearly necessary. Three forms of heat input into the cooling flow region have been discussed: (i) heating by the energy output of the central AGN (e.g. Pedlar et al. 1990, Binney & Tabor 1993, McNamara et al. 2000,

(ii) heating by heat conduction from the hotter gas outside the cooling flow (e.g. Tucker & Rosner 1983, Bertschinger & Meiksin 1986), and (iii) heating by magnetic fields, basically through some form of reconnection (e.g. Soker & Sarazin 1990, Makishima et al. 2001). The latter two processes depend on poorly known plasma physical conditions and are thus more speculative. The energy output of the central AGN, however, can be determined as shown below.

A heating scenario can only successfully explain the observations if among others the two most important requirements are met: (i) The energy input has to provide sufficient heating to balance the cooling flow losses, that is about 10^{60} to 10^{61} erg in 10 Gyr or on average about $3 \cdot 10^{43} - 3 \cdot 10^{44}$ erg s^{-1}, and (ii) The energy input has to be fine-tuned. Too much heating would result in an outflow from the central region and the central regions would be less dense than observed. Too little heat will not reduce the cooling flow by a large factor. Therefore the heating process has to be self-regulated: mass deposition triggers the heating process and the heating process reduces the mass deposition.

Further constraints are discussed by Böhringer et al. (2001b). The total energy input into the ICM by the relativistic jets of the central AGN can be estimated by the interaction effect of the jets with the ICM by means of the scenario described in Churazov et al. (2000). It relies on a comparison of the inflation and buoyant rise time of the bubbles of relativistic plasma which are observed e.g. in the case of NGC 1275 (Böhringer et al. 1993, Fabian et al.

Table 1. Estimated energy output from the central AGN in M87, NGC1275, and the central galaxy of the Hydra A cluster. The input parameters are the bubble radius, r_B, the ambient pressure, P_{th}, and the Keplerian velocity at the bubble location, v_K.

system	r_B	P_{th}	v_K	L_{kin}
M87	8kpc	10^{-10}erg cm^{-3}	460 km s^{-1}	1.2$10^{44}$erg s^{-1}
NGC1275	15kpc	2 10^{-10}erg cm^{-3}	600 km s^{-1}	1 10^{45}erg s^{-1}
Hydra A	15kpc	2.8 10^{-10}erg cm^{-3}	550 km s^{-1}	2 10^{45}erg s^{-1}

2001b) and requires as observational input parameters the bubble size, r_B, the ambient pressure, P_{th}, and the Keplerian velocity at the bubble radius in the cluster, v_K. The parameters and the estimated total energy output is given in Table 1 for three examples, M87, Perseus, and Hydra A. These values for the energy input have to be compared with the energy loss in the cooling flow, which is of the order of 10^{43} erg s^{-1} for M87 and about 10^{44} erg s^{-1} for Perseus. Thus in these cases the energy input is larger than the radiation losses in the cooling flow for at least about the last 10^8 yr. We have, however, evidence that this energy input continued for a longer time with evidence given by the outer radio halo around M87 with an outer radius of 35 - 40 kpc (e.g. Kassim et al. 1993, Rottmann et al. 1996). Owen et al. (2000) give a detailed physical account of the halo and model the energy input into it. They estimate the total current energy content in the halo in form of relativistic plasma to $3 \cdot 10^{59}$ erg and the power input for a lifetime of about 10^8 years, which is also close to the lifetime of the synchrotron emitting electrons, to the order of 10^{44} erg s^{-1}, consistent with our estimate. The very characteristic sharp outer boundary of the outer radio halo of M87, noted by Owen et al. (2000), has the important implications, that this could not have been produced by magnetic field advection in a cooling flow.

Thus, we find a radio structure providing evidence for a power input from the central AGN into the halo region of the order of about ten times the radiative energy loss rate over at least about 10^8 years (for this representative example of M87). The energy input could therefore balance the heating for at least about 10^9 years. The observation of active AGN in the centers of cooling flows is a very common phenomenon. E.g. Ball et al. (1993) find in a systematic VLA study of the radio properties of cD galaxies in cluster centers, that 71% of the cooling flow clusters have radio loud cDs compared to 23% of the non-cooling flow cluster cDs. Therefore we can safely assume that the current episode of activity was not the only one in the life of M87 and its cooling flow.

The mechanism for a fine-tuned heating of the cooling flow region should most probably be searched for in a feeding mechanism of the AGN by the cooling flow gas. The most simple physical situation would be given if simple Bondi type of accretion from the inner cooling core region would roughly provide the order of magnitude of the power output that is observed and required. Using the classical formula for spherical accretion from a hot gas by Bondi (1952) we can obtain a very rough estimate for this number. For the proton density near the M87 nucleus ($r \leq 15$ arcsec) of about 0.1 cm^{-3}, a temperature of about 10^7 K (e.g. Matsushita et al. 2001), and a black hole mass of $3 \cdot 10^9$ $M_\odot$ (e.g. Ford et al. 1994) we find a mass accretion rate of about 0.01 $M_\odot$ yr^{-1} and an energy output of about $7 \cdot 10^{43}$ erg s^{-1}, where we have assumed the canonical value of 0.1 for the ratio of the rest mass accretion rate to the energy output. The corresponding accretion radius is about 50 pc (~ 0.6 arcsec). This accretion rate is more than a factor of 1000 below the Eddington value and thus no reduction effects of the spherical accretion rate by radiation pressure has to be expected. Small changes in the temperature and density structure in the inner cooling core region will directly have an effect on the accretion rate. Therefore we have all the best prospects for building a successful self-regulated AGN-feeding and cooling flow-heating model.

5. Conclusions

Several observational constraints have let us to the conclusion that the mass deposition rates in galaxy cluster cooling cores are not as high as previously predicted. The new X-ray spectroscopic observations with a lack of spectral signatures for the coolest gas phases expected for cooling flows and the lower mass deposition rates indicated at other wavelength bands than X-rays are more consistent with mass deposition rates reduced by one or two orders of magnitude below the previously derived values. This can, however, only be achieved if the gas in the cooling flow region is heated. The most promising heating model is a self-regulated heating model powered by the large energy output of the central AGN in most cooling flows.

Most of the guidance and the support of the heating model proposed here (based on concepts developed in Churazov et al. 2000, 2001) is taken from the detailed observations of a cooling core region in the halo of M87 and to a smaller part from the observations in the Perseus cluster. These observations show that the central AGN produces sufficient heat for the energy balance of the cooling flow, that the most fundamental and classical accretion process originally proposed by Bondi (1952) provides an elegant way of devising a self-regulated model of AGN heating of the cooling flow, and that most of the further requirements that have to be met by a heating model to be consistent with the observations can most probably be fulfilled. Since these ideas are mostly developed to match the conditions in M87, it is important to extent such detailed studies to most other nearby cooling flow clusters.

In this new perspective the cooling cores of galaxy clusters become the sites where most of the energy output of the central cluster AGN is finally dissipated. Strong cooling flows should therefore be the locations of AGN with the largest mass accretion rates. While in the case of M87 with a possible current mass accretion rate of about 0.01 $M_\odot$ y^{-1} the mass addition to the black hole (with an estimated mass of about $3 \cdot 10^9$ $M_\odot$) is a smaller fraction of the total mass, the mass build-up may become very important for the formation of massive black holes in the most massive cooling flows, where mass accretion rates above 0.1 $M_\odot$ y^{-1} become important over cosmological times.

References

Allen, S.W., 2000, MNRAS, 315, 269

Allen, S.W., Fabian, A.C., Johnstone, R.M., et al., 2001, MNRAS, 322, 589

Ball, R., Burns, J.O., Loken, C., 1993, AJ, 105, 53

Bertschinger, E. & Meiksin, A., 1986, ApJ, 306, L1

Böhringer, H., Nulsen, P.E.J., Braun, R., Fabian, A.C., 1995, MNRAS, 274, L67

Böhringer, H., Belsole, E., Kennea, J., et al., 2001a, A&A, 365, L181

Böhringer, H., Matsushita, K., Ikebe, Y., et al. 2001b, A&A, submitted

Bondi, H., 1952, MNRAS, 112, 195

Churazov, E., Forman, W., Jones, C., Böhringer, H., 2000, A&A, 356, 788

Churazov, E., Brüggen, M., Kaiser, C.R., Böhringer, H., & Forman, W., 2001, ApJ, 554, 261

Fabian, A.C., 1994, ARA&A, 32, 277

Fabian, A.C., Nulsen, P.E.J., 1977, MNRAS, 180, 479

Fabian, A.C., Nulsen, P.E.J., & Canizares, C.R., 1984 Nature, 310, 733

Fabian, A. C., Sanders, J. S., Ettori, S., et al. 2000, MNRAS 318, 65

Fabian, A.C., Mushotzky, R.F., Nulsen, P.E.J., Peterson, J.R., 2001, MNRAS, 321, 20

Ford, H.C., et al. 1994, ApJ, 435, L27

Ikebe, Y., Makishima, K., Ezawa, H., et al., 1997, ApJ, 481, 660

Ikebe, Y., Makishima, K., Fukazawa, Y., et al., 1999, ApJ, 525, 58

Kassim, N., Perley, R.A., Erickson, W.C., Dwarakanath, K.S., 1993, AJ, 106, 2218

Makishima, K., Ezawa, H., Fukazawa, Y., et al., 2001, PASJ, 53, 401

Matsushita, K., Belsole, E., Finoguenov, A., Böhringer, H., 2001, A&A, submitted

McNamara, B., Wise, M., Nulsen, P.E.J., et al. 2000, ApJ, 534, L135

Molendi, S. & Pizzolato, F., 2001, ApJ (in press), astro-ph/0106552

Nulsen, P.E.J., 1986, MNRAS, 221, 377

Owen, F.N., Eilek, J.A., Kassim, N.E., 2000, ApJ, 543, 611

Pedlar, A., Ghataure, H. S., Davies, R. D., et al., 1990, MNRAS, 246, 477

Peterson, J.R., Paerels, F.B.S., Kaastra, J.S., et al., 2001, A&A, 365, L104

Rottmann, H., Mack, K.-H., Klein, U., Wielebinski, R., 1996, A&A, 309, L9

Silk, J., 1976, ApJ., 208, 646

Soker, N. & Sarazin, C.L., 1990, ApJ, 348, 73

Stewart, G.C., Canizares, C.R., Fabian, A.C., & Nulsen, P.E.J., 1984, ApJ, 278, 536

Tabor, G. & Binney, J., 1993, MNRAS, 263, 323

Tamura, T., Kaastra, J.S., Peterson, J.R., et al., 2001, A&A, 365, L87

Thomas, P.A., Fabian, A.C., & Nulsen, P.E.J., 1987, MNRAS, 228, 973

Tucker, W.H. & Rosner, R., 1983, ApJ, 267, 547

Tracing Cosmic Evolution with Galaxy Clusters
ASP Conference Series, Vol. 268, 2002
S. Borgani, M. Mezzetti, and R. Valdarnini, eds.

The X-Ray Virial Relations for Relaxed Lensing Clusters Observed with Chandra

S.W. Allen, R.W. Schmidt and A.C. Fabian

Institute of Astronomy, Madingley Road, Cambridge CB3 0HA, UK

Abstract. We examine the relations linking mass, X-ray temperature and bolometric luminosity for a sample of luminous, relatively relaxed clusters of galaxies observed with the Chandra Observatory, for which independent confirmation of the mass results is available from gravitational lensing studies. Within radii corresponding to a fixed overdensity $\Delta = 2500$ with respect to the critical density at the redshifts of the clusters, the observed temperature profiles, scaled in units of T_{2500} and r_{2500}, exhibit an approximately universal form which rises within $r \sim 0.3\, r_{2500}$ and then remains approximately constant out to r_{2500}. We obtain best-fit slopes for the mass-temperature and temperature-luminosity relations consistent with the predictions from simple scaling arguments i.e. $M_{2500} \propto T_{2500}^{3/2}$ and $L_{2500} \propto T_{2500}^{2}$, respectively. We confirm the presence of a systematic offset of ~ 40 per cent between the normalizations of the observed and predicted mass-temperature relations for both SCDM and ΛCDM cosmologies.

1. Introduction

A major goal of studies with the new generation of X-ray missions including the Chandra Observatory and XMM-Newton, which permit the first direct spatially-resolved X-ray spectroscopy of hot, distant clusters, is the verification and accurate calibration of the virial relations for galaxy clusters. In particular, detailed studies of systems for which precise mass measurements have been made using other, independent methods are required. Here we use new Chandra observations to determine the X-ray virial relations for a sample of luminous, relatively relaxed clusters spanning the redshift range $0.1 < z < 0.45$, for which lensing mass measurements are available and have been shown to be in good agreement with the Chandra results (Allen, Ettori, & Fabian 2001; Schmidt, Allen, & Fabian 2001; Allen, Schmidt & Fabian 2001a; Allen et al. in prep.). We present gas mass-weighted temperatures, bolometric luminosities and total mass measurements within radii corresponding to a fixed overdensity $\Delta = 2500$ at the redshifts of the clusters, and compare the observed scaling relations between these quantities with those predicted by simulations. Results are given for two cosmologies: SCDM with $h = H_0/100$ km/s/Mpc $= 0.5$, $\Omega_{\rm m} = 1$ and $\Omega_\Lambda = 0$, and ΛCDM with $h = 0.7$, $\Omega_{\rm m} = 0.3$ and $\Omega_\Lambda = 0.7$. A full discussion of the analysis, results and implications is presented by Allen, Schmidt & Fabian (2001b).

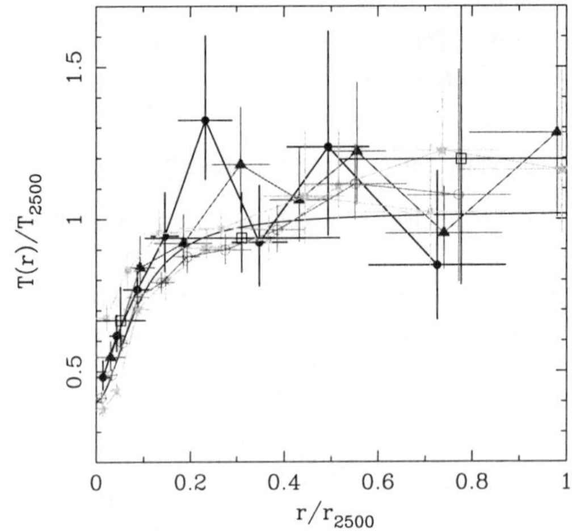

Figure 1. The observed (projected) spectrally-determined temperature profiles in the clusters, scaled in units of kT_{2500} and r_{2500}, for the ΛCDM cosmology. PKS0745-191: open circles, Abell 2390: dark filled triangles, Abell 1835: grey filled stars, MS2137-2353: grey filled squares, RXJ1347-1145: dark filled circles, 3C295: open squares. The best-fit to the combined data set using the functional form in Section 2 is shown as the thin solid line.

2. A universal temperature profile for relaxed clusters

Fig. 1 shows the observed (projected), spectrally-determined temperature profiles for the clusters in our sample in units of the mean gas mass-weighted temperature, kT_{2500}, with the radial axis scaled in units of r_{2500}. The ΛCDM cosmology is assumed. The clusters exhibit similar scaled-temperature profiles which rise within $r \sim 0.3 r_{2500}$ and then remain approximately isothermal out to r_{2500}. The combined data set can be modelled using a simple function of the form $T(r)/T_{2500} = T_0 + T_1[(x/x_c)^\eta/(1+(x/x_c)^\eta]$ where $x = r/r_{2500}$, $T_0 = 0.40 \pm 0.02$, $T_1 = 0.61 \pm 0.07$, $x_c = 0.087 \pm 0.011$ and $\eta = 1.9 \pm 0.4$.

3. The mass-temperature relation

Fig. 2 shows the $M_{2500} - kT_{2500}$ relation for the ΛCDM cosmology. Fitting only the data for those clusters for which independent confirmation of the X-ray mass results is available from lensing studies (*i.e.* excluding 3C295 from the present sample) using a power-law model of the form

$$E(z)\left(\frac{M_{2500}}{1\,M_\odot}\right) = A\left(\frac{kT_{2500}}{10\,\mathrm{keV}}\right)^\alpha \tag{1}$$

and a χ^2 estimator, we obtain $A = 6.91 \pm 0.68 \times 10^{14}$, $\alpha = 1.43 \pm 0.26$ for SCDM ($h = 0.5$) and $A = 5.38 \pm 0.52 \times 10^{14}$, $\alpha = 1.51 \pm 0.27$ for ΛCDM ($h = 0.7$).

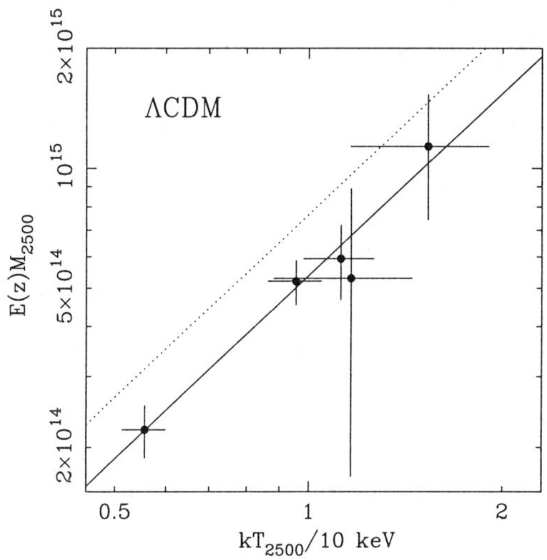

Figure 2. The observed mass-temperature relation for the ΛCDM cosmology with M_{2500} in $M_{\odot}$ and kT_{2500} in keV. The solid line is the best-fitting power-law model $E(z)M_{2500} = A[kT_{2500}/10]^{\alpha}$ with $A = 5.38 \pm 0.74 \times 10^{14}$ and $\alpha = 1.52 \pm 0.36$. The dotted curve is the predicted result from the simulations of Mathiesen & Evrard (2001).

Fixing the slope at $\alpha = 1.5$, we measure normalizations of $A = 6.99 \pm 0.60 \times 10^{14}$ (SCDM) and $A = 5.38 \pm 0.46 \times 10^{14}$ (ΛCDM). Uncertainties are quoted at the 68 per cent confidence level.

Using the modified least-squares estimator of Fasano & Vio (1988), which accounts for errors in both axes, we obtain values $A = 7.02 \pm 0.98 \times 10^{14}$, $\alpha = 1.47 \pm 0.36$ for SCDM and $A = 5.38 \pm 0.74 \times 10^{14}$, $\alpha = 1.52 \pm 0.36$ for the ΛCDM cosmology. We conclude that the measured slope is consistent with the expected value of $\alpha = 1.5$ in all cases.

The dotted curve in Fig. 2 shows the predicted (zero-redshift) relation for the ΛCDM cosmology $[E(z)M_{2500} = 7.7 \pm 0.6 \times 10^{14} kT^{1.52 \pm 0.03}]$ from the hydrodynamical simulations of Mathiesen & Evrard (2001). For both SCDM and ΛCDM, the predicted normalization lies approximately 40 per cent above the observed value. The implications of these results are explored by Allen et al. (2001b).

4. The temperature-luminosity relation

Fig. 3 shows the $kT_{2500} - L_{2500}$ relation for the ΛCDM cosmology. Fitting the $kT_{2500} - L_{2500}$ data for all six clusters using a power-law model of the form

$$\left(\frac{kT_{2500}}{10\,\mathrm{keV}}\right) = B\left(\frac{L_{2500}}{10^{45}\mathrm{erg/s}E(z)}\right)^{\beta} \tag{2}$$

and a χ^2 estimator, we obtain $B = 0.43 \pm 0.05$, $\beta = 0.45 \pm 0.09$ for SCDM, and $B = 0.42 \pm 0.05$, $\beta = 0.56 \pm 0.10$ for ΛCDM. Using the BCES$(X_2|X_1)$ estimator

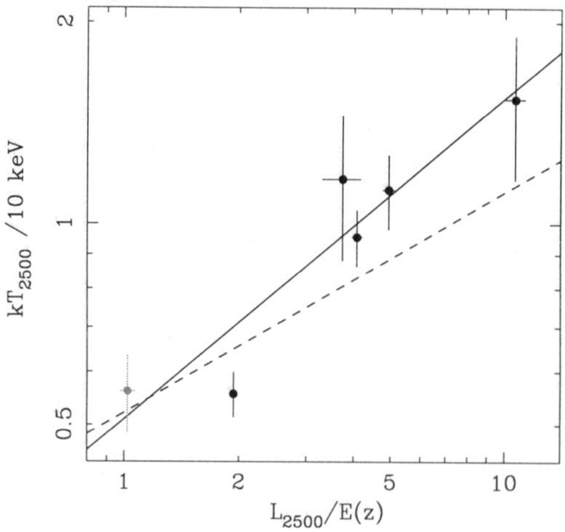

Figure 3. The observed temperature-luminosity relation for the ΛCDM cosmology with kT_{2500} in keV and L_{2500} in units of 10^{45} erg/s. The solid line is the best-fitting power-law model of the form $kT_{2500}/10 = B[L_{2500}/10^{45}E(z)]^{\beta}$ using the BCES estimator, for which $B = 0.51 \pm 0.05$ and $\beta = 0.48 \pm 0.06$. The data for 3C295 are in lighter shading. The dashed line shows the best-fit curve with $\beta = 0.33$ fixed which provides a poor description of the data.

of Akritas & Bershady (1996), which accounts for errors in both axes and the presence of possible intrinsic scatter, we obtain $B = 0.48 \pm 0.07$, $\beta = 0.46 \pm 0.08$ for SCDM and $B = 0.51 \pm 0.05$, $\beta = 0.48 \pm 0.06$ for ΛCDM.

We conclude that the slope of the temperature-luminosity relation for the present sample of hot, relaxed clusters is consistent with the predicted value of $\beta = 0.5$ (Section 1). Fixing $\beta = 0.33$ results in a poor fit: $\chi^2 = 12.5$ for 5 degrees of freedom, as opposed to $\chi^2 = 6.7$ with $\beta = 0.5$ (ΛCDM).

References

Akritas M.G., Bershady M.A., 1996, ApJ, 470, 706

Allen S.W., Ettori S., Fabian A.C., 2001, MNRAS, 324, 877

Allen S.W., Schmidt R.W., Fabian A.C., 2001a, MNRAS, submitted

Allen S.W., Schmidt R.W., Fabian A.C., 2001b, MNRAS, in press

Evrard A.E., Metzler C.A., Navarro J.F., 1996, ApJ, 469, 494

Fasano G., Vio R., 1988, Bull. Inf. Cnt. Données Stellaires, 35, 191

Mathiesen B.F., Evrard A.E., 2001, ApJ, 546, 100 MNRAS, 317, 1029

Schmidt R.W., Allen S.W., Fabian A.C., 2001, MNRAS, in press (astro-ph/0107311)

Tracing Cosmic Evolution with Galaxy Clusters
ASP Conference Series, Vol. 268, 2002
S. Borgani, M. Mezzetti, and R. Valdarnini, eds.

Discovery of Holes in the Core of the Distant Galaxy Cluster RBS797 in a CHANDRA Observation

E. De Filippis

Astrophysics Research Institute (Liverpool John Moores University)

S. Schindler (1), A. Castillo-Morales (1), A. Schwope (2), J. Wambsganss (3)

(1) Astrophysics Research Institute (Liverpool John Moores University), (2) Astrophysikalisches Institut Potsdam, (3) Universität Potsdam, Institut für Physik

Abstract. We present CHANDRA X-ray observations of the hot ($T = 7.7^{+1.2}_{-1.0}$), distant ($z = 0.354$) galaxy cluster RBS797. The most striking features are two spectacular minima in the X-ray emission in the core of the cluster; this suggests an interaction of radio lobes with the intracluster gas. This is the first time such depressions have been observed in a distant cluster.

1. Introduction

In this poster we present the final results of the analysis of the galaxy cluster RBS797. RBS797 has been observed by the CHANDRA Advanced CCD Imaging Spectrometer (ACIS) I detector on October 20th, 2000 for a total exposure time of 13.3 *ksec*.
Throughout this poster we have used: $H_0 = 50\ km \cdot s^{-1} \cdot Mpc^{-1}$ and $q_0 = 0.5$ and errors, if not otherwise specified, are 90% confidence levels.

2. General characteristics

A CHANDRA image of the centre of the cluster in the $0.5 - 7.0\,keV$ energy band, smoothed with a Gaussian filter of $\sigma = 0.75\,arcsec$, is shown in the left panel in Figure 1.
RBS797 shows a slightly elliptical X-ray morphology. The most spectacular features are two minima in the X-ray emission of about $20 - 30\,kpc$ in size; these minima are located opposite to each other with respect to the cluster centre (ENE and WSW direction), and at a distance of about $5\,arcsec$ from the cluster centre.
The central morphology of the cluster, with its depressions in the X-ray surface brightness, is similar to the one of Perseus cluster (Fabian et al. 2000) and of Hydra-A (McNamara et al. 2000), with the difference that in RBS797 the two holes are even more pronounced. Cuts in direction of the minima in the

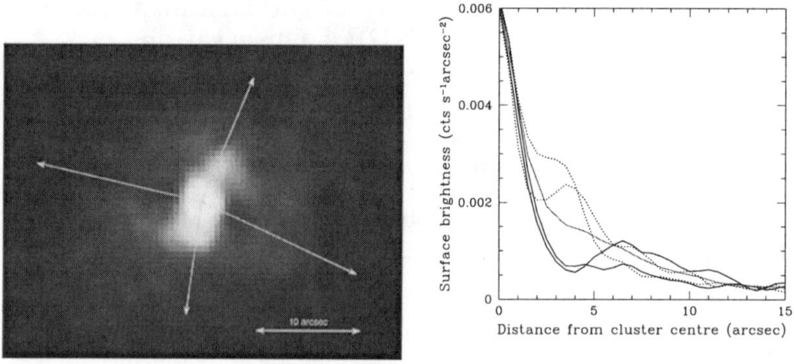

Figure 1. **Left panel**: CHANDRA image of the central 35 $arcsec^2$ of the cluster RBS797. **Right panel**: Traces of the X-ray emission from the cluster centre in direction of the minima (solid lines, position angles 77° and 246°), in the perpendicular direction (dotted lines, position angles 172° and 337°) and the average for the whole cluster (red line).

X-ray emission (solid lines) and in direction perpendicular to it (dotted lines) are plotted in the right panel in Fig. 1 (for reference in the directions of the cuts also see the left panel); the red line in the right panel shows an X-ray emission profile averaged for the whole cluster centre. We can estimate an X-ray deficit of a factor of approximately $3 - 4$ in the holes, compared to the perpendicular directions where the gas is probably compressed. These features suggest an interaction of the central cluster galaxy with the intra-cluster medium. This is the first discovery of such holes in a *distant cluster*.

3. Spectral analysis

The CHANDRA data have then been used to derive the temperature and metallicity of RBS797. The results of the spectral analysis are summarised in Table 1. We treat the central point-like source (radius $\leq 2\,arcsec$), probably an AGN, separately, because it would distort the thermal spectrum of the intra-cluster gas. The source has a count rate of about $0.026\,cts/s$ and a flux $F(2-10\,keV) = 2.2 \times 10^{-13} erg/s/cm^2$. The spectrum is vary flat; a power law yields a good fit with a very law index of 1.2 ± 0.2.
The cluster overall emission is well fit by a model with a relative high temperature of $T = 7.7keV$ and a metallicity of 0.26 in solar units. The X-ray luminosity is $L_X(0.5 - 7.0keV) = 4.0 \times 10^{45} erg \cdot s^{-1}$ and $L_X(bol) = 6.7 \times 10^{45} erg \cdot s^{-1}$.
The cluster has then been subdivided into three annuli in order to get an estimation of its radial temperature profile. The result is shown in the left panel in Figure 2; a trend is observed of increasing temperature with the radius, but the results are almost consistent with constant values. In the figure are also shown, in dotted lines, the confidence limits of the mean ambient cluster temperature.

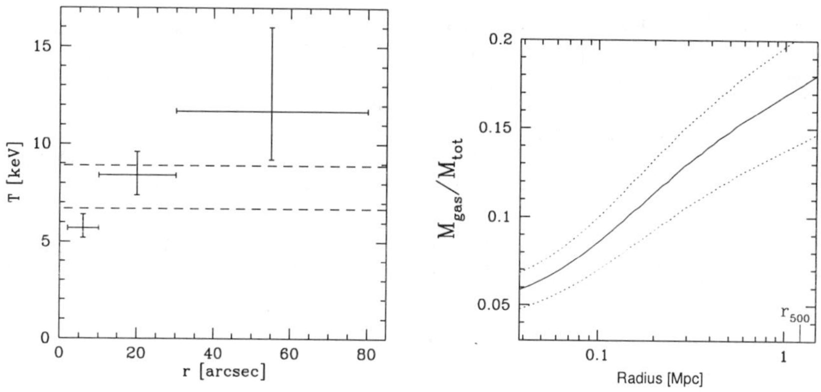

Figure 2. **Left panel**: Radial projected X-ray gas temperature profile. Horizontal error bars show the boundaries of the annuli. The dotted lines mark the 90% confidence region of the mean ambient cluster temperature. **Right panel**: ratio of the X-ray gas mass to total gravitating mass as a function of radius. The three lines show the best-fit value (solid line) and conservative confidence limits (dotted lines).

A spectral analysis of the bright *arms* surrounding the minima has been performed separately: their emission comes mainly from *soft* X-rays. With 1387 source counts extracted from these regions, a temperature of $4.4\,keV$, lower than for the rest of the cluster, is found, implying that these can not be shocked regions.

The X-ray surface brightness profile of RBS797 can be fitted well by a β model (Cavaliere & Fusco-Femiano 1976) when ignoring the central source: $\beta = 0.63 \pm 0.01$, $r_c = 8.1 \pm 0.6\,arcsec$ and a central surface brightness $S_0 = 6.6\,cts/s/arcmin^2$. Assuming a spherical symmetry, this corresponds to a gas mass $M_{gas}(r_{500}) = 1.13(\pm 0.09) \times 10^{14} M_\odot$, with $r_{500} = 1.22(\pm 0.08)$Mpc. Assuming also hydrostatic equilibrium, the total mass of the cluster can be estimated. For an isothermal cluster of $T = 7.7\,keV$ the total mass is $M_{tot}(r_{500}) = 6.5^{+1.6}_{-1.2} \times 10^{14} M_\odot$. This corresponds to a gas mass fraction $f_{gas}(r_{500}) = 0.17^{+0.06}_{-0.05}$ which is in good agreement with the gas mass fractions found in samples of nearby and distant clusters (Ettori & Fabian 1999; Schindler 1999). For this cluster we find an increasing gas mass fraction (see Figure 2, right panel) implying that the gas distribution is more extended than the dark matter.

4. Summary and conclusions

RBS797 appears as a slightly elliptical, hot cluster with the gas temperature slightly increasing outward with the radius. Two strongly pronounced depressions are present in the X-ray emission, having diameters of $20 - 30\,kpc$ and with a factor of $3 - 4$ less emission that in the other directions. The low temperature of the regions surrounding the holes is a clear indication that these cannot be

Table 1. Column (1) shows the radii, in arc-seconds, of the regions from which the photons have been extracted. Col.(2): model used for the fit: MeKaL = Kaastra & Mewe (1993), Liedahl, Osterheld & Goldstein (1995), PL = power law. Col.(3): energy range in keV. In columns (4-6) the fit parameters are given: temperature in keV (for PL fit is the power law index), metallicity in solar units, redshift. Col.(7): reduced χ^2. The hydrogen column density is fixed to the Galactic value (Dickey & Lockman 1990).

Radius	Model	En. range	kT	Metallicity	Redshift	$\frac{\chi^2}{d.o.f.}$
$2-80$	MeKaL	$0.5-10$	$7.7^{+1.2}_{-1.0}$	0.26 ± 0.10	0.39 ± 0.02	1.2
$2-80$	MeKaL	$2-10$	$7.1^{+1.6}_{-1.5}$	0.25 ± 0.12	0.39 ± 0.02	1.0
$2-80$	MeKaL	$2-10$	$7.4^{+1.7}_{-1.1}$	0.17 ± 0.11	0.354(fixed)	0.9
$2-10$	MeKaL	$0.5-10$	$5.7^{+0.7}_{-0.5}$	$0.38^{+0.17}_{-0.16}$	0.39(fixed)	1.6
$10-30$	MeKaL	$0.5-10$	$8.4^{+1.2}_{-1.0}$	0.25 ± 0.18	0.39(fixed)	1.0
$30-80$	MeKaL	$0.5-10$	$11.7^{+4.3}_{-2.5}$	$0.20^{+0.35}_{-0.20}$	0.39(fixed)	1.2
arms	MeKaL	$0.5-10$	$4.4^{+0.7}_{-0.6}$	0.26(*fixed*)	0.39(fixed)	1.2
$0-2$	PL	$0.5-10$	1.2 ± 0.2	-	0.39(fixed)	0.9
$0-2$	0.59× MeKaL +0.41 *PL*	$0.5-10$	7.7 (fixed) 1.1 ± 0.2	0.26 (fixed)	0.39(fixed)	0.9

shock regions; it is likely that the intra-cluster gas has been subsonically pushed away from the areas of low X-ray emission to the areas of higher emission by the pressure of the relativistic particles in radio lobes.

In the next months planned follow-up radio data will hopefully give a clearer picture of the internal structure of this cluster and of the interaction of the cluster active galaxy with the intracluster medium.

References

Cavaliere, A. & Fusco-Femiano, R. 1976, A&A, 49, 137

Dickey, J.M. & Lockman, F.J. 1990, ARA&A, 28, 215

Ettori, S. & Fabian, A.C. 1999, MNRAS, 305, 834

Fabian, A.C. et al. 2000, MNRAS, 318, 65

Kaastra, J.S. & Mewe, R. 1993, A&AS, 97,443

Liedahl, D.A., Osterheld, A.L. & Goldstein, W.H. 1995, ApJ, 438, 115

McNamara, B.R. et al. 2000, ApJ, 534, 135

Schindler, S. 1999, A&A, 349, 435

Schindler, S. et al. 2001, accepted by A&AL

Tracing Cosmic Evolution with Galaxy Clusters
ASP Conference Series, Vol. 268, 2002
S. Borgani, M. Mezzetti, and R. Valdarnini, eds.

The Coma cluster of galaxies observed by XMM-Newton

Ulrich G. Briel

Max-Planck-Institut fuer extraterrestrische Physik, Garching, Germany

J. Patrick Henry

Institute for Astronomy, Honolulu, Hawaii, USA

Monique Arnaud and Doris Neumann

CEA/DSM/DAPNIA Saclay, Service d'Astroph., Gif-sur-Yvette, France

Abstract. The Coma cluster of galaxies was observed with XMM-Newton in 13 partially overlapping pointings, with three of them re-observed because of a high proton background level. We present here the resulting X-ray maps in different energy bands and discuss the large scale structure of this cluster. Many point sources were found throughout the observed area, for some of them we give tentative identifications. We also present a temperature map and a hardness ratio map at the so far highest angular resolution. We discuss the variation of the temperature across the central cluster region in the light of recent merger activities.

1. Introduction

The Coma cluster of galaxies was one of the performance verification targets of XMM-Newton (Jansen et al. 2001) with the aim to verify XMM-Newton's ability to map large extended X-ray sources. Coma was chosen because it is a well studied source at all wavelength ranges and has a diameter which is several times larger than the FOV of the EPIC cameras of about 26 arcmin. Although Coma was long believed to be the archetype of a relaxed cluster of galaxies, the X-ray image taken during the Rosat All Sky Survey revealed the complex large scale structure, especially the subgroup around NGC4839 probably falling into the main cluster (Briel, Henry, & Böhringer 1992). Subsequent pointed observations with Rosat showed that Coma has irregular structure on different scales and was most probably formed by hierarchical clustering of several subunits (White, Briel, & Henry 1993; Vikhlinin, Forman, & Jones 1994 & 1997). ASCA observations and the pointed Rosat observation showed evidence of temperature structure in Coma (Honda et al. 1996; Briel & Henry 1998; Donnelly et al. 1999; Watanabe et al. 1999). Its interpretation favors the ongoing merging of the subgroup around NGC4839 with the main cluster. This is consistent with the large scale filamentary structure of the Coma-A1367 supercluster: there is striking alignment between the optical large scale filaments (see Fig.1 of West, Jones, & Forman 1995) and the NGC4839 group – main cluster orientation.

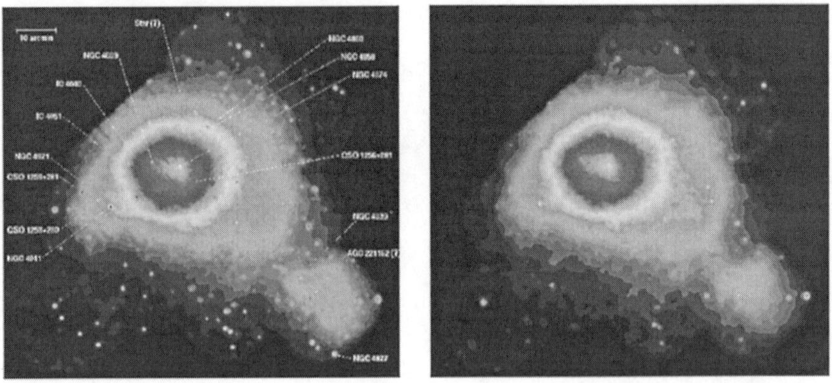

Figure 1. Mosaic of Coma in the energy range from 0.3 to 2 keV (left) and 2 to 5 keV (right), measured with the XMM-Newton pn camera. The images are smoothed with a two-dimensional Gauss function of varying width. Clearly seen is the enhanced emission between the NGC4839 group and the main cluster in the "hard" image when comparing it to the "soft" image: evidence for a hotter region, found also by spectral fitting (Arnaud et al. 2001) and in the hardness ratio map of the inner region (Figure 3, left).
Probably a hundred point sources are seen in the soft image; indicated are tentative identifications of some of them. Many bright galaxies seem to be X-ray sources including: NGC4827, 4839, 4858, 4860, 4874, 4889, 4911, & 4921, and IC4040 & 4051. At least three sources are quasars.

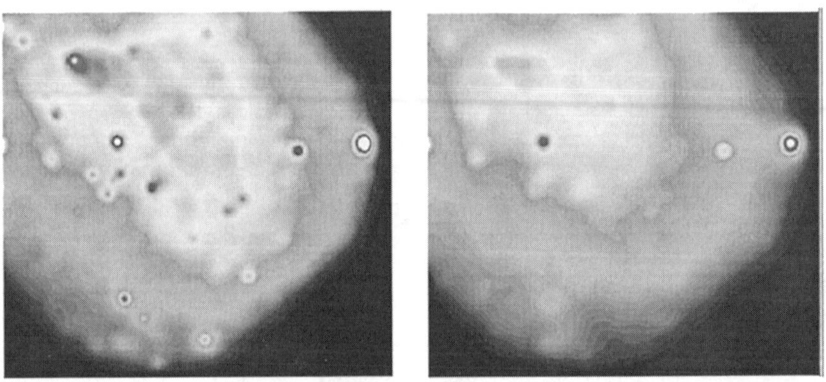

Figure 2. Zoom of the Coma-NGC4839 region from the above images, again in the energy range from 0.3 to 2 keV (left) and 2 to 5 keV (right). The contrast was enhanced to show the lower surface brightnes region more clearly. Note the swept back gas out of the NGC4839 galaxy and the displacement of the center of the group ICM and NGC4839. Both findings indicate, and are consistent with simulations, that the group is falling into the main cluster.

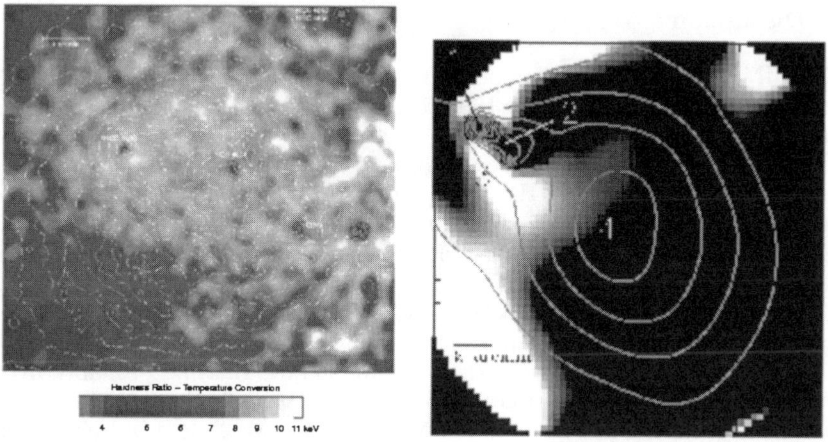

Figure 3. Hardness ratio maps of the central part of Coma (left, from Briel et al. 2001) and the Coma-NGC4839 region (right, from Neumann et al. 2001). The central map is obtained using pn data, the NGC4839 map using MOS data. Overlaid are isointensity contours of the corresponding smoothed soft (0.3 to 2 keV) X-ray images. Spectral fitting of the MOS data of different regions of the NCG4839 group revealed different temperatures: 3.1 keV in the core of NCG4839, 4.8 keV in the tail and 8.0 in the region south of the tail. The gas of the main body of the sub group is at 4.4 keV (from Neumann et al. 2001).

2. Observations and data analysis

Between May and June 2000, the Coma cluster was observed with the pn-camera (Strüder et al. 2001) and the MOS-camera (Turner et al. 2001) on bord XMM-Newton in 12 partially overlapping pointings. During these observations the proton induced background often varied and reached count rates of up to several hundred counts/sec in the energy band from 0.15 to 10 keV which led to a considerable loss of effective observing time in some of the pointings. Because of that three fields plus one additional field have been re-observed in December 2000. In Figure 1 (which is an update of the figures, shown in Briel et al. 2001) we show the new mosaic image of Coma, using the data from the pn-camera in the energy range from 0.3 to 2 keV and 2 to 5 keV. The images are vignetting corrected and a corresponding background is subtracted, using the procedure described in detail in Briel et al. (2001). In addition, to visualize better the low surface brightness of the cluster, we smoothed the images with two-dimensional Gaussian filters of variable widths.

We produced a hardness ratio map of the central Coma region, using the pn data, and of the NGC4839 group, using the MOS data. They are shown in Figure 3 with overlaid isointensity contours of the smoothed soft (0.3 to 2 keV) X-ray images.

3. Discussion

The XMM-Newton observation of the Coma cluster confirmed the scenario that the NGC4839 group is indeed in the process of falling into the main cluster, as it was originally proposed by Briel et al. (1992). The main arguments in favor of the infall are: a) the X-ray surface brightness distribution of the group, showing swept-back material around the group galaxy 4839 as well as the displacement of the galaxy and the center of the group ICM, probably also caused by ram-pressure, b) the complex temperature structure around the NCG4839 galaxy with the indication of a bow-shock.

Furthermore, there might be a second sub-cluster, located between the main cluster and the infalling group which is clearly visible in the harder surface brightness map of Figure 1 as enhanced emission, originating probably from a hot bubble. In both, the hardness ratio map of the inner Coma region (Figure 2, left) and the temperature map, obtained with the MOS data by Arnaud et al (2001), we see this hotter region, most likely to be due to adiabatic compression, caused by the merging of this second sub-cluster with the main cluster.

Acknowledgements: XMM-Newton is an ESA Science Mission with instruments and contributions directly funded by ESA Member States and the USA (NASA). The XMM-Newton project is supported by the Bundesministerium für Bildung und Forschung (BMFT/DLR), the Max-Planck Society and the Heidenhain-Stiftung, and also by PPARC, CEA, CNES, and ASI.
J. P. Henry thanks Prof. J. Trümper for the hospitality at the MPE.

References

Arnaud M., Aghanim N., Gastaud R., et al. 2001, A&A, 356, L67

Briel U.G., Henry J.P., & Böhringer H. 1992, A&A, 259, L31

Briel U.G., Henry J.P., Lamb, D., et al. 2001, A&A, 356, L60

Briel U.G. & Henry J.P. 1998, in "A Vision of an old Cluster: Untangling Coma Berenices", Eds. A. Mazure, F. Casoli, F. Durret and D. Gerbal, World Scientific Publishing, 170

Donnelly R., Markevitch M., Forman W., et al. 1999, ApJ, 513, 690

Honda H., Hirayama M., Watanabe M., et al. 1996, ApJ, 473, L71

Jansen F., Lumb D., Altieri B., et al. 2001, A&A, 365, L1

Neumann D., Arnaud M., Gastaud R., et al. 2001, A&A, 356, L74

Strüder L., Briel U.G., Dennerl K., et al. 2001, A&A, 356, L18

Turner M.J.L., Abbey A., Arnaud M., et al. 2001, A&A, 356, L27

Vikhlinin A., Forman W., & Jones C. 1994, ApJ, 435, 162

Vikhlinin A., Forman W., & Jones C. 1997, ApJ, 435, L7

Watanabe M., Yamashita K., Furuzawa A., et al. 1999, ApJ, 527, 80

West M., Jones C., & Forman W. 1995, ApJ, 451, L5

White S., Briel U.G., & Henry J.P. 1993, MNRAS, 261, L8

Tracing Cosmic Evolution with Galaxy Clusters
ASP Conference Series, Vol. 268, 2002
S. Borgani, M. Mezzetti, and R. Valdarnini, eds.

Magnetic Fields in Clusters of Galaxies

Federica Govoni

Dip. di Astronomia, Universita' degli studi di Bologna, Via Ranzani 1, 40127 Bologna, Italy

Gabriele Giovannini

Dip. di Fisica, Universita' degli studi di Bologna, Via B. Pichat 6/2, 40127 Bologna, Italy

Luigina Feretti

Istituto di Radioastronomia CNR, Via P. Gobetti 101, 40129 Bologna, Italy

Daniele Dallacasa

Dip. di Astronomia, Universita' degli studi di Bologna, Via Ranzani 1, 40127 Bologna, Italy

Gregory B. Taylor

National Radio Astronomy Observatory, Socorro, NM 87801, USA

Abstract. Here we present the analysis of the magnetic field in the cluster A2255 through the study of two radio galaxies belonging to the cluster. Since the radio galaxies sample different lines-of-sight across the cluster, an analysis of their Faraday Rotation Measures, together with the X-Ray data, provides information on the strength and the structure of the cluster magnetic field. A comparison between magnetic field strengths estimated using different techniques will also be discussed.

1. Introduction

The presence of cluster magnetic fields is now well established, although different methods of analysis give somewhat discrepant results for the field strength.

Direct observational evidence of cluster magnetic fields is demonstrated by the existence of diffuse cluster-wide synchrotron radio emission (radio halos and relics). Assuming energy equipartition between the magnetic field and the relativistic particles, measurements of the strength of the magnetic fields in some radio halos have been obtained. From Very Large Array (VLA) observations of six diffuse sources we have found magnetic fields of about 1 μG [1] (Govoni et

[1]We adopt H_0=50 km $sec^{-1}Mpc^{-1}$.

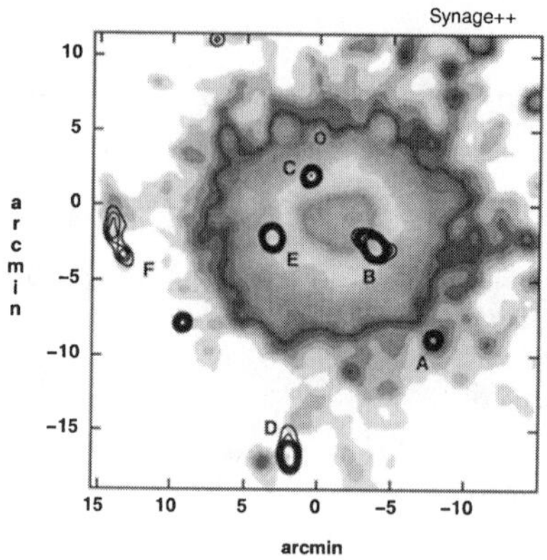

Figure 1. The 1.4 GHz radio image obtained from the NVSS (contour) superimposed to the X-Ray image of Abell 2255 taken from Rosat PSPC archive. The radio contour levels are 5, 10, 20, 30, 50, and 70 mJy/beam. The radio image has an angular resolution of 45″. The Rosat image has an angular resolution of 25″ (FWHM) and has been smoothed with a Gaussian of $\sigma = 30''$.

al. 2001a). In the Coma cluster, the equipartition magnetic field is $\simeq 0.4\mu$G (Giovannini et al. 1993). These values are consistent with those suggested from the recent detection of Inverse Compton (IC) hard X-ray emission in clusters with halos or relics. By comparing the non-thermal hard X-ray emission detected in the Coma cluster with the diffuse radio emission (the halo Coma C) Fusco-Femiano et al. (1999) obtained a magnetic field of 0.15 μG.

Indirect measurements of the magnetic field strength can also be determined in conjunction with X-ray observations of the hot gas, through the study of the Faraday Rotation Measure of radio sources located inside or behind the clusters. Strong magnetic fields ($\simeq$ 5-30 μG) have been found in cooling flow clusters (e.g. 3C 295, Perley & Taylor 1991 and Allen et al. 2001; Hydra A, Taylor & Perley 1993) where extremely large Faraday rotations have been revealed. Significant magnetic fields ($\simeq$ 2-7 μG) have also been detected in clusters without a cooling flow (e.g. Coma, Feretti et al 1995; A119, Feretti et al. 1999; 3C129, Taylor et al. 2001; A514, Govoni et al. 2001b). From a statistical study of several clusters without cooling flows, Clarke et al. (2001) obtained magnetic fields of 4-8 μG, in agreement with previous results obtained from individual clusters.

2. A2255

Here we present the analysis of the magnetic field in the cluster A2255. Fig. 1 shows the radio image at 1.4 GHz obtained from the NRAO Very Large Array Sky Survey (NVSS, Condon et al. 1998), overlaid on the X-Ray image obtained from the Rosat PSPC archive. This cluster (z=0.0806) does not posses a cooling flow. It is known to contain two diffuse radio sources classified as a radio halo and a relic (Jaffe & Rudnick 1979, Harris et al. 1980; Burns et al. 1995, Feretti et al. 1997) and it is characterized by the presence of several radio sources located at various distances from the cluster center. The sensitivity of the NVSS image is not enough to show the halo and the relic sources in Fig. 1. Feretti et al. (1997) estimated an equipartition magnetic field strength of 0.5 μG from the study of the extended halo source.

To estimate the cluster magnetic field from the Rotation Measure (RM), we performed a study in total intensity and polarization of all the radio galaxies indicated by letters in Fig. 1 using the Very Large Array (VLA) at 3.6 and 6 cm. No polarized flux was detected in sources A, C, D, and F, therefore only the two sources B and E are suitable for a Rotation Measure study. In Fig. 2 we present the total intensity images of the two sources at 6 cm obtained with a resolution of $2''(\simeq 4$ kpc). The RM images were obtained by performing a least-square fit of the polarization angle images at each pixel as a function of λ^2.

In the Table 1, along with the distance from the cluster center, we report the maximum absolute value, the mean, and the σ of the RM values for the two sources.

Table 1. Rotation Measure

Name	Dist. center ($arcmin$)	$\|RM_{max}\|$ (rad/m^2)	$< RM >$ (rad/m^2)	σ_{RM} (rad/m^2)
A2255B	3.2	-225	-89	196
A2255E	5.0	+176	94	212

The determination of the strength of the magnetic field in a cluster depends on several assumptions, including the model for the X-ray gas distribution and the magnetic field structure. Following Felten et al. (1996), we assumed in A2255 a randomly oriented magnetic field in cells of uniform size and strength, and a gas density distribution given by the hydrostatic isothermal β-model. From the RM images we estimated that the magnetic field is tangled on scales of about 10 kpc. Considering the β-model parameters obtained by Feretti et al. (1997), the values of σ_{RM} obtained for the sources fit fairly well with Felten's model and a magnetic field strength of about $5 - 7$ μG.

3. Conclusions

We have estimated the cluster magnetic field in A2255 through the Rotation Measure study of two cluster galaxies. Even allowing for the uncertainties related to the model used in the computation, a plausible value for the magnetic field strength is about 5-7 μG with coherence length on scale of about 10 kpc. This result is in agreement with previous results obtained on clusters without cooling flows. By studying the radio halo in A2255 Feretti et al. (1997) found an equipartition magnetic field of about 0.5 μG. This cluster is the second one, after Coma, where magnetic fields have been evaluated both from equipartition arguments and from RM measurements. As in the Coma cluster, in A2255 there is a significant difference in the magnetic field strength calculated with the two different methods of about an order of magnitude. Some explanations for the discrepancy obtained using different methods of analysis have been proposed in literature (Carilli & Taylor 2002 and the references therein) however this open problem it will be investigated in more detail in the future.

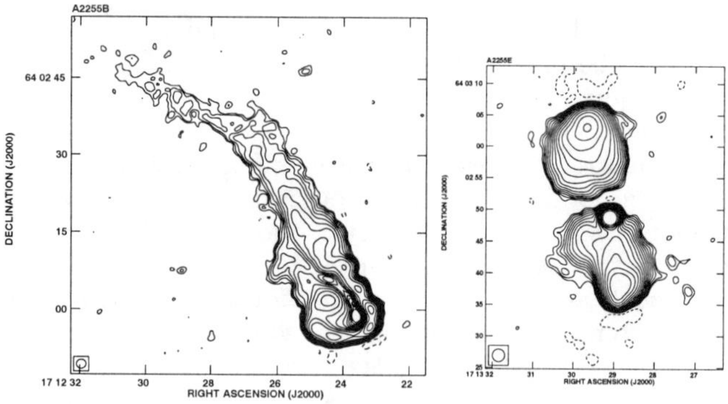

Figure 2. Total intensity radio images of B and E sources obtained at 6 cm with a resolution of 2″. Contour levels are: -0.06, 0.06, 0.08, 0.12, 0.17, 0.24, 0.34, 0.48, 0.68, 0.96, 1.36, 1.92, 2.71, 3.84, 5.43 mJy/beam.

References

Allen, S.W., Taylor, G.B., Nulsen, P.E.J., Johnstone, R.M. et al., 2001, MNRAS 324, 842

Burns, J.O., Roettiger, K., Pinkney, J., et al. 1995, ApJ 446, 583

Carilli, C.L., & Taylor G.B., 2002, ARA&A, in press

Clarke, T.E., Kronberg, P.P., & Böhringer, H. 2001, ApJ 547, L111

Condon J.J., Cotton W.D., Greisen E.W. et al., 1998, AJ 115, 1693

Felten, J.E., 1996, Clusters, Lensing, and the Future of the Universe, ASP Conference Series, Vol. 88, Eds. V.Trimble and A.Reisenegger, p. 271

Feretti, L., Dallacasa, D., Giovannini, G., & Tagliani, A. 1995 A&A 302, 680

Feretti, L., Böhringer, H., Giovannini, G., & Neumann, D. 1997, A&A 317, 432

Feretti, L., Dallacasa, D., Govoni, F., Giovannini, G., Taylor, G.B., Klein, U., 1999, A&A 344, 472

Fusco-Femiano, R., Dal Fiume, D., Feretti, L., Giovannini, G., Grandi, P., Matt, G., Molendi, S., & Santangelo A. 1999 ApJ, 513, L21

Giovannini, G., Feretti, L., Venturi, T., Kim, K.T., Kronberg, P.P 1993, ApJ 406, 399

Govoni, F., Feretti, L., Giovannini, G., Böhringer, H., Reiprich, T.H., & Murgia, M. 2001a, A&A in press astro-ph/0107275

Govoni, F., Taylor, G.B., Dallacasa, D., Feretti, L., & Giovannini, G. 2001b, A&A submitted

Harris, D.E., Kapahi, V.K., Ekers, R.D., 1980, A&AS 39, 215

Jaffe, W.J., Rudnick, L., 1979, ApJ 316, 113

Perley R.A., Taylor G.B., 1991, AJ 101, 1623

Taylor, G.B., Perley R.A., 1993, ApJ, 416, 554

Taylor G.B., Govoni F., Allen S., Fabian A.C., 2001, MNRAS in press, astro-ph/0104223

THEORY AND SIMULATIONS

Tracing Cosmic Evolution with Galaxy Clusters
ASP Conference Series, Vol. 268, 2002
S. Borgani, M. Mezzetti, and R. Valdarnini, eds.

Frequently Asked Questions on the Luminosity–Temperature relation in Groups and Clusters

Fabio Governato

Osservatorio Astronomico di Brera, Milan, Italy

Abstract. Efforts to understand the deviation of the L–T relation from a simple scaling law valid for clusters and groups have triggered a number of studies on the subject. Techniques and approaches differ widely but most works agree on the important role played by gas cooling and heating sources like AGNs and SNs. Observations set useful constraints on the evolution of the intra cluster medium (ICM): a 100KeV/cm^2 entropy floor in the core of groups and about 5–10% of baryons being converted into stars. However, essential details like the nature of the dominant heating mechanism and the quantitative importance of cooling still need to be addressed. I suggest that a new generation of high resolution N-body simulations and a quantitative comparison of results between different approaches is required to improve results and increase our understanding of the problem.

1. Q: What are the Theoretical Expectations for the L–T relation?

The first attempt to model the ICM in the framework of the hierarchical scenario assumed its thermodynamical properties to be entirely determined by gravitational processes, like adiabatic compression during collapse and shock heating (Kaiser 1986). If no characteristic scales are present in the underlying cosmology (i.e., Einstein–de-Sitter cosmology and power–law shape for the power spectrum of density perturbations), this model should predict hot gas within rich clusters to look the same as within poor groups, since gravity in itself does not have characteristic scales. Under the assumptions of emissivity dominated by free–free bremsstrahlung and of hydrostatic equilibrium of the gas, this model predicts $L_X \propto T^2(1+z)^{3/2}$ for the shape and evolution of the relation between X–ray luminosity and ICM temperature (also Eke, Navarro & Frenk 1998). Furthermore, if we define the gas entropy as $S = T/n_e^{2/3}$ (n_e: electron number density; e.g., Eke et al. 1998), then the self–similar ICM has $S \propto T(1+z)$.

2. Q: What are the observational Constraints?

The simple model was soon recognized to fail at accounting for several observational facts: (a) the L_X–T relation for nearby clusters is steeper than predicted, with $L_X \propto T^{\sim 3}$ for $T > 2$ keV clusters (e.g., David et al. 1993; White, Jones & Forman 1997; Allen & Fabian 1998; Markevitch 1998); with a possible further steepening at the group scale, $T < 1$ keV (Ponman et al. 1996; Helsdon & Ponman 2000); (b) no evidence of evolution for its amplitude has been detected out to $z > 1$ (e.g., Fairley et al. 2000; Della Ceca et al. 2001; Borgani et al. 2001a); (c) the gas density profiles in central regions of cooler groups is relatively softer than for hotter cluster and, correspondingly, the entropy is higher (e.g., Ponman, Cannon & Navarro 1999; Lloyd–Davis et al. 2000). Observational evidence provides a fairly robust measure of the entropy floor for the diffuse hot gas in groups (Lloyd–Davis et al. 2000) of 100 KeV/cm^2.

A related, very important measurement is the fraction of baryons that has been converted into a "cold", non X–ray emitting phase (i.e stars and neutral hydrogen). This constraint is crucial as it defines the amount of energy available from SNs and the importance of cooling in groups and clusters (Renzini 2000). Balogh et al. (2001) estimate that only a small fraction (5–10%) of baryons has been converted into stars, independently of the total virial mass of the cluster.

3. Q: What are the physical processes that could affect the L–T relation?

• *Gas cooling.* At the center of groups the cooling time is much shorter than a Hubble time. Cooling would remove low entropy gas transforming it into stars and possibly originate the entropy floor observed. While advocated on the ground of its simplicity (Muanwong et al 2001 also Voit & Bryan 2001), the amount of gas involved cannot exceed the observed fraction of 5–10% of the total baryon fraction.
• *Energy injection from AGNs* (e.g Böhringer, this proceedings) would be an important source of heating. Most of the energy would come from jets that would carve "bubbles" in the ICM (Mc Namara et al 2001, Quilis et al. 2001), while the X-ray emission from the central engine in radio quiet QSOs would likely have a small effect due to the small cross section for the process.
• *Energy injection from SNs.* Winds propelled by SNs explosions (an hefty 10^{51}erg/SN) would transfer heat to the ICM, possibly escaping the halos of individual galaxies in the forming protocluster and enrich the ICM with metals. (Note: substantial metal enrichment could originate by ram pressure and tidal stripping of galaxies, see Moore, Quilis & Bower 2001).

4. Q: What are the most used theoretical tools to study the L–T relation?

• *Analytical methods* allow a fast exploration of parameter space, but must sometimes rely on the parametrization of the end results of complicated physical processes. While this approach requires studying the physics of gas in a somewhat idealized situation (spherical collapse, accretion from a uniform background) the amount of insight gained is impressive, although conclusions sometimes differ widely (Kaiser 1991,Valageas & Silk 1999, Tozzi & Norman 2000, Bryan 2000, Balogh, Babul & Patton 1999, among many). The most refined methods are an improvement of the standard machinery developed to link gravitational clustering in CDM models and gas physics (Sommerville & Primack 1999, Bower et al. 2001, Menci & Cavaliere 2000, Cavaliere, Giacconi & Menci 2000).

•*N-body simulations* are a powerful method, able to treat the highly non–linear problem of the formation of cosmic structures with very few assumptions. Recent observational evidence for a mono phase ICM (Böhringer et al 2001) remove some potential worries about the use of SPH methods, which do not describe well a multiphase medium. Some recent works on the L–T relation including feedback and star formation (Metzler & Evrard 1994, Bialek, Evrard & Mohr 2001, Borgani et al 2001b, Valdarnini 2001) provide a good starting point and an extensive list of references to the subject of simulations of galaxy clusters.

5. Q: Heating, Cooling or both?

As mentioned above, several papers suggested that simply the addition of cooling would remove enough baryons to reconcile models with the observed L–T relation. This would likely be in contrast with observational evidence that points to a limited fraction of baryons in clusters in a "cold" non X–ray emitting phase. Moreover it would be difficult to hide the large quantity of cooled gas in "dark baryons" as they would cause detectable deepening of the central potential well at the center of clusters (e.g Lewis et al. 1999, Valdarnini 2001)

On the other hand recent theoretical works, both analytical and numerical seem to (perhaps slowly) converge to a required energy input comparable to 100% of what available from SNs (Kravtsov & Yepes 2000, Bower et al 2001, or even a few times higher (Borgani et al. 2001b, Valageas & Silk 1999 , Wu, Fabian & Nulsen 2000) to satisfy the L–T relation.

While semi-analytical models estimate the energy available from SNs to be of the order of ~ 0.25 keV per particle , Renzini (2000) suggested an even smaller energy budget available from SNs: just 0.1 keV. Taken together these works suggest that the required energy be provided both by SNs and AGNs, perhaps active as major mergers form the early type population of cluster galaxies. Unfortunately none of the N–body simulations cited above (some analytical works did) treated *both* cooling and heating in a fully consistent way or included the effect of AGNs on the ICM. This leaves the quantitative contribution of cooling and heating rather uncertain.

6. Q: Is there a preferred epoch for heating the ICM?

Not really. As the entropy S depends on a inverse power of ρ it might seem more efficient to heat the ICM at lower z (say z<5) as suggested in Tozzi & Norman (1999). However, in their model the authors heated the gas *before* accretion, while at the background density. In a more realistic case gas will be heated when inside halos that will later merge to form the cluster. While this requires more energy for a given entropy S, (ρ is higher by a factor ~ 200) this is more than compensated by the fact that halos collapsed at high z will form the core of the cluster at lower z, i.e. you dump energy right where it is needed. Borgani et al (2001b, 2001c) and Bialek, Evrard & Mohr (2000) tried a large range of z (1 to 5 and 20 respectively) finding a weak dependence on results for the argument given above. Of course, a redshift too close to z $\sim$1 for heating would cause the L–T to evolve significantly at moderate redshift, contrary to observations.

To wrap up these results: it is very likely that a combination of cooling & feedback will give the right, non evolving L–T relation, although only stronger observational constraints and more refined simulations will allow us to evaluate the *quantitative* role of cooling, SNs and AGNs.

7. Future Improvements:

Numerical simulations including a treatment of hydrodynamical processes have reached a high degree of maturity, with several codes giving comparable results (Frenk et al 1999). However, the number of particles used to simulate *individual* halos is still systematically lower than in DM only runs (e.g Ghigna et al. 2001). The hot debate on the shape of inner density profiles in DM halos has started a race to higher resolution runs (Governato, Ghigna & Moore 2001). While the quest for the ultimate density profile of DM halos might ultimately prove futile it has sparkled a number of improvements on codes and, equally important, increased the robustness of results. A key technique is the so called "renormalization technique" (Katz & White 1993) where a halo taken from a cosmological simulation is resimulated at much higher resolution, while sparsely resampling the surrounding region. Fig.1 shows the crucial effect of increasing resolution on a sample of SPH simulations of renormalized halos. (Borgani et al 2001c). The L–T obtained from simulations of large cosmological volumes (where particle number and ϵ/R_{vir} for each halo depends on its mass) likely suffer of systematic biases due to the worsening resolution for less massive halos: namely smaller L_x and slower cooling due to lower central densities.

This is my very personal wish list for future theoretical and observational studies on the subject:

• Estimates of L_X and T in groups that exclude contamination from AGNs.
• Estimates of the amount of metals in clusters to a significant fraction of R_{vir} as another method to infer the energy release from SNs (Renzini 2000, De Grandi & Molendi 2001).
• A robust estimate of the fraction of "cold" baryons which keeps into account the amount of intracluster stars. Recent claims (see Arnaboldi et al 2001) based on intracluster PNs would almost double the the efficiency of star formation in clusters.
• Simulations of sufficient resolution to treat cooling and star formation processes in a robust way. (note: robust does not mean realistic, but it's a start).
My checklist for simulating an individual cluster:

$N > 10^5$
$\epsilon/R_{vir} < 1\%$
$N_{gas}(r > R_{vir}) > 7\text{–}10\ 10^4$
Simulation box $\sim$ 100Mpc
$N_{steps} >> 10^{3-4}$

• an exploration of the role of QSO's and AGNs in heating the ICM in a cosmological context.

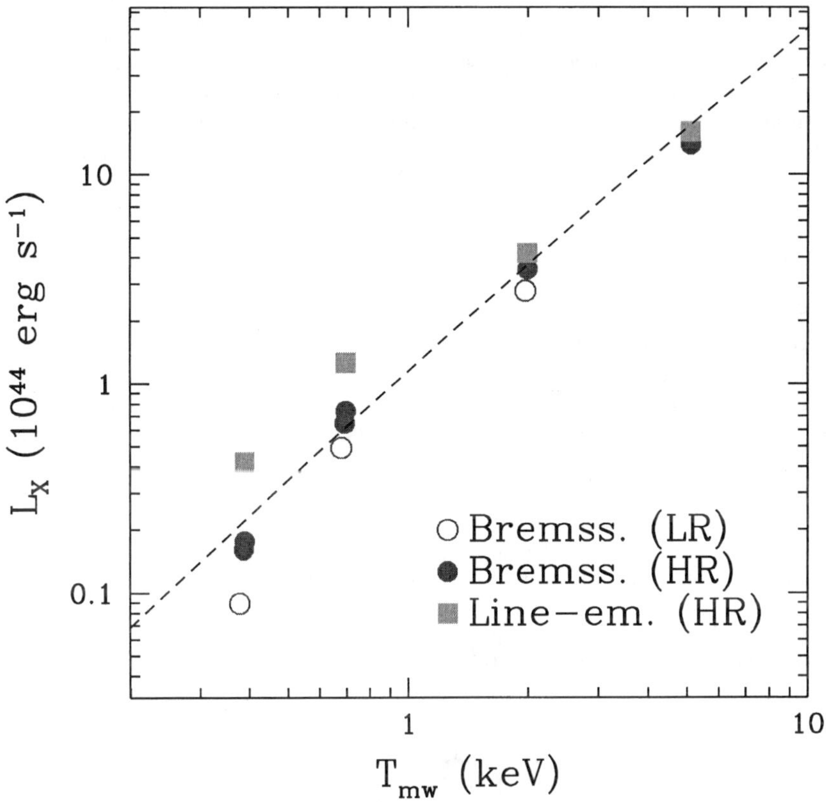

Figure 1. X-ray luminosity vs mass weighted Temperature: effects of resolution. Empty (red) dots are runs with ~ 0.5–2×10^4 gas particles within R_{vir}. Full dots are halos simulated with $N_{gas}(r > R_{vir}) \sim 10^5$. Softening is $1\% R_{vir}$. The slightly higher filled (red) dots show the effect of having instead $\epsilon = 0.5\%$ R_{vir}: L_x grows by 15–20%. (Green) filled squares show the large effect of including line emission in the estimate of X–ray luminosities for the high res runs. The dashed line is the scaling predicted by Kaiser (1986) assuming NFW (Navarro, Frenk & White 1995) profiles. (data from Borgani et al. 2001 in prep.)*Bottom line: use at least 10^5 gas particles per halo to avoid nasty systematic effects like reproducing a bend in the L–T relation at $T <$ 0.3 KeV. Correct for line emission if you compare vs. real data.*

• Introduction of heating schemes that closely follow the time evolution of the physical process behind it.
• Resolving the discrepancy between adiabatic runs with mesh and SPH codes (e.g Kravtsov et al. 2001, also Springel & Hernquist 2001 and Cen & Ostriker 1999) : why SPH simulations do not have an entropy floor and mesh codes do?

Acknowledgments. I wish to thank Stefano Borgani, Marino Mezzetti and Riccardo Valdarnini for organizing such a interesting and lively meeting, and Uros Seljak and Bepi Tormen for stimulating discussions (although they made me skip a coffee break or two). I also thank my collaborators Stefano Borgani, Nicola Menci, Tom Quinn, Joachim Stadel, Paolo Tozzi & James Wadsley for allowing me to show results in Fig.1. in advance of publication.

8. References

Allen S.W., Fabian A.C., 1998, MNRAS, 297, 63
Arnaboldi et al, 2001 AJ in press
Balogh M.L., Babul A., Patton D.R., 1999, MNRAS, 307, 463
Balogh et al., 2001, MNRAS, 326, 1228
Bialek, J.J., Evrard, A.E., Mohr, J.J, 2001, ApJ, 555, 597
Böhringer, H. et al 2001, A&A, in press
Borgani et al. 2001a ApJ, 561, 13
Borgani et al. 2001b, ApJ, 559, 71
Borgani et al. 2001c, in preparation
Bower R.G., Benson A.J., Bough C.L., Cole S., Frenk C.S., 2001 MNRAS325,497
Bryan G.L. 2000, ApJ, 544, 1
Cavaliere A., Giacconi R., Menci N., 2000, ApJ, 528, L77
Cen R., & Ostriker J.P., 1999 ApJ517, 31
David L.P., Slyz A., Jones C., Forman W., Vrtilek S.D., Arnaud K.A., 1993, ApJ, 412, 479
De Grandi S. & Molendi S., 2001 ApJ, 551, 153
Della Ceca R. et al. 2000, A&A, 353, 498
Eke, V., Navarro J.F. & Frenk C.S. 199,8 MNRAS, 503, 569
Fairley B.W. et al., 2000, MNRAS, 315, 669
Frenk, C.S. et al. 1999, ApJ, 525, 554
Ghigna S. et al. 2000, ApJ, 544, 616
Governato, F., Ghigna, S. & Moore, B. 2001, proceedings of "Astrophysical Ages and Timescales, ed. Ted Von Hippel
Helsdon S.F., Ponman T.J., 2000, MNRAS, 315, 356
Kaiser N., 1986, MNRAS, 222, 323
Kaiser N., 1991, ApJ, 383, 104
Katz, N. & White, S.D.M., 1993 ApJ412, 455
Kravtsov A.V., Yepes G., 2000, MNRAS, 318, 227
Kravtsov A.V. et al. ApJsubmitted, astro–ph/0109077
Lewis G.F., Babul A., Katz N. Quinn T., Hernquist L., 2000, ApJ, 536, 623
Lloyd-Davies E.J., Ponman T.J., Cannon D.B., 2000, MNRAS, 315, 689
Markevitch M., 1998, ApJ, 504, 27
Mc Namara, B.R. et al 2001, ApJ, 562, 149
Muanwong et al 2001, ApJ2001 552, 27

Menci N., Cavaliere A., 2000, MNRAS, 311, 50
Metzler C.A., Evrard, A.E. 1994 ApJ437, 564
Moore B., Quilis V. & Bower, R. 2000, ASP Conf. Sr. 197, 363
Ponman T.J., Bourner P.D.J., Ebeling H., Böhringer H., 1996, 293, 690
Ponman T.J., Cannon D.B., Navarro J.F., 1999, Nature, 397, 135
Quilis, V., Bower, R.G. & Balogh, M.L. 2001, astro-ph/0109022
Renzini A., 2000 astro-ph/0001312, proccedings of "LSS in the X–ray Universe, 1999, ed Atlantisscience.
Somerville, R.S., & Primack, J.R. 1999, MNRAS, 310, 1087
Springel, V. & Hernquist L. 2001 astro-ph/0111016 Tozzi P., Norman C. 2001, ApJ, 546, 63
Valageas P., Silk J., 1999, A&A, 350, 725
Valdarnini R., 2001, ApJ, in press astro–ph/0110545
Voit G.M., & Bryan G.L., 2001, Nature, in press
White D.A., Jones C., Forman W., 1997, MNRAS, 292, 419
Wu K.K.S., Fabian A.C., Nulsen P.E.J., 2000, MNRAS, 318, 889 (preprint astro-ph/9907112)

Tracing Cosmic Evolution with Galaxy Clusters
ASP Conference Series, Vol. 268, 2002
S. Borgani, M. Mezzetti, and R. Valdarnini, eds.

Simulations of Galaxy Clusters observed by Chandra

Alessandro Gardini

Dipartimento di Astronomia, Università di Padova, vicolo dell'Osservatorio 2, 35122 Padova - ITALY

Abstract. A software package able to simulate imaging observations of galaxy clusters by the Chandra X-ray telescope is here presented. We start from high resolution N-body hydrodynamical simulations of galaxy clusters and assign to each gas particle a spectrum of emissivity, after assuming the MeKaL model. We then construct spatial images of the source differential flux which are used to create lists of incoming X-ray photons, preserving information on photon direction and energy. The photon lists are passed on to the Chandra simulator (MARX) to produce the final observation events. Background events are added to complete the simulation. Data analysis is currently in progress and simulated observations by other telescopes will become available in the Future.

1. Introduction

We present a software package that, starting from high resolution hydrodynamical simulations of galaxy clusters, reproduces observations of the ICM by the Chandra X–ray telescope. Syntetic data can then be processed and analysed using the same tools and procedures as real observations. This enables us to directly compare the results of the simulations with the observational data and furthermore to compare the intrinsic properties of clusters, as given by simulations, with the same quantities obtained at the end of the process of data reduction and analysis. In this way, we are able to check the reconstructed properties of clusters as well as to get indication on the quantity of information retained or lost in the observations. On the other hand it has to be remembered that the reliability of the comparison between a simulated quantity and the real counterpart depends on how well the relevant physics inside simulations is modelled.

2. Cluster Simulations

The package can be easily adapted to run on every kind on hydro-simulations. For our input we used a set of cluster simulations realized by G.Tormen via the resimulation technique of regions of interest inside a cosmological simulation (see e.g. Tormen, Bouchet, & White 1996).

This starting simulation reproduces a ΛCDM model of Universe, with parameters $\Omega_0 = 0.3$, $\Omega_\Lambda = 0.7$, $H = 70\, km\, s^{-1}\, Mpc^{-1}$ and baryon fraction ~ 0.1.

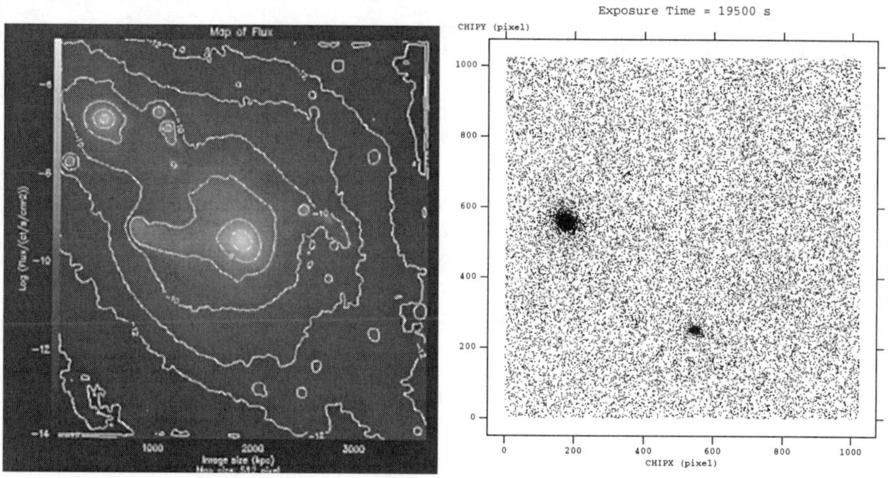

Figure 1. Left: map of projected flux ($photon\, s^{-1}\, cm^{-2}$) of a cluster at redshift $z \simeq 0.778$. Right: simulated observation of the same object by the Chandra ACIS-S3 detector given an exposure time of 19500 s. The image is turned upside-down and the main clump is set at the Chandra ACIS-S aimpoint.

It describes a box of size equal to 479 $Mpc\, h^{-1}$ with 512^3 particles of Dark Matter[1].

Clusters are identified in its final output using a spherical overdensity algorithm and resimulated singularly from the beginning using the GADGET Tree-SPH code. Each resimulation makes use of $\sim 10^6 + 10^6$ particles of Dark Matter and gas to model the cluster and the surrounding region, while the external tidal field is reproduced by a smaller number of more massive particles. The mass of the Dark Matter particles ranges between 2 and $5 \cdot 10^9 M_\odot\, h^{-1}$ depending on the cluster, and the gas particle mass scales accordingly. The softening related to the cubic spline kernel is $\sim 5\, kpc\, h^{-1}$. At the end of the simulation the cluster is described by 2 to $3 \cdot 10^5$ particles for each species inside the virial radius. A set of twelve clusters has been simulated and for each one 50 outputs were retained between $z = 10$ and $z = 0$.

3. Method : (1) general

In order to describe how the package works, it can be useful to distinguish between a first general part of the method and a second one related to Chandra. For a cluster at a given redshift z, the first part accounts for the construction

[1] http://www.mpa-garching.mpg.de/Virgo/VLS.html

of a map of the incoming flux (or of some other related quantity) that preserves the energy information.

We start by considering the position $\vec{x}_i$, mass m_i, density ρ_i and temperature T_i for each gas particle i. We then assign to the particle a volume $V_i = m_i/\rho_i$ that is assumed cubic and centered on $\vec{x}_i$.

We then assign to each particle the related spectral emissivity $\epsilon_i^\nu = \epsilon_i^\nu(\rho_i, T_i)$ computed by XSPEC using the MeKaL spectral model. The metallicity enters as a parameter; here we present results obtained assuming $Z = 0.3\, Z_\odot$, but any other value is possible. The luminosity of each particle becomes then $L_i^\nu = \epsilon_i^\nu \cdot V_i$ and is assumed uniformly distibuted on V_i.

In general, we consider emissivity and derived quantities (luminosity, flux, surface brightness) in terms of incoming photons instead of energy because this quantity is more easily related to observations.

Given an observational direction, we get a bidimensional map of the cluster spectral luminosity projecting each L_i^ν on a pixellized perpendicular image and summing on particles. The information is then stored in a 3D array, where two dimensions are spatial and the third one contains the energy channels. The transformation from the cluster luminosity at redshift z to the flux per pixel at $z = 0$ is then direct: in term of incoming photons, it holds

$$F_\nu = \frac{(1+z)^2 L_{\nu(1+z)}}{4\pi D_L^2},$$

being D_L the luminosity distance. Note that the frequency ν in the flux is related to $\nu(1+z)$ in the cluster luminosity.

Tipically, the number of pixels ranges from 256^2 to 1024^2, depending on the field of view, while the energy ranges from 0.1 to $10\, keV$ in bins of $20\, eV$, even if $100\, eV$ should be sufficient according to the energy resolution of Chandra (Chandra Proposers' Observatory Guide[2]).

4. Method : (2) Chandra

It has to be remembered that a simulator of Chandra already exists: it is called MARX[3] and performs the ray-tracing of incoming photons of a given source inside the telescope up to the possible final detection. It returns a list of binary data files describing the detection events that can be assembled to reproduce a CXC Level 1 FITS data file.

However, only some kind of sources of simple form and uniform spectral flux are implemented in MARX, while users are allowed to construct their specific source writing some C routines and linking it dynamically to the code. So did we, as described in the following.

For a given cluster, we generate a list of incoming photons by random extraction from the cumulative distribution of the spectral flux array described above. Each photon hence owns a position in a pixel and an energy in the energy

---

[2] http://asc.harvard.edu/udocs/docs/docs.html

[3] http://space.mit.edu/ASC/MARX

band, with the given resolution. Other random extractions define the position of photon inside the pixel and its energy in the bin. The number of photons extracted is set to 10^6.

We then enable MARX to accept as input this photon list and the total flux of the cluster (in $photon\, s^{-1}\, cm^{-2}$): photons are randomly extracted from the list according to the total flux and the exposure time, and usually processed up to the possible detection.

No background is introduced by MARX. We add background events randomly extracting them from Chandra ACIS observations of empty fields available in the home page of M.Markevitch[4]. An example of a complete simulated observation is shown in the Fig. 1.

5. Data analysis and future developments

As written above, the detection events generated by MARX have the same caracteristics of the CXC Level 1 observational data file and can be analysed using the same procedures and CIAO tools. This analysis is currently in progress and will concern the reconstucted properties of clusters, the cluster morphology (maps and profiles) and its relation to the dynamical state of the object. Detectability of clusters at high redshift is another possible application.

We plan to extend this package to the XMM-Newton telescope, using the related simulator SCISIM, in the next future.

6. Acknowledgments

The author wish to thank L.Moscardini, G.Tormen and S.DeGrandi, that collaborate to this project.

References

Tormen, G., Bouchet, F.R., & White, S.D.M., 1997, MNRAS, 286, 865

[4]http://hea-www.harvard.edu/~maxim

Tracing Cosmic Evolution with Galaxy Clusters
ASP Conference Series, Vol. 268, 2002
S. Borgani, M. Mezzetti, and R. Valdarnini, eds.

Evolution of Cluster Scaling Relations

Peter Thomas

Astronomy Centre, University of Sussex, Brighton, BN1 9QJ, UK

Abstract. I look at the evolution of scaling relations in a simulation of clusters that includes radiative cooling of the intracluster medium. The temperature-mass relation for the brightest clusters approximately follows the self-similar theoretical prediction, but that for the luminosity-temperature relation does not. Individual clusters show mainly temperature evolution but with occasional large excursions in both temperature and luminosity during merging events.

1. Introduction

Predictions of the evolution of cluster properties often make use of the theoretical scaling relations for self-similar systems. However, these do not apply to systems in which there is non-gravitational heating and/or cooling, as is required to reproduce the present-day luminosity-temperature relation. In this paper I look at the evolution of the X-ray properties of clusters in a cosmological simulation that includes radiative cooling.

The simulation is described in Muanwong et al. (2001; see also her contribution in this volume). It has 160^3 particles each of gas and dark matter in a box of comoving size $100\,h^{-1}$ Mpc, and the cosmology is ΛCDM. By the end of the simulation approximately 20 percent of the baryons have cooled and formed stars in the box as a whole, but much more than this in clusters. Despite this overcooling, the clusters reproduce the observed luminosity-temperature relation rather well.

2. Temperature-mass evolution

The theoretical scaling relation between temperature and mass is

$$T \propto \frac{M}{R} \propto M^{\frac{2}{3}}\rho^{\frac{1}{3}} \propto M_{200}^{\frac{2}{3}}(1+z), \qquad (1)$$

where M_{200} is measured within a fixed overdensity relative to the critical density.

As a population, the lower-mass clusters are more affected by radiative cooling and this makes them hotter (by removal of low-entropy core gas) and raises the constant of proportionality in Equation 1. However, this effect is small compared to the change in temperature over a cluster's lifetime and, as Figure 1 shows, the emission-weighted temperatures of clusters as they age do approximately follow the self-similar prediction.

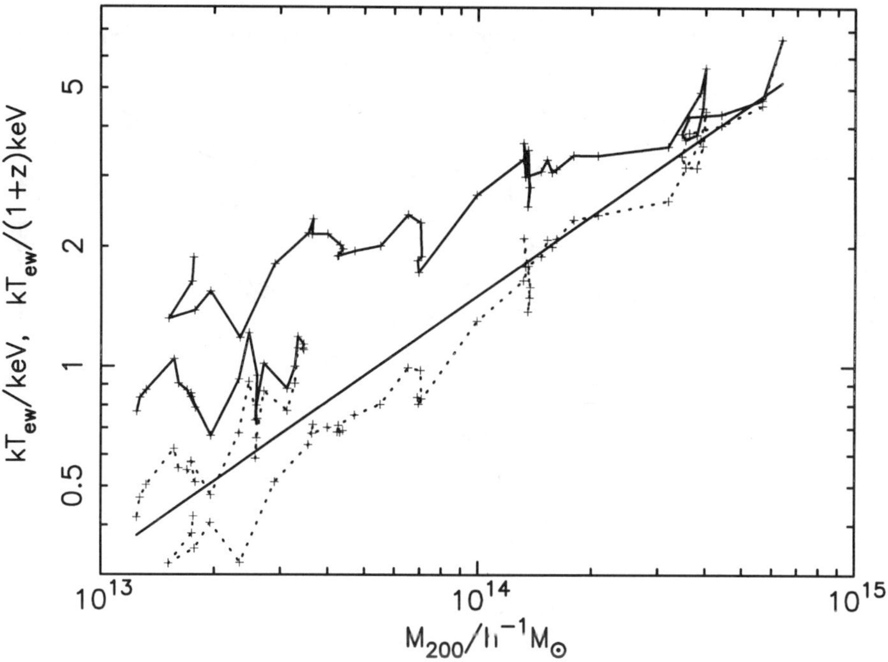

Figure 1. The solid curves show the evolution in the $T_{\rm ew}$-M_{200} plane of two example clusters. The dotted curves show the temperature multiplied by the scale factor of the Universe; for self-similar evolution this should lie parallel to the straight solid line.

3. Luminosity-temperature evolution

Assuming isothermal, self-similar clusters radiating via bremsstrahlung ($kT \gtrsim 1\,\mathrm{keV}$) then the luminosity scales as

$$L \propto \int \rho^2 T^{\frac{1}{2}} \mathrm{d}V \propto M\rho T^{\frac{1}{2}} \propto (1+z)^{\frac{3}{2}} T^2. \tag{2}$$

However, the observed relation is steeper than this self-similar prediction (see e.g. Xue & Wu 2000) because the luminosity is dominated by the core emission that is strongly affected by heating and cooling; nor is there much evidence for redshift evolution below $z = 1$ (Borgani et al. 2001). For this reason, the results from simulations may be regarded as an empirical prediction.

As Figure 2 shows, individual clusters evolve mainly in temperature. The high-mass clusters maintain a roughly constant luminosity over time, whereas lower-mass clusters show a gradual decline in luminosity. Overall, the evolution is stronger than is seen in observations but less than predicted by the self-similar theory.

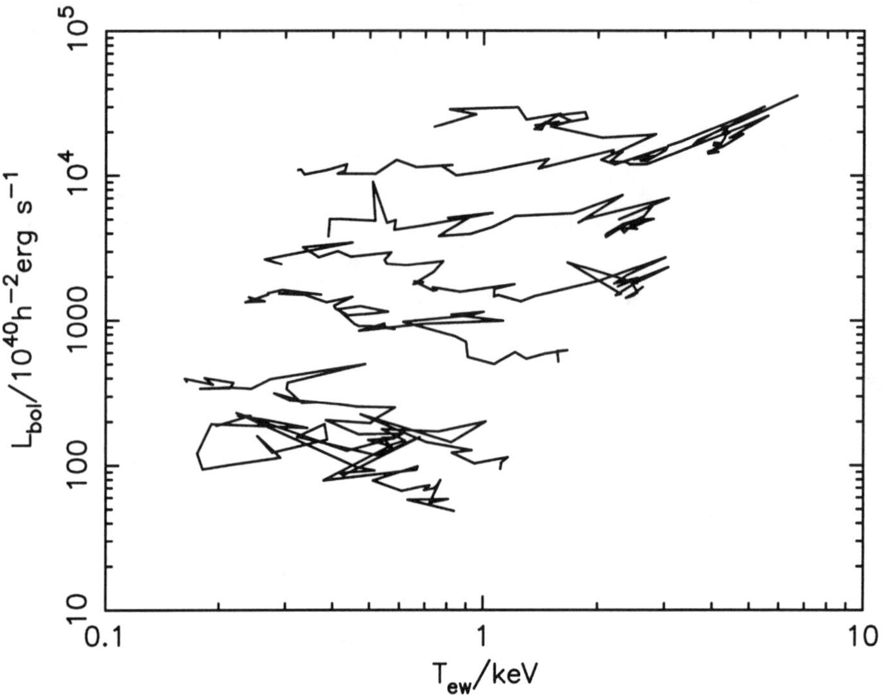

Figure 2. Evolution in the $L_{\rm bol}$-$T_{\rm ew}$ plane of a selection of clusters.

4. Mergers

Many of the trajectories in Figure 2 show large excursions in temperature and luminosity which are associated with merger events. High-resolution simulations of mergers of pairs of clusters (for a detailed description see Ritchie & Thomas 2001) show that the temperature and luminosity of merging systems can temporarily increase by factors of 3 and 10 or more, respectively. However, as shown in Figure 3, these fluctuations are roughly parallel to the observed L-T relation and so do not add considerable scatter.

5. Conclusions

Cluster simulations show that cooling breaks the expected self-similarity between temperature and mass: in low-mass clusters $T_{\rm ew} > T_{\rm vir}$, but in high-mass clusters $T_{\rm ew} \sim T_{\rm vir}$ and shows the correct redshift scaling.

Both observations and simulations give a steeper $L_{\rm bol}$-$T_{\rm ew}$ relation than is predicted by self-similar theory. However the simulated clusters show more evolution in $L_{\rm bol}$-$T_{\rm ew}$ than is observed.

Individual clusters maintain an almost constant luminosity while their temperature changes by a factor of 10, however large excursions in temperature and

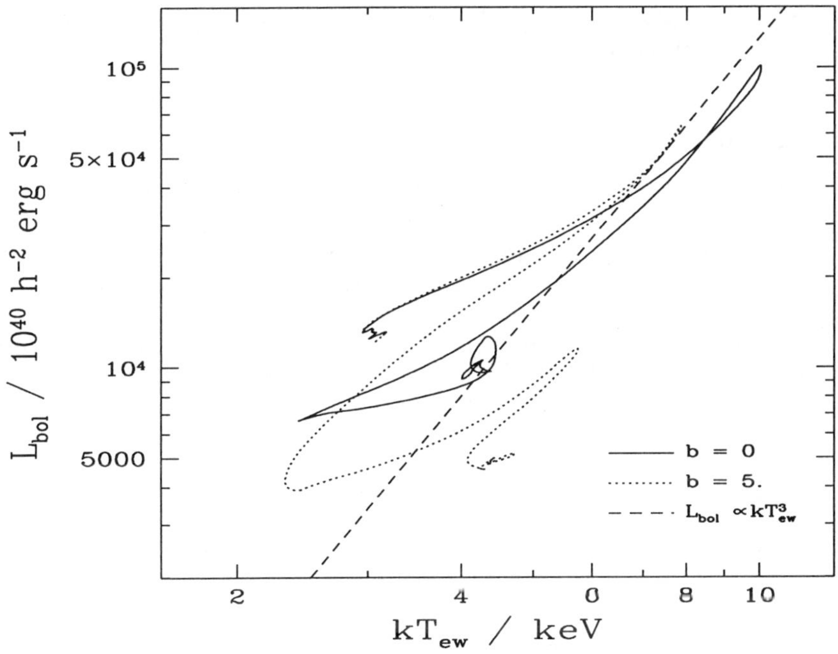

Figure 3. Evolution in the $L_{\rm bol}$-$T_{\rm ew}$ plane of two merging systems: one head-on, one with an impact parameter of 5 core radii.

luminosity are seen during merger events. This adds scatter to the $T_{\rm ew}$-M relation, but the excursions are roughly parallel to the observed $L_{\rm bol}$-$T_{\rm ew}$ relation.

For a copy of the transcripts of the conference talk, including many more results than presented in this paper, see
http://astronomy.sussex.ac.uk/~petert/public_html/index.html

Acknowledgments. PAT is a PPARC Lecturer Fellow. The simulations analysed in this paper were carried out as part of the programme of the Virgo Supercomputing Consortium (http://star-www.dur.ac.uk/~frazerp/virgo/) using computers based at the Computing Centre of the Max-Planck Society in Garching and at the Edinburgh Parallel Computing Centre.

References

Borgani et al. 2001, ApJ, submitted, (astro-ph/0106428)

Muanwong, O., Thomas, P. A., Kay, S. T., Pearce, F. R., Couchman, H. M. P. 2001, ApJ, 552, L27

Ritchie, B. W., Thomas, P. A. 2001, MNRAS, submitted, (astro-ph/0107374)

Xue, Y.-J., Wu, X.-P. 2000, ApJ, 538, 65

Tracing Cosmic Evolution with Galaxy Clusters
ASP Conference Series, Vol. 268, 2002
S. Borgani, M. Mezzetti, and R. Valdarnini, eds.

The Effect of Heating and Cooling on Scaling Laws of X-Ray Clusters

Orrarujee Muanwong

Astronomy Centre, University of Sussex, Brighton, BN1 9QJ, UK

Abstract. We present results of including heating and cooling in cosmological simulations of the ΛCDM cosmology and demonstrate their effects on scaling laws of galaxy clusters. The scaling relations when radiative cooling is included are in good agreement with observations but the fraction of cooled gas is on the upper limit allowed by observations. On the contrary the preheating model has a more realistic cooled fraction but the scaling relations are less well reproduced.

1. Introduction

The predicted X-ray luminosity and temperature relation from self-similarity ($L_X \propto T_X^2$) is shallower than the observed one ($L_X \propto T_X^{\sim 3}$). The steepening of the relation implies that low temperature clusters are less luminous than they would have been if self-similarity were to hold. A physical explanation of this effect is that as the luminosity is roughly proportional to the gas density squared, there must exist processes which lower the density preferentially in low temperature clusters, hence the steepening of the relation. Such processes are for example preheating and radiative cooling of the gas.

Ponman, Cannon, & Navarro (1999) have shown that the observed core entropies of low temperature clusters are in excess of what can be achieved by gravitational collapse alone. They suggest that preheating can raise the entropy before the infall and that level of entropy is maintained during gravitational collapse. As the entropy scales as $T/n_e^{2/3}$, where n_e is the electron gas density, and the temperature of the gas is fixed by the virial temperature of the halo, the excess entropy implies a decrease in density, hence a decrease in luminosity. Energy sources which can inject energy into the intergalactic medium (IGM) are for example AGN, quasars, hypernova explosions, supernova explosions and so on. However, it is still debatable whether the energy injection efficiency of the heating sources is sufficient to provide the required entropy level.

On the other hand, radiative cooling is a process which is also important throughout the history of cluster formation. Low entropy gas, i.e. dense and cold gas, is removed from the ICM, leaving behind high entropy material. Pearce et al (2000) have shown that when radiative cooling is included, the gas is slightly heated and the luminosity is greatly reduced thus acting so as to reconcile simulations and observations.

In these Proceedings, we present results from cosmological simulations of the ΛCDM cosmology in which 3 different models are considered. In the first, 'non-radiative' model, non-adiabatic heating comes about only via shocks following gravitational collapse and the gas is not allowed to cool. The second 'cooling' model introduces radiative cooling such that 20 percent of the gas in the box has turned into stars by the current day - a value that is higher than suggested by observations (Balogh et al 2001). Finally, the third 'preheating' model has both radiative cooling and an impulsive heating event in which 0.1 keV per particle is injected into the IGM at a redshift of 4. This lowers the cooled fraction down to 10 percent.

The simulations consist of 160^3 particles each of gas and dark matter within boxes of side 100 h^{-1} Mpc. The cosmological parameters are : density parameter, $\Omega_0 = 0.35$; cosmological constant, Λ_0=0.65, Hubble parameter, $h = H_0/100\text{km s}^{-1}\text{ Mpc}^{-1} = 0.71$ and baryon density parameter, $\Omega_b h^2 = 0.019$. The metallicities are varied with time in the form of $Z = 0.3(t/t_0)Z_\odot$, where t/t_0 is the age of the Universe in units of the current time. The objective of this is to mimic the gradual enrichment of the ICM by stars. Cluster catalogues are constructed at redshift zero. The catalogues are complete down to a cluster mass of $1.18 \times 10^{13} h^{-1}$ M$_\odot$ which corresponds to 500 particles of each species. Details of cluster selection criteria can be found in Muanwong et al (2001).

2. Results

In this section, cluster properties are compared to those observed. We calculate the X-ray properties (temperature and luminosity) in a soft X ray band (0.3-1.5 keV) by using the cooling tables of Raymond & Smith (1977). The luminosity is then corrected to a bolometric one by assuming that the gas is isothermal. The main effect of using band-limited rather than bolometric fluxes is to raise the emission-weighted temperature of low-mass clusters slightly.

2.1. $T_X - M_{200}$ Relations

We compare the derived $T_X - M_{200}$ relations of the simulated catalogues with a compilation of observations made by Horner, Mushotzky, & Scharf (1999). By using mass estimates from galaxy velocity dispersions, X-ray temperature profiles, the isothermal β-model and the surface brightness deprojection method, they obtain different scaling laws shown as dashed lines in Figure 1.

The normalisation of the cooling simulation (triangles) is greater than that of the non-radiative simulation (squares) by about a factor of 2 at the mass of $10^{14} h^{-1}$ M$_\odot$. For clarity, we omit the preheating results but they lie roughly between the two. Most of the radiative cooling and preheating clusters, particularly the small systems, are hotter than their virial temperatures (solid line), in accord with expectation. Non-radiative cooling clusters are cooler compared to the virial temperatures due to emission-weighting of cold and dense gas remaining in the cluster core. The slopes of all 3 simulated catalogues are roughly 0.54. We find a good agreement between the relations from the radiative cooling model and observations, given the spread of the 4 estimates from observations.

2.2. $L_{bol} - T_X$ Relations

The effect of radiative cooling is demonstrated dramatically in terms of luminosity. The $L_{bol} - T_X$ relations of the non-radiative cooling, radiative cooling and preheating simulations are shown as squares, triangles and circles respectively

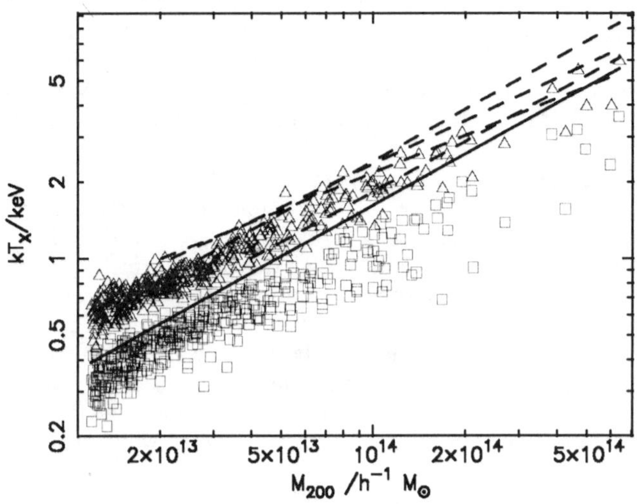

Figure 1. The temperature-mass relation. The dashed lines are from observations (Horner et al 1999). The solid line is the relation between the virial temperature and mass. Non-radiative and radiative cooling clusters are shown as squares and triangles, respectively.

in Figure 2. The large decrease in density as a result of radiative energy loss of the gas manifests in a significant reduction in luminosity. At a temperature of 1 keV, the luminosity is reduced by a factor of 100. The radiative cooling model again is found to have a good agreement with observations as shown as solid lines (Xue & Wu 200). The different solid lines are different fits to their subsamples, i.e. groups (roughly below 1 keV), clusters (above 1 keV) and a mixture of the two subsamples.

The preheating model produces a relation which lies between that of the other two models. Preheating still results in an increase in entropy, causing the normalisation of the relation to be a factor of 10 smaller than that of the non-radiative cooling simulation, but the agreement with observations is not as good as the radiative cooling simulation. The non-radiative clusters are consistent with self-similarity. The relations in the other two models are significantly steeper than 2, even more so in low temperature clusters.

3. Conclusions

We have demonstrated the effect of heating and cooling on X-ray properties of clusters in cosmological simulations. The radiative cooling model can reproduce both the temperature-mass and luminosity-temperature relations in good agreement with observations. However, the fraction of cooled gas is considerably higher than observed. In contrast, the preheating model results in a lower cooled fraction but gives a poor match to the observed scaling relations. As a possible resolution, we are currently undertaking a simulation which includes energy feedback from supernovae.

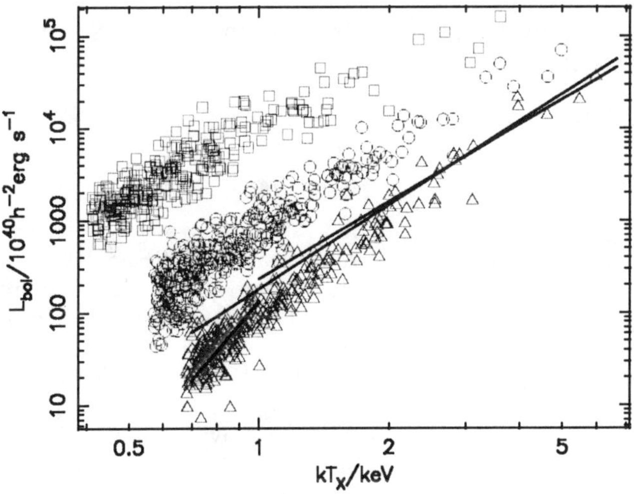

Figure 2. The luminosity-temperature relation. The solid lines are from observations (Xue & Wu 2000). Symbols are as same as those used in Figure 1. Circles represent preheated clusters.

4. Acknowledgement

I am grateful for the financial support from the Thai Government. I wish to thank Peter Thomas and Scott Kay for reading this manuscript and their feedbacks. Simulations presented here were carried out at Edinburgh Parallel Computing Centre as part of the Virgo Supercomputing Consortium.

5. References

References

Balogh, M. L., Pearce, F. R., Bower, R. G., & Kay, S. T. 2001, MNRAS, in press (astro-ph/0104041)

Horner, D. J., Mushotzky, R. F., & Scharf, C. A. 1999, ApJ, 520, 78

Muanwong, O., Thomas, P. A., Kay, S. T., Pearce, F. R., & Couchman, H. M. P. 2001, ApJL, 552, L27

Pearce, F. R., Thomas, P. A., Couchman, H. M. P., & Edge, A. C. 2000, MNRAS, 317, 1029

Ponman, T. J., Cannon, D. B., & Navarro, J. F. 1999, Nature, 397, 135

Raymond, J. C., & Smith, B. W. 1997, ApJS, 35, 419

Xue, Y.-J. & Wu, X.-P. 2000, ApJ, 538,65

Tracing Cosmic Evolution with Galaxy Clusters
ASP Conference Series, Vol. 268, 2002
S. Borgani, M. Mezzetti, and R. Valdarnini, eds.

Radiative cooling in SPH hydrodynamical simulations of X-ray clusters

R. Valdarnini

SISSA Via Beirut 2-4 34014, Trieste, Italy

Abstract. The results from hydrodynamical TREESPH simulations of galaxy clusters are used to investigate the dependence of the final cluster X-ray properties upon the numerical resolution and the assumed star formation models for the cooled gas. When cold gas particles are allowed to convert into stars the final gas profiles show a well defined core radius and the temperature profiles are nearly flat. Star formation methods based on a local density threshold, as in Navarro and White (1993), are shown to give stable results. Final X-ray luminosities are found to be numerically stable, with uncertainties of a factor ~ 2.

1. Introduction

Hydrodynamical simulations have been widely used to predict for different theoretical frameworks the time evolution of the temperature distribution of X-ray clusters. With increasing availability of computational power numerical hydrodynamical simulations have been attempted to model the effects of radiative cooling on the gas in the formation and evolution of cluster galaxies (Katz & White 1993; Anninos & Norman 1996; Yoshikawa, Jing & Suto 2000; Pearce et al. 2000; Lewis et al. 2000). In this contribution I will show the preliminary results that have been obtained from a series of hydrodynamical SPH simulations of galaxy clusters. The simulations have different numerical resolution and include the effects on the gas component of radiative losses, star formation and energy feedback from SN. Final profiles are compared in order to assess the effects of numerical resolution, or different star formation prescriptions, on the cluster X-ray variables.

2. Simulations

In a previous paper (Valdarnini et al. 1999, hereafter VGB) a large set of hydrodynamical simulations was used to study the global X-ray cluster morphology and its evolution. The simulations were run using a TREESPH code with no gas cooling or heating. I refer to VGB for a detailed description of the simulations. In Valdarnini (2001) I have used the same cluster sample to study the effects of including in the simulations additional physics such as gas cooling and star formation. I will report here the results for the cluster with label ΛCDM00 in VGB. This is the most massive cluster ($M_v \simeq 1.5 \cdot 10^{15} M_{\odot}$) extracted from a cosmological ΛCDM N-body simulation with size $L = 200h^{-1}Mpc$, matter density $\Omega_m = 0.3$ and Hubble constant $H_0 = 70 Km\ sec^{-1}\ Mpc^{-1}$. In order to check the effects of radiative cooling for this cluster a set of TREESPH simulations was performed, with initial conditions provided by the cosmological simulation. The effects of radiative cooling are modelled in these simulations by adding to the SPH thermal energy equation an energy-sink term represented by the radiative cooling function. The simulation parameters are: number of gas particles, $N_g \sim 22,500$; gas particle softening parameter $\varepsilon_g \sim 30Kpc$; gas particle mass, $m_g \sim 10^{10} M_\odot$; number of dark particles, $N_d \sim 45,000$ (including those in an external shell surrounding the cluster, cf. VGB).

Allowing the gas to cool radiatively will produce dense clumps of gas at low temperatures ($\simeq 10^4\ °K$). In these regions the gas will be thermally unstable and will likely meet the physical conditions to form stars. In TREESPH simulations star formation (SF) processes have been implemented using different approaches . According to Katz, Weinberg & Hernquist (1996, hereafter KWH) a gas particle is in a star forming region if the flow is convergent and the local sound crossing time is larger than the dynamical time (i.e. the fluid is Jeans unstable). In a simplified version Navarro & White (1993 , hereafter NW) assume as a sufficient condition that a gas particle must be in a convergent flow and its density exceeds a threshold ($\rho_g > \rho_{g,c} = 7 \cdot 10^{-26} grcm^{-3}$). If a gas particle meets these criteria then it is selected as an eligible particle to form stars. The local star formation rate (SFR) obeys the equation

$$d\rho_g/dt = -c_\star \rho_g/\tau_g = -d\rho_\star/dt\ , \tag{1}$$

where ρ_g is the gas density, $\rho_\star$ is the star density , $c_\star$ is a characteristic dimensionless efficiency parameter, τ_g is the local collapse time and it is the maximum of the local cooling time τ_c and the dynamical time τ_d. At each step the probability that a gas particle will form stars in a time step Δt is compared with a uniform random number. If the test is successful then a mass fraction $\varepsilon_\star$ of the gas mass is converted into a new collisionless particle. This star particle has the position, velocity and gravitational softening of the original gas particle. Typical assumed values are $\varepsilon_\star = 1/3$ and $c_\star = 0.1$ (KWH). For the local SFR, NW adopt Eq. 1 with $c_\star = 1$, $\tau_g = \tau_d$ and $\varepsilon_\star = 1/2$ when a gas particle can convert part of its mass into a star particle. The numerical tests have been performed following the NW prescriptions for selecting gas particles which can form stars. Once a star particle is created it can release energy into the surrounding gas through supernova (SN) explosions. All the stars with mass above $8M_\odot$ end as a SN, leaving a $1.4M_\odot$ remnant. The SN energy ($\simeq 10^{51} erg$) is released gradually into the gas according to the lifetime of stars of different masses.

3. Results

Three simulations were performed with the same numerical parameters, but using different SF methods or parameters. This in order to test the final dependence of the cluster X-ray variables on the assumed star formation model for the cooled gas. In simulation I cold gas particles have been converted into stars according to the NW method. In the other two runs (II and III) the KWH prescription is adopted for converting gas particles into stars, but with different values of the star formation efficiency parameter $c_\star$: 0.1 and 1.. As is shown in

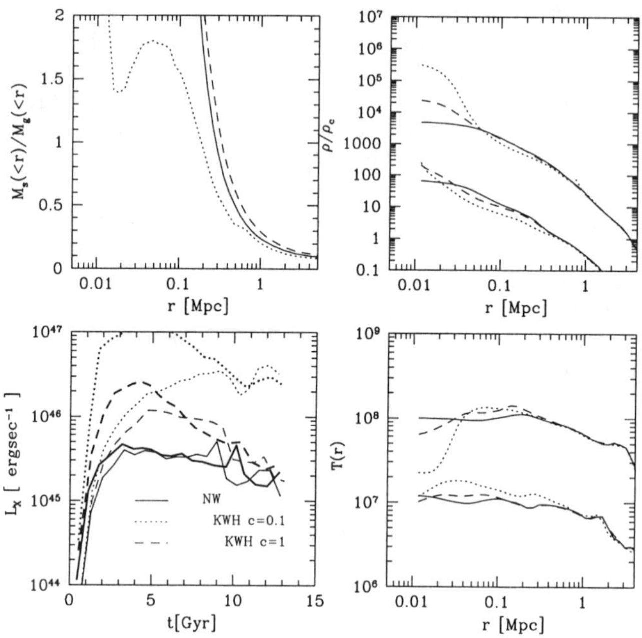

Figure 1. Plots showing several cluster properties in simulation runs with different SF prescriptions. The continuous line refers to the NW method, the others to KWH with different $c_\star$ (c in the bottom left panel). *Top left*: ratio of the star mass within the radius r over the gas mass within r. *Top right*: final radial density behavior for the gas component. The simulation results are compared with the corresponding high-resolution runs (see text). To facilitate a comparison the radial profiles of the high resolution results have been shifted downward by 10^k, $k = 2$ for densities and $k = 1$ for temperatures. *Bottom right*: Radial temperature profiles. *Bottom left* : X-ray luminosity versus time, the thick lines correspond to the high resolution simulations.

Fig. 1 the most important differences are found between the KWH simulations with different $c_\star$. The differences are dramatic in the final X-ray luminosities, which differ by a factor $\simeq 40$. The source of this discrepancy relies in the different gas density profiles, which have substantial differences in the cluster core regions for $r \lesssim 100Kpc$. These differences are localized at the cluster center, beyond $r \simeq 100Kpc$ all the profiles converge, as it is shown in the plots of Fig. 1. The temperature profiles have a peak value of $\simeq 10^8\,°K$ at $r \simeq 100Kpc$ and thereafter decline outward by a factor ~ 2 out to r_v. Below $\sim 100Kpc$ the profiles instead show large differences. Compared to the NW run the simulation with $c_\star = 0.1$ has gas temperatures which decline inwards by a factor ~ 10 from $\sim 100Kpc$ down to $r \sim 10Kpc$. These radial decays follow because of the less efficient conversion of the cooled gas into stars compared to the NW run. There is a remarkable agreement for the ratio of cluster mass locked into stars to the gas mass, which is $\simeq 10\%$ at r_v for all the runs considered. In order to assess the effects of numerical resolution upon final results simulations I, II and III have been run again but with a number of particles increased by a factor $\simeq 3$. These simulations will be referred as IH, IIH and IIIH, respectively. The simulation results are shown in Fig. 1. For simulations IH there are not appreciable differences in the radial profiles. The profiles of simulation IIH are instead different from those of run II at $r \lesssim 50Kpc$. The strong drop in $T(r)$ has been removed and the gas density profile is much closer to the NW one. Simulation IIIH yields final profiles very similar to the ones of the parent simulation. The bottom left panel of Fig. 1 shows that high resolution runs have final X-ray luminosities which can differ within a factor ~ 2 from the parent simulations.

To summarize, the above results demonstrate that simulations I and III have an adequate numerical resolution to reliably predict X-ray cluster properties, such as the X-ray luminosity. For simulation II (KWH with $c_\star = 0.1$) there are large differences at the cluster core between the final profiles when the numerical resolution is increased.

References

Anninos, P. & Norman, M.L. 1996, ApJ, 459, 12

Katz, N. & White, S.D.M. 1993, ApJ, 412, 455

Katz, N., Weinberg, D.H. & Hernquist, L. 1996, ApJS, 105, 19

Lewis, G.F., Babul, A., Katz, N., Quinn, T., Hernquist, L. & Weinberg, D.H. 2000, ApJ 536, 623

Navarro, J. & White, S.D.M. 1993, MNRAS, 265, 271 (NW)

Navarro, J., Frenk C.S. & White, S.D.M. 1995, MNRAS, 275, 720

Pearce, F.R., Thomas, P.A., Couchman, H.M.P. & Edge, A.C. 2000, MNRAS 317, 1029

Valdarnini, R., Ghizzardi, S. & Bonometto, S. 1999, New Astr, 4, 71 (VGB)

Valdarnini, R. 2001, in preparation

Yoshikawa, K., Jing, Y.P. & Suto, Y. 2000, ApJ, 535, 593 (YJS)

Tracing Cosmic Evolution with Galaxy Clusters
ASP Conference Series, Vol. 268, 2002
S. Borgani, M. Mezzetti, and R. Valdarnini, eds.

Group Scaling Relations From a Cosmological Hydrodynamic Simulation: No Pre-heating Required?

Romeel Davé[1], Neal Katz[2], Lars Hernquist[3], David Weinberg[4]

[1] *Steward Observatory, 933 N. Cherry Ave., Tucson, AZ 85721*
[2] *Astronomy Dept., Univ. of Massachussetts, Amherst, MA 01003*
[3] *Harvard/Smithsonian Center for Astrophyiscs, Cambridge, MA 02138*
[4] *Astronomy Dept., Ohio State Univ., Columbus, OH 43210*

Abstract.
We investigate the X-ray vs. optical scaling relations of poor groups to small clusters ($\sigma \approx 100-700$ km/s) identified in a cosmological hydrodynamic simulation of a ΛCDM universe, with cooling and star formation but no pre-heating. We find that the scaling relations between X-ray luminosity, X-ray temperature, and velocity dispersion show significant departures from the relations predicted by simple hydrostatic equilibrium models or simulations without cooling, having steeper $L_X-\sigma$ and L_X-T_X slopes and a "break" at ≈ 200 km/s (≈ 0.3 keV). These departures arise because the hot (X-ray emitting) gas fraction varies substantially with halo mass in this regime. Our predictions roughly agree with observations. Thus radiative cooling is a critical physical process in modeling galaxy groups, and may present an alternative to *ad hoc* models such as pre-heating or entropy floors for explaining X-ray group scaling relations.

1. Introduction

The simplest view of a cluster is as a sphere of Virial-temperature gas punctuated by old galaxies, with perhaps a cooling flow onto the cD galaxy. Within this model, the gas cooling time is longer than a Hubble time everywhere except near the center, and thus it is predicted that clusters should follow simple "self-similar" scaling relations derived from the Virial theorem combined with an assumption of hydrostatic equilibrium, namely: $L_X \propto T_X^2$, $L_X \propto \sigma_{\rm gal}^4$, and $T_X \propto \sigma_{\rm gal}^2$, where L_X is the total X-ray luminosity, T_X is the gas temperature (presumed to be the halo Virial temperature), and $\sigma_{\rm gal}$ is the velocity dispersion of cluster galaxies (presumed to trace the system's total mass).

While observations of the most massive clusters follow these relations, as one progresses to smaller systems, the self-similar model fails. The failure is most striking when one extends observations to poor groups, where Helsdon & Ponman (2000; HP) found that the luminosity of their poorest groups declined much faster than expected ($L_X \propto T_X^{4.9}$), a fact also interpreted as an "entropy floor" of ~ 100 keV cm^{-2} (Ponman, Cannon & Navarro 1999). One model that successfully accounts for these observations postulates that the gas is "pre-heated" by some unknown physical process (possibly supernovae or winds) with

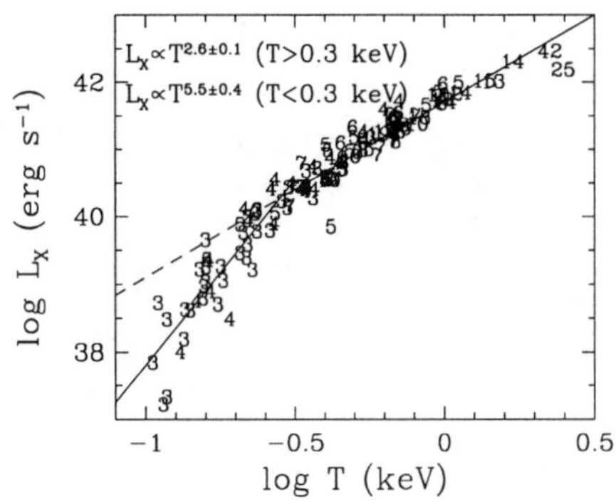

Figure 1: $L_X - T_X$ relation, showing a break at $T \approx 0.3$ keV.

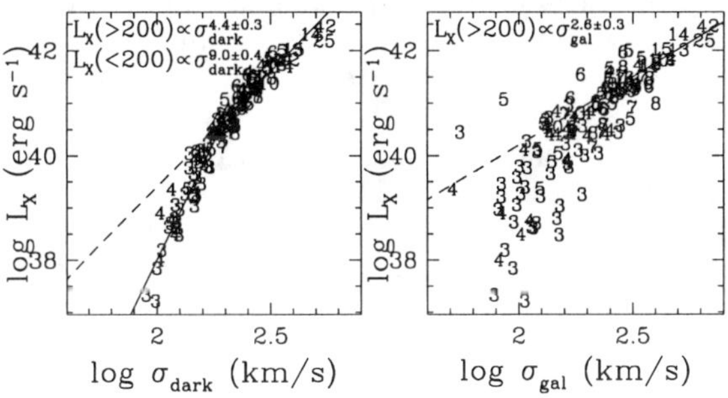

Figure 2: $L_X - \sigma_{\rm DM}$ (left) and $L_X - \sigma_{\rm gal}$ (right) relations.

an energy of ~ 1 keV/baryon. However, the success of this model relies on the assumption that groups are self-similarly scaled-down versions of large clusters.

In these proceedings we present a preliminary test of whether self-similar scaling relations arise naturally in a cosmological hydrodynamic simulation of galaxy formation. We find that on group scales, the self-similar scaling relations are not followed, and that features can arise that quantitatively mimic a pre-heating model. The breaking of self-similarity from clusters to groups is due to an increased cooling efficiency in smaller systems, and does not imply additional heat or entropy input at early times (see also Bryan 2000).

2. Simulation and Group Identification

Our simulation is of a $50h^{-1}$Mpc random volume in a ΛCDM universe with $\Omega_m = 0.4$, $H_0 = 65$ km s^{-1} Mpc^{-1}, and $\sigma_8 = 0.8$. There are 144^3 dark matter and 144^3 SPH particles. It includes cooling, star formation, and thermal feedback. We identify galaxies using SKID, and halos using a spherical-overdensity criterion on friends-of-friends halos, as described in Murali et al. (2001). Our 64-particle galaxy mass resolution limit ($5 \times 10^{10} M_\odot$) corresponds to $\approx L_\star/4$. Halos that contain three or more galaxies are identified as "groups"; we find 128 at $z = 0$.

The X-ray luminosity of a group is computed from all particles with $T > 10^{4.5}$K that are members of that group. Only Bremmstrahlung emission is used;

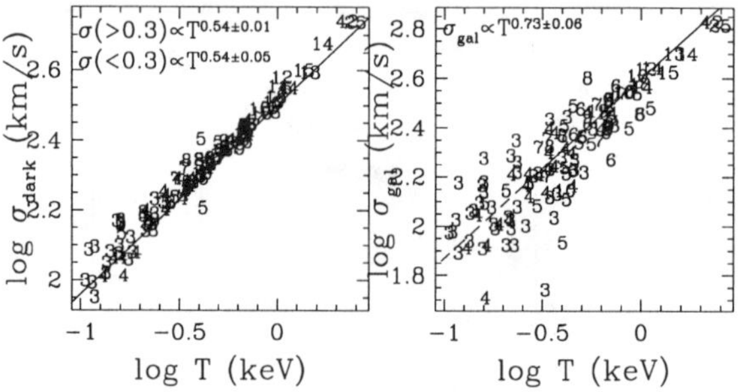

Figure 3: $T_X - \sigma_{\rm DM}$ (left) and $T_X - \sigma_{\rm gal}$ (right) relations.

metal emission is not included in this preliminary analysis to facilitate a straightforward comparison with the self-similar model, and also because intragroup gas metallicities are uncertain. In order to avoid the well-known problem of overestimation of X-ray luminosity due to SPH oversmoothing, we recalculate gas densities using only hot particles, ignoring all particles with $T < 10^{4.5}$K. This explicit decoupling of the hot and cold phases is shown to reproduce the correct X-ray luminosity in analytic cases (Croft et al. 2001). The X-ray temperature is given by the average luminosity-weighted temperature of group particles.

3. Results

Figure 1 shows the $L_X - T_X$ relation for our simulated groups. The plot symbols indicate the number of group members. Best fit power law relations above and below T = 0.3 keV are shown in the upper left. The slope, for groups with $T_X > 0.3$ keV, is steeper than predicted by the self-similar model, and steepens even more below $\approx$ 0.3 keV (the dashed line shows a continuation of the high-temperature slope). The latter slope is consistent with the HP data, and the former with results from clusters (White, Jones & Forman 1997). The break appears at a somewhat lower temperature than is suggested by HP, but more detailed modeling (i.e. including metal emission and aperture effects) may improve agreement. The crucial point is that a break occurs at all, when no physics has been input into the simulation that specifically picks out this temperature scale. This suggests that a straightforward extrapolation of cluster scaling relations (i.e. self-similarity) is an inappropriate model for poor groups.

Figure 2 shows the $L_X - \sigma_{\rm DM}$ (left panel) and $L_X - \sigma_{\rm gal}$ relations (right panel). The dark matter velocity dispersion faithfully traces the group mass. A break is seen in L_X at $\sigma_{\rm DM} \approx 200$ km/s. The slopes above and below the break are consistent with a compilation of $L_X - \sigma$ for clusters down to galaxies by Mahdavi & Geller (2001). Their break occurs at $\sigma \approx 350$ km/s, but given observational uncertainties it is roughly consistent with our prediction.

The galaxy velocity dispersion does *not* trace the group mass when small numbers of galaxies are used. The right panel of Figure 2 shows smaller groups shifted towards lower $\sigma_{\rm gal}$ compared to the left panel, and the best-fit slope is altered significantly. Thus we confirm Zabludoff & Mulchaey's (1998) result that a large number of galaxies ($\gtrsim$ 10) must be used to accurately estimate group velocity dispersions. HP find a slope consistent with our $L_X - \sigma_{\rm DM}$, but claim that their σ is lowered due to misestimation from using small numbers of galaxies. However, Zimer, Zabludoff & Mulchaey (these proceedings) find a similar slope using more than 20 galaxies per group.

Figure 3 shows the $T_X - \sigma$ relation for $\sigma_{\rm DM}$ (left panel) and $\sigma_{\rm gal}$ (right panel). Interestingly, the $T_X - \sigma_{\rm DM}$ relation shows no break at 200 km/s, indicating that the drop in L_X is not due to an additional heat source such as shock heating on filaments. The slope is in reasonable agreement with the self-similar model, indicating that these systems are Virialized. However, the misestimation of σ from using too few galaxies results in a steeper slope (right panel), bringing it more into agreement with HP; this can be misinterpreted as excess heat injection.

Since there is no excess heat, we conclude that the break in L_X arises from a drop in the hot fraction due to an increase in efficiency of galaxy formation in smaller systems. The fraction of hot ($T > 10^5$ K) gas for our simulated groups drops linearly from 50% at $\sigma > 500$ km/s to 20% at $\sigma \approx 100$ km/s. When placed on a logarithmic scale (cf. the L_X plots), the decline appears to steepen at lower σ; hence the appearance of a "break". While the large cold gas + stellar fraction may appear to contradict baryon estimates in clusters, there are indications that the hot gas fraction decreases in smaller systems (see Bryan 2000 for summary). In any case, the trend for a rapid decline of the hot fraction in this mass regime seems to be a generic feature of CDM-based galaxy formation, arising in semi-analytic models as well (Bower, these proceedings).

4. Summary

Using a cosmological hydrodynamic simulation, we have shown that group X-ray scaling relations are not expected to follow simple extrapolations from clusters. For rich groups, the scaling relations for the X-ray luminosity are somewhat different than predicted by self-similarity arguments, while for poor groups the scaling relations deviate dramatically. The trends of our predictions are in agreement with observations. The location of the predicted break predicted is slightly different than that observed, but more careful modeling and better data may improve the agreement. We conclude that cooling is responsible for breaking self-similarity, and hence it is premature to use observed poor group scaling relations as evidence for an excess injection of heat or entropy into these systems.

References

Bryan, G. L. 2000, ApJ, 544, L1

Croft, R. A. C., et al. 2001, ApJ, 557, 61

Helsdon, S. F. & Ponman, T. J. 2000, MNRAS, 315, 356

Mahdavi, A. & Geller, M. J. 2001, ApJ, 554, L129

Murali, C., et al. 2001, ApJ, submitted, astro-ph/0106282

Ponman, T. J., Cannon, D. B., & Navarro, J. F. 1999, Nature, 397, 135

White, D. A., Jones, C., & Forman, W. 1997, MNRAS, 292, 419

Zabludoff, A. I. & Mulchaey, J. S. 1998, ApJ, 496, 39

Tracing Cosmic Evolution with Galaxy Clusters
ASP Conference Series, Vol. 268, 2002
S. Borgani, M. Mezzetti, and R. Valdarnini, eds.

The Thermal State of the ICM Probed with the SZ Effect

A. Cavaliere[1], A. Lapi[1] and N. Menci[2]

[1] *Astrofisica, Dip. Fisica 2a Università, Roma I-00133*

[2] *Osservatorio Astronomico di Roma, Monteporzio, I-00044*

Abstract. Origins, levels and effects of non-gravitational energy inputs into the ICM pervading groups and clusters of galaxies. The role of the SZ effect as a probe of these processes.

1. Introduction

Groups and clusters of galaxies are dominated by cold dark matter (DM), with "halo" masses $M \sim 10^{13} \div 10^{15}\, M_\odot$ and radii from a few 10^{-1} to a few Mpcs. These condense and virialize under the action of self-gravity; then they do not remain secluded, rather they grow hierarchically by repeated and stochastic merging with comparable clumps and by accretion of much external matter.

Closely scale-invariant structures with nearly self-similar profiles are expected (Kaiser 1986) as long as only weak gravity is active, and no strong features are imprinted by the initial conditions (power spectrum of the initial fluctuations) or pressed on by the boundary conditions (cosmology at large). The scales are provided by the density $\rho \approx 2\,10^2 \rho_u \propto (1+z)^3$ and the virial radius $R \approx (M/\rho)^{1/3}$; the depths of the potential wells scale as the 1-D velocity dispersions $\sigma^2 \propto GM/R$.

The halos contain considerable amounts of baryons in the form of a diffuse plasma (ICM) at temperatures $kT \approx kT_v = \mu m_p \sigma^2 \sim 1 \div 10$ keV and at densities $n \sim 10^{-4} \div 10^{-3}$ cm^{-3}; the ICM emits by Bremsstrahlung strong X-ray luminosities $L_X \propto n^2 T^{1/2} R^3 \sim 10^{42} \div 10^{45}$ erg/s.

Such emissions should follow $L_X = L_{grav} \propto T_v^2$ if the ICM passively followed the DM behavior. However, it has been stressed by Ponman, Cannon & Navarro (1999) that in moving from clusters to groups L_X is observed to progressively deviate toward lower values. In other words, the baryons *break* the DM scale-invariance by active, non-gravitational processes.

These data may be rephrased in the form of a progressive upward deviation of the central ICM entropy $S \propto \ln kT/n^{2/3}$ from the self-similar behavior $S = S_{grav} \propto \ln T_v$. In fact, in poor groups relatively high "floor" values $e^S \propto kT/n^{2/3} \approx 140$ keV cm^2 are attained (Lloyd-Davies, Ponman & Cannon 2000, see fig. 1b); such values require a temperature increase combined with a density decrease, quite a non adiabatic behavior.

On the other hand, doubts have been raised about the data and their interpretation in terms of broken scale-invariance; several authors (e.g., Mahdavi et al. 1997; Arnaud & Evrard 1999; Roussel, Sadat & Blanchard 2000) have stressed the observational uncertainties at the group scales, the systematics especially due to single galaxy subtraction, and the selection effects.

This has motivated Cavaliere & Menci (2001) to propose the SZ effect (see Sunyaev & Zel'dovich 1980) as an independent probe of the same physics. The SZE is observed at μwave and submm wavelengths; along a line of sight l its amplitude is governed by the Componization parameter

$$y = 2\,\frac{k\sigma_T}{m_e c^2} \int_0^R dl\, T(r)\, n(r) \propto n\, T\, R\ , \tag{1}$$

which causes a temperature decrement of the CMB photons $(\Delta T/T)_{CMB} \approx -2\,y$ at μwave wavelengths. In rich clusters y attains values around 10^{-4} corresponding to $\Delta T_{CMB} \approx -1/2$ mK (see fig. 2a). Scale-invariance would imply $y = y_{grav} \propto T_v^{3/2}$; will the actual behavior deviate in poor clusters and groups?

2. Model-independent relations

The deviations of y, L_X, and S from their gravitational scaling are related by the following *model-independent* relations

$$y/y_{grav} = (L_X/L_{grav})^{1/2}\,(T/T_v)^{3/4} \tag{2}$$

$$e^{S-S_{grav}} = (L_X/L_{grav})^{-10/9}\,(y/y_{grav})^{14/9}\ . \tag{3}$$

These are easily derived from the definitions, and show that an entropy *excess* is to correspond to systematic and correlated *deficits* of both L_X and y.

These correlations are bound to hold whenever the observations refer to a single-phase ICM. They link observables constituted by different combinations of n and T; in particular, y weights n and T in a balanced way, while L_X stresses n. So they provide two quantitative probes of the ICM mass and thermal state, which are complementary as to the goal, but are fully independent as for instrumental bands, object selection, systematics and uncertainties.

The amplitude as a function of T_v of either deficit in L_X or y probes the strength and possibly the nature of the process causing scale-*variance*. Such analysis requires some modeling, which we will focus on key features and basic issues in view of the increasingly complex ICM substructures revealed by the X-ray observations at increasing sensitivity and resolutions (see Forman 2001).

3. Modeling: is hydrostatics enough?

The high level of the entropy "floor" requires a thermal input combined with a considerable density decrease; may we have them both as parallel outcomes of an energy injection into the ICM? We will start from the canonical heating (recalled in § 6) $\Delta T \approx 0.3$ keV provided by SN explosions within the member galaxies, and discuss how sensitive is the ICM density to ΔT.

An answer may be sought in terms of hydrostatic equilibrium of the ICM in the DM potential wells, a good approximation in between major merging events.

As is well known (Cavaliere & Fusco-Femiano 1976; Jones & Forman 1984), this yields

$$n(r)/n_2 = e^{\beta\,\Delta\phi} \tag{4}$$

in isothermal conditions $T(r) \approx T_2$; the subscript 2 denotes values at the boundary to be discussed in § 5, $\Delta\phi = [\Phi_2 - \Phi(r)]/\sigma^2$ is the scaled potential difference, and the ratio of the DM/ICM scale heights reads $\beta = \mu m_p \sigma_2^2/kT = T_v/T$.

In moving from rich clusters to groups, a given energy input causes a relatively larger temperature raise $a\,\Delta T/T_v$, say; in turn, this will *lower* the exponent β from the asymptotic value β_{grav} in rich clusters to

$$\beta(T) \approx \frac{\beta_{grav}}{1 + a\Delta T/T_v}\ . \tag{5}$$

Smaller values of β in eq. (4) imply flatter density profiles in groups, a trend which accords with the resolved data of Ponman et al. (1999).

Fig. 1a shows how the same thermal input ΔT with $a = 1.3$ (discussed in § 5) lowers the central density from a cluster to a group; this may be referred to as the *de-amplification* factor from hydrostatics. On the other hand, the lower curve of fig. 1b shows this alone to be insufficient to account for the observed entropy floor; the deficit observed in the relation $L_X - T$ is also not accounted for, as implied by eq. (3).

Note that both a polytropic relation $T(r) \propto n^{\Gamma-1}(r)$ ($\Gamma \lesssim 1.2$) and the adoption of the potential proposed by Navarro, Frenk & White (1997, NFW) further reduce the sensitivity of the density to a given thermal input. In a given structure either change lowers the central n, while the variations from clusters to groups are squeezed (see Cavaliere, Lapi & Menci 2001).

4. Internal heating/cooling

In looking for stronger effects and further amplification, one possibility is to rely on processes *internal* to the given structure, focusing on boundary conditions that occlude it from the infall of external gas and prevent dilution.

For example, Bryan (2000) adopts $p_2 = 0$, and focuses on fast central cooling of the low entropy gas, which leaves behind higher values of S. The floor is approached even with $\Delta T = 0$, at the cost of accruing a large fraction f_c of baryons in cooled form and of producing excess star formation.

At the other extreme, Bower et al. (2001) require a rather large $\Delta T \approx 1$ keV to cause substantial ICM outflows. In terms of eq. (5), their model results in a combined value $a\,\Delta T \gtrsim 0.3$ keV, while cooling plays here a side role.

More on internal processes will be found in these Proceedings. We just note that when heating or cooling (with the related star formation) dominate, either process has to be tuned high to yield the excess entropy required. Can we strike an intermediate course and tune ΔT and f_c to realistic values?

5. Hierarchical heating

To achieve this, Cavaliere, Menci & Tozzi (1997) started from the tenet of the hierarchical clustering, that is, the infall into a given structure of baryons together

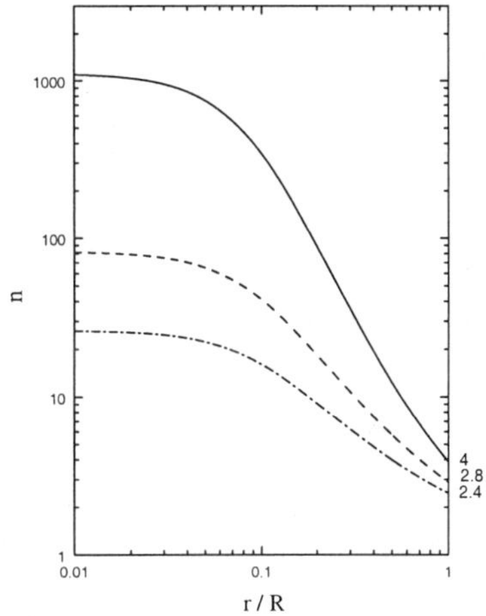

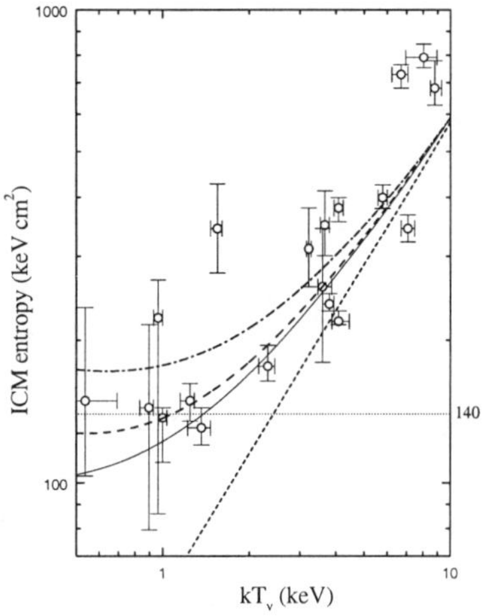

Figure 1. *Left*: density profiles of the isothermal ICM in the King (1972) potential. Solid line: a cluster with $T_v = 8$ keV; dashed line: a group with $T_v = 0.5$ keV; both are computed for $\Delta T = 0.3$ keV. The dot-dashed line refers to the same group with $\Delta T = 0.6$ keV. The boundary values n_2 for the two lower curves are specified in § 5. *Right*: the S-T relation for the ICM within $10^{-1}\,R$. Dashed line: $\Delta T = 0.3$ keV with the named boundary condition; solid line: same with fixed n_2. Dot-dashed line: $\Delta T = 0.6$ keV. The short-dashed line represents the scaling $S_{grav} \propto \ln T_v$. Data from Lloyd-Davies et al. (2000).

with DM during the frequent merging/accretion of other lumps. As a bonus, these events involve smaller clumps which have higher L/M and are more effective at producing and feeding back star-related energy. Since under any model warmer external gas less easily flows into a shallow potential well, an enhanced density de-amplification from clusters to groups is expected for a given ΔT.

In particular, a less supersonic inflow during merging/accretion implies decreased strength of the accretion shocks that are located at about R, see Knight & Ponman (1997); Gheller, Pantano & Moscardini (1998); Takizawa & Mineshige (1998). This means decreasing the jumps at the shock of the temperature and the density that provide boundary conditions for the inner ICM settled to hydrostatic equilibrium.

One such condition concerns the temperature. With $\Delta T = 0.3 \div 0.6$ keV the shocks are still strong around rich clusters where $\Delta T/T_v \ll 1$ holds; then the post-shock value reads $T \approx 1.1\, T_v + 1.5\, \Delta T$, and this gives

$$\beta \approx \frac{0.9}{1 + 1.3\, \Delta T/T_v} \, . \tag{6}$$

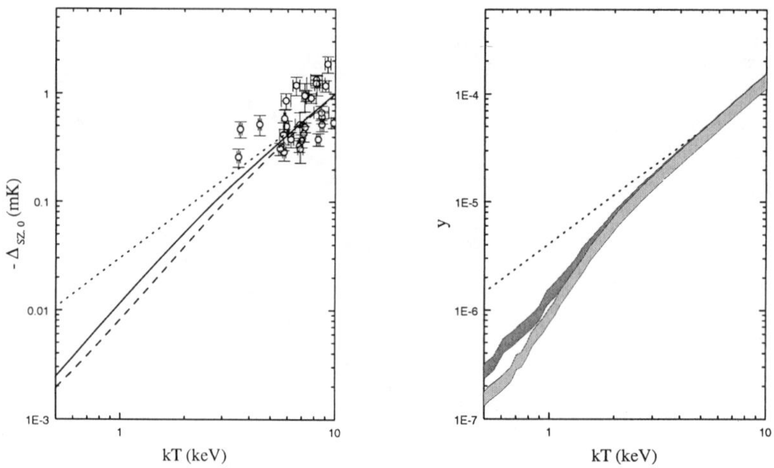

Figure 2. *Left*: the SZ decrement from isothermal ICM in the King potential for $\Delta T = 0.3$ keV (solid line), and $\Delta T = 0.6$ keV (dashed line). The dotted line represents the scaling $T^{3/2}$. Data from Zhang & Wu (2000). *Right*: the $y - T$ relation (with its 2-sigma variance) from ICM grafted to SAMs with $\alpha_h = 2$ (heavy shaded region) and $\alpha_h = 5$ (light shaded region). Polytropic $\Gamma = 1.2$, NFW potential, and ΛCDM cosmology, which concur to minimize the deviations from scale-invariance (dotted line).

Values $T \approx 2\,T_v$ obtain in groups where $\Delta T/T_v \approx 1$ holds, to yield $\beta \lesssim 0.5$ (see Cavaliere et al. 2001 for details). In sum, the post-shock boundary condition for the temperature directly specifies the first *de-amplification* factor in eq. (4).

By the same token, a second *de-amplification* arises from the density jump n_2/n_1 across the shock, to be inserted in eq. (4). The jump again depends on $\Delta T/T_v$; it asymptotes to 4 in very rich clusters and goes to 1 for groups with $\Delta T/T_v \approx 1$. The combined effect of such de-amplifications is illustrated in figs. 1 and fig. 2a.

Next, one may take full advantage of the feature intrinsic to this approach, i.e., the groups build up from the smaller, galactic structures with enhanced star formation rates and larger energies fed back onto the associated baryons. Such a *link* among different scales is implemented on grafting the above ICM modeling onto the SAM (semi-analitic model) treatment of star and galaxy formation, see Menci & Cavaliere (2000); this is because the *hot* baryonic phase appearing in the SAMs is just the ICM that pervades groups and clusters.

In addition, the SAMs are just based on the DM hierarchical clustering; they specifically embody the "merging history" of a given structure, that is, its buildup through stochastic merging events with a range of partners. In fact, the "window" of such progenitors effectively contributing to the growth of a structure marked by given values of M and T is in the range $M' \approx M/2 \div M/20$, that is, $T'_v \approx 0.6 \div 0.15\, T_v$.

The feedback effects, namely, gas heating and ejection are also embodied in the SAMs, often in terms of the parameterization $\Delta m_h/\Delta m_\star \propto \sigma^{-\alpha_h}$ ($\alpha_h = 2 \div 5$). This enhances in the smaller progenitor galaxies the amount of cool gas ejected and reheated by stellar winds and by SN explosions following the formation of a star mass $\Delta m_\star$; so, near the window's lower end it is easier to have higher energy production and substantial gas heating to yield $\Delta T/T'_v \gtrsim 1$. Such gas eventually falls into the forming groups.

Two specific outcomes result from this approach combining SAMs with hydrostatics. First, the convolutions associated with the merging histories introduce variance (e.g., $\delta y/y \approx 15\%$ at the two sigma level, see fig. 2b), while they also smooth out the deviations from self-similarity at the scales of rich clusters.

Second, moving toward groups such smoothing is overwhelmed by the window effect, to the point of enhancing the deviations; this is because the smallest progenitors still effective for the mass inflow are also most effective at producing stellar feedback and higher ΔT. The result is to lower not only β, but also the average $\langle n_2/n_1 \rangle$ values; the latter condition is localized at the boundary, independently of the inner potential shape and of the isothermal or polytropic runs of $T(r)$, so it blurs the difference between the ICM dispositions.

Note that the SAMs also include the moderate density decrease in all progenitors due to amount of cooling related to realistic cool gas fractions and to star formation rates fitting the optical observations.

In sum, in the SAM framework the preheating is *distributed* over the range of cosmic times and scales envisaged by the hierarchical clustering, with overall de-amplification optimized and *linked* to the stellar observables.

6. Energy inputs

Obvious first candidates for such preheating are type II SN explosions following the star formation in the galaxian members of groups and clusters. Such events release $E_{SN} \approx 10^{51}$ erg of initial kinetic energy, coupled to the surrounding gas in the form of SN Remnants. While isolated SNRs are subject to strong radiative losses after some 10^5 yr, nearly coeval starbursts produce cooperative SNR propagation and galactic winds (see Pettini et al. 2001 for recent evidence), to the effect of discharging and mixing their high internal entropy into the ICM.

The energy input per unit solar mass condensed into stars comes to $E_\star = f_{SN}\, \eta_{SN}\, E_{SN} \approx 3\, 10^{48}$ erg/M$_\odot$, considering also the contribution of the strong winds from young hot stars. The related thermal energy of the affected gas mass is given by $E_\star\, m_p\, \Delta m_\star / 3\, \Delta m_h = 0.5\, \Delta m_\star/\Delta m_h$ keV; in groups where $\Delta m_\star/\Delta m_h \approx 1/2$ holds the stellar heating attains $\Delta T \approx 0.3$ keV, and higher values in galaxies.

However, it has been contended (see Wu, Fabian & Nulsen 2000) that such feedback levels produced by stars may be marginally (in)sufficient to cause the observed deviations of $S - T$ from the self-similar behavior. Additional feedback is conceivably provided by the outputs of AGNs powered by those galactic baryons that condense under strong gravity onto a central, massive black hole.

Such a use of small amounts of baryons to build up black hole masses M_{BH} introduces another scale of order $\eta c^2 M_{BH}$, the energy radiated with efficiency $\eta \sim 10^{-1}$. This converts to $E_{AGN} = f_{AGN}\,\eta c^2\, M_{BH}/M_{bulge} \approx 4\, f_{AGN}\, 10^{50}$ erg/M$_\odot$ per unit solar mass condensed into stars, considering the black hole/bulge relation $M_{BH}/M_{bulge} \approx 2\,10^{-3}$ (Gebhardt et al. 2000, Merritt & Ferrarese 2001); but the coupling efficiency f_{AGN} to the ISM throughout the host galaxy is even more uncertain than for SNe. The radio-loud AGNs, where the output is dominantly in kinetic energy of relativistic jets, have shown to now evidence of limited current heating of the surrounding ICM (see McNamara et al. 2001).

As to the 90% radio-quiet AGNs, a small coupling f_{AGN} is expected in the optically unobscured ones, in view of the low photon momentum and of the flat spectra. In fact, the energy deposited in the ISM troughout the host galaxy over times of order 10^8 yr should not exceed the gas binding energy, to yield

$$f_{AGN}\,\eta c^2\, M_{BH} \lesssim 10^{-1}\, GM^2/R \tag{7}$$

(Silk & Rees 1998); for masses $M_{BH} \sim 10^8 M_\odot$ this translates into $f_{AGN} \lesssim 10^{-2}$.

Smaller values would yield an irrelevant AGN contribution, but larger ones would cause evacuation of the ISM; such may have been the case for some of the EROS (Cimatti et al. 2001), which to now appear to constitute a limited galaxy population. In less extreme cases the above relation is closer to equality, and then it yields as a specific signature the *steep* correlation $M_{BH} \propto \sigma^5$ with the galaxian dispersion following from eq. (7) as discussed by Cavaliere & Vittorini (2001, in preparation). Interactions of the host galaxies with their companions in groups easily destabilize gas amounts matching or exceeding those allowed by condition (7), and are likely to induce massive starbursts; the result is increased feedback levels but minimal variation of the input rate relative to SNe.

7. Relics and probes

The ICM constitutes an archive of the energy inputs, mainly driven by weak gravitation on large scales, but also affected by nuclear explosions and by activity due to strong gravity around black holes. To these additional inputs the ICM responds with readjustments that amount to density *de-amplification* from clusters to groups. These imply: decreased ratio β of the hydrostatic scale heights; lower densities at the boundary associated with warmer inflows of external gas; self-regulated cooling and ejection related to star formation.

The *relics* of the ICM thermal history include not only star related quantities (L_B, L_K and f_c) in the member galaxies, but also at smaller scales relic black holes (with the imprinted $M_{BH} - \sigma$ correlation), and on larger scales deficits in X-ray luminosities ($L_X - T$) and in the targeted SZ effect ($y - T$).

For all its merits, the *hierarchical* preheating approach calibrated to stellar observables yields an $L_X - T$ relation arguably marginal relative to the current, debated data. The other probe, *complementary* and observationally *independent*, is provided by the targeted $y-T$ relation, see fig. 2b. A strong, systematic deficit observed also in the latter will reinforce the case for an energy input exceeding that from SNe.

The related measurements currently constitute a challenging proposition. They ought to discern levels $\Delta T_{CMB} \approx 50 \div 25\,\mu K$; this will be feasible with

PLANCK over scales of several arcminutes matched to nearby groups; over scales of tens of arcseconds matched to distant groups, such deficits will emerge from mapping performed with dedicated interferometers like *AMiBA* or eventually with *ALMA*, and compared with optical and X-ray surveys.

Such levels stand out from integrated values $\langle y \rangle \lesssim 10^{-6}$ due to groups and clusters (Valageas & Silk 1999, Cavaliere & Menci 2001); hardly a factor of 2 is conceivably contributed by AGN energy escaping from the host galaxies out to scales larger than groups and clusters.

References

Arnaud, M., & Evrard, A.E. 1999, MNRAS, 305, 631

Bower, R.G., Benson, A.J., Lacey C.G., Baugh, C.M., Cole, S., & Frenk, C.S. 2001, MNRAS, 325, 497

Bryan, G.L. 2000, ApJ, 544, 1

Cavaliere, A., & Fusco-Femiano, R. 1976, A&A, 49, 137

Cavaliere, A., Menci, N., & Tozzi, P. 1997, ApJ, 484, L21

Cavaliere, A., & Menci, N. 2001, MNRAS, 327, 488

Cavaliere, A., Lapi, A., & Menci, N. 2001, in Proc. *Chemical Enrichment of Intracluster and Intergalactic Medium*, Vulcano, in press

Cimatti, A. et al. 2001, astro-ph/0111527

Forman, W. 2001, in Proc. *Lighthouses in the Universe*, Garching, in press

Gebhardt, K. et al. 2000, ApJ, 539, L13

Gheller, M., Pantano, O., & Moscardini, L. 1998, MNRAS, 296, 85

Jones, C., & Forman, W. 1984, ApJ, 276, 38

Kaiser, N. 1986, MNRAS, 222, 323

King, I.R. 1972, ApJ, 174, L123

Knight, P.A., & Ponman, T.J. 1997, MNRAS, 289, 955

Lloyd-Davies, E.J., Ponman, T.J., & Cannon, D.B. 2000, MNRAS, 315, 689

Mahdavi, A., Boehringer, H., Geller, M.J., & Ramella, M. 1997, ApJ, 483, 68

McNamara, B.R. et al. 2001, astro-ph/0110554

Menci, N., & Cavaliere, A. 2000, MNRAS, 311, 50

Merritt, D., & Ferrarese, L. 2001, ApJ, 547, 140

Navarro, J.F., Frenk, C.S., & White, S.D.M. 1997, ApJ, 490, 493

Pettini, M. et al. 2001, ApJ, 554, 981

Ponman, T.J., Cannon, D.B., & Navarro, J.F. 1999, Nature, 397, 135

Roussel, H., Sadat, R., & Blanchard, A. 2000, A&A, 361, 429

Silk, J., & Rees, M.J. 1998, A&A, 331, 1

Sunyaev, R.A., & Zel'dovich, Ya.B. 1980, ARA&A 18, 537

Takizawa, M., & Mineshige, S. 1998, ApJ, 499, 82

Valageas, P., & Silk, S. 1999, A&A, 350, 725

Wu, K.K.S., Fabian, A.C., & Nulsen, P.E.J. 2000, MNRAS, 318, 889

Zhang, T., & Wu, X. 2000, ApJ, 545, 141

Tracing Cosmic Evolution with Galaxy Clusters
ASP Conference Series, Vol. 268, 2002
S. Borgani, M. Mezzetti, and R. Valdarnini, eds.

The Heating of the ICM: Energy Crisis and viable solutions

Paolo Tozzi

Oss. Astronomico di Trieste, via G.B. Tiepolo 11, 34100 Trieste – Italy

Abstract. X–ray observations indicate that non–gravitational processes play a key role in the thermodynamics of the Intra Cluster Medium (ICM). The effect of non–gravitational processes is imprinted in the ICM as an entropy minimum, whose effects are visible in the Luminosity–Temperature relation and in the Entropy–Temperature relation. However, the X–ray emission alone cannot discriminate between different mechanisms and sources of heating. There are no answers at present to the following questions: how much non–gravitational energy per baryons is present in the ICM? When was this energy injected? Which are the sources of heating? The embarrassment in front of these questions is amplified by the fact that the most viable sources of heating, SNae and stellar winds, seem to be inefficient in bringing the ICM to the observed entropy level. We may call it the *energy crisis*. Here we review the main aspects of this crisis, listing possible solutions, including other sources, like AGNs and Radio Galaxies, or other mechanisms, like large scale shocks and selective cooling.

1. Evidences for non–gravitational heating

Observations in the X–ray band provided convincing evidences for non–gravitational heating of the diffuse baryons in the potential wells of groups and clusters of galaxies (ICM). Among them: the shape of the L–T relation (steeper than the self–similar behaviour $L \propto T^2$ predicted in the case of gravitational processes only) and the entropy excess in the center of groups, recently found by Ponman, Cannon & Navarro (1999). Non–gravitational heating of the ICM is expected also on the basis of observations of an average metallicity $Z \simeq 0.3\ Z_{\odot}$. In fact, SNae and stellar winds are the most viable source of heating and metal enrichment of the ICM. It should be understood how much of the ejected energy goes into the ICM and if it is enough to generate the observed entropy plateau.

2. How the entropy works

If the excess entropy is present in the baryons before collapse (*external heating*, or preheating, see Tozzi & Norman 2001, hereafter TN), it will be preserved in the cores of dark matter halos after virialization. In virtue of the extra pressure support, in fact, the gas is accreted adiabatically without shock heating. The excess entropy also decreases the density in the central regions. This, in turn,

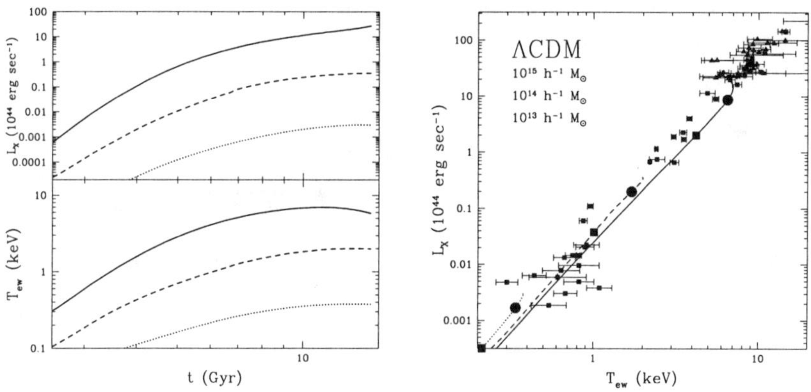

Figure 1. Left: The evolution of the bolometric luminosity L_X and of the emission–weighted temperature T_{ew} is shown as a function of time for a final mass of $10^{15} - 10^{14} - 10^{13} h^{-1} M_\odot$ (solid, dashed and dotted lines respectively) for a ΛCDM cosmology, with an initial adiabat $K_{34} = 0.3$ (in units of 10^{34} erg cm^2 g$^{-5/3}$). Right: The evolutionary tracks along the L–T relation for the halos on the left. Large squares and circles mark $z = 1$ and 0.5 respectively. Local data from Allen & Fabian (1998), Arnaud & Evrard (1999), Ponman et al. (1996).

rapidly decreases the X–ray luminosity which is proportional to the square of the density. The effect is stronger in small groups, where the energy responsible for the entropy is comparable to the gravitational one, while clusters, where gravity is dominant, are mostly unaffected. This produces a steepening of the L–T relation and breaks the scale invariance in the density profiles of clusters and groups. The entropy has also the effect of suppressing the radiative cooling in the central regions. This model allows to trace the evolution of X–ray luminosity and temperature of groups and clusters after assuming a Press & Schechter–like law for the accretion rate of baryons (see TN). In Figure 1 we show that the evolution corresponds to tracks moving along the local L–T relation. Thus, a constant L–T is predicted up to $z \simeq 1$, in agreement with observations (Mushotzky & Scharf 1997, Borgani et al. 2001). The external entropy level that satisfies the observations is in the range $K_* = 0.2 - 0.3 \times 10^{34}$ erg cm^2 g$^{-5/3}$ (see TN).

3. The energy crisis

The external heating scenario is appealing, but the excess entropy cannot be spread uniformly into the cosmic baryons at high redshifts, because this would make the Ly$_\alpha$ forest to disappear. Recently, it has been estimated that the level of preheating in the Ly$_\alpha$ clouds is of the order of few $10^4 K$ (Cen & Brian 2001), corresponding to an entropy level at least one order of magnitude lower than that observed in groups. A possible solution is that only the baryons that end up in the cores of groups and clusters are heated by a biased distribution of

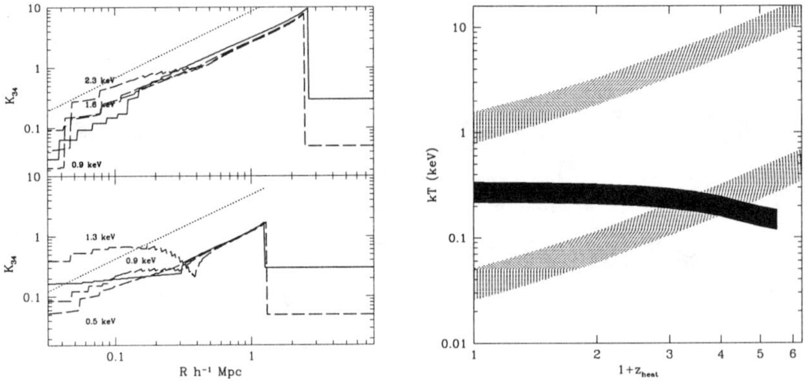

Figure 2. Right: The entropy profiles in the external (solid) and the internal (dashed lines) scenario for a rich ($M = 10^{15}h^{-1}M_{\odot}$, top) and a small cluster ($M = 10^{14}h^{-1}M_{\odot}$, bottom). The external entropy level is $K_{34} = 0.3$. In the internal scenario, a total energy of 1–2 keV per particle is released, as shown by the labels. Left: The upper and lower stripes show the required energy per particle needed to obtain an excess entropy of $K_{34} = 0.2 - 0.3$ in virialized structures and in the background baryons respectively. The solid stripe is the energy per particle dumped in the ICM by TypeII and TypeIa SNae after Pipino et al. (2001).

sources. Such a warm, low density gas would be unobservable at high z, and can be detected as OVI absorption systems at low (Tripp, Savage & Jenkins 2001) or at intermediate z (Reimers et al. 2001).

If OVI systems cannot account for the pre–heated baryons, the ICM must be heated after the collapse. Of course, for a given entropy level, the much higher density of the collapsed regions implies a much higher energy input. An energy budget of 1–2 keV per particle (3–10 times higher than that in the external scenario) seems to be required to reproduce the entropy floor (see also Wu, Fabian & Nulsen 1999, Valageas & Silk 1999) and entropy profiles similar to the ones predicted in the external heating scenario (see Figure 2, left).

Despite the SNae can provide a large amount of energy, their efficiency in heating the gas is unknown. In particular, if the heated gas has high density, the thermal energy received from SNae is rapidly radiated away, with a small net increase in the gas entropy. A recent study has been made by Pipino et al. (2001), starting from the observed luminosity function of cluster galaxies. The efficiency of TypeII and TypeIa SNae is computed assuming a spherical gas distribution around galaxies. The energy per particle dumped in the ICM as a function of redshift is plotted in Figure 2 (right, black stripe). It turns out that SNae can contribute a substantial amount of the required energy ($\simeq 0.3$ keV/particle), which is, however, lower than that needed in virialized region (upper stripe). The first attempts to include the stellar feedback in hydrodynamical simulations, seem to indicate a low efficiency from SNae (see, e.g., Borgani et al.

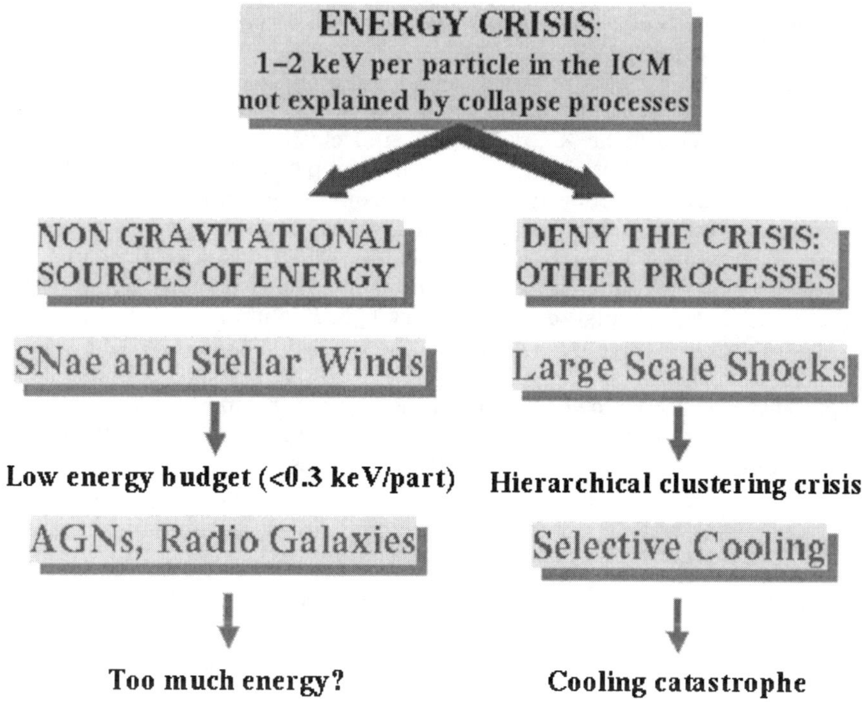

Figure 3. A sketch of the current situation. The description may be not exhaustive.

2001). Following the natural predisposition of cosmologists towards the use of the word *crisis*, I would call this empasse the *energy crisis* of the cosmic baryons.

4. Solutions and possible way out

If SNae will be shown unadequate for this job, the energy crisis can be solved by other sources of non–gravitational energy. Several studies already considered AGNs as the main source of heating. Radio Galaxies can provide an energy output two orders of magnitude larger than stellar sources, and they can heat the baryons at large distance, reaching low densities for which the radiative cooling is negligible (see Inoue & Sasaki 2001). On the other hand, AGNs can heat the baryons so efficiently as to exceed the upper limit of the comptonization parameter of the CMB (Yamada & Fujita 2001).

As in political life, a possible way to solve the crisis is to deny it: some viable scenario avoids the intervention of non–gravitational heating to generate the entropy plateau. The extra energy could be due to gravitational shocks produced by large scale structure (filaments and sheets) before the collapse of group–sized regions. However, this requires large scale shocks to occur at redshifts much larger than the typical formation epoch of the cores of clusters or groups, which can be as high as $z \simeq 3$ in ΛCDM universes. These large–scale effects have never been noticed in N–body simulations with current CDM power spectra, for which there is a large consensus.

Another way consists in simply eliminating the low entropy gas. Some simulations suggest that the cooling itself, mostly efficient in the higher density region, can reproduce the X–ray properties of clusters, like the L–T relation (Muanwong et al. 2001). In this case, the entropy plateau is created by the removal of the lowest entropy gas, which cools out of the diffuse, emitting phase. However, the majority of present–day simulations, in absence of any feedback, find an unacceptably high fraction of cooled baryons (Balogh et al. 2001). We must stress that the cooling is extremely difficult to treat numerically and there is consensus among simulations only at very high resolutions.

To summarize, denying the crisis can lead to a worst crisis, like the inversion of the clustering hierarchy or the cooling catastrophe, as sketched in Figure 3. At the present stage there are no obvious solutions and it is important to investigate all the possible way out. Both X–ray observations and theoretical modelling will be crucial in the next years to get out of this empasse.

References

Allen, S.W., & Fabian, A.C. 1998, MNRAS, 297, 63L

Arnaud, M., & Evrard, A.E. 1999, MNRAS, 305, 631

Balogh, M., Pearce, F., Bower R., Kay S. 2001, MNRAS, 326, 1228

Borgani, S. et al. 2001, ApJ in press, astro-ph/0108329

Cen, & Bryan 2001, ApJ, 546, 81L

Inoue, S., & Sasaki S. 2001, ApJ in press, astro-ph/0106187

Muanwong et al. 2001, ApJL, 552, L27

Mushotzky, R. F., & Scharf, C. A. 1997, ApJ, 482, 13

Ponman, T. J., et al. 1996, MNRAS, 283, 690

Ponman, T. J., Cannon, D. B., & Navarro, F.J. 1999, Nature, 397, 135

Reimers, D., et al. 2001, A&A submitted, astro-ph/0106097

Tozzi, P., & Norman, C. 2001, ApJ, 546, 63 (TN)

Tozzi, P., Scharf, C., & Norman, C. 2000, ApJ, 542, 106 (TSN)

Tripp, T.M., Savage, B.D., & Jenkins, E.B. 2001, ApJL in press, astro-ph/0003277

Valageas, P., & Silk, J. 1999, A&A, 350, 725

Yamada, M., & Fujita, Y. 2001, ApJL, 553, 14

Wu, K. K. S., Fabian, A. C., & Nulsen, P. E. J. 1999, MNRAS, 318, 889

Tracing Cosmic Evolution with Galaxy Clusters
ASP Conference Series, Vol. 268, 2002
S. Borgani, M. Mezzetti, and R. Valdarnini, eds.

Blowing bubbles in the ICM

Vicent Quilis, Richard Bower and Michael Balogh

Department of Physics, University of Durham, South Road, Durham DH1 3LE, U.K.

Abstract. We use a three dimensional hydrodynamical code to simulate the effect of energy injection on cooling flows in the intra-cluster medium. Specifically, we compare a simulation of a $10^{15}\,M_{\odot}$ cluster with radiative cooling only, with a second simulation in which thermal energy is injected. The heat injection forms a hot, low density bubble which quickly rises, dragging behind it material from the cluster core. The main result is that the cooling flow is disrupted for up to $\sim 50\,Myr$ after the injection of energy ceases. Thus, this mechanism provides a very efficient method for regulating cooling flows, if the injection events occur with 1:1 duty cycle.

1. Introduction

Radiative cooling in clusters of galaxies makes the distribution of gas unstable. As the gas in the centre of the cluster radiates its thermal energy, the pressure support in the outer parts is reduced and a flow is established. In cluster centres, the cooling time is short compared to the age of the cluster, which makes this an important mechanism for accreting gas onto a central object. The astrophysical puzzle is to understand how the flow rates than have been estimated for some clusters (up to 2000 $M_{\odot}$ per year, e.g. Allen 2000) can be compatible with the relatively small masses of cold gas and young stars seen in central cluster objects (Crawford et al. 1995, Edge et al. 1999), or with the lack of gas seen at temperatures below about 2 KeV (eg., Peterson et al. 2001, Tamura et al. 2001). A popular solution to these problems is the suggestion that gas cooling might trigger heating processes (feedback) which regulate the cooling flow (eg., White & Frenk 1991).

Although there are many physical processes known which may be viable sources of feedback energy, it is unknown exactly how such injections of energy interact with the surrounding gas. In particular, there are three possible outcomes of an energy injection event: (1) the energy may serve to increase the total emissivity of the gas (for example, by compressing surrounding material), resulting in all the input energy radiated away; (2) the energy may be concentrated in a small amount of gas that then escapes form the cluster core but has little effect on the cooling flow; (3) the injected energy may disrupt the cooling flow, either temporarily or permanently. The focus of this work is to determine what happens to energy injected in clusters with substantial cooling flows. All of the recent simulations find that the energy injection results in buoyant bubbles which displace and disturb the surrounding gas.

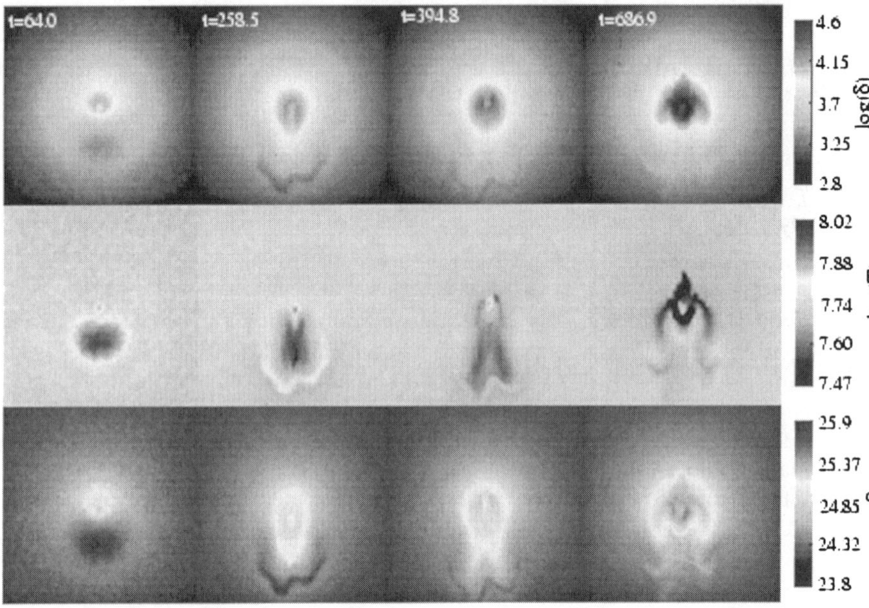

Figure 1. A time sequence of two dimensional slices taken through the fully 3 dimensional simulations with the heat injection. The box shown is 250 Kpc on each side; the full simulation is 1 Mpc. The different rows show evolution in the ICM (a) density contrast; (b) temperature; (c) entropy. The panels are labeled with the time in Myr since the start of the simulation

2. Simulations

We use the three dimensional Eulerian fixed-grid hydrodynamical code described in Quilis et al. (1996). We run two simulations of a $10^{15}\,M_\odot$ cluster, using a grid of 256^3 cells in a cube 1 Mpc per side, corresponding to a cell size of 3.9 Kpc. The simulations were carried out on an Origin 2000 parallel computer, and follow the cluster evolution over a simulated time of 900 Myr. We assume $H_o = 70\,km\,s^{-1}Mpc^{-1}$, $\Lambda = 0.7$, $\Omega_o = 0.3$, and $\Omega_b = 0.039$. The initial cluster profile is constructed assuming that isothermal gas – KT=4.75 – sits in a $10^{15}\,M_\odot$ dark matter potential well in pressure-supported hydrostatic equilibrium. For the dark matter we assume a NFW profile. The bolometric X-ray luminosity of this model cluster is $4.9 \times 10^{44}\,erg\,s^{-1}$. Considering the range of temperatures and densities of the ICM , we model the ICM gas as a non-relativistic monoatomic fluid with $\gamma = 5/3$.

Our model is not intended to investigate the interaction of the AGN jet with the surrounding material (Reynolds et al. 2001), so we simply model the heating process by injecting energy centred in a cell which is displaced 8 cells (31 kpc) from the cluster center. The heating is distributed over the surrounding pixels using a spherical (SPH-like) kernel with smoothing length 8 cells. The heating rate was chosen to be of the same order as the cooling rate of the cluster: $4.9 \times 10^{44}\, erg\, s^{-1}$; this is expected to be typical of the energy injection from radio jets(e.g. Owen et al. 2000). This choice is conservative in the sense that it is less likely to be successful at disrupting the cooling flow than if we had deliberately targeted the heating of the lowest entropy gas. The heat input is continued for 50 Myrs, by which time a near empty bubble has been formed in the ICM. It is then switched off so that the energy can be distributed around the cluster. This is motivated by the desire to model a scenario in which the AGN (or other feedback source) is "woken-up" by gas deposited in the cooling flow. This disrupts the flow which turns off the supply of fresh material and hence limits its lifetime.

3. Results

We run simulations where energy is injected into a single cell over 50 Myr. The evolution of the gas properties is shown in several two-dimensional slices in Figure 1. The three rows show the gas density contrast, temperature and entropy, at four different times[1]. The energy injected heats a relatively small amount of mass, $3.5 \times 10^{10}\, M_{\odot}$, to $10^{8}\, K$, creating a hot but almost empty bubble. In effect, the bubble is generated by a subsonic wave expanding into the ICM (a sonic boom in the terminology of Reynolds et al. 2001). In the simulation, the energy injection is sufficiently weak that bubble expands subsonically, pushing gas out of the cluster centre. Since this gas is moving adiabatically from a high pressure region to one of lower pressure, it expands and cools. This results in a shell of gas around the bubble which is cooler that the surrounding material, this is qualitatively similar to *Chandra* observations in some cluster (e.g. Fabian et al. 2000).

Once the heat source is turned off (at 50 Myrs), the bubble ceases to expand, but does not collapse. The buoyancy of the bubble compared to the surrounding material takes over as the dominant force, pushing it outward from the centre of the simulation. This causes the bubble to adopt a cap-like geometry; roughly a semi-spherical shell of gas with a low density, low pressure interior. This is similar to the geometry found in the simulations of Churazov et al. (2001). The surface of the shell in our simulation experiences turbulent mixing with the surroundings. The shell rises until its entropy becomes comparable to the surrounding medium, at which point it dissolves into the surrounding ICM. This occurs well within the core region of the clusters (< 250 kpc).

As a result of the low pressure region create behind the bubble as it rises, a plume of low entropy material (dark blue color) is drawn out of the core. This gas, although not heated directly, becomes mixed into the higher entropy gas surrounding the cluster, slightly depressing the entropy on scales of 50–100 kpc. But the entropy of the very central gas (within 10 Kpc) is higher in the feedback simulation, relative to the cooling-only simulation. This is because, in the cooling-only simulation, the dense core gas has radiated its energy and

[1] A movie displaying the formation and evolution of the bubble can be seen at the web site http://star-www.dur.ac.uk/~quilis/movies/bubble_den.mpg. The movie shows the time evolution (in units of 10^{6} years) of the logarithm of the overdensity.

dropped in entropy; this cooling has been prevented in the simulation with feedback, resulting in a relatively higher core entropy.

4. Conclusion

We have presented a heating model which has been successful at regulating the cooling of the ICM. We have seen that the energy injected is not immediately radiated. Although only a small amount of gas is involved in the formation of the bubble, its effect is felt through the cluster core. The aim of these simulations has been to quantify the effectiveness of this form of energy injection, and we summarize our finding as follows:

(1)The effect of the delay to the cooling lasts $\sim$ 50 Myr after the initial energy injection event finishes, i.e., an amount of time equivalent to the time over which energy was supplied. Therefore, a 1:1 duty cycle of heating events would be very effective in regulating the radiative cooling process.

(2)There is still a net radiation of energy in both simulations. Measuring energy relative to the cooling-only simulation shows that the cooling flow has been disrupted, and the injected energy has been efficiently mixed into the surrounding ICM: the radiated energy never exceeds the energy radiated in the "nonfeedback" case, and none of the heated bubble material is convected out of the cluster. Although our simulations have not directly reduced the flow rate over timescales of $\sim$500 Myr, the efficiency and timescale over which the cooling flow is disrupted makes it seem likely that multiple events will be very effective at regulating the cooling rate.

Acknowledgements. We would like to thank Sebastian Heinz for useful discussions. VQ is a Marie Curie Fellow (grant HPMF-CT-2000-00052). MB acknowledges support from a PPARC rolling grant for extragalactic cosmology at Durham. Simulations were carried out as part of the Virgo consortium on COSMOS an Origin 2000.

References

Allen S.W., 2000, MNRAS, 315, 269

Crawford C. S., Edge A. C., Fabian A. C., Allen S. W., Bohringer H., Ebeling H., McMahon R. G., Voges W., 199,MNRAS, 274, 75

Churazov E., Brueggen M., Kaiser C.R., Boehringer H., Forman W., 2001, ApJ, astro-ph/0008215

Edge A. C., Ivison R. J., Smail I., Blain A. W., Kneib J.P., 1999, MNRAS, 306, 599

Fabian A. C., Sanders J. S., Ettori S., Taylor G. B., Allen S. W., Crawford C. S., Iwasawa K., Johnstone R. M., Ogle P.M. 2000, MNRAS, 318, L65

Owen F.N., Eilek J.A., Kassim N.E., 2000, ApJ, 543, 611

Peterson, J.R., Paerels, F. B. S., Kaastra, J. S., et al., 2001, astro-ph/0010658

Quilis V., Ibáñez J.M^{a}., Sáez D., 1996, ApJ, 469, 11

Reynolds, C.S., Heinz, S., Begelman, M.C., 2001, astro-ph/001104

Tamura, T., Kaastra, J.S., Peterson, J.R., et al., 2001, astro-ph/0010362

White S. D. M., & Frenk C. S., 1991, ApJ., 379, 52

Tracing Cosmic Evolution with Galaxy Clusters
ASP Conference Series, Vol. 268, 2002
S. Borgani, M. Mezzetti, and R. Valdarnini, eds.

Making the Connection: Feedback, X-rays and Galaxy Formation

Richard G. Bower

Department of Physics, University of Durham, South Road, Durham, DH1 3LE, UK

Abstract. The feedback process is key to understanding how galaxies are formed: without feedback, too much gas is able to cool in small dense objects. But what is the nature of the additional physics is needed? In this talk, I take a look at progress that is being made in using observations of galaxies and the X-ray universe to understand the elusive feedback phenomenon.

1. The need for Feedback

This is an exciting epoch for cosmology. We are moving away from simply measuring basic cosmological parameters and beginning to really understand how galaxies are formed. Advances in large galaxy redshift surveys and cosmic microwave background experiments have defined the most important cosmological parameters (Ω_0, Λ and the mass fluctuation power spectrum) have become increasingly well defined (eg., Peacock et al. 2001). However, "galaxy formation" is still very much an unsolved problem. The theoretical framework sounds simple: dark matter haloes grow due to gravitational attraction; gas cools within them to form galaxies; the galaxies grow in mass both through additional gas cooling and by merging with each other. The model suggests that the basic galaxy building blocks are disks which can be disrupted by the mergers to give the observed diversity of morphology.

But what regulates cooling? In the scenario above far too much gas is able to cool compared with observations. It is straightforward to theoretically show that most of the baryonic content of the universe will condense into cold, small fragments at early times (eg., Cole 1991). Yet in the real universe, only a small fraction of the available baryons have been locked into galaxies. This puzzling inefficiency means that the most fundamental ingredient of galaxy formation (commonly referred to as "feedback" or "sub-grid physics", White & Frenk, 1991, Cen & Ostriker, 1999) is completely missing from the simple scheme. We can guess at what is required, but the physical mechanism by which this "feedback" operates is currently an open issue.

2. What is the Cold Fraction?

The first step we need to make is to observationally determine the fraction of baryons that are locked into galaxies. This is mainly the stellar mass of the galaxy, but we have a good handle on the contribution from cold neutral and hydrogen too. I referee to this as the "cold fraction", f_{cold}. Fukugita, Hogan & Peebles were able to show that this fraction is small, and their numbers can now be updated to set an even harder limit.

In Balogh et al., 2001, we set out to compile the latest available information on f_{cold}. It can be robustly estimated in clusters, groups and as an average value in the universe as a whole. We obtain remarkably consistent answers in each case.

For the universe as a whole the technique is to estimate the the cold fraction by combining the K-band luminosity density with an estimate of the mass to light ratio. This gives a robust estimate since M/L_K is less sensitive to star formation history, varying by only a factor of about two due to stellar population age differences. Fortunately, Cole et al. 2001 have computed the mass function using over 17000 galaxies with redshifts from the 2dF galaxy redshift survey and magnitudes from the 2-micron all sky survey (2MASS). Furthermore, dynamical measurements of early type galaxies (Bell & de Jong, 2001) favour $M/L_K \sim 1.1h$; this is an upper limit to the global average because later type galaxies have lower M/L_K. This leads to the estimate of $\Omega_{\rm stars} = 0.0014h^{-1}$.

Cooled gas is also present in the form of neutral and molecular gas, but these make only small contributions. The density of neutral gas is $\Omega_{\rm atomic} \approx 0.000188h^{-1}$, as recently calculated by the HI Parkes All Sky Survey (Kilborn et al. 1999). For the amount of gas in molecular form, we adopt the relation $\rho_{H_2}/\rho_{HI} = 0.81$. Together, the neutral and molecular gas only represent $\sim 10\%$ of the stellar mass so that $\Omega_{\rm cold} = 1.1\Omega_{\rm stars}$.

The total baryon content of the universe is unknown observationally, because baryons in the warm plasma phase believed to occupy normal galaxy haloes are very difficult to see . So we must base the total baryon density, Ω_b, on deuterium abundances. The best current estimate, $\Omega_b = 0.019h^{-2}$, therefore implies $f_{cold} = \Omega_{\rm cold}/\Omega_b = 0.073h$.

Combining these estimates, only about 5% of the available baryons in the universe have cooled (for $h \approx 0.7$ and a Kennicutt IMF). This result is slightly lower than the minimum of the range $0.062 < f_{cold} < 0.107$ found by FHP from optical data, but much more certain as it does not rely on knowing the relative abundances of different galaxies (disks, spheroids and irregulars) with very different (and uncertain) M/L_B ratios.

3. X-ray Feedback

There are three different physical schemes that could be used to regulate galaxy formation: (1) photo-ionisation by the UV background; (2) ejection of gas from the disks of galaxies; (3) heating baryons in the dark matter halo so that their cooling rate is less. In this talk I will concentrate on the third of these options. Recent reviews of the first two schemes can be found in Gnedin et al. 2001 and Cole et al. 2000 respectively.

The reason why we think the third from of feedback may be relevant is the observed luminosities of groups of galaxies (Ponman et al., 1999), and the lack of detectable X-ray emission from the haloes of normal spiral galaxies (Benson et al., 2000). These can be explained pre heating the intracluster medium before

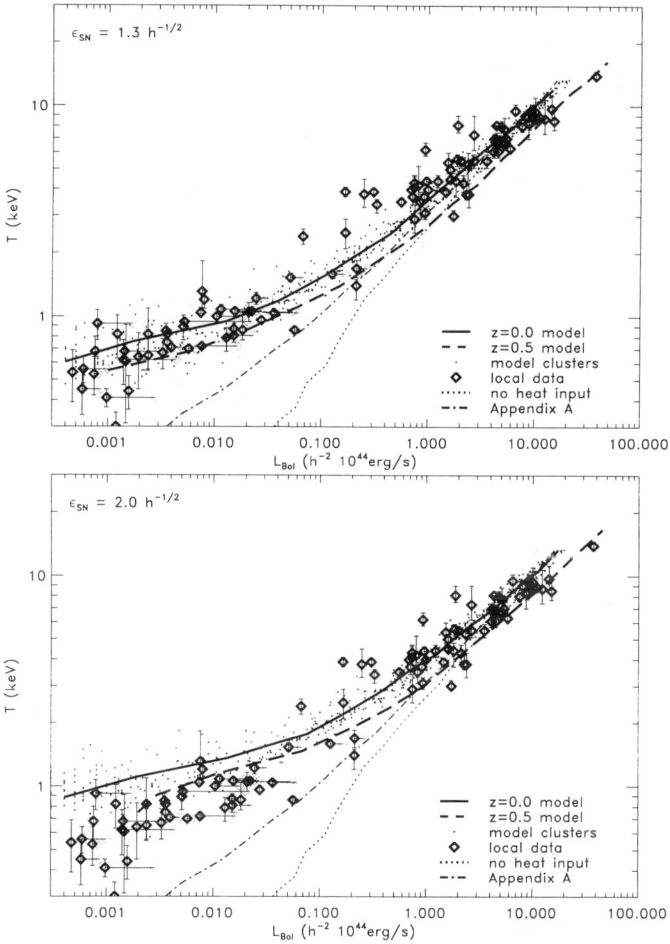

Figure 1. A comparison of the predicted and observed T-L relations for heating efficiencies of $\epsilon_{\rm SN} = 1.3\,h^{-1/2}$ *(upper panel)* and $\epsilon_{\rm SN} = 2\,h^{-1/2}$ *(lower panel)*. The distribution of model clusters at $z = 0$ is shown as points, with the thick solid line showing the median T at each L. The thick dashed line shows the median T-L relation in this model at $z = 0.5$. Bold diamonds are observational data points for clusters and groups with $z < 0.1$ taken from a variety of sources (see Bower et al., 2001). The dotted line shows the median T-L relation from a model with $\epsilon_{\rm SN} = 0$, i.e. in which heat input from galaxy formation is turned off. The dot-dashed line shows an estimate of the T-L relation when there is no heat input from galaxies, but additional gas is removed.

it falls into galaxy haloes, or else by accounting for the energy released from supernovae during the collapse of galaxies (Bower et al., 2001). In this paper, we developed a new model for the evolution of the masses and density profiles of the hot gas in galaxy, group and cluster haloes, using the semi-analytic galaxy formation scheme of Cole et al. (2000) to predict the evolution of the supernova heating rate in dark haloes of different masses, and calculating the response of the gas profile to this heating using a method related to those of Wu et al. 1998 and Cavaliere et al. 2000.

Heat input into the intracluster gas by processes associated with the formation of cluster galaxies, such as supernovae and/or AGN winds, succeeds in flattenning the slope of the temperature-luminosity (T-L) relation, but it is necessary to introduce very substantial heating of the ICM. In the upper panel of Fig 1, we show the T-L relation at $z = 0$ in a model with a heating efficiency $\epsilon_{\rm SN} = 1.3h^{-1/2}$. This value of $\epsilon_{\rm SN}$ is already larger than can be accounted for by supernova feedback alone, if the IMF has the conventional solar neighbourhood form, even if there are no radiative energy losses. This suggests that a significant contribution from AGN is probably also required. If the heating produced by galaxies is smaller than this, the model predictions at the bright end fall too steeply with decreasing luminosity. Even with an efficiency of $\epsilon_{\rm SN} = 1.3h^{-1/2}$, the predicted T-L relation seems somewhat too steep for the most luminous clusters ($L_{\rm X} > 10^{44}h^{-2}$ergs s^{-1}). These clusters are not much affected by this amount of heating and tend to follow the self-similar slope. To bring the most luminous clusters into line with the observed T-L slope requires that the energy injection efficiency be increased to $\epsilon_{\rm SN} = 2.0h^{-1/2}$. However, the model then fails to reproduce the presence of X-ray luminous clusters with temperatures below 1 keV (Fig. 1, lower panel). Thus, it seems that the heating efficiency needs to be slightly greater in the progenitor haloes of the most massive clusters, in order to produce the best match to the T-L relation.

4. Next Steps

The treatment in Bower et al. 2001 is not self-consistent since the feedback from galaxy formation is not used to modify the rate at which gas cools; rather we have taken the successful GALFORM model of galaxy formation, with the same parameters as in Cole et al. (2000), and used it to predict the energy injection and thus the evolution of cluster and group X-ray properties. In practice, we should use the methods developed here to calculate the gas density profiles of all haloes at each epoch, as modified by the energy injection, compute gas cooling rates using these modified profiles, and then calculate the energy injection from the resultant star formation histories. This represents a large computational overhead on the standard GALFORM model, but this is clearly an important step to take.

However, there are clearly some problems that this type of feedback model must overcome.

- Will any massive galaxies form at all? With a large value of the feedback parameter $\epsilon_{\rm SN}$, there is a danger of heating the halo so strongly that a small amount of star formation prevents all subsequent cooling. This would still allow large galaxies to grow by mergers, but they would not have gas disks.

- Is the process stable? Care is needed in the way that cooling and heating interact. Small dense high redshift objects have very short cooling times. Thus to avoid the problem described above, the halo must be frequently checked to see if sufficient energy has been released to eject all the remaining halo gas. Frequent recalculataion should allow a stable regime to be established in which the fraction of cold gas grows slowly with halo mass.

- The largest galaxies? A balance must also be struck between degree of cooling in cluster haloes and the galaxy luminosity function. If there is insufficient heat input on group and cluster scales, cooling remains efficient in these systems leading to the formation of a few very bright galaxies and a luminosity function that has too soft a break.

A paper summarising these models is currently in preparation (Benson et al., 2002). The new model provides exciting prospect and potential a very different way of forming galaxies. We wait to see if it is possible to obtain as good a match to the data as that seen in Cole et al. 2000.

Acknowledgments. I'll take this opportunity to thank my collaborators on the various projects that have made this summary possible: Andrew Benson, Shaun Cole, Carlton Baugh, Cedric Lacey, Carlos Frenk (Bower et al., 2001 and Benson et al. 2002) and Mike Balogh, Fraser Pearce and Scott Kay (Balogh et al., 2001).

References

Balogh M., Pearce F.R., Bower R.G. & Kay S., 2001, MNRAS, 326, 1228

Bell E.F. & de Jong R.S., 2001, astro-ph/0011493

Benson A. J., Bower R. G., Frenk C. S., White S. D. M., 2000, MNRAS, 314, 557

Bower, Benson, Lacey et al., 2001, MNRAS, 325, 497

Cavaliere A., Giacconi R., Menci N., 2000, ApJ, 528, 77

Cen R. & Ostriker J., 1999, ApJ, 514, 1

Cole S., 1991, ApJ, 367, 45

Cole S., Lacey C. G., Baugh C. M., Frenk C. S., 2000, MNRAS 319, 168

Peacock J.A., Cole S., Norberg P. et al., 2001, astro-ph/0103143

Fukugita M, Hogan C.J. & Peebles P.J.E., 1998, ApJ, 503, 518

Gnedin N.Y., 2000, ApJ., 542, 535

Kilborn. K., Webster R. L., Staverly-Smith L., 1999, PASA, 16, 8

Ponman T. J., Cannon D. B., Navarro J. F., 1999, Nature, 397, 135

Wu K.K.S., Fabian A.C., Nulsen P.E.J., 1998, MNRAS, 301, L20

White S. D. M., Frenk C. S., 1991, ApJ, 379, 52

GALAXY POPULATION IN CLUSTERS

Tracing Cosmic Evolution with Galaxy Clusters
ASP Conference Series, Vol. 268, 2002
S. Borgani, M. Mezzetti, and R. Valdarnini, eds.

Formation and Evolution of Galaxies in Clusters

Pieter G. van Dokkum

California Institute of Technology, MS 105-24, Pasadena, CA 91125

Abstract.

Elliptical and S0 galaxies dominate the galaxy population in nearby rich clusters such as Coma. Studies of the evolution of the colors, mass-to-light ratios, and line indices of early-type galaxies indicate that they have been a highly homogeneous, slowly evolving over the last $\sim 65\,\%$ of the age of the Universe. On the other hand, recent evidence suggests that many early-type galaxies in clusters have been transformed from spiral galaxies since $z \sim 1$. Arguably the most spectacular evidence for such transformations is the incidence of red merger systems in several high redshift clusters. Due to this morphological evolution the sample of early-type galaxies at high redshift is only a subsample of the sample of early-type galaxies at low redshift. This "progenitor bias" results in an overestimate of the mean formation redshift if simple models without morphological transformations are used. Models which incorporate morphological evolution explicitly can bring the homogeneity, slow evolution, and morphological transformations into agreement. The modeling shows that the corrected mean formation redshift of the stars in early-type galaxies may be as low as $z \approx 2$ in a Λ dominated Universe.

1. Introduction

The galaxy population in rich clusters is dominated by early-type galaxies (S0 and elliptical galaxies). The study of these objects gives insight in the formation of the most massive disk- and spheroidal galaxies in the Universe, and in the processes governing star formation at early times. Furthermore, cluster galaxies provide critical tests of the hierarchical paradigm for galaxy formation. In currently popular semi-analytical galaxy formation models in a CDM Universe the descendants of the Ly-break population are massive galaxies in groups and clusters (Baugh et al. 1998). These models also predict that significant differences should exist between early-type galaxies in clusters and those in the general field (Kauffmann 1996).

It has been known for a long time that early-type galaxies in clusters form a very homogeneous population: at a given luminosity, they show a very small scatter in their colors, M/L ratios, and line indices (e.g., Bower et al. 1992). The simplest interpretation of this high degree of homogeneity is a small spread in age, although it has been argued that a larger age spread could be "masked" by correlated metallicity variations (e.g., Trager et al. 2000)[1]. Determining the *mean* age of nearby early-type galaxies has proven to be a formidable challenge. The main reason is the well known age-metallicity degeneracy in fitting early-type galaxy spectra (e.g., Worthey 1994). Furthermore, the observed abundance ratios of early-type galaxies cannot be reproduced with simple stellar population synthesis models, which makes absolute determinations of age and metallicity even more uncertain. It is therefore not surprising that most of our understanding of the formation and evolution of cluster early-type galaxies has come from studies of clusters at large lookback times.

Since the seminal work by Butcher & Oemler (1978) on the colors of galaxies in two distant clusters this field has witnessed great progress: measurements of redshifts, morphologies, colors, M/L ratios, and line indices of cluster galaxies currently span $\sim 65\,\%$ of the age of the Universe. Examples of successful ongoing programs are the MORPHS collaboration (Smail et al. 1997), who obtained deep HST images of the cores of ~ 10 clusters at $0.3 < z < 0.5$, the CNOC group (Yee et al. 1996), who obtained extensive wide field spectroscopy and imaging of X-ray selected clusters at $0.2 < z < 0.5$, the work by Lubin, Postman, & Oke on optically selected clusters at $z \sim 0.8$, and our wide field HST imaging and extensive spectroscopy of X-ray clusters[2].

As in many other fields, Hubble Space Telescope imaging and spectroscopy with large ground-based telescopes have been instrumental in this advance. An additional factor has been the success of surveys of increasing sophistication to find ever more distant clusters. The progress in this area that was demonstrated at the meeting is encouraging, and should lead to a better understanding of the interplay between the selection of clusters and the derived evolution of the galaxies within them (see, e.g., the review by Marc Postman).

2. Evolution of early-type galaxies

Studies of the evolution of early-type galaxies in clusters are in remarkable agreement. Studies of their colors (e.g., Ellis et al. 1997, Stanford et al. 1998), M/L ratios (van Dokkum & Franx 1996, Bender et al. 1998, van Dokkum et al. 1998, Kelson et al. 2000), and line indices (Bender et al. 1998, Kelson et al. 2001) show that they remain a very homogeneous population and evolve only slowly all the way from $z = 0$ to $z \sim 1$.

2.1. Evolution of the Fundamental Plane

The strongest constraints on the mean star formation epoch have come from the evolution of the Fundamental Plane (FP) relation. The Fundamental Plane (Djorgovski & Davis 1987) is a relation between the effective radius r_e, effective surface brightness μ_e, and central velocity dispersion σ, such that $r_e \mu_e^{0.8} \propto \sigma^{1.25}$ in the B band. The implication of the existence of the FP is that M/L ratios of galaxies correlate strongly with their structural parameters: $M/L \propto r_e^{0.2} \sigma^{0.4} \propto$

[1]Note that this interpretation requires that we observe early-type galaxies at a special time, when age and metallicity variations exactly cancel.

[2]Possibly dubbed AWACS, for A Wide Angle Cluster Survey.

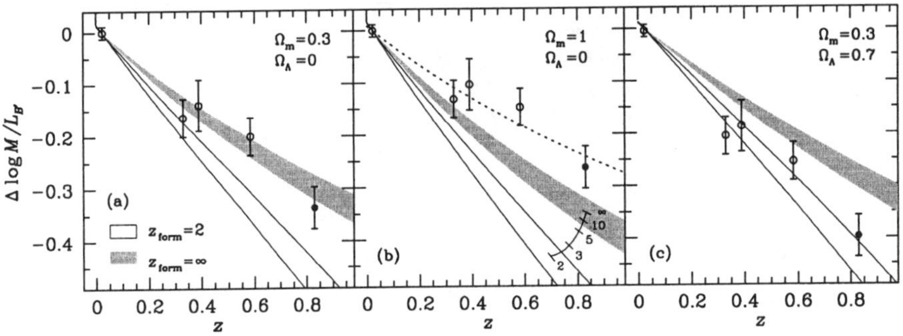

Figure 1. Evolution of the mean M/L_B ratio of early-type galaxies, as determined from the FP relation. The observed evolution is slow, indicating an early formation of the stars.

$M^{0.2}$ (Faber et al. 1987). The power of the FP lies in this relation to M/L ratios, and its usefulness for galaxy evolution studies stems from its small scatter and the fact that it applies to both elliptical and S0 galaxies.

The M/L ratios of galaxies are expected to evolve because the luminosity of their stellar populations decreases as they age ("passive evolution"). The *rate* of evolution depends on the time that has elapsed since the population was formed: the light of young stellar populations is dominated by massive stars which have a short life time on the main sequence, whereas the light of old stellar populations is dominated by long lived, low mass stars. In the rest frame B band, the expected evolution is $M/L \propto (t - t_{\rm form})^{0.91}$ for a Salpeter IMF; the coefficient depends on the passband and the IMF, but is only weakly dependent on the metallicity. As a result, the rate of evolution of the intercept of the FP gives a strong constraint on the mean stellar age of early-type galaxies.

The measured evolution of M/L_B to $z = 0.83$ is shown in Fig. 1, from van Dokkum et al. (1998). The evolution is surprisingly low, $\ln M/L_B \propto -z$, indicating stellar formation redshifts of $z > 2.8$ for $\Omega_m = 0.3$, $\Omega_\Lambda = 0$, and a Salpeter IMF. Studies of the evolution of colors (e.g., Stanford et al. 1998) and line indices (Bender et al. 1998; Kelson et al. 2001) have yielded very similar results. In general, color evolution (effectively the difference between the luminosity evolution in each of two passbands) can not be measured to the same precision as evolution in M/L ratios.

It is difficult, but not impossible, to extend the FP measurements to even higher redshift. The practical limit probably lies around $z \sim 1.3$: 12.5 hr Keck spectra of "Extremely Red Objects" in the cluster RXJ 0848+4453 at $z = 1.27$ are just sufficient to measure velocity dispersions (van Dokkum & Stanford, in prep).

2.2. Evolution of the scatter in the color-magnitude relation

The color-magnitude (CM) relation provides important additional constraints on the star formation epoch of early-type galaxies. Because spectroscopy is not required it is relatively straightforward to obtain large samples, enabling studies of the evolution of the scatter and slope of the relation as well as its zeropoint.

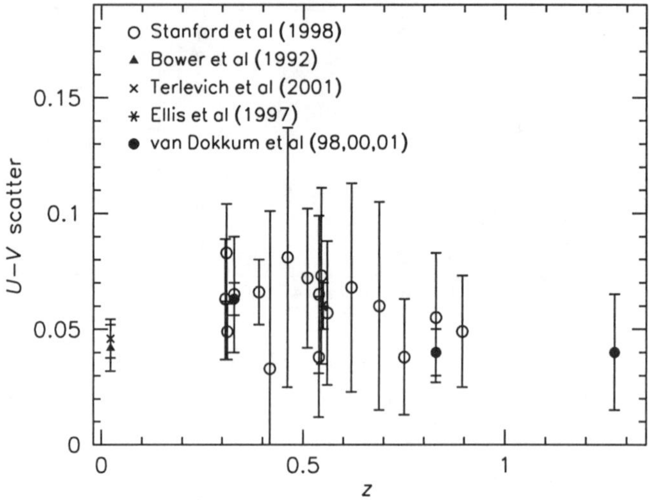

Figure 2. Evolution of the scatter in the color-magnitude relation with redshift, from literature data (see text). The scatter shows little evolution, implying that the scatter in age is small at all times.

The scatter is of particular interest, because it measures the spread in stellar age among early-type galaxies. The rest frame $U-V$ color evolution of a stellar population can be described by $L_V/L_U \propto (t - t_{\rm form})^{\kappa_U - \kappa_V}$, with $\kappa_U \approx 1.08$ and $\kappa_V \approx 0.81$. It can be shown that the scatter in the CM relation at any time t is proportional to the scatter in luminosity weighted age divided by the mean age (e.g., van Dokkum et al. 2000).

The observed evolution of the scatter in the CM relation is shown in Fig. 2. Ground based data are from Bower et al. (1992) and Terlevich et al. (2001) for Coma and from Stanford et al. (1998) for high redshift clusters. HST measurements have smaller errorbars, and are from Ellis et al. (1997) and van Dokkum et al. (1998, 2000, 2001). The scatter remains very small all the way from $z = 0$ to $z = 1.3$. This result is quite surprising, because the mean age of galaxies should be *at least* a factor 2 smaller at $z = 1$. Therefore, if the scatter in the CM relation at $z = 0$ is caused by age variations, one might expect the scatter to increase by at least a factor 2 from $z = 0$ to $z = 1$. Following a similar line of reasoning it has been argued that the scatter in the CM relation of nearby clusters is mainly due to metallicity variations, and that the formation of early-type galaxies is even more synchronized than implied by the tight scaling relations observed at low redshift (e.g., Stanford et al. 1998). Indeed, when taken at face value, the scatter observed at $z = 1.27$ implies a spread in age of only $\sim 5\,\%$ at the present epoch!

In summary, studies of the observed evolution of early-type galaxies indicate that they have been a highly homogeneous, slowly evolving population over at least the latter $\sim 65\,\%$ of the age of the Universe. To satisfy the tightest observational constraints it appears that early-type galaxies would have to have formed at very high redshift ($z > 3$) in a very short time ($\leq 500\,$Myr).

3. Assembly time of cluster galaxies

3.1. The Butcher-Oemler effect

It has been known for a long time that something must be amiss with the simple picture of early formation presented above. The earliest evidence for significant recent evolution in cluster environments was the discovery of the Butcher-Oemler effect: the increase with redshift of the fraction of blue galaxies in clusters (Butcher & Oemler 1978, 1984). Figure 3 is a compilation showing the evolution of the blue fraction with redshift, with data from Butcher & Oemler (1984), Smail et al. (1998), Fabricant et al. (1991), van Dokkum et al. (2000), and Ellingson et al. (2001).

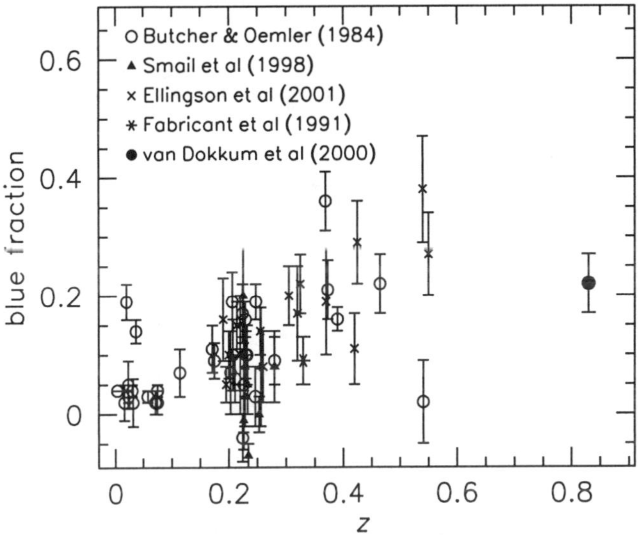

Figure 3. Evolution of the blue galaxy fraction in clusters. The trend has significant scatter.

There is clearly a trend in Fig. 3, albeit with a large scatter. Much of the work on cluster galaxies in subsequent decades – too much to do justice in this short review – was aimed at understanding the nature of these blue galaxies. It has become clear that most of the blue galaxies are not early-type galaxies but low mass spirals and irregulars (e.g., Smail et al. 1997). Some show ongoing star formation; others are currently not forming stars but have enhanced Balmer absorption lines indicating a recent star burst (e.g., Dressler & Gunn 1983). Among the more persistent ideas is that the Butcher-Oemler effect is driven by infall of blue, late-type galaxies from the field, which subsequently lose their fuel for star formation in interactions with other galaxies and/or the hot X-ray gas (e.g., Abraham et al. 1996, Ellingson et al. 2001). In this picture, the redshift dependence of the blue fraction may be the result of a decreasing infall rate with time, and/or reflect the well established overall decrease of the star formation rate in the field population (e.g., Kauffmann 1995).

It is usually assumed that a sizable fraction of the blue population are progenitors of (low mass) red early-type galaxies in nearby clusters (e.g., Kodama & Bower 2001), in apparent conflict with the early formation of early-type galaxies inferred from studies of their color and luminosity evolution.

3.2. Evolution of the early-type galaxy fraction

Dressler et al. (1997) report a high fraction of spiral galaxies in clusters at $0.3 < z < 0.5$. These galaxies are much rarer in nearby rich clusters, and hence must have transformed into early-type galaxies between $z = 0.5$ and $z = 0$. Other studies (e.g., Couch et al. 1998, van Dokkum et al. 2001) have confirmed this trend, and extended it to $z = 1.3$. The evolution of the early-type galaxy fraction is shown in Fig. 4. The early-type fraction decreases by a factor ~ 2 from $z = 0$ to $z \sim 1$, although the trend has significant scatter.

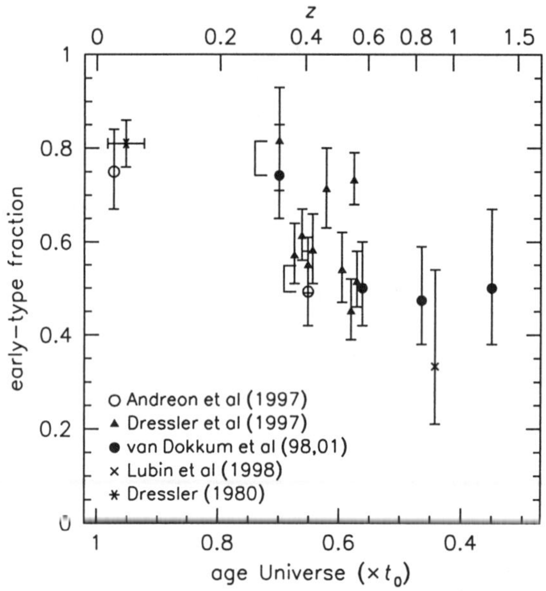

Figure 4. Evolution of the early-type galaxy fraction compiled from various studies, and taken from van Dokkum et al. (2001).

Dressler et al. found that the increased fraction of spiral galaxies at high redshift is accompanied by a low fraction of S0 galaxies, and concluded that the $z \approx 0.4$ spiral galaxies transform into S0 galaxies. They postulate that the formation of elliptical galaxies predated the virialization of the clusters in which they now live. However, there is some controversy over the relative numbers of elliptical and S0 galaxies in distant clusters[3]. Dressler et al. claim that S0 galaxies are virtually absent at $z \sim 0.4$, suggesting a factor ~ 4 evolution over

[3]Importantly, there is no such controversy over the *combined* number of Es and S0s, i.e., the early-type galaxy fraction.

the past ~ 4 Gyr. Others have debated this, and find a much milder evolution in the E/S0 ratio (e.g., Andreon et al. 1997).

One of the problems is the difficulty in distinguishing elliptical and S0 galaxies at high redshift (see Fabricant et al. 2000 and references therein). Spatially resolved kinematics may offer an elegant solution. An example is shown in Fig. 5. From the HST image alone it is difficult to determine whether this $z = 0.83$ galaxy is an elliptical or an S0. The kinematics reveal a rapidly rotating cold disk in addition to a hot bulge, and show that this object is a massive S0.

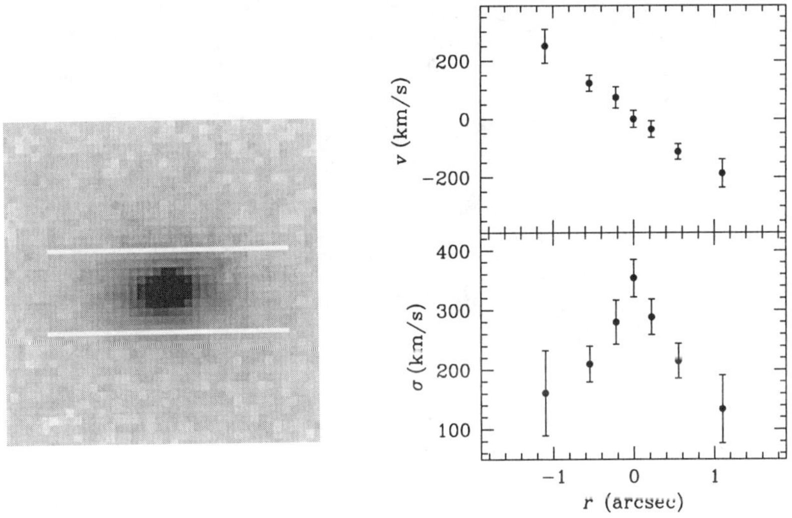

Figure 5. Based on the HST image shown at left it is difficult to determine whether this $z = 0.83$ galaxy is an E or an S0. The kinematics reveal a rapidly rotating disk, demonstrating that this is an S0.

3.3. Mergers in $z \sim 1$ clusters

The discovery of a large number of red merger systems in the cluster MS 1054–03 at $z = 0.83$ is arguably the most spectacular evidence for recent formation of massive early-type galaxies (van Dokkum et al. 1999). We obtained deep, multicolor images of this cluster at 6 pointings with WFPC2 on HST. Redshifts of galaxies in this field were obtained with the Keck Telescope; 89 of those are cluster members. The survey is described in van Dokkum et al. (2000).

We found that 17 % of the galaxies in MS 1054–03 are merger systems. Most of the mergers are very luminous ($M_B \sim -22$ in the rest frame, or $\sim 2L_*$ at $z = 0.83$), and a striking way to display our result is to show a panel with the 16 brightest confirmed cluster members (Fig. 6). Five were classified as mergers.

The mergers are generally red, with a few exceptions. Similarly, the spectra of most of the mergers do not show strong emission lines. These results suggest that the bulk of the stars was formed well before the merger. Hence the stellar age of the merged galaxies will be significantly different from the "assembly age". The physical reason for the low star formation is unknown: it is possi-

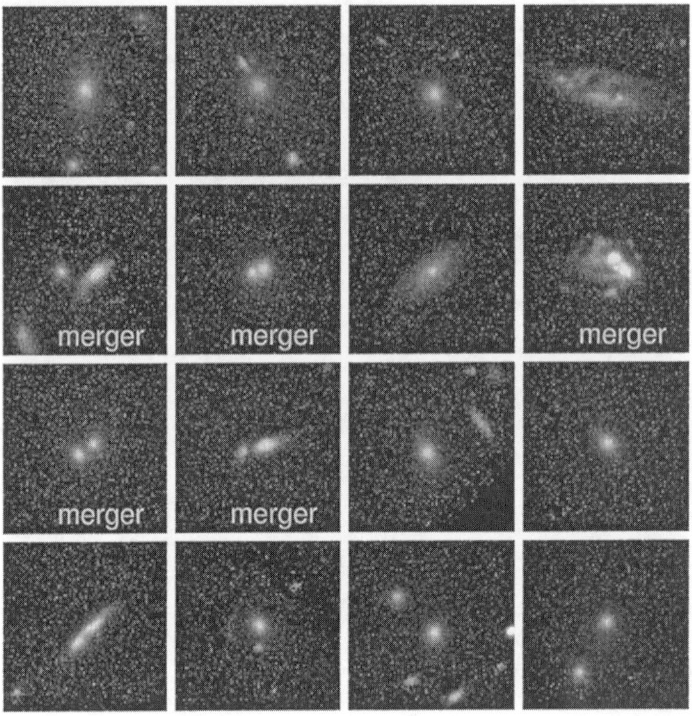

Figure 6. The 16 brightest confirmed members of MS 1054–03 at $z = 0.83$, ordered by I magnitude. Note the large number of mergers. A color version of this figure can be found at http://www.astro.caltech.edu/ pgd/ms1054.

ble that the massive precursor galaxies had already lost their cold gas due to internal processes (such as super winds, or winds driven by nuclear activity). Alternatively, the cold gas may have been stripped by the cluster X-ray gas, or exhausted in an earlier phase of merging in dense, infalling groups.

It is not yet known whether the galaxy population of MS 1054–03 is typical for its redshift. One possibility is that such a phase of enhanced merging occurs at different redshifts for different clusters. In MS 1054–03 the mergers probably occur in infalling subclumps, and its high merger fraction could be related to its overall unvirialized state (van Dokkum et al. 1999).

We recently completed a morphological study of an even higher redshift cluster, RXJ 0848+4453 at $z = 1.27$ (van Dokkum et al. 2001). The Brightest Cluster Galaxy has an asymmetric outer envelope, demonstrating that it recently experienced a merger or strong tidal interaction. The second brightest galaxy is yet another red merger system, in this case between three galaxies of comparable brightness. This remarkable system shows that red mergers are not unique to MS 1054–03, and may even be common at $z \sim 1$. Studies of more clusters are required to better quantify the role of these red mergers in the formation of massive galaxies.

4. Effects of morphological evolution: the progenitor bias

The existence of the Butcher-Oemler effect, the evolution of the early-type galaxy fraction, and the presence of mergers in distant clusters imply that simple models for the evolution of early-type galaxies in clusters are insufficient. Therefore, we need to consider more complex models that incorporate morphological transformations. The problem was first addressed in Franx & van Dokkum (1996), and worked out in van Dokkum & Franx (2001).

4.1. Complex models

We assume that early-type galaxies have two phases in their history: first a relatively long phase in which they were forming stars, at a rate which can be constant, or variable with time. Second, they are transformed into galaxies without star formation, through a merger, gas-stripping, or other mechanism. During a period of ≈ 1 Gyr they are classified as post-starburst galaxy, merger galaxy, or other "special type". After that, they are classified as normal early-type galaxies. As a result, the set of early-type galaxies evolves, and the set of galaxies classified as early-types at $z = 0$ is not the same as the set of galaxies classified as early-types at high redshift.

The evolution of the M/L ratio of individual is shown in Fig. 7c. The evolution of the M/L ratio of each galaxy is indicated with a dotted line when it is not yet classified as early-type, and with a continuous curve when it is classified as early-type galaxy. As is obvious from the plot, the continuous addition of young early-type galaxies to the sample has a significant effect on the evolution of the mean M/L ratio: the newly added galaxies pull the mean M/L ratio to lower values. As a result the evolution of the mean M/L ratio of the full sample is very slow – much slower than the evolution of the M/L ratio of "typical" individual galaxies. This effect is called "progenitor bias": as we compare the M/L ratios of early-type galaxies at different redshifts, we compare between different sets of galaxies, and the evolution of their properties can be misinterpreted when morphological evolution is ignored.

The evolution of the mean M/L ratio of the early-types is shown in Fig. 7d. As can be seen, the mean evolution is slow. The slope of the $M/L - z$ relation is comparable to the slope for a single galaxy which formed at very high redshift, even though the mean formation redshift of all early-type galaxies at $z = 0$ is low at $z_{\mathrm{form}} = 2$. The long dashed curve in the figure indicates the evolution of the mean M/L ratio of all galaxies classified as early-types at $z = 0$. The evolution is much faster, as expected. The difference between these two curves is caused by the progenitor bias, and it can be quite substantial.

The scatter for the early-types at any redshift is indicated by the shaded region in Fig. 7d. Because the youngest galaxies drop out of the sample at higher redshift, the scatter remains constant, or even decreases slightly at higher redshift. The same holds for the scatter in the CM relation. This counter-intuitive result can be explained by the fact that the models are approximately scale free in time, and the relative age differences between the early-type galaxies are similar at all times.

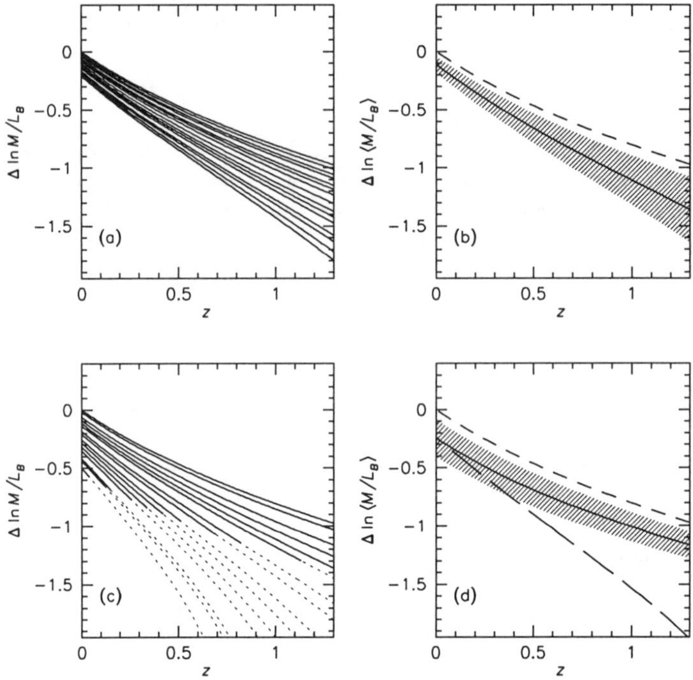

Figure 7. Model predictions for the evolution of the M/L ratio for simple models (a,b), and complex models with morphological transformations (c,d). See text and van Dokkum & Franx (2001).

4.2. Application to data

Figure 8 shows two models which fit the evolution of the M/L ratio measured from the Fundamental Plane, and the evolution of the scatter in the color-magnitude relation. The morphological transformations are described by a simple function which provides a good fit (panel a). Two models are explored for the star formation rate during the phase when the galaxies are spirals: the solid line indicates a model with constant star formation rate, and the dotted line shows a model with declining star formation rate. They both fit the evolution of the M/L ratio well (panel b). The model with the declining star formation ratio underpredicts the scatter in the color magnitude relation and M/L ratios. Hence for this model the scatter is not produced entirely by age differences, but also by scatter in the metallicity-magnitude relation, or other effects.

The progenitor bias for the two models is different, as might be expected from the difference in the predicted scatter due to age variations. When the evolution of the M/L ratio is fitted with a simple model without morphological transformations, we obtain a mean formation redshift $z_{\rm form} = 6.5$. The complex model with constant star formation produces an estimate of $z_{\rm form} = 3$. The model with a declining star formation rate produces $z_{\rm form} = 4$. These values apply for a cosmology with $\Omega_m = 0.3$. They change to 2.6, 2.0, and 2.2, respec-

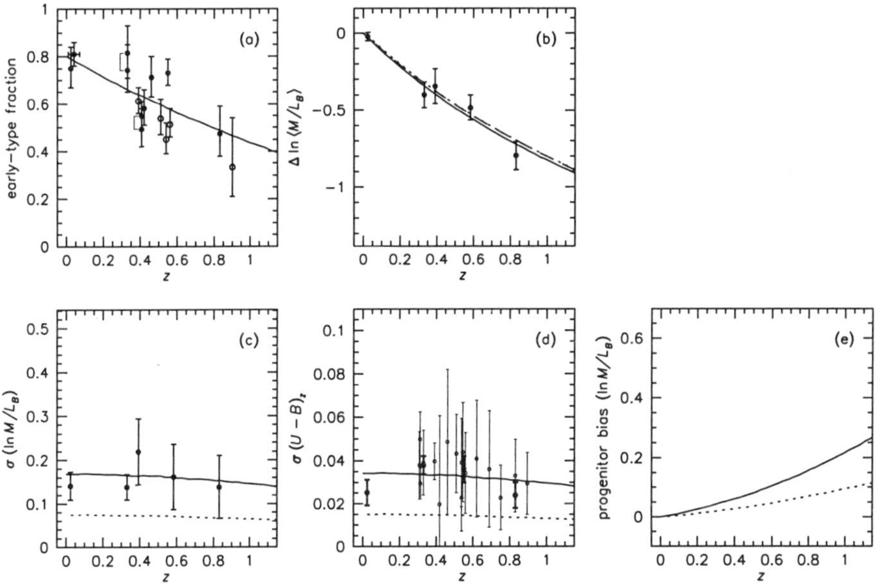

Figure 8. Complex models fitted to observations. Models which include morphological transformations can fit the evolution of the early-type galaxy fraction (a), the mean M/L ratio (b), and the scatter in the FP (c) and the CM relation (d) simultaneously. The error that is made when using simple models (i.e., without morphological transformations) is shown in (e). See van Dokkum & Franx (2001) for details.

tively, for a flat universe with $\Omega_m = 0.3$. Hence the effects of the progenitor bias are modest, but should not be ignored.

5. Conclusions

Models which include morphological transformations can reconcile two apparently contradictory lines of evidence: the low scatter in the color magnitude relation and the slow evolution of the M/L ratio on one hand, and the morphological evolution observed in rich clusters on the other. A basic framework for galaxy evolution in clusters, which includes infall and morphological transformations, seems to be developing and future observations will be aimed at refining and testing these ideas. The available data at $z \sim 1$ are still sparse, and the clusters that have been studied so far may not be typical progenitors of "run of the mill" nearby clusters. It is expected that ACS on HST will make it much easier to study clusters at this epoch; the future of this field can therefore be considered bright!

It is a pleasure to thank the organizers for an interesting and lively meeting in beautiful Sesto Pusteria, and for financial support.

References

Abraham, R. G., et al. 1996, ApJ, 471, 694
Andreon, S., Davoust, E., & Heim, T. 1997, A&A, 323, 337
Baugh, C. M., Cole, S., Frenk, C. S., & Lacey, C. G. 1998, MNRAS, 297, 427
Bender, R., et al. 1998, ApJ, 493, 529
Bower, R. G., Lucey, J. R., & Ellis, R. S. 1992, MNRAS, 254, 601
Butcher, H., & Oemler, A. 1978, ApJ, 219, 18
Butcher, H., & Oemler, A. 1984, ApJ, 285, 426
Couch, W. J., et al. 1998, ApJ, 497, 188
Djorgovski, S., & Davis, M. 1987, ApJ, 313, 59
Dressler, A. 1980, ApJ, 236, 351
Dressler, A., & Gunn, J. E. 1983, ApJ, 270, 7
Dressler, A., et al. 1997, ApJ, 490, 577
Ellingson, E., Lin, H., Yee, H. K. C., & Carlberg, R. G. 2001, ApJ, 547, 609
Ellis, R. S., et al. 1997, ApJ, 483, 582
Faber, S. M., et al. 1987, Nearly Normal Galaxies. Springer, New York, p. 175
Fabricant, D. G., McClintock, J. E., & Bautz, M. W. 1991, ApJ, 381, 33
Fabricant, D., Franx, M., & van Dokkum, P. 2000, ApJ, 539, 577
Franx, M., & van Dokkum, P. G. 1996, in New Light on Galaxy Evolution (IAU 171), R. Bender & R. L. Davies, Eds., Kluwer, p. 233
Kauffmann, G. 1995, MNRAS, 274, 161
Kauffmann, G. 1996, MNRAS, 281, 487
Kelson, D., Illingworth, G., van Dokkum, P., & Franx, M. 2000, ApJ, 531, 184
Kelson, D., Illingworth, G., Franx, M., & van Dokkum, P. 2001, ApJ, 552, L17
Kodama, T., & Bower, R. G. 2001, MNRAS, 321, 18
Lubin, L. M., et al. 1998, AJ, 116, 584
Smail, I., et al. 1997, ApJS, 110, 213
Smail, I., Edge, A. C., Ellis, R. S., & Blandford, R. D. 1998, MNRAS, 293, 124
Stanford, S. A., Eisenhardt, P. R., & Dickinson, M. 1998, ApJ, 492, 461
Terlevich, A. I., Caldwell, N., & Bower, R. G. 2001, MNRAS, 326, 1547
Trager, S. C., Faber, S. M., Worthey, G., Gonzalez, J. J. 2000, AJ, 120, 165
van Dokkum, P. G., & Franx, M. 1996, MNRAS, 281, 985
van Dokkum, P., Franx, M., Kelson, D., Illingworth, G. 1998, ApJ, 504, L17
van Dokkum, P. G., et al. 1999, ApJ, 520, L95
van Dokkum, P. G., et al. 2000, ApJ, 541, 95
van Dokkum, P. G., et al. 2001, ApJ, 552, L101
van Dokkum, P. G., & Franx, M. 2001, ApJ, 553, 90
Worthey, G. 1994, ApJS, 95, 107
Yee, H. K. C., Ellingson, E., & Carlberg, R. G. 1996, ApJS, 102, 269

Tracing Cosmic Evolution with Galaxy Clusters
ASP Conference Series, Vol. 268, 2002
S. Borgani, M. Mezzetti, and R. Valdarnini, eds.

A wide field survey of the distant rich cluster Cl0024+1654

T. Treu, R.S. Ellis, P. Trivedi

California Institute of Technology, MS 105-24, Pasadena, CA 91125

J.-P. Kneib

Observatoire Midi-Pyrénées, UMR5572, 14 Av. Édouard Belin, 31400 Toulouse, France

A. Dressler, A. Oemler

Carnegie Observatories, 813 Santa Barbara Street, Pasadena, CA 91101

P. Natarajan

Department of Astronomy, Yale University, P.O. Box 208101, New Haven, CT 06250

I. R. Smail

Department of Physics, University of Durham, South Road, Durham DH1 3LE, UK

Abstract. We describe the first results from a comprehensive study of the distant cluster Cl0024+1654 ($z = 0.39$) based upon a pattern of 38 mosaiced HST-WFPC2 images extending to radii ~ 5 Mpc. These are being analysed in conjunction with extensive spectroscopy conducted with the CFHT, WHT, and Keck Telescopes. The overall goal is to understand the morphological transformations and associated short-term star formation histories of representative numbers of infalling field galaxies in the context of the cluster potential as defined by weak lensing studies. Our HST database contains over 2000 galaxy morphologies to $I = 22.5$. Spectroscopic data and HST morphologies are currently available for about 215 members over an unprecedented range of environments. We confirm the existence of a well-defined morphology-density relation over a large dynamic range within a single system at a significant look-back time. Tentative trends in the E/S0 fraction as a function of radius are discussed. A weak lensing signal in the background galaxies has been detected at the cluster periphery and its inversion demonstrates only marginal substructure. A statistically-significant galaxy-galaxy lensing signal has also been seen for the cluster members. Further work will relate radial dependencies in the dark matter and halo masses in the context of spectroscopic and morphological diagnostics of truncated star formation.

1. Introduction

The ability of the Hubble Space Telescope (HST) to resolve galaxy morphologies at cosmologically-significant look-back times has enabled great progress in understanding the origin of the morphology density relation and, *inter alia*, the influence of the environment on galaxy morphology. On the basis of HST images of several clusters at $z \sim 0.3 - 0.5$, the Morphs team (Dressler et al. 1997) determined a remarkable decline with redshift in the fraction of S0s with a corresponding rise in that for spirals. They deduced that spirals were transformed to S0 galaxies remarkably recently, most likely because their gas was removed by processes such as tidal effects and ram pressure stripping.

Alongside this evolutionary signal, strong evolution is also found in star formation and morphological characteristics of field galaxies (Glazebrook et al. 1995; Lilly et al. 1996). Understanding the role that gas-rich field galaxies play in fueling the transformations inferred in cluster cores remains an interesting issue. As field galaxies most likely fall into clusters at all epochs, the origin of the recent demise of spirals is puzzling.

A promising way forward is to study, in considerable detail, the transformations occurring *in situ* in a single system from the virialised core to the periphery where field galaxies infall on radial orbits. This is preferable to drawing deductions based on studies of many clusters, observed at different redshifts, each at an unknown stage in its evolutionary history.

Because of the small field of view of WFPC-2, most morphological studies have been limited to the very central regions of distant clusters (e.g. Smail et al. 1997), and correspondingly little is known about the properties of the infalling galaxy population at large radii. Ground based studies (Abraham et al. 1996a) have explored the *integrated* properties of galaxies to large cluster radii, finding evidence for radial trends in diagnostics of recent star formation. It seems likely that star formation is suppressed as galaxies infall, although the physical mechanisms and timescales are still quite controversial (Poggianti et al. 1999; Balogh et al. 2000). These latter uncertainties may be overcome if the timescales of stellar activity/truncation can be linked to those determined dynamically in a known gravitational potential (e. g. from weak lensing signal).

We present the first results from a wide field HST survey of the rich cluster Cl0024+1654 (hereafter 0024; $z = 0.39$) whose goal is to address many of the above questions. In the following, the Hubble constant is $H_0 = 50 h_{50}$kms^{-1} Mpc^{-1}, assuming $h_{50} = 1.3$ when necessary. The matter density of the Universe and the cosmological constant are $\Omega = 0.3$ and $\Omega_\Lambda = 0.7$ respectively.

2. Overview of Database

0024 ($z = 0.39$) is an ideal cluster with which to begin a detailed dynamical, morphological and stellar population study of this type. The results can be contrasted with similar pioneering ground-based studies of the Coma cluster by Kent & Gunn (1982) and Colless & Dunn (1996) and of Abell 2390 ($z = 0.23$) by Abraham et al. (1996a). The distance to 0024 is sufficient that only a relatively modest number (38) of WFPC2 images is needed for a well-sampled study almost to the turn-around radius. The cluster appears regular and symmetric in

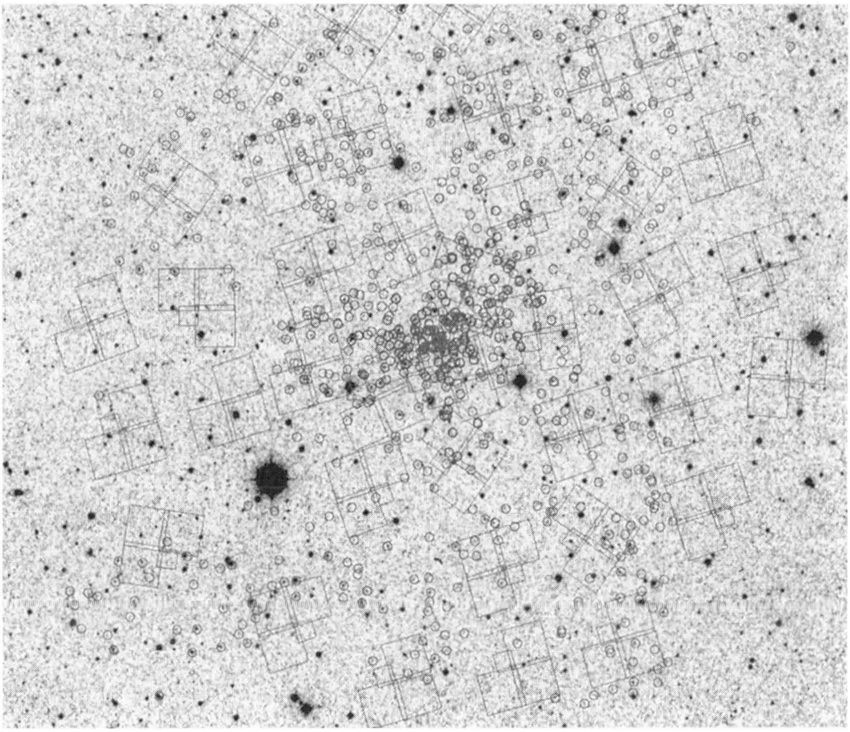

Figure 1. WFPC2 pointings and available redshifts (open circles). The working database of cluster 0024 ($z = 0.39$) contains $\sim$ 2000 HST morphologies and $\sim$ 750 spectroscopic redshifts.

both its galaxy distribution and X-ray morphology. The availability of over 600 spectroscopic redshifts from the CFHT/WHT (Czoske et al. 2001a) provides a major starting point for a comprehensive dynamical study. As is often the case with such extensive sampling of the velocity field (c.f. Coma), complications emerge. The redshift distribution of confirmed members already suggested the existence a foreground system of modest mass. However, recent work by Czoske et al. (2001b) suggests the dynamical distribution of the innermost galaxies is consistent, although not uniquely so, with a line-of-sight merger of two systems. In addition, the central mass distribution can be independently reconstructed from the quintuple gravitational arc system (Broadhurst et al. 2000).

The first component of our dataset is the wide-field WFPC2 data, providing galaxy morphology and the weak shear to $\sim$ 15'($\sim$ 5Mpc) from the cluster center. The primary imaging was accomplished using 2 orbit exposures using the F814W filter, with parallel images taken using STIS in clear CCD mode. As of July 2001 when the conference was held, only 22/38 planned pointings had been observed. However, the remainder arrived soon after and this report presents the first preliminary analysis based on the entire dataset (see Fig. 1).

Via comparisons with both the Morphs and Medium Deep Survey data taken in the same F814W filter, it was determined that morphological classifications in 0024 could be reliably performed within the MDS classification system (Glazebrook et al. 1995, Abraham et al. 1996b) to a limiting magnitude of $I = 22.5$. The slightly deeper Morphs exposure in the cluster core permits a similar classification to $I = 23$ (Smail et al. 1997). Systematic classifications performed by one of us (RSE) yields a sample of 1800 objects ($\sim$ 600 E/S0 $\sim$ 800 S, $\sim$ 400 Irr), in addition to $\sim$ 250 already classified in the core. Independent classification is currently being performed by a second collaborator (AD).

The second component of our survey is the redshift survey. In the outer fields (>1-2 Mpc from the center), the galaxy excess above background per WFPC2 field to $I \sim 22.5$ is within the observed fluctuations in the field counts (Abraham et al. 1996a); to identify infalling members, spectra are crucial. So far, we have collated 718 redshifts (Dressler et al. 1999; Czoske et al. 2001a; Owen 2001; Metevier & Koo 2001) including 334 members. Further spectroscopy in the periphery is necessary because of the incomplete sampling by earlier studies.

The third component is a spectroscopic campaign with the Keck Telescopes, following up known members to $I = 21$ and extending the completeness of the redshift surveys to $I = 23$. These data represent a crucial part of the study of the infalling population providing diagnostics of recent star formation and dynamical masses either through stellar velocity dispersions or resolved emission line rotation curves. A first attack in Oct 2001 yielded spectra for $\sim$ 100 objects.

3. First results

3.1. Galaxy population

Our dataset provides the first comprehensive and uniform set of morphological data across the range of environments from the cluster core out to $\sim$ 5 Mpc ($\simeq$ "field"). This has enabled us to examine the suggestion that there are strong radial gradients in the morphological mix and, in future, to tie these to spectroscopic diagnostics which can be used to understand the interplay between dynamical infall and the truncation of star formation.

Whereas photometric techniques are effective in addressing the statistical properties of background-subtracted galaxies in low densities (Kodama et al. 2001), we seek here to identify members individually so the dynamical and spectral diagnostics can be brought into play. As an illustration of the difficulties associated with statistical subtraction techniques, we show in Fig. 2, the run of morphological type with radius from the cluster core to regions of very low density. The left panel shows the $I < 22.5$ galaxy density as a function of cluster radius divided by broad morphological type. As expected, E/S0 are more centrally concentrated than spirals, while Irr/Unclass/Mergers show little evidence of any concentration. At radii $\gtrsim$ 2 Mpc, the observed density lies within the random fluctuations of the average field density (dashed line; Abraham et al. 1996b), and the background contamination can no longer be reliably removed statistically. We can obtain a smoother morphology-radius relation by averaging azimuthally, as shown in the right panel of Fig. 2. The peaked distribution of E/S0s is now more prominent and an excess in both E/S0s and spirals is seen to the largest radius probed. However, this excess is critically dependent on the

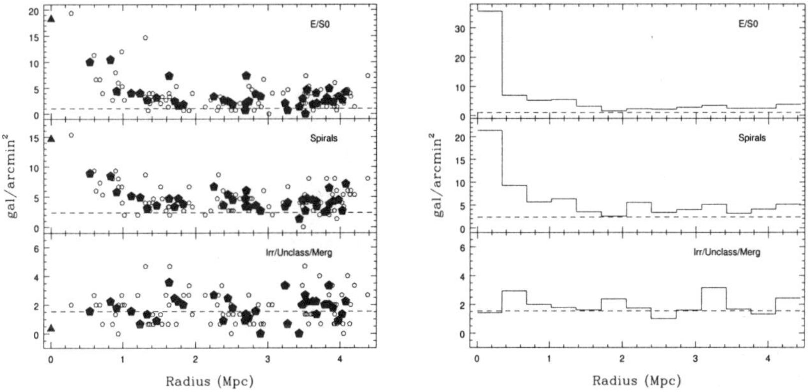

Figure 2. *Left:* Galaxy surface density to $I < 22.5$ as a function of cluster-centric radius. Open symbols represent individual WF chips, the filled symbols represent the average of a entire WFPC2 pointing. The dashed line is the average field density (as inferred from Abraham et al. 1996b). *Right:* As above, after azimuthal averaging.

assumed background count and its morphological distribution. Astonishingly, the morphological counts (Glazebrook et al. 1995, Driver et al. 1995) remain uncertain simply by virtue of the limited area sampled by HST. This emphasises the importance of a redshift survey (to a depth equivalent to that of local determinations) to construct a reliable morphology-density relation.

What does the extensive spectroscopic data tell us about the distribution of confirmed members in Figure 2? As discussed, the kinematic structure of 0024 is more complex than that of a single relaxed system. In particular, the distribution of radial velocities is double peaked, with a secondary peak blueshifted by 3000 kms^{-1} with respect to the main peak (Czoske et al. 2001b). If we assign membership based on the redshift limits adopted in Dressler et al. (1999), there are currently 215 confirmed members within our WFPC2 survey, 115 of which are E/S0s. The left panel of Figure 3 shows the fraction of E/S0s in 0024 ($I < 21.1$ equivalent to the luminosity limit of Dressler 1980) as a function of radius in comparison to a local relation and values reported at intermediate redshift by Couch et al. ($z = 0.31$) and van Dokkum et al. (2000; $z = 0.83$). Examining the trends across the various clusters, the fraction of E/S0s in the cluster cores increases with redshift, consistent with earlier findings by Dressler et al (1997) and van Dokkum et al. (2000). However, perhaps surprisingly, in the outer regions of 0024, the fraction of E/S0s remains relatively high and consistent with the local value. Of course, the uncertainties are large as the number of spectroscopic redshifts beyond 2 Mpc remains small. If this result is borne out by enlarged spectrosopic samples, one possible explanation is that the growth of 0024 is fuelled from merging substructures containing spheroidals rather than solely via an abundant population of infalling isolated field galaxies. Alternatively, in the model proposed by Czoske et al (2001b), some members of

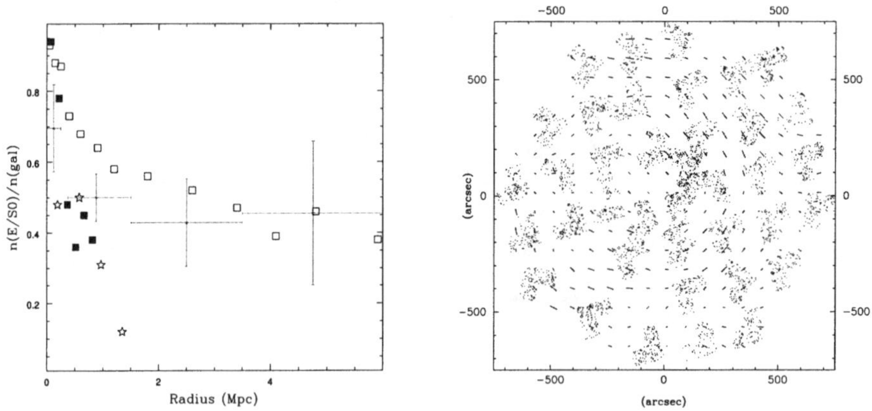

Figure 3. *Left:* The fraction of E/S0s as a function of radius (points with error bars) as compared to a local relation (open squares, Whitmore, Gilmore & Jones, 1993). Also shown are values reported at $z = 0.31$ (filled squares Couch et al. 1998) and at $z = 0.83$ (open stars; van Dokkum et al. 2000). *Right:* Shear map (solid sticks) computed from the faint galaxy catalogue (small ellipses). The cluster shear can be detected right to the periphery of our HST data.

the core of the merging foreground component will be scattered to large radii; in this case we would expect a different dynamical distribution.

3.2. Weak lensing

In classical studies of cluster dynamics, galaxies (or X-ray gas) have been regarded as tracers of the gravitational potential under a number of assumptions. Foremost it has been assumed that they are not biased in any respect with respect to the dominant dark matter potential. A significant advantage of studying in detail a cluster at $z \simeq 0.4$ is that it lies $\simeq$half-way to the faint background population reached with a 2-orbit WFPC2 exposure.

It is possible, therefore, to contemplate combining the galaxy dynamical data with the dark matter distribution as revealed by the weak lensing induced by the cluster on the background population. As an illustration of where we have reached with the recently arrived data, Figures 3 and 4 show the first results arising from this important additional physical diagnostic.

The right panel of Figure 3 shows the weak shear as a function of position using a simple interpolation scheme to handle the sparse-sampling. Remarkably, the shear is detected right to the periphery of the HST mosaic where its value is typically $\sim 3\%$. Interestingly there is some discrepancy between the centre of the shear pattern and that of the visible light, the multiply-imaged arcs or the X-ray gas. This may be an artefact of the preliminary reduction or it could represent contamination from the foreground component. In the context of the model proposed by Czoske et al. (2001b) this seems unlikely.

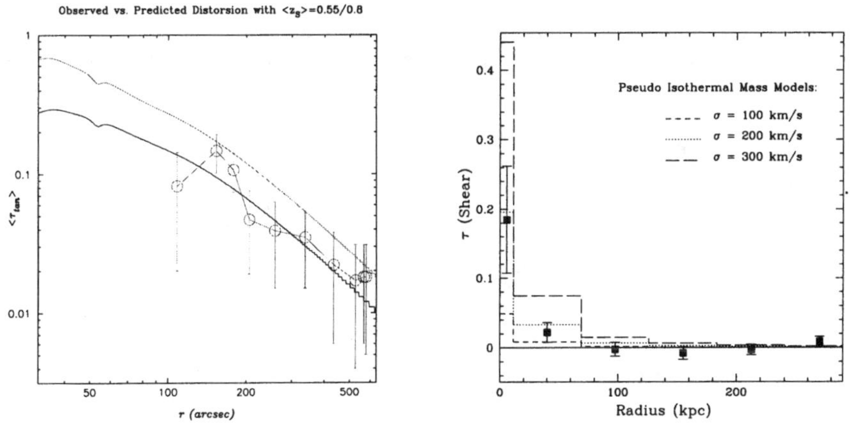

Figure 4. *Left:* Mean shear as a function of cluster radius. An extrapolation of the strong lensing model by Smail et al. (1996) is shown, for different values of the average redshift of the background objects ($1'' = 5.7$ kpc). *Right:* Average galaxy-galaxy lensing tangential shear τ versus radial distance from cluster (lens) galaxy. Histograms represent the expected shear for pseudo-isothermal mass distributions, assuming a core radius $r_c = 0.15$ kpc, truncation radius $r_t = 30$ kpc and a source redshift distribution peaked at $z_s = 1$

3.3. Galaxy-galaxy lensing

A final application of this rich dataset is in determining the halo properties of individual galaxies as probed by weak lensing induced by individual cluster members. In dense environments, dark matter halos of galaxies are likely to be truncated or stripped by environmental effects (e.g. Moore et al. 1996). Previous studies in the cluster AC 114 at $z = 0.31$ (Natarajan et al. 1998) suggested an apparent halo cutoff for cluster members not seen in the equivalent field population. While comprehensive measures of galaxy-galaxy lensing by field galaxies have been carried out recently (e.g. McKay et al. 2001), in clusters such studies have been limited by small sample sizes and sample inhomogeneities.

The 0024 dataset can, in principle, probe halo properties in a statistical sense from the high density core to the low density field thereby securing a direct an unbiased test of the result claimed by Natarajan et al. A preliminary analysis of the 38 WFPC2 pointings yielded 137 cluster galaxies (71 with confirmed redshifts, 66 selected on the color-magnitude diagram) and a carefully constructed background galaxy sample of $\approx 11,000$ objects. The shear obtained by direct averaging (Natarajan & Kneib 1997) is plotted in Fig. 4.

A shear detection of $\tau = 0.18 \pm 0.08$ ($1\ \sigma$) is observed in the innermost $2''$. Despite significant statistical uncertainties, the observed value is consistent with that expected for an isothermal mass distribution with $\sigma \sim 200$ km s^{-1} (see caption for detail). Compared to Natarajan et al. (1998) who detected $\tau = 0.16^{+0.12}_{-0.13}$ ($1\ \sigma$) in the innermost $1.5''$, the galaxy-galaxy lensing signal is more clearly seen and to further radial distance (see plot). Although still a weak signal, the improved selection criteria and larger sample of spectroscopically confirmed E/S0 galaxies are major advances over the AC114 study. The signal will improve further when data from the Morphs field is used. Employing a detailed cluster mass model which incorporates strong lensing features, a maximum likelihood method (Natarajan & Kneib 1997) will be used to robustly extract halo parameters and look for radial trends in halo mass and size.

References

Abraham R. G. et al. 1996a, ApJ, 471, 694

Abraham R. G. et al. 1996b, ApJS, 107, 1

Balogh M., et al. 2000, ApJ, 540, 113

Broadhurst T., et al. 2000, 534, L15

Colless M. & Dunn A. 1996, ApJ, 458, 435

Czoske O., et al. 2001a, A&A, 372, 391

Czoske O., et al. 2001b, A&A, submitted, astro-ph/0111118

de Blok et al. 2001, ApJ, 552,L23

Dressler A. 1980, ApJ, 236, 351

Dressler A. et al. 1997, ApJ, 490, 577

Dressler A. et al. 1999, ApJS, 122, 51

Driver, S.P. et al. 1995, ApJ, 449, L23

Kent, S. & Gunn, J.E. 1982, AJ, 87, 945

Kodama T., et al. 2001, ApJ, 562, L9

McKay T.A. et al. 2001, submitted to ApJ, astro-ph/0108013

Moore B. et al. 1996, Nature, 379, 613

Natarajan P. & Kneib J.-P 1997, MNRAS, 287, 833.

Poggianti B., et al. 1999, ApJ, 518, 576

Smail I. R. et al. 1997, ApJS, 479, 70

van Dokkum P. G. et al. 2000, ApJ, 541, 95

Tracing Cosmic Evolution with Galaxy Clusters
ASP Conference Series, Vol. 268, 2002
S. Borgani, M. Mezzetti, and R. Valdarnini, eds.

Tracing Cosmic Evolution with the Coma Cluster

Bianca M. Poggianti

Osservatorio Astronomico, vicolo dell'Osservatorio 5, 35122 Padova -I

Terry J. Bridges

Anglo-Australian Observatory, PO Box 296, Epping, NSW 1710, Australia

Bahram Mobasher

Space Telescope Science Institute, 3700 San Martin Drive, Baltimore, MD 21218, USA

Dave Carter

Liverpool john Moores University, Twelve Quays House, Egerton Wharf, Birkenhead, Wirral, CH41 1LD, UK

Abstract. We summarize here the first results of a new spectroscopic survey of galaxies in the Coma cluster with $-20.5 < M_B < -14$. Differences between the last epoch of star formation (SF) activity in S0 and elliptical galaxies are discussed. Furthermore, we show that about half of all galaxies without present-day SF display signs of activity at $z < 1 - 2$. Systematic trends of most recent SF epoch with galaxy luminosity are presented.

1. Introduction

In this contribution we report on the first results of a new spectroscopic survey of galaxies in the Coma cluster. This survey was designed to have three main characteristics:
1) It extends over almost 7 mag in galaxy luminosity ($-20.5 < M_B < -14$), well into the dwarf regime. In fact, this is by far the largest spectroscopic sample of cluster dwarf galaxies to date.
2) It covers two areas of $33' \times 50'$ (1×1.5 Mpc) towards the center and the southwest region of Coma. The photometry (B and R) has been obtained with the Japanese Mosaic CCD camera at the William Herschel Telescope (WHT) in collaboration with M. Doi, M., Iye, N.Kashikawa, Y. Komiyama, S. Okamura, M. Sekiguchi, K., Shimasaku, M. Yagi, N. Yasuda.
3) It is essentially a magnitude limited sample, without significant color or morphological bias, therefore it includes galaxies of all Hubble types.
We obtained multifiber spectra with the WYFFOS fibre spectrograph at the WHT for about 300 members of Coma and 200 field galaxies. The spectra cover the central 2.7 arcsec (1.3 kpc) of the galaxies and have a mean S/N ranging

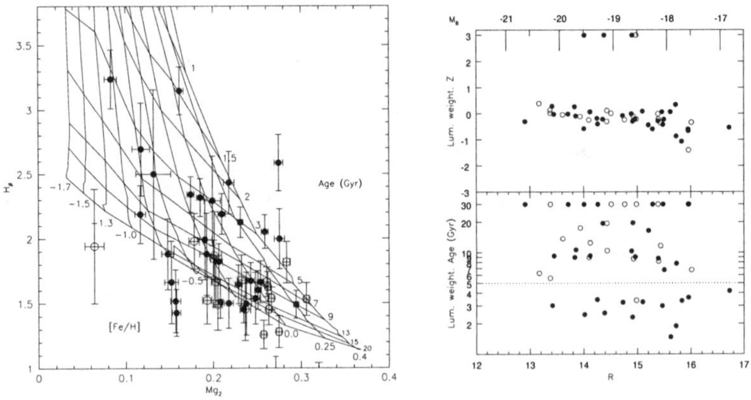

Figure 1. *Left.* Hβ versus Mg_2 index strength for ellipticals (empty circles) and S0 galaxies (filled circles). *Right.* Luminosity weighted metallicities (top) and ages (bottom) of ellipticals (empty circles) and S0 galaxies (filled circles) as a function of the R-band magnitude.

from 16 for the brightest subset to 8 for the faintest subset. Hence, the errorbars on the spectral line measurements are large for the faint galaxies, and for them only broad conclusions regarding ages and metallicities can be reached. *Here we consider only the* $\sim$ 250 *galaxies without emission lines in their spectra.* For them, we have measured the absorption line indices in the Lick system. The stellar properties (luminosity weighted ages and metallicities) have been derived from index-index diagrams, comparing with spectrophotometric models based on the Padova isochrones (http://astro.sau.edu/~worthey/dial/dial_a_pad.html, Worthey). A full description of the results presented here can be found in Papers III and IV. The dataset is described in Papers I and II.

2. S0 versus elliptical galaxies

There are 19 ellipticals and 33 S0s in our sample with a secure morphological classification from Dressler (1980). All but one of the ellipticals are consistent with luminosity weighted ages ≥ 9 Gyr, while 13 S0s have ages < 5 Gyr (Fig. 1, left). Thus, a significant fraction ($> 40\%$) of the S0s show evidence for some recent SF. This is consistent with the hypothesis that they evolved from star-forming spirals that were accreted onto the cluster and had their SF truncated at intermediate redshifts (Dressler et al. 1997, Fasano et al. 2000, see also van Dokkum and Treu in these proceedings). This difference in age spread between Es and S0s is similar to the results of Kuntschner & Davies (1998) in the Fornax cluster and Smail et al. (2001) in A2218. In contrast, other works (e.g. Jones, Smail & Couch 2000, Ellis et al. 1997) have not found a statistically significant difference between the ages of the stars in S0s and Es. It is obviously important to understand whether S0s with recent SF are a *common* phenomenon in clusters or not. In this respect, it is interesting to look at the age distributions

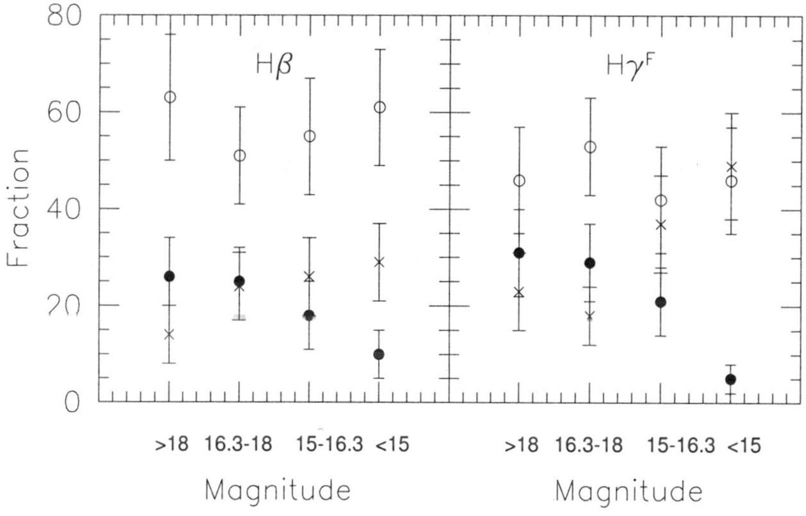

Figure 2. Fraction of young (filled dots, age < 3 Gyr), intermediate-age (crosses, 3 to 9 Gyr) and old (empty dots, age > 9 Gyr) galaxies within each magnitude bin as derived from the Hβ/Mg$_2$ diagram (left) and the Hγ_F/<Fe> diagram (right).

of galaxies as a function of magnitude.[1] In Fig. 1 (right), the population of S0s with recent SF is noticeable below the dotted line: only 4 of the "young" S0s are brighter than $M_B = -19$. In fact, considering galaxies brighter and fainter than $M_B = -19$, bright and faint Es have similar age distributions, while the fraction of S0s with recent SF is higher at fainter luminosities. Therefore, a possible reason for the discrepant results found so far (recent SF in cluster S0s, yes or no) could be the luminosity range explored by the different studies. Given the luminosity function of spirals at $z \sim 0.5$ and the expected fading if their SF

[1]We note that S0s and Es follow a broadly similar metallicity-luminosity relation (top in the right panel of Fig. 1), although the relation for the S0s is slightly steeper and more scattered than that of the Es.

is halted, the great majority of the spirals evolved into S0s should be typically fainter than $M_B = -20/-19$.[2]

3. The infall history of the Coma cluster

In the following we will consider the complete spectroscopic sample with no emission lines, including galaxies of all magnitudes and Hubble types. This sample represents the total cluster population of galaxies without current SF. There is a broad range of ages, from younger than 3 Gyr to older than 9 Gyr, among galaxies of any magnitude. However, there are systematic trends of age with galaxy magnitude. Dividing all galaxies into four magnitude bins, ~ 50 % of galaxies in any bin are "old", i.e. with no sign of SF during the last 9 Gyr (at $z < 1-2$). As shown in Fig. 2, the proportion of *luminous* galaxies that are passive at z=0 and that experienced a SF activity at intermediate redshift ($0.35 < z < 1$) is 30 to 50 %, and it is higher than the fraction of *dwarf* galaxies that were active at that epoch. This is consistent with the observations of current/recent SF in a large number of galaxies in intermediate-z clusters (e.g. Dressler et al. 1999). At low redshift, instead, SF involved a higher fraction of faint than bright galaxies. These results can be used to trace the infall history of galaxies onto Coma, and the consequent star formation history of its present-day galaxies. The luminosity(mass) dependence of the most recent SF epoch points to a "down-sizing effect", and can be compared with the accretion history of clusters predicted by cosmological models.

References

Dressler, A. 1980, ApJS, 424, 565

Dressler, A., Oemler, A. Jr., Couch, W. J., Smail, I. et al. 1997, ApJ, 490, 577

Dressler, A., Smail, I, Poggianti, B. M., Butcher, H. et al. 1999, ApJS, 122, 51

Ellis, R. S., Smail, I., Dressler, A., Couch. W. et al. 1997, ApJ, 483, 582

Fasano, G., Poggianti, B. M., Couch, W. J. et al. 2000, ApJ, 542, 673

Jones, L. A.,, Smail, I., Couch, W. J. 2000, ApJ, 528, 118

Komiyama, Y., Sekiguchi, M., Kashikawa, N. et al. 2001, ApJS, in press (Pap I)

Kuntschner, H., Davies, R. L. 1998, MNRAS, 295, L29

Mobasher, B., Bridges, T. J., Carter, D. et al. 2001, ApJS, 137, 230 (Paper II)

Poggianti, B.M., Bridges, T.J., Mobasher, B., Carter, D. 2001a, ApJ, 561 in press (Paper III, astro-ph 0107158)

Poggianti, B.M., Bridges, T.J., Carter, D., Mobasher, B. 2001b, ApJ, 563, 1 in press (Paper IV, astro-ph 0108212)

Smail, I., Kuntschner, H., Kodama, T. et al. 2001, MNRAS, 323, 839

[2]Those S0s with recent SF have an asymmetric distribution in the sky, being preferentially located in a region east/north-east of NGC4874 (the cluster centre), while older S0s and ellipticals are spread out both north and south of NGC4874.

Tracing Cosmic Evolution with Galaxy Clusters
ASP Conference Series, Vol. 268, 2002
S. Borgani, M. Mezzetti, and R. Valdarnini, eds.

Brightest Cluster Galaxy Formation in Low-Redshift Galaxy Clusters

Wayne A. Barkhouse and H.K.C. Yee

Department of Astronomy & Astrophysics, University of Toronto, 60 St. George Street, Toronto, ON, M5S 1A7, Canada

O. López-Cruz

INAOE-Tonantzintla, Tonantzintla, Pue., Mexico

Abstract. We present the composite luminosity function of a sample of 17 low-redshift galaxy clusters. Our luminosity functions have been measured for an inner region ($r \leq r_{200}$) and an outer region ($r_{200} \leq r \leq 2r_{200}$) centred on each Brightest Cluster Galaxy. The inner region luminosity function has a significantly flatter faint-end slope ($\alpha_{inner} = -1.81 \pm 0.02$) than the outer region ($\alpha_{outer} = -2.07 \pm 0.02$). These results are consistent with the hypothesis that a large fraction of dwarf galaxies near cluster centres are being tidally disrupted.

1. Introduction

Galaxy clusters provide a laboratory in which one can study the effects of galaxy formation and evolution over time. A detailed understanding of low-redshift clusters would furnish us with a benchmark in which to compare observations of higher redshift galaxy clusters. Cluster properties, such as luminosity functions, have the potential to yield valuable clues concerning the dynamical evolution of cluster galaxies, including the formation of Brightest Cluster Galaxies (BCGs).

Recent studies have concentrated on the merging process as an efficient way of creating BCGs (Merritt 1984; Schombert 1988). The formation of cD-like galaxies (we define cD galaxies as those with an extended halo (e.g., Schombert 1988) as well as those galaxies which are halo-dominated; Brown 1997) has been problematic since the merging process cannot account for the presence of an extended halo (Merritt 1984; Tremaine 1990).

From a survey of 45 low-redshift ($0.04 \leq z \leq 0.18$) galaxy clusters, López-Cruz et al. 1997 proposed that the envelope of cD galaxies was created through the process of tidal disruption of dwarf galaxies (e.g., dwarf spheroidals). Evidence for this hypothesis was obtained by examining cluster luminosity functions (LFs). López-Cruz et al. (1997) found that clusters which contain a cD galaxy (Bautz Morgan types I, I-II; Rood-Sastry type "cd") had flatter faint-end slopes (as parameterized by the Schechter function; Schechter 1976) than those which did not. Further support for this model was found by examining the dwarf-to-giant galaxy ratio (D/G), which was found to increase with increasing Bautz Morgan type (López-Cruz 1997).

2. Cluster Survey

In order to test the dwarf galaxy disruption model, we have conducted an imaging survey of low-redshift ($0.02 \leq z \leq 0.04$) Abell clusters using the KPNO 0.9m telescope + 8k mosaic camera. We have obtain R band images for 27 clusters, of which a sub-sample of 11 were also observed in the B band. The 8k mosaic camera has a field-of-view of approximately one square degree and allows us to cover a region 2 to 4 Mpc in length, centred on each cluster.

The purpose of this survey is to test the dwarf galaxy disruption hypothesis by; 1) sampling cluster luminosity functions to a depth adequate to measure the faint-end slope, 2) measuring cluster LFs over a wide area in order to sample the LF as a function of cluster-centric radius, 3) determining the colour gradient of the BCG envelope and compare it to the radial colour gradient of the dwarf galaxy population, and 4) measuring the properties (LFs, spatial distributions, etc.) of morphologically selected galaxy sub-populations.

In this paper we concentrate on results based on the analysis of cluster luminosity functions (we use $H_o = 50\ km\ s^{-1}\ Mpc^{-1}$ and $q_o = 0$ throughout).

2.1. Data Reductions and Photometry

Object detection, classification, and photometry was performed using PPP (Yee 1991). All bright, early-type galaxies were modeled and subtracted from their parent image using the same procedure as in Brown (1997). This allowed us to accurately measure the magnitudes of overlapping fainter galaxies which becomes a serious problem in the cluster core. Complete photometric catalogs of each cluster contain approximately 25,000 objects.

Since we do not have redshifts for every cluster galaxy, luminosity functions are created from the statistical subtraction of background galaxies. The background correction was determined by observing six controls fields, covering a total of six square degrees in area.

3. Luminosity Functions

The direct comparison of LFs for various surveys has been problematic. Many studies have relied on photographic plates and have compared cluster LFs measured using a different radial cutoff from the BCG (e.g., Lugger 1986). Since we expect that environment may have an influence on the overall galaxy population mixture (morphology-density relation; Dressler 1980), it is prudent to compare cluster LFs using the same dynamical radius. For this study, we have chosen to measure our LFs using the r_{200} radius (the radius within which the average density is 200 times the critical density). Since we lack radial velocity measurements, we use the correlation between r_{200} and Bgc (a richness indicator; Yee & López-Cruz 1999) as measured for the CNOC 1 cluster survey (Carlberg, Yee, & Ellingson 1997). We find that as cluster richness increases, the value of r_{200} also increases. Using this relationship, we can estimate a value of r_{200} for a given value of Bgc (which we can calculate from our photometry). In this way, we are able to make a more direct and robust comparison of cluster LFs.

In an effort to improve the signal-to-noise of cluster galaxies over fore/background galaxies, we impose a colour criteria from which we select cluster members. Cluster colour-magnitude diagrams clearly show the well known colour-magnitude relation (CMR; Baum 1959) for early-type cluster galaxies. The colour-magnitude diagram can reveal the presence of background contaminating clusters (López-Cruz 1997). We have used the colour-magnitude diagram to establish an upper envelope in which to reject redder background galaxies.

3.1. Composite Luminosity Function

We computed the composite LF for two radial bins centred on the BCG. The inner area was chosen to have a radius of r_{200}, and the outer bin had an inner radius of r_{200} and an outer radius $2r_{200}$. Clusters whose photometry was 100% complete to $M_R = -16.0$, and which had complete imaging out to a radius of $2r_{200}$, were included in the final sample. This sample consisted of 17 clusters, 7 of which have colour information which allowed us to impose a colour selection criteria based on the cluster's CMR. All magnitudes were K corrected based on Coleman, Wu, & Weedman (1980) and corrected for galactic extinction (Burstein & Heilis 1982).

Individual cluster LFs were constructed through the statistical subtraction of background galaxy counts as measured from six control fields. The rejection criteria used for each cluster was also applied to the background counts (including normalization with respect to cluster area). Each cluster LF was then summed to produce a composite LF for each radial bin.

Figure 1 shows the composite LF computed for the inner and outer radial regions. It is clear from this figure that a single Schechter function is inadequate to describe the composite LFs. We have therefore used the sum of two Schechter functions to fit the LFs in both cases. For the inner radial bin ($r \leq r_{200}$), we find a faint-end slope of $\alpha_{inner} = -1.81 \pm 0.02$. The faint-end slope for the outer radial bin ($r_{200} \leq r \leq 2r_{200}$) was measured to be $\alpha_{outer} = -2.07 \pm 0.02$. It is clear from this result that the inner region has a significantly flatter faint-end slope than the outer region.

4. Discussion and Conclusions

Schechter (1976) described an empirical function which was shown to provide an adequate fit to a sample of 14 cluster LFs. Many studies have investigated the possibility that the LF is "universal" (e.g., Lugger 1986). The results of several recent studies (e.g., López-Cruz et al. 1997; Driver, Couch, & Phillipps 1998), clearly demonstrate that cluster LFs are not universal but are instead a function of environment (this is really just a re-statement of the density-morphology relation).

The comparison of the composite LFs in this study support the conclusion that cluster LFs are not universal and are in fact dependent upon cluster-centric radius (ie., environment). This result is also consistent with the dwarf galaxy disruption model since one expects that tidal forces acting on dwarf galaxies would be greater near the central regions of clusters. A reduction in the fraction of dwarf galaxies, relative to the remaining galaxy cluster population, would be expected for the inner cluster regions.

References

Baum, W.A. 1959, PASP, 71, 106
Brown, J.P. 1997, Ph.D. thesis, University of Toronto
Burstein, D., & Heiles, C. 1982, AJ, 87, 1165
Carlberg, R.G., Yee, H.K.C., & Ellingson, E. 1997, ApJ, 478, 462
Coleman, G.D., Wu, Chi-Chao, & Weedman, D.W. 1980, ApJS, 43, 393
Dressler, A. 1980, ApJ, 236, 351
Driver, S.P., Couch, W.J., & Phillipps, S. 1998, MNRAS, 301, 369
López-Cruz, O. 1997, Ph.D. thesis, University of Toronto
López-Cruz, O., Yee, H.K.C., Brown, J.P., Jones, C., & Forman, W. 1997, ApJ, 476, L97
Lugger, P.M. 1986, ApJ, 303, 535
Merritt, D. 1984, ApJ, 276, 26
Schechter, P.L. 1976, ApJ, 203, 297
Schombert, J.M. 1988, ApJ, 328, 475
Tremaine, S.D. 1990, in Dynamics and Interactions of Galaxies, Wielen, R., Ed. (Heidelberg FRG: Springer-Verlag), 394
Yee, H.K.C. 1991, PASP, 103, 396
Yee, H.K.C., & López-Cruz, O. 1999, AJ, 117, 1985

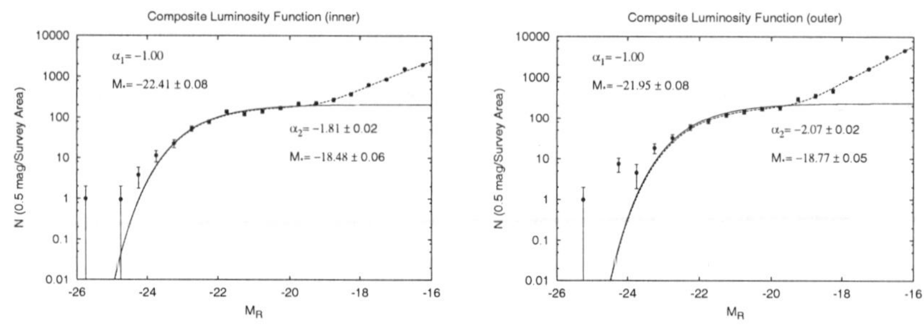

Figure 1. This figure depicts the composite LF for the inner ($r \leq r_{200}$) and outer ($r_{200} \leq r \leq 2r_{200}$) radial bins. The faint-end slope of the inner region is significantly flatter ($\alpha_{inner} = -1.81 \pm 0.02$) than the faint-end slope of the outer region ($\alpha_{outer} = -2.07 \pm 0.02$).

Tracing Cosmic Evolution with Galaxy Clusters
ASP Conference Series, Vol. 268, 2002
S. Borgani, M. Mezzetti, and R. Valdarnini, eds.

NIR and Optical Structural Parameters of Galaxies in the Cluster AC118 at z=0.31

G. Busarello, P. Merluzzi, M. Massarotti

Osservatorio Astronomico di Capodimonte, Napoli, Italy

F. La Barbera, M. Capaccioli

Physics Department, Università Federico II, Napoli, Italy

G. Theureau

Observatoire de Paris, Paris, France

Abstract. We present some preliminary results related to a project aimed at studying the evolution of the galaxy population in rich environments by means of the Color-Magnitude relation and of the Fundamental Plane. We derive the NIR and optical structural parameters for a sample of galaxies in the cluster AC118 at z=0.31. We prove that reliable structural parameters of galaxies at z~0.3 can be still derived from ground–based observations. The NIR effective radii, measured for the first time at this redshift, turn out to be significantly smaller than those derived from the optical data, providing new insight into the evolution of colour gradients in galaxies.[1]

1. Introduction

Numerous photometric surveys are currently done on wide areas of the sky with ground-based telescopes. They are providing an enormous output of data that, in particular, are quickly improving our knowledge of distant galaxies. In addition to integrated quantities like magnitudes and colours, the possibility of extracting galaxy structural information would significantly increase their scientific outcome.

Structural parameters have been recently derived for distant galaxies from HST photometry in optical wave bands (Kelson et al. 2000, and references therein). Jørgensen et al. (1999) proved that structural parameters can be still derived from ground–based data up to z=0.18. We now extend the redshift range and present the first measurement of NIR effective radii at z=0.31.

[1]Based on observations at ESO NTT (OAC guaranteed time) and HST Data Archive.

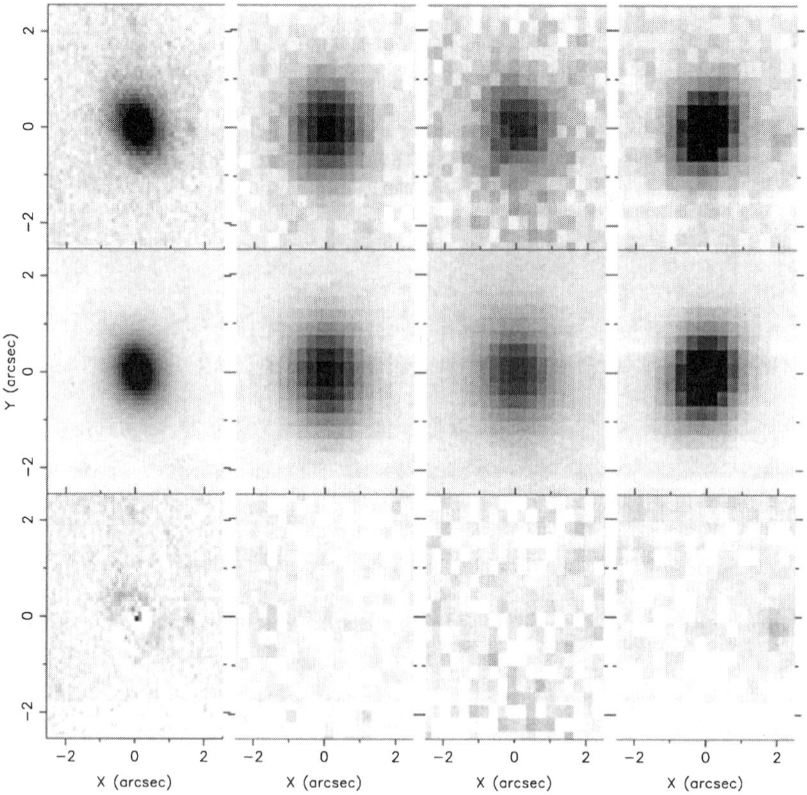

Figure 1. *Upper panel:* image of a galaxy (from left to right: HST–F702W, NTT R, V, and K). *Middle panel:* model used to fit the galaxy surface brightness distribution. *Bottom panel:* residuals of the fit. The galaxy orientation changes in the different wave bands.

2. The Data

We obtained photometry in the V, R, I, K bands for the cluster of galaxies AC118 with the ESO NTT telescope (EMMI and SOFI) during four observing runs (October 1998 – September 2000). Here we will make use of the optical and K-band images, and HST (WFPC2-F702W) archive data to derive surface photometry for a sample of cluster members.

The data reduction and photometric calibration were performed following the standard procedures and are described elsewhere. The sample of cluster members were selected via photometric redshifts.

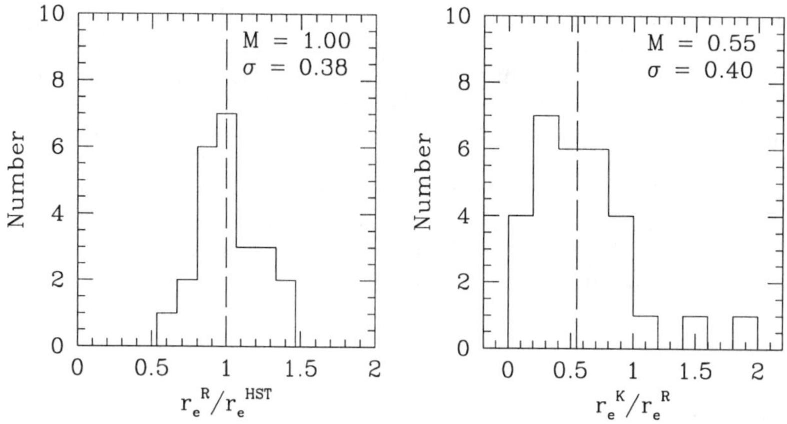

Figure 2. *Left panel*: comparison between the effective radii estimated from ground–based (R) and HST (F702W) data. *Right panel*: comparison between the effective radii estimated from R and K data. The median and standard deviation of the distributions are indicated.

3. Structural Parameters

Galaxy sizes at intermediate redshifts are comparable with the typical seeing of ground-based images ($\sim 1''$). To obtain reliable estimates of the galaxy structural parameters, it is therefore crucial to take into account the effects of the Point Spread Function (PSF).

This is usually performed 1) by a one dimensional fitting approach (see Saglia et al. 1997) and 2) by two–dimensional fitting methods (see e.g. van Dokkum & Franx 1996). In case 1) the integrated light curve or the galaxy intensity profile are constructed by an elliptical fit of the galaxy isophotes and fitted with a proper model. Due to the small galaxy size in ground-based images, the isophotal fit has to be done with few pixels and the brightness profiles consist of very few data points. In the two–dimensional approach, a direct fit of the galaxy brightness distribution is performed without any intermediate step. Because of these reasons, we adopted the two-dimensional approach (see Figure 1).

4. Optical and NIR Effective Radii

Structural parameters were derived in the optical wave bands for galaxies brighter than R=21 mag. The effective radii derived from the HST (F702W) and from the NTT (R) images are compared in Figure 2 (left panel). The fitted parameters are fully consistent, with a dispersion of $\sim$40%, that is the typical uncertainty on effective radii measurements (see Kelson et al. 2000). This result shows that reliable structural parameters of galaxies can be still derived from ground–based observations at z$\sim$0.3.

The right panel of Figure 2 shows the ratios of the NIR to the optical effective radii. The K–band (rest-frame H) effective radii turn out to be ~50% smaller than those measured in the R–band (rest-frame V). This implies the presence of strong internal (optical-NIR) colour gradients in galaxies at z=0.31. Following Sparks & Jørgensen (1993, eq. 21) we obtain Δ(R-K)/Δlog(r)~-0.28 at z=0.31. When compared with the local (z~0) value Δ(V-H)/Δlog(r)~-0.18 (Scodeggio et al. 1998, see also Peletier et al. 1990), this result implies that the colour gradients of galaxies decreased by a factor ~1.5 since z=0.31 (or ~4.5 Gyr). In a forthcoming paper we will make a more reliable comparison between optical and NIR structural parameters by extending the sample to ~100 cluster galaxies.

References

Jørgensen, I., Franx, M., Hjorth, J., van Dokkum, P.G. 1999, MNRAS, 308, 833

Kelson, D.D., Illingworth, G.D., van Dokkum, P.G., Franx, M. 2000, ApJ, 531, 137

Peletier, R. F., Valentijn, E. A., Jameson, R. F. 1990, A&A, 233, 62

Saglia, R.P., et al. 1997, ApJS, 109, 79

Scodeggio, M., et al. 1998, MNRAS, 301, 1001

Sparks, W.B., & Jørgensen, I. 1993, AJ, 105, 1753

van Dokkum, P.G., & Franx, M. 1996, MNRAS, 281, 985

Tracing Cosmic Evolution with Galaxy Clusters
ASP Conference Series, Vol. 268, 2002
S. Borgani, M. Mezzetti, and R. Valdarnini, eds.

The dark side of star formation in galaxy clusters: spectroscopic follow-up of clusters observed with ISOCAM

Dario Fadda

Instituto de Astrofisica de Canarias, Via Lactea S/N, E-38200 La Laguna, Tenerife - Spain

Pierre-Alain Duc

CNRS (URA 2052), CEA Service d'astrophysique, F-91191 Gif-sur-Yvette Cedex, France

Abstract. The evolution of galaxies in cluster environments can be studied using mid-IR observations which trace star forming regions hidden by dust. The optical follow-up of A1689 (z=0.2), observed at 6.7 and 15 μm by ISOCAM, have revealed a systematic excess of the B-[15 μm] galaxy color distribution with respect that of Coma and Virgo clusters. This result suggests the existence of a dark side of the Butcher-Oemler effect measured in the optical. We present an analysis of the optical/mid-IR properties of the mid-IR emitters in A1689, comparing in particular the star formation rates based on mid-IR and optical data. Morover, we present preliminary result for J1888, a cluster at $z = 0.56$ deeply observed with ISOCAM, based on recent VLT/FORS and NTT/SOFI observations.

1. Introduction

Since the classical study of Butcher & Oemler (1984) who put in evidence an increasing fraction of blue, presumably star-forming, galaxies as a function of cluster redshift, several authors have tried to estimate the importance of the cluster environment on the evolution of cluster galaxies. Recent studies with sample of distant clusters (Dressler et al. 1999, Balogh et al. 2000) pointed out that the star formation rate (SFR) per cluster galaxy appears to be lower than that in similar types of galaxies in the surrounding field. The main drawback of these studies is that the SFR rely on measurements of optical line fluxes (usually [OII] or in a few cases Hα) which suffer from strong dust extinction. Since dust obscuration depends on object types and on environmental conditions, it is impossible to correct it on average. On the other hand, although galaxies affected by extinction show typical spectral features (see Dressler et al. 1999), it is quite impossible to quantify the total SFR on the basis of the optical data alone. Mid-IR surveys, along with radio centimetric fluxes, provide the best estimates of the dust obscured SFR (see e.g. Chary & Elbaz 2001) when far-IR are not available. For this reason a sample of 10 clusters at different redshift $(0.1 < z < 0.9)$ has been observed with the ISOCAM camera on-board ISO. We

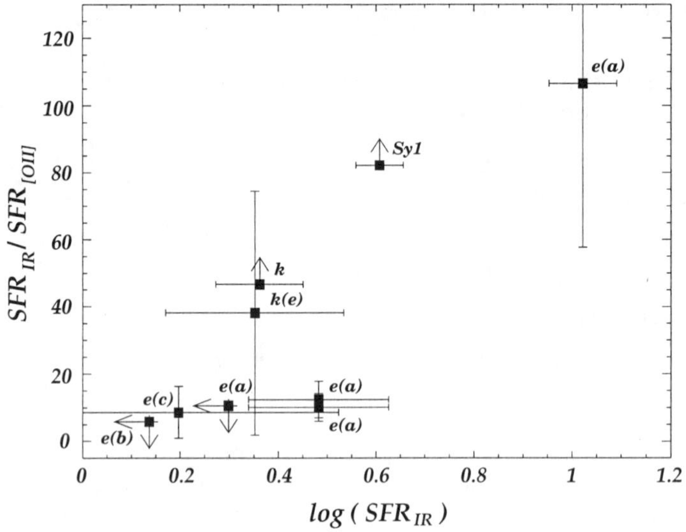

Figure 1. Ratio of the mid-IR to optical SFR estimates versus the mid-IR based SFR estimate for A1689. Note the high dust extinction in some apparently passive galaxies ('k' type).

present here the optical/mid-IR properties of the galaxies detected at 15 μm in the A1689 cluster (see Fadda et al. 2000 for ISOCAM data) and preliminar results of the VLT follow-up program of J1888, a cluster at $z = 0.56$ deeply observed with ISOCAM.

2. Hidden star formation in A1689

Using available images (HST and NTT) and about 100 spectra measured with the NTT (Duc et al. 2001), we studied membership, morphologies and spectral types of the objects detected by ISOCAM in A1689. Our analysis confirms the high fraction of blue galaxies in the cluster (already detected by Butcher & Oemler 1984) relying on a secure cluster membership of the galaxies (spectroscopic and photometric redshifts are known for all the galaxies observed). Most of the 15 μm sources are luminous, blue, emission-line or morphologically disturbed galaxies, i.e. a population of galaxies usually associated with the "Butcher–Oemler" effect. However, ~30% of the 15 μm sources do not show any sign of star-formation activity in their optical spectrum. More of 70% of the emission-line galaxies in our spectroscopic sample are detected at 15 μm and all the galaxies classified as dusty starbursts ("e(a)" type in Dressler et al. 1999) are 15 μm sources. On the contrary, none of the galaxies with a post-starburst optical spectrum have been detected at 15 μm.

Since the AGN activity is very low in A1689, the 15 μm flux is a reliable tracer of the dust-obscured star formation activity (see Chary & Elbaz 2001). Comparing the SFR based on the mid-IR (15 μm flux) and optical ([OII] line flux), we found that for galaxies with mid-IR emission the ratio

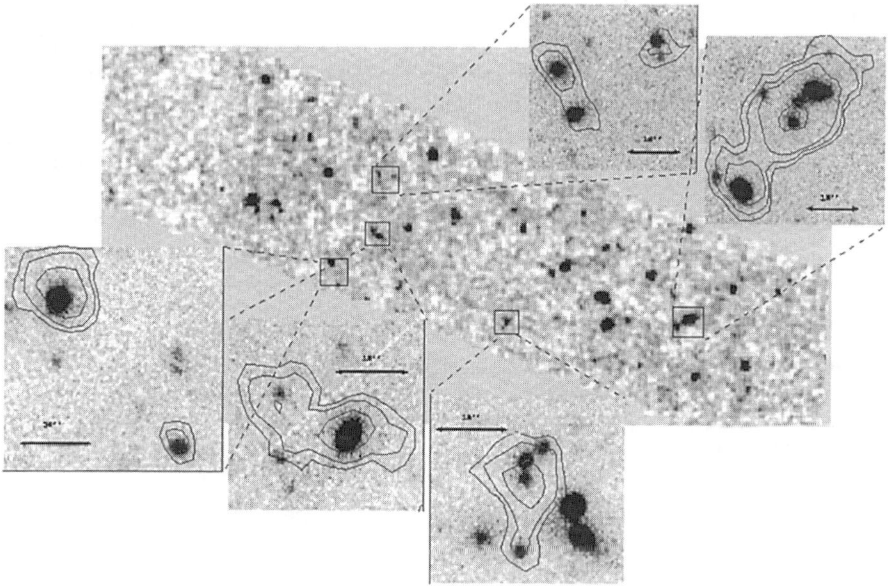

Figure 2. 15μm map of J1888 with more than 100 sources detected. Subsets show 15μm overlays on optical images. Note the excellent astrometric precision of ISOCAM and the clear optical identifications.

SFR(IR)/SFR(Opt) is very high and ranges between 10 and 100, being the highest among "e(a)" galaxies (see Figure 1). The median SFR(IR) is 2 $M_{\odot}yr^{-1}$, while the median SFR(Opt) of the [OII] detected galaxies is 0.2 $M_{\odot}yr^{-1}$. Although we are not observing luminous infrared galaxies (the highest total IR luminosity 6.2×10^{10} $L_{\odot}$ is measured in a "e(a)" galaxy), we conclude that a significant portion of the star formation activity is visible only using mid-IR data. At least 90% of the star formation is missed when estimated from the [OII] line. A comparison with the field, essential to investigate the effects of the environment on the star formation of the cluster galaxies, will be possible only when a statistical sample of mid-IR galaxies for the coeval field of A1689 will be available.

3. Preliminary results in J1888

J1888, a cluster at $z = 0.56$, has been surveyed by ISOCAM at the same depth of the Hubble Deep Field covering a strip of $14.5' \times 3.5'$ crossing the cluster center in order to explore possible variation of the mid-IR properties with the cluster-centric distance (see Figure 2). We have recently obtained about 120 spectra with VLT/FORS1 and a Ks image with NTT/SOFI, which complements B and R images obtained with the 2.2m ESO telescope. So far we have redshifts for 55 out of 70 sources detected at 15 μm and 22 out of 47 sources detected at 6.7 μm.

The redshift distribution of the cluster, which appears very loose in the optical images, peaks at $z = 0.56$ but the field is highly polluted by background galaxies. Considering the range of magnitudes of the cluster members (21.4 $< B <$25.3), 55% of the galaxies do not belong to the cluster. Considering only the 15 μm sources, only 30% of the ISOCAM sources are cluster members.

The color-magnitude relation is not obvious to determine even for cluster members. Quite surprisingly there is a large number of "red outliers" ($B - R > 2$) which belong to the cluster and most interestingly several of them are detected at 15 μm. The presence of "red outliers" is not uncommon in clusters at $z \sim 0.5 - 0.6$ (see e.g. Margoniner & de Carvalho 2000) but in this case any contamination by field galaxies is ruled out. Contrary to the case of A1689, we have almost no mid-IR emitters among blue cluster members.

On the basis of VLT spectra we did a classification of galaxies according to the scheme of Dressler et al. (1999). As in the case of A1689, we detect at 15 μm all the dusty starbursts ("e(a)" type) and no post-starburst galaxies. Compared to A1689, we detect a larger fraction of spiral-like spectra which is not unexpected for clusters showing the "Butcher-Oemler" effect.

A detailed reanalysis of the J1888 ISOCAM data is now in progress using the new method developed by Lari et al. (2001) in parallel with the analysis of the deep ISOCAM surveys in the region of the Lockman Hole. The comparison of the data from this survey, whose redshift distribution peaks at redshift of 0.6 close to that of J1888, will be fundamental to study the influence of cluster environment on the evolution of galaxies.

References

Balogh, M. L., Morris, S. L., Yee, H. K. C. et al. 1999, ApJ, 527, 54
Butcher, H. & Oemler, A. Jr. 1984, ApJ, 285, 426
Chary, R. & Elbaz, D. 2001, ApJ, 556, 562
Dressler, A., Smail, I., Poggianti, B. M. et al., 2000, ApJS, 122, 51
Duc, P.-A., Poggianti, B.-M., Fadda, D. et al., 2001, A&A, submitted
Fadda, D., Elbaz, D., Duc, P.-A. et al. 2000, A&A, 361, 827
Lari, C., Pozzi, F., Gruppioni, C. et al., 2001, MNRAS, 325, 1173
Margoniner, V. E. & de Carvalho, R. R. 2000, AJ, 119, 1562

Tracing Cosmic Evolution with Galaxy Clusters
ASP Conference Series, Vol. 268, 2002
S. Borgani, M. Mezzetti, and R. Valdarnini, eds.

History of Mass Assembly and Star Formation in Clusters

T. Kodama[1], Ian Smail[2], F. Nakata[1], S. Okamura[1], R. G. Bower[2]

[1]*Dept of Astronomy, Univ of Tokyo, Bunkyo-ku, Tokyo 113–0033, Japan*
[2]*Dept of Physics, Univ of Durham, South Road, Durham DH1 3LE, UK*

Abstract. We present deep, panoramic multi-color imaging of the distant rich cluster A 851 ($z = 0.41$) using Suprime-Cam on Subaru. These images cover a $27'$ field of view ($\sim 11h_{50}^{-1}$ Mpc), and by exploiting photometric redshifts, we can isolate galaxies in a narrow redshift slice at the cluster redshift. Using a sample of ~ 2700 probable cluster members ($< M_V^* + 4$), we trace the network of filaments and subclumps around the cluster core. The depth of our observations, combined with the identification of filamentary structure, gives us an unprecedented opportunity to test the influence of the environment on the properties of low luminosity galaxies. We find an abrupt change in the colors of faint galaxies ($> M_V^* + 1$) at a local density of 100 gal. Mpc^{-2}. The transition in the color-local density behavior occurs at densities corresponding to subclumps within the filaments surrounding the cluster. Identifying the sites where the transition occurs brings us much closer to understanding the mechanisms which are responsible for establishing the present-day relationship between environment and galaxy characteristics.

1. Introduction

Clusters of galaxies are continuously growing through the accretion of galaxies and groups from the field. The star formation activity in the accreting galaxies must be quenched during the assimilation of the galaxies into the cluster. This transformation is a key process in creating the environmental dependence of galaxy properties, and may also underpin the observed evolution of galaxy properties in distant clusters [e.g. 1.2] However, the physical mechanism which is responsible for these changes has not yet been identified [3,4,5].

The advent of Suprime-Cam, a revolutionary wide-field camera on the Subaru telescope, has opened a new window in this field. Its $27'$ field of view can trace the variation of galaxy properties from the cluster cores out to the surrounding field in an attempt to identify the environment where the decline in the star formation in accreted galaxies begins.

2. Large scale structure

As a first step towards a systematic study of distant clusters with Subaru and Suprime-Cam, we obtained deep ($\sim M^* + 4$) $BVRI$ imaging of the rich cluster

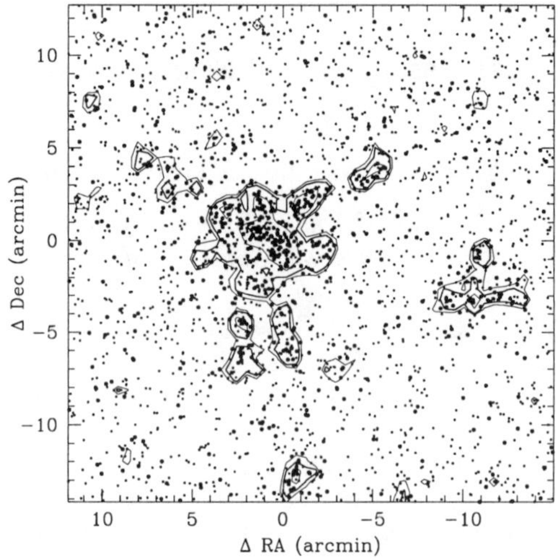

Figure 1. A wide-field map of A 851 from our Suprime-Cam *BVRI* imaging. The field is 11 Mpc on a side. The points identify cluster members based on photometric redshifts. The remaining field contamination is not subtracted. The large and small circles separate the colors of galaxies at the color 0.5 mag bluer than the color-magnitude sequence in $B - I$, where large ones are in red. The contours indicate the local surface density of the member galaxies, corresponding to $\log_{10} \Sigma = 1.8$, 2.0 (thick) and 2.5 Mpc^{-2} (after correcting for field contamination), respectively. The thick contour traces the boundary where the color distribution of galaxies changes dramatically. Large scale structure is clearly visible. The overall shape resembles an 'octopus', with a round head and many legs.

A 851 at z=0.4. We constructed an I-band selected sample which contains 15,055 galaxies brighter than $I = 23.4$.

In order to assign cluster membership across our large field and to faint magnitudes, we applied photometric redshift technique [6] to isolate the galaxies in a narrow redshift slice around the cluster redshift. By applying the redshift cut of $0.32 < z_{\mathrm{phot}} < 0.48$, we can reduce the field contamination by factor 10 at $I = 23.4$, while keeping most of the cluster members (80%, from the estimate using the spectroscopic members [7]).

After this photometric redshift selection, it is straightforward to map out the distribution of galaxies which are likely to be associated with the cluster. As shown in Fig. 1, several large scale structures are visible around the cluster core, comprising many clumps and filamentary extensions coming out directly from the core. Importantly, most of these extensions from the core are aligned with the surrounding subclumps.

The structures identified in this region are qualitatively similar to those seen in cosmological simulations of the growth of clusters which exhibit the filamentary/clumpy substructures on similar scales [e.g. 8] It appears therefore

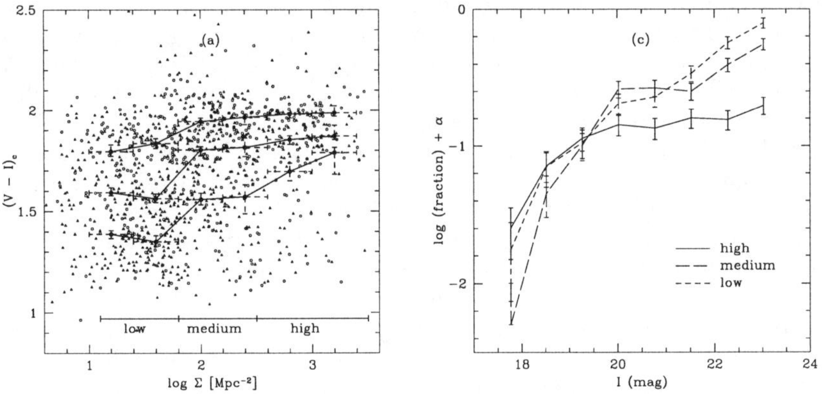

Figure 2. (a) The variation in color versus local galaxy density, Σ, for cluster members brighter than $I = 23.4$. The open circles and filled triangles show the galaxies brighter or fainter than $I = 21.4$ (M_V^*+2), respectively. The three red lines represent the loci of the 25, 50, and 75th percentile colors. The colors are corrected for the slope of the color-magnitude relation. We also correct for the residual field contamination using the blank field data. (c) The field-corrected luminosity function for cluster galaxies in the three density slices from Panel (a). The distributions are normalized at $I = 19.4$ (M_V^*) to match the high density curve. A clear steepening of the luminosity function is visible at lower densities.

that we are witnessing A 851 as it assembles through the accretion of galaxies and groups along the filaments onto the cluster core from the surrounding field.

2.1. Environmental dependence of galaxy properties

By exploiting the striking large scale structure around this cluster, we can investigate the influence of environment on the photometric properties of galaxies. We define the environment for each galaxy using the local surface number density, Σ, of members, calculated from the 10 nearest neighbors. We statistically subtract residual field contamination using a Subaru blank field data [9].

We divide the density distribution into three regimes as indicated in Fig. 2a. There is a close correspondence between the local density and structure: the high density region corresponds to the cluster core within $\lesssim 1$ Mpc; the medium density region includes the structures defining the filaments surrounding the cluster; and the low density region comprising the rest of the volume (Fig. 1).

As shown in Fig. 2a, the color distribution in the high density region is strongly peaked at $(V - I)_c \sim 2$, the color of an early-type cluster member. However, as we move to lower densities, the distribution becomes dramatically bluer. Notably, this color transition with local density occurs quite abruptly at $\log_{10} \Sigma \sim 2$, indicating a threshold effect in transforming galaxy properties. The boundary corresponding to this critical density is highlighted in Fig. 1. Surprisingly, this transition occurs in subclumps well outside the core.

Our ability to pin-point this environment is a fundamental step towards identifying the dominant mechanism behind the environmental dependence of galaxy properties. Since the ram-pressure stripping is largely supressed in groups [4], the effective mechanism at this hierarchy is either galaxy–galaxy collisions (which cause cold disk gas to be driven to the galaxy center creating a star burst [3]); and 'suffocation' (where *warm* gas in the galaxy's halo is shock heated by the intra-cluster medium so that it can no longer cool and replenish the cold gas in the disk [10]). Our data do not distinguish directly between these possibilities, but they can be distinguished by tracing the variation in the individual components of galaxies, bulges and disks, between the cluster and the field.

It is important to note that the changes in galaxy properties as a function of local density are most prominent in faint galaxies ($> M_V^* + 1$). This is mirrored to the dramatic change in the shape of the luminosity function with local galaxy density: where we see a steepening of the faint end slope of the luminosity function with decreasing density (Fig. 2c). Assuming that their star formation effectively ceases, the relative absence of low-luminosity galaxies at projected densities of $\log_{10}\Sigma > 2$ can be understood as they will fade by $\gtrsim 1\,\mathrm{mag}$ as their star formation declines [2]. This will put many of their descendents below our magnitude limit. This is also qualitatively consistent with the presence of large numbers of very low luminosity, $\gg M_V^* + 3$, passive dwarf galaxies in local clusters [11]

Acknowledgments. We are grateful to Dr Y. Komiyama for his assistance during our observations. We acknowledge the Suprime-Cam team for allowing us to use the blank field data. We also thank Drs. M. Balogh, N. Arimoto and K. Shimasaku for helpful discussion. TK and FN acknowledge the Japan Society for the Promotion of Science for support through its Research Fellowships for Young Scientists. IRS acknowledges support from the Royal Society and the Leverhulme Trust.

References

[1] Butcher, H., Oemler, A., 1984, ApJ, 285, 426

[2] Kodama, T., Bower, R. G., 2001, MNRAS, 321, 18

[3] Moore, B., Katz, N., Lake, G., Dressler, A., Oemler, A., 1996, Nature, 379, 613

[4] Adabi, M. G., Bower, R. G., Navarro, J. F., 2000, MNRAS, 314, 759

[5] Balogh, M. L., Navarro, J. F., Morris, S. L., 2000, ApJ, 540, 113

[6] Kodama, T., Bell, E. F., Bower, R. G., 1999, MNRAS, 302, 152

[7] Dressler, A., Smail, I., Poggianti, B. M., Butcher, H., Couch, W. J., Ellis, R. S., Oemler, A., 1999, ApJS, 122, 51

[8] Ghigna, S., Moore, B., Governato, F., Lake, G., Quinn, T., Stadel, J., 1998, MNRAS, 300, 146

[9] Ouchi, M., et al., 2001, ApJ, 558, L83

[10] Larson, R. B., Tinsley, B. M., Caldwell, C. N., 1980, ApJ, 237, 692

[11] Binggeli, B., Sandage, A., Tammann, G. A., 1988, ARA&A, 26, 509

Tracing Cosmic Evolution with Galaxy Clusters
ASP Conference Series, Vol. 268, 2002
S. Borgani, M. Mezzetti, and R. Valdarnini, eds.

The Butcher-Oemler Effect in High Redshift X-ray Selected Clusters

David A. Wake, Chris A. Collins

Astrophysics Research Institute, Liverpool John Moores University, Twelve Quays House, Egerton Wharf, Birkenhead CH41 1LD, UK.

Bruce W. Fairley, Laurence R. Jones

School of Physics and Astronomy, University of Birmingham, Birmingham, B15 2TT, UK.

Doug J. Burke

Harvard-Smithsonian Center for Astrophysics, 60 Garden Street, Cambridge, MA 02138, USA.

Robert C. Nichol, Kathy A. Romer

Physics Department, Carnegie Mellon University, Pittsburgh, PA 15213, USA.

Abstract. We are engaged in a wide-field, multi-colour imaging survey of X-ray selected clusters at intermediate and high redshift. We present blue fractions for the first eight out of 29 clusters, covering almost a factor of 100 in X-ray luminosity. We find no correlation of blue fraction with redshift or X-ray luminosity. The lack of a correlation with L_X, places strong constraints on the importance of ram-pressure stripping as a driver of the Butcher-Oemler effect.

1. Introduction

Butcher & Oemler (1984) found the fraction of blue galaxies (f_b), in their sample of centrally concentrated clusters, increased with redshift . The Butcher-Oemler (BO) effect was the first evidence for rapid galaxy evolution in clusters, and as such led to an extensive follow up program. Further photometric studies have generally confirmed the presence of this effect, although to varying degrees (e.g. Smail et al. 1998; Margoniner & de Carvalho 2000; Kodama & Bower 2001). Spectroscopic studies of the blue galaxies in question have revealed that they are cluster members, and have spectral features indicative of star-formation, often recently truncated (e.g. Dressler & Gunn 1992). Morphological studies with HST indicate that these galaxies are disc dominated, with some evidence of enhanced interaction rates (Couch et al. 1994). These morphological studies also revealed that although the luminous elliptical galaxies present in local clusters remain in their high redshift counterparts, the large population of S0 galaxies are absent at high redshift. This has led to the association of these

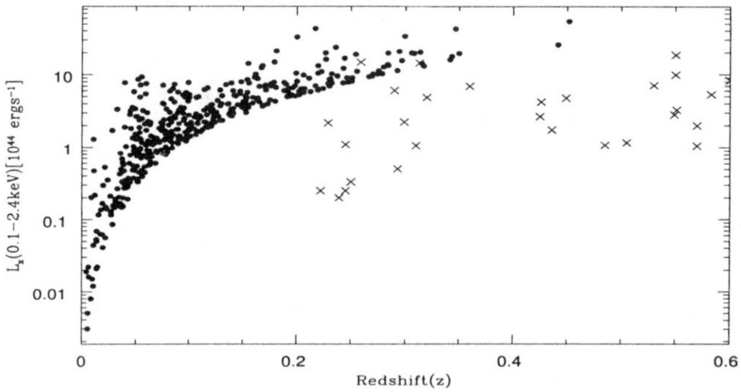

Figure 1. Redshift versus X-ray luminosity for our 29 clusters (crosses) and for the total REFLEX sample (filled circles).

two effects with many mechanisms, such as ram-pressure stripping, harassment and tidal interaction, being suggested for the transformation of the BO galaxies to S0s. However a number of problems with these studies have been noted. In particular selection effects may be driving the observed BO effect (e.g. Andreon & Ettori 1999). The clusters observed by Butcher and Oemler and by most of the additional BO studies have used optically selected cluster catalogues. These have tended to be selected in a single photometric band, so that as clusters are selected to higher redshift, bluer galaxies will become more important in the selection. This will bias clusters at high redshift to contain bluer galaxies. Another potential problem suffered by all studies to date, is caused by the tendency to select more and more rich/massive clusters with increasing redshift, leading to only the richest and most massive clusters being selected at high redshift . It is not yet clear whether f_b depends on the cluster mass and how much this dependence may contribute to the BO effect.

2. A Wide-Field Optical Imaging Survey of High Redshift X-ray Selected Clusters

We are presently engaged in a broad-band photometric study of a well controlled sample of X-ray selected clusters, and present results on the blue fractions of the first eight observed here. Our clusters come from 4 catalogues, EMSS (Gioia et al. 1990), SHARC (Burke et al. 1997), BSHARC (Romer et al. 1999) and WARPS (Scharf et al. 1997). The flux limits of these surveys vary, with WARPS and SSHARC having very low flux limits allowing us to select clusters with low X-ray luminosities to high redshift. We have imaged a total of 29 clusters over two redshift intervals, $z \sim 0.25$ and $z \sim 0.5$, with a similar range in X-ray luminosities. Figure 1 shows a plot of the X-ray luminosities and redshifts of our cluster sample alongside those of the REFLEX survey (Bohringer et al. 2001). It is clear that our clusters have X-ray luminosities consistent with typical clusters in the local universe, allowing for easy and meaningful evolutionary comparisons to local samples. We have imaged each of our clusters over a wide field in BVR for the $z \sim 0.25$ sample, and VRI for the $z \sim 0.5$ sample. These filters approximately map to rest-frame UBV for the chosen redshifts. The wide field allows an individual background to be calculated for each cluster taking

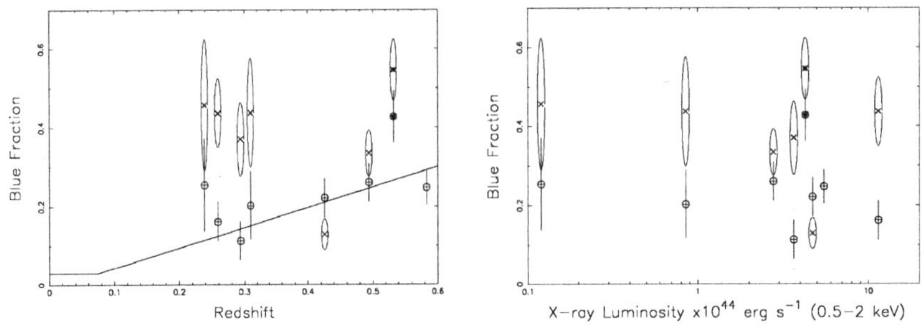

Figure 2. Left plot shows blue fractions versus reshift and the right plot blue fraction versus X-ray luminosity for the eight clusters presented to $M_V = -20$, within R_{30}. See text for description of symbols.

into account line-of-sight large-scale structure variations. The imaging was done using the wide field mosaic cameras of the INT 2.5m, the CTIO 4m and the UH 2.2m, which have fields of view of 0.29, 0.1, and 0.36 deg^2 respectively.

3. Results and Conclusions

Here we present blue fractions for 8 clusters imaged with the INT. We selected 4 in each redshift interval, with the lower redshift clusters chosen to have a wide range in L_X and the high redshift sample very similar L_X. Our photometric accuracy is ~0.03 mag, with completeness typically 24.2, 24.0, 22.7 mag for the three filters, with seeing between 0.79 and 1.52 arcsec. Figure 2 shows both f_b versus z and f_b versus L_X, where f_b is defined as the fraction of galaxies, within a characteristic radius R_{30}, 0.2 mag bluer than the early-type red-sequence, to a limiting magnitude $M_V = -20$ (the standard BO definition). The open circles shows the rest-frame $B-V$blue fraction, and the crosses show the rest frame $U-B$blue fraction. The solid circle and square are for RXJ2146.0+0423 which appears to be contaminated by a foreground group. Immediately apparent from both graphs is the significant difference in the blue fractions measured in the two different colours, with the $U-B$blue fraction tending to be higher. This illustrates the importance of consistency when comparing blue fractions, and also illustrates the potential biases that may arise when selecting clusters in a single optical band. No trend with either redshift or X-ray luminosity is apparent, although the small sample and large errors may be masking the trend. The original trend line of BO is plotted on the f_b-z graph and is not inconsistent with our data. The lack of a correlation with L_X agrees with previous studies (e.g. Andreon & Ettori 1999), and is shown to continue over a much wider range in L_X. This has particular consequences for the importance of ram-pressure stripping as a mechanism in the BO effect. The gas density in our poorest clusters, even in the core, is unlikely to provide effective stripping, whereas galaxies in our most X-ray luminous clusters should be rapidly stripped far from the core. As we see no dependence of blue fraction on L_X, this mechanism does not seem to be important in determining the blue fraction. Figure 3 shows the blue fraction

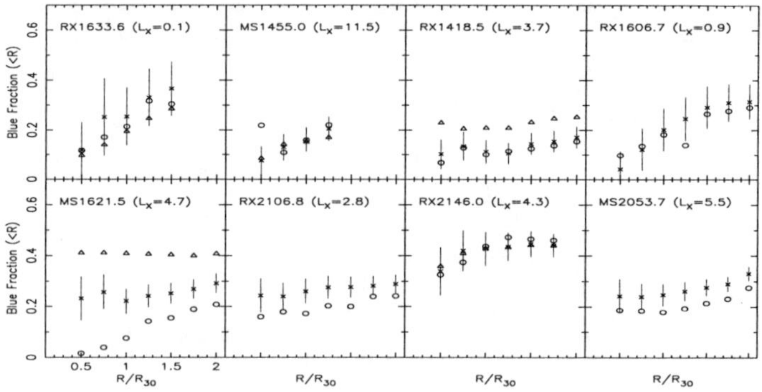

Figure 3. Blue fraction versus radius for all eight clusters in rest frame $B-V$. Open circles show blue fractions for $M_V < -21$, crosses for $M_V < -20$, and triangles for $M_V < -19$

in rest frame $B-V$as a function of radius and absolute magnitude for our 8 clusters. We confirm the widely noted trend of increasing blue fraction with radius, although the rate of change varies greatly from cluster to cluster. For most, but not all, of our clusters we see an increasing blue fraction with fainter magnitude cut in agreement with Kodama & Bower (2001). The radial trend of increasing f_b is consistent with the idea of in-falling field galaxies being somehow transformed as they fall into the cluster. However, as previously mentioned it seems ram-pressure stripping is an unlikely mechanism for this transformation. Further details can be found in Fairley et al. (2001).

References

Andreon, S., & Ettori, S. 1999, ApJ, 516, 647

Bohringer, H., Schuecker, P., Guzzo, L., Collins, C. A., Voges,W., Schindler, S., Neumann, D. M., Cruddace, R. G., DeGrandi, S., Chincarini, G., Edge, A. C., MacGillivray, H. T., Shaver, P. 2001, A&A, 369, 826

Burke, D. J., Collins, C. A., Sharples, R. M., Romer, A. K., Holden, B. P., & Nichol, R. C. 1997, ApJ, 488, L83

Butcher, H., & Oemler, A. 1984, ApJ, 285, 426

Couch, W. J., Ellis, R. S., Sharples, R. M., & Smail, I. 1994, ApJ, 430, 121

Dressler, A., & Gunn, J. E. 1992, ApJS, 78, 1

Fairley, B. W., Jones, L. R., Wake, D. A., Collins, C. A., Burke, D. J., Nichol, R. C., & Romer, A. K. 2001, MNRAS, submitted

Gioia, I. M., & Luppino, G. A. 1994, ApJS, 94, 582

Kodama, T., & Bower, R. G., 2001, MNRAS, 312, 18

Margoniner, V. E., & de Carvalho, R. R. 2000, AJ, 119, 1562

Romer, A. K., Nichol, R. C., Holden, B. P., Ulmer, M. P., Pildis, R. A., Merrelli, A. J., Adami, C., Burke, D. J., Collins, C. A., Metevier, A. J., Kron, R. G., Commons, K. 2000, ApJS, 126, 209

Smail, I., Edge, A. C., Ellis, R. S., Blandford, R. D. 1998, MNRAS, 310, 663

CONFERENCE SUMMARY

Tracing Cosmic Evolution with Galaxy Clusters
ASP Conference Series, Vol. 268, 2002
S. Borgani, M. Mezzetti, and R. Valdarnini, eds.

Conference Summary

Richard S Ellis
Astronomy Department, California Institute of Technology, Pasadena CA 91125, USA

1. Introduction

Clusters of galaxies are of wide interest in the astronomical community both as laboratories where baryons and dark matter can be conveniently studied and, through their statistical properties at various look-backtimes, as tracers of cosmic structure and evolution. The Local Organising Committee wisely narrowed the topics of this meeting to four broad themes and so I will organise my summary remarks under these headings.

First I think I should address the plea "Why bother to study clusters?", raised provocatively by Colin Norman at one stage during the proceedings[1]. We have learnt enough about clusters and the growth of structure to know that two of the traditional motivations require much more careful consideration.

For years on time allocation committees, I read observers repeat a *mantra* in the first sentence of their proposals: *"Clusters of galaxies represent the largest bound structures in the Universe.."* in justification of their role as tracers of large scale structure. A number of speakers at this meeting reminded us that clusters no longer uniquely occupy this role. We have other tracers of large scale structure and, moreover, it seems we need to be much clearer of what exactly we mean by a cluster before we can convincingly use them as cosmological probes.

A second, traditional, motive for studying clusters has been observational convenience, e.g. in studying constituent populations such as galaxies at various look-back times. At first sight, it's an attractive proposition for an observer to study a few rich clusters at various redshifts each containing hundreds of accessibly-luminous galaxies and then to "join the results" in a timeline to make some evolutionary claim. But, at most redshifts of interest, we can expect to find a wide range of overdensities represented (groups, merging systems, virialised clusters). Put simply, rich cluster A at z=1.2 is unlikely to evolve into rich cluster B at z=0.5 and neither may necessarily become a present-day Coma.

Both these worries indicate a high degree of rigour is needed in using clusters. We need very large samples spanning ranges of mass at any redshift of interest and perhaps it would be foolish to adopt one single selection criterion for their study. A similar "panchromatic" theme has emerged over the past decade in understanding how to address the question of evolution using samples of galaxies.[2]

---

[1] Fortunately, I have forgotten to which speaker he made this remark!

[2] Witness the controversy surrounding the use of UV/optical and far-infrared probes of the cosmic star formation history. No single technique wins: several are required.

The motivation for studying clusters that emerged at this meeting focused broadly along the following themes:

- Testing gravitational instability by measuring the number density of massive clusters at high redshift, viz. $\Phi(M_{cluster}, z)$.

- Breaking degeneracies in estimates of the the cosmological parameters by examining the local population of clusters as a probe of the mean mass density and the normalisation of the mass power spectrum, Ω_M and σ_8.

- Verifying hierarchical structure and the nature of the dark matter and keeping those theorists in check who predict a universal mass profile with central cusps, $\rho_M(r) \propto r^{-1.5}$.

- Determining the origin of the heating of the intracluster medium, its enrichment history and examining whether non-gravitational processes are involved.

- Examining the history of spheroidal galaxies and role of the environment, for example in understanding the origin of the morphology-density relation and the destiny of the infalling component of gas-rich field galaxies.

Let me make a disclaimer in what follows. I am not an expert in any of the areas I was asked to summarise and so I submit these concluding remarks only as someone who tried to listen carefully to most talks, dutifully avoiding the lunchtime mountain hikes to make sense of what I heard[3]. With $\simeq$70 talks and $\simeq$80 posters, inevitably I have had to be very selective in discussing results.

If one could summarise the meeting in two paragraphs, I would say the following:

- There is an explosion of terrific data, from large surveys (X-ray and ground-based) which offer qualitatively new ways in which we find and do statistical studies of clusters, and also in *resolved* data within clusters which opens up new opportunities for understanding the detailed astrophysics of dense environments.

- The subject is moving from exploratory surveying to detailed astrophysics. As part of this "growing up" there is a need to admit defeat on some of the old methods and to embrace new ones, particularly on the statistical questions. For many years cluster workers had something of a monopoly in the study of early galaxies and large scale structure, but it is time to take advantage of other datasets being delivered and view cluster astrophysics as only one part of a larger body of information.

---

[3]I apologise if I missed some talks but I tried to secure powerpoint files or transparencies for most of those.

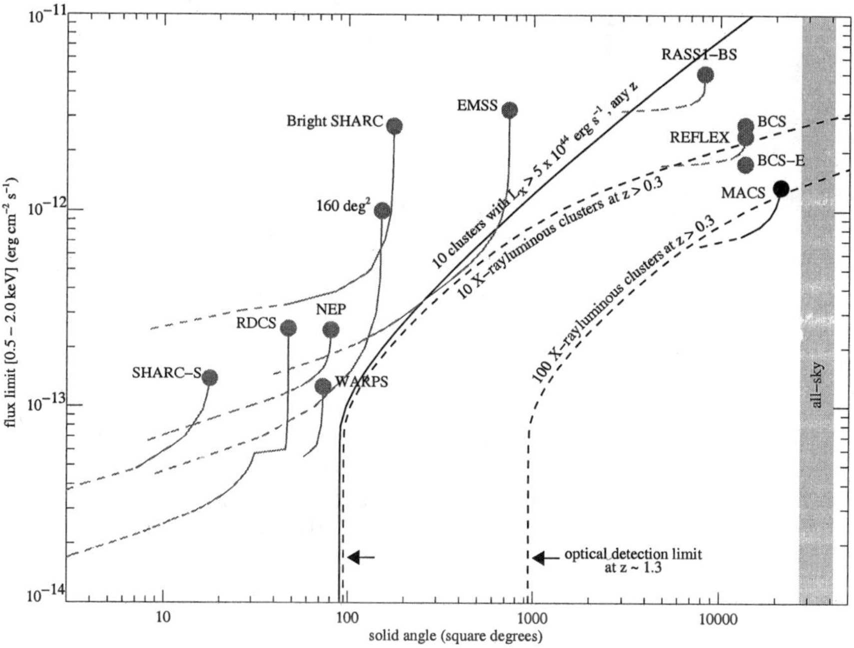

Figure 1. The impressive effort to track down the evolution of X-ray luminous clusters to high redshifts: survey parameters taken from the compilation of Harald Ebeling.

2. Searching for Distant Clusters

During the first part of our meeting we listened to progress made in finding clusters, mostly (but not exclusively) at high redshift. We heard about a bewildering number of deep surveys (MACS, REFLEX, WARPS, LCDCS, EIS, RDCS to name but a few..Figure 1) each requiring significant follow-up with the world's most powerful facilities. Excellent reviews of this active field have been given by Borgani (2001) and Borgani & Guzzo (2001).

The motivation is a sound one: testing gravitational instability in the context of a known cosmological model, or alternatively, by assuming a structure formation model such as CDM, constraining the cosmological parameters. The fraction of collapsed large massive structures ($M \simeq 10^{14} M_{\odot}$) at, say, $z \simeq 1$ for a given cosmology is one of the most robust predictions of the dark matter models (Figure 2). The necessary ingredients are a large well-defined sample and robust estimates of the cluster masses.

An additional motivation for locating high z clusters, is to undertake evolutionary studies. For galaxies, this is less appealing now we can find statistically-complete field samples at similar redshifts by independent photometric means (Adelberger 2000, Daddy et al 2001, McCarthy et al 2001), many of which are extensive enough for locating clustered objects in a self-consistent way. Marc

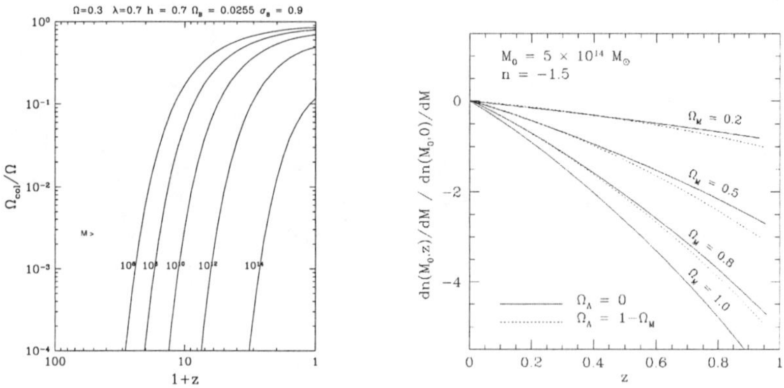

Figure 2. (Left) The fraction of gravitationally collapsed objects (measured in terms of Ω_M) with mass greater than a given value (in solar mass units) as a function of redshift, $1+z$, is one of the simplest and most robust predictions of CDM theory given a set of cosmological parameters (ΛCDM here, after Fukugita 2000). (Right) The wanted evolutionary signal, a decline in the number of massive clusters with redshift, as a function of cosmological model (after Voit 2000).

Postman stressed the dangers here by contrasting differences found between conclusions drawn from galaxies studied in the X-ray selected CNOC and optically-selected Morphs samples. If clusters are found by a manner which relates to the properties of their constituent galaxies (e.g. by only searching for ones with prominent red sequences as discussed by Mike Gladders, or finding ones associated with powerful radio galaxies as reviewed by Philip Best), one can get different views of what is going on in dense regions.

The quest for high z clusters has a chequered history. Most of the early work was motivated by a quest for the deceleration parameter based on brightest cluster galaxies (Gunn & Oke 1975, Kristian et al 1976). Denis Zaritsky gave us a new twist to this story with his evidence for varying accretion onto brightest cluster galaxies in the Las Campanas Distant Cluster Survey (see also Aragón-Salamanca et al 1998). Claims for number evolution in cluster counts with redshift have also swung to and fro over the years. Until recently only a handful of optical or X-ray selected clusters were available[4]. The most exciting aspect about this meeting is the enormous advance in the numbers of distant clusters.

So how best to find distant clusters? Marc Postman nicely summarised the various search techniques, their advantages and biases. The most useful techniques are those that minimise projection effects. Faint X-ray surveying in the 0.1-10 keV range is cerainly an expensive method in telescope time (but good things don't come cheap): the background is low, we expect $L_x \propto n_e^2$ and thus,

[4]On more than one occasion, challenges to a standard paradigm have been made on the basis of the existence of a single high z, assumed massive, cluster.

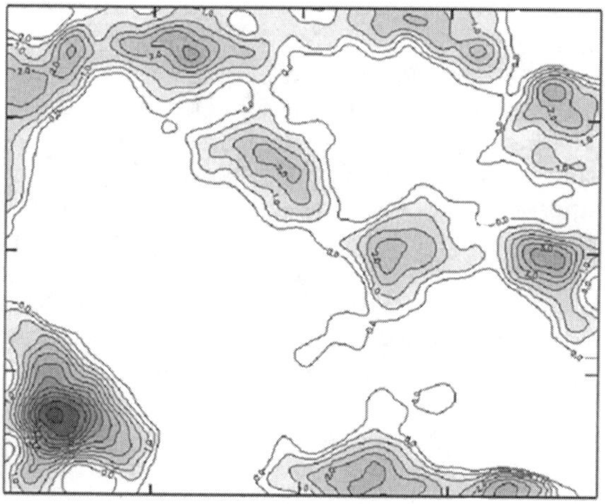

Figure 3. The projected distribution of dark matter in a 40 arcmin field determined from the shear of faint galaxies by Wittman et al (2001). The tangential shear as a function of the photometric redshift of the background galaxies places the 4.5σ concentration in the lower left corner at a redshift of $z \simeq 0.3 \pm 0.08$. Spectroscopy of the cluster in this region yields z_c=0.28 with a dispersion σ_c=615 km s^{-1}.

so long as massive clusters have hot intracluster media, this is the method of choice. Kathy Romer emphasized how, even with the higher EPIC background, XMM promises to extend the depth of the ROSAT surveys to mean redshifts well beyond $z \simeq$1.

Whilst we have yet to reap the benefits of SZ effect as a survey tool, James Bartlett showed us that this method has the great advantage of being distance independent (modulo evolution) with a signal $\propto n_e$. Instruments such as Planck Surveyor, and the proposed Arcminute Microkelvin Imager which Rüdiger Kneissl described, should also readily probe beyond $z \simeq$1.

In a similar category to the SZ effect is weak lensing. As this technique wasn't explicitly reviewed at the meeting, I reproduce the nice discovery of a z=0.28 cluster by Wittman et al (2001) in Figure 3. Although lensing offers the most direct route to the total mass, it still suffers from projection effects, the effect of the mass sheet degeneracy[5] and, as a high density of background galaxies is needed, the technique will most likely be limited to the study of clusters with 0.1< z <0.8.

Concerning the optical searches, Marc showed us, from the analysis of Goto et al (in preparation) how, even locally within the SDSS, changing the search algorithms can deliver very different samples. This is worrying to me as I imagine this kind of uncertainty can only be worse in faint surveys. While smart ideas such as "matched filters" designed to minimise projection and maximise contrast,

[5]Mass estimates based on weak shear of a background population are insensitive to an additional unclustered component of dark matter.

are of course to be welcomed there is a natural concern that, like Maximum Entropy restoration techniques, such methods work reliably only when we have a clear idea of what we are searching for. Do we?

The Red Cluster Sequence method described by Mike Gladders seems the most promising of the optical methods. Indeed the authors claim it to be "comparable or superior to X-ray methods" (Gladders & Yee 2000) - a bold assertion! Certainly panoramic detectors sensitive in R and z' make this a much more efficient way to locate red colour-magnitude sequences to $z \simeq 1$ and, at CFHT, there is the prospect of coupling lower z detections with weak lensing constraints to directly get masses. Simulations suggest this search technique should be complete to $z \simeq 1$, even allowing for some dispersion and bluing in the color-magnitude relation. Unless there is a (perverse) population of massive clusters devoid of a uniform population of spheroidals, the technique seems a sound one.

But the next thorny issue is how to get a *reliable mass*, the second important requirement to test Figure 2 and one worth thinking about before embarking on any time-consuming survey! The mass of a distant cluster has traditionally been inferred from the X-ray luminosity (assuming a $L_x - T_x$ relation), an optical richness (very crude) or a velocity dispersion based on a few members. Clearly none of these is really adequate. This is, I think, where the X-ray surveys win hands down even if "smart" algorithms rescue the optical searches. We can fully expect, with adequate investment, X-ray temperatures for large numbers of distant clusters in the coming years. Brian Holden showed new Chandra data that indicates little evolution in the $L_x - T_x$ relation to $z \simeq 0.8$. Whilst assumptions are necessary to link T_x and virial mass (Voit 2000), the prospects look good, particularly with cross-checks from the SZ effect. Lauro Grego showed us the promise of deriving masses independently from her interferometric SZ techniques.

The major problem in my opinion is that the optical searchers have no real route to cluster mass (except by recourse to independent X-ray techniques or perhaps weak lensing over a limited z range); projection also remains an issue. Multiple systems, often separated by only modest velocity differences, would be very hard to detect without a huge spectroscopic investment. A good example in showing the pitfalls is the cluster surrounding the z=1.206 galaxy 3C324 (Smail & Dickinson 1995, Figure 4) now known from spectroscopic evidence to consist of two components separated by only $\delta z \simeq 0.06$. Although radio-selected, not only would its richness have been overestimated without careful spectroscopy, it seems unlikely the Red Sequence method *applied to a similar system at any redshift* would be able to photometrically-separate the two systems.

In the case of the local X-ray cluster data essential as a basis for all high z comparisons, Luigi Guzzo and Hans Böhringer demonstrated the remarkably precise cluster luminosity function obtained from the REFLEX survey. Few would disagree that this is a major achievement in the subject not just in statistical terms but also because of the considerable care taken to ensure homogeneity in the survey.

Turning then to the cosmological results and necessarily being somewhat selective, Harald Ebeling (MACS) and Piero Rosati (RDCS) discussed a fairly modest decline in the abundance of massive clusters to $z \simeq 1$ (Figure 5); this is a fairly convincing conclusion for $z < 0.5$ but beyond I suspect there is still some

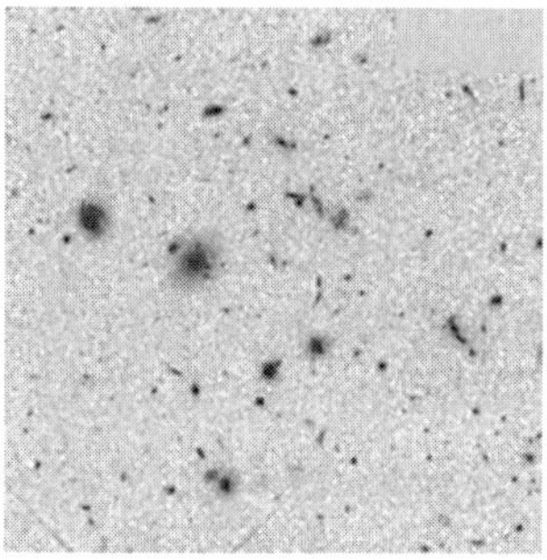

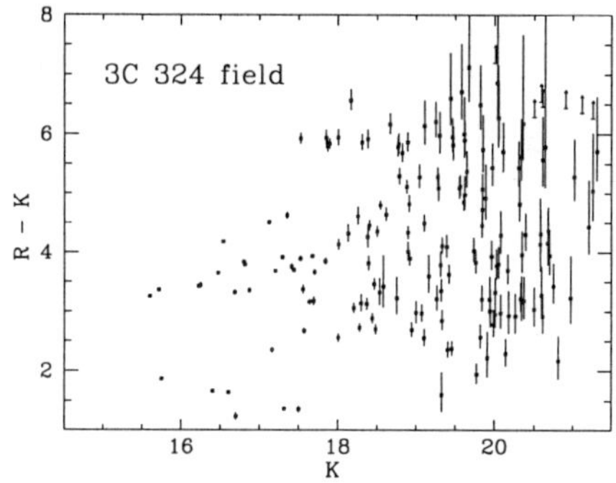

Figure 4. 3C324: an example of a complex system whose significance would most likely be overestimated in any optical or infrared-based search. The colour-magnitude relation (right) masks the presence of two separate systems (with $R-K \sim 6$) revealed only with spectroscopy: one at z=1.206 containing the radio galaxy and a second, separate system at z=1.115 (Courtesy of Mark Dickinson)

room for manoever. Qualitatively, this is similar to the 4.7σ decline claimed by Gioia (2001) in her NEP sample (not discussed at this meeting). Stefano Borgani has modelled the decline with redshift in the RDCS-3 sample with available $L_x - T_x$ data and claims the data is consistent with popular choices of Ω_M and σ_8 (see §3).

Can we say there is "concordance" between the observers? Not yet but good progress is being made. Most of any agreement refers to evolution in luminous, presumed massive, clusters. A number of niggly issues remain including the effects of incompleteness, cosmic variance for the smaller field surveys, AGN/cooling flow contamination, and how to compare surveys with minimal L_x, z overlap. For example, at the lower luminosities probed by the WARPS survey, the situation is quite unclear with some arguing for no evolution at all. Much of the dispersion in the inferred evolution arises from uncertainties in the number of *local* massive clusters against which comparisons are needed. WARPS and RDCS appear to disagree on the "local" abundance to an extent that seriously affects their respective analyses. However, the prospects for extending these tests to higher redshift and clarifying the temperatures with Chandra appear very promising, even before the new XMM surveys are underway.

As an outsider to the subject of distant X-ray cluster surveys, I was struck in both Harald and Marc Postman's reviews of the large number of competing surveys underway. If the labour involved in constructing these surveys was not enough, remember that the time required to fully exploit them is even larger. For example, the construction of Ebeling's ROSAT-based MACS sample of 840 clusters involved two colour CCD imaging and spectroscopic verification of a significant sample. To exploit all 109 clusters beyond z >0.3 with multi-slit spectrographs, a weak lensing deep imaging study, Sunyaev-Zel'dovich (SZ) de-

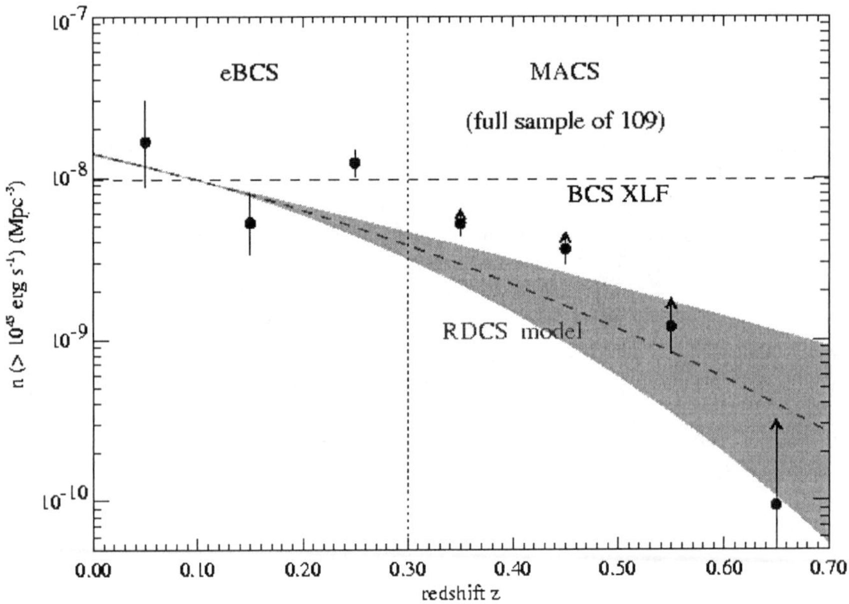

Figure 5. Evolution in the comoving space density of the most X-ray luminous clusters ($L_x > 10^{45}$ cgs) from the MACS survey (Ebeling), a result which indicates a slightly more modest decline in number to $z \simeq 0.5$ than that adopted by the RDCS team discussed at the meeting. Broadly comparable results have been obtained for the NEP sample (Gioia 2001).

tections and other applications would require almost dedicated access to a huge range of facilities (a tall order even by Harald's standards!). When one realizes this is just *one* of the many ambitious surveys we heard about with more in the pipeline, I worry has the planning of all these surveys seriously taken into account the necessary investment for a full exploitation?

3. Mapping Large Scale Structure with Clusters

Bob Nichol gave us a very balanced review of the changing role of clusters as new probes of large scale structure became available. He also posed some controversial questions, e.g: "in the era of galaxy and lensing surveys, who needs clusters?" He also stressed many of the complications that arise from the frequent merging of clusters on our assumption of relaxed systems in virial equilibrium.

I think it fair to say the traditional role of using clusters selected in various way (optically from the Abell/ACO catalogues or from X-rays) as the most efficient way to get to the mass power spectrum $P(k)$ on large scales is being overtaken by the large redshift surveys. In the era of 2dF and SDSS, huge volumes populated by over 100,000 galaxies are publically available[6] and delineate the power spectrum more reliably than the earlier Abell/ACO cluster samples whose integrity is still being debated. In the case of the X-ray cluster samples, Luigi Guzzo's analysis of the REFLEX survey is comparable in importance to 2dF and SDSS at the time of writing. A particularly impressive achievement in that sample is the detection of infall in the $\xi_v(\sigma, \pi)$ plane convincingly demonstrating the quality of the sample. However, the redshift surveys are continuing apace and from them one can create far more reliable spectroscopically-based cluster samples. Of particular interest is the volume to $z \simeq 0.55$ probed by SDSS Luminous Red Galaxies survey. The message is clear: wholesale galaxy mapping has arrived..use the data!

Bob drew attention to possible "baryon wiggles" in both the power spectrum of Abell/ACO clusters seen by Miller et al (2001) and referred to one tentatively in the 2dF data (Percival et al 2001). It must be remembered that, in such plots, the redshift-space power spectrum is significantly affected by aliasing introduced by the window function of the survey volume. Without careful simulation of this effect it is very hard to be sure whether bumps are artefacts or genuine baryon oscillations. Neither the REFLEX not the 2dF surveys claim to have detected these features. Indeed, the 2dF team estimates unless the baryon density is much higher than expected, they will remain undetected in the completed SDSS survey.

Elena Pierpaoli described a second aspect of local cluster statistics, namely constraining the *amplitude* of the mass power spectrum. The abundance of clusters of known mass provides a joint constraint on the variance, σ_8 on $8h^{-1}$ Mpc scales and the mean mass density, Ω_M in a form dependent on $\sigma_8\,\Omega_M^{0.5}$. By adding X-ray temperature measurements and improving the mass-T_x connection, an improved constraints has been determined. In the past year, the same

[6]See http://www.mso.anu.edu.au/2dFGRS/ & http://archive.stsci.edu/sdss/edr_main.html.

constraint has also been probed in a completely independent way via weak lensing studies of randomly-chosen fields (so called "cosmic shear" surveys) (eg. van Waerbeke et al 2001, Bacon et al 2000, 2001 in prep.). Reassuringly both methods get fairly similar answers with $\sigma_8\,\Omega_M^{0.6}$=0.40-0.50 although it is important to remember that cosmic variance and the redshift distribution of the background faint galaxies may affect uncertainties in the lensing estimates[7]

The weak lensing results are independent of any assumed cluster population and do not rely on Gaussian fluctuations in the mass spectrum. Accordingly, comparisons of lensing and cluster-based methods are an important way to verify the assumptions we may take for granted, not only in the cosmological framework but also in cluster physics.

4. Physical Processes Within Clusters

We had an interesting session on the internal mass distributions within cluster where Andrea Biviano gave a very comprehensive review of years of effort invested in estimating orbital anisotropies and mass distributions in clusters. The number of clusters which have been comprehensively surveyed spectroscopically is increasing rapidly and the new field surveys (SDSS, 2dF) will generate lots more. Cluster galaxy dynamics will remain a critical tool in determining the orbital anistropy of cluster galaxies of different types. As Andrea showed, there is convincing evidence that spheroidal galaxies retain an isotropic velocity field, consistent with their long-standing membership, whereas late type systems show a detectable radial anisotropy reflecting their continuous infall.

The peripheries of clusters at moderate redshifts are largely unstudied regions important in linking cluster-based environmental evolution to that in field galaxies. Tommaso Treu and Taddy Kodama described complementary dynamical/HST and photometric imaging programs to address this question. Kodama finds a potentially important result whereby the colours of field galaxies undergo a sharp transition from blue-to-red at some galaxian surface density, presumably reflecting some process associated with their arrival at the cluster. Treu is embarking on a longer term project to understand the mechanistic details of infall in a single well-studied cluster. Both routes are important ones to pursue.

Andrea was cautious about deriving mass profiles from galaxy velocities. This has been a long standing problem in individual clusters, even Coma, because of the inevitable degeneracies arising from unknown orbital anisotropies as a function of radius. If a functional form is adopted for the mass profile (e.g. NFW), solutions can be found but the present motivation is surely to determine these forms directly, as is possible in a more competitive way with X-ray data in the inner regions and weak lensing to the periphery. Steve Allen showed how much progress has been made in reconciling strong lensing and X-ray based mass estimates following improved Chandra data on 7 clusters and found no significant deviations from the universal mass profile seen in numerical simulations.

---

[7] Since the conference, estimates about 20% lower for the combination of σ_8 and Ω_M have been published, both from cluster studies and large scale structure results. Lahav et al (astro-ph/0112162 gives a good summary)

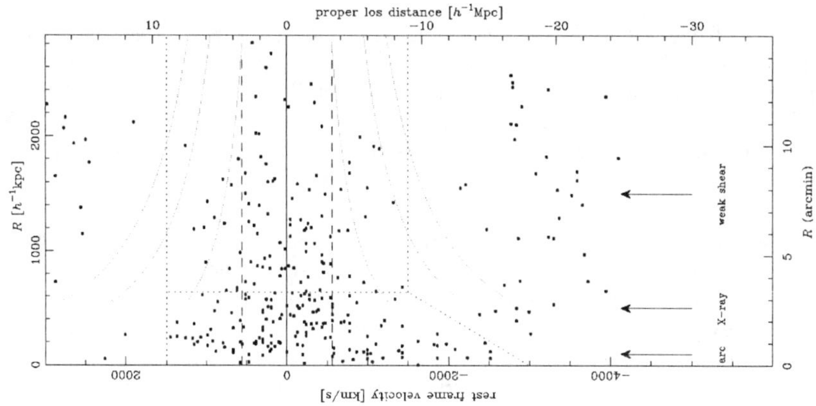

Figure 6. Velocities versus cluster-centric radius for the rich cluster Cl0024+16 from the CFHT survey of Czoske et al (2001). For many years this cluster was considered to have a very high velocity dispersion and a surprisingly low X-ray luminosity. The authors now propose that the innermost velocity distribution indicates a head-on collision of two systems of comparable mass. Although not an unique explanation, the cluster serves as a warning to those interested in cooadding velocity data over many assumed regular clusters.

In an effort to overcome some limitations, it has become common practice to co-add data from many clusters in the hope of improving signal to noise and erase asymmetries. Peter Katgert and Roland van der Marel showed us results from coadded samples based on the ENACS and CNOC surveys respectively, and concluded mass traces light to a reasonable approximation. However, is a composite cluster really a physical entity? A number of critics were unconvinced because of the dangerous effect one or two "complex" systems might have on the final conclusion. Even with a substantial number of redshifts, it seems one has to be cautious in interpreting dynamical data in terms of a simple gravitational potential.

A salutory lesson can be learnt from the comprehensive survey of the regular cluster Cl0024+16 (see Treu et al, Metevier et al, these proceedings) by Czoske et al (2001). For many years this was regarded as a classic regular virialised system at z=0.4 and its mass profile even formed the basis for promoting self-interacting dark matter as discussed by Oleg Gnedin. However, when 650 redshifts were gathered (of which 295 are members), Czoske et al resolved a foreground system (whose mass was initially thought to be quite modest but now is claimed to be significant) undergoing an end-on merger with the main body of the cluster (Figure 6). This discovery may explain its low X-ray luminosity c.f. the original high velocity dispersion. With less than 100 member velocities, Cl0024+16 would surely have been considered a representative cluster for coaddition but now we know it is a more complex beast. Bootstrap-style experiments might be helpful to clarify the robustness of the coaddition procedure.

Soon after the first X-ray detections of hot gas from the intracluster medium (ICM) were detected, it was realised that most of the baryons in clusters are in gaseous form. Assuming an isothermal distribution, the surface brightness profiles offer a valuable probe of cluster mass disribution, and the X-ray morphology indicates the evolutionary state. Clusters are expected to continuously assemble in popular hierarchical models and we can expect to directly witness this growth via shock heating events.

XMM and Chandra are transforming this field and this conference has provided one of the first opportunities for us to discuss the assumptions made in analysing earlier data. Monique Arnaud illustrated some of this progress in her excellent review. Although merging substructures produce shock-heating Chandra reveals unforeseen complexities, for example in the "cold fronts" seen in some clusters which suggest the associated timescales may have been underestimated. Monique demonstrated several lines of evidence for the continuing assembly of clusters (X-ray substructure, anisotropic accretion etc). However, even in the well-studied case of the outskirts of the Coma cluster, Uli Briel demonstrated significant uncertainties in understanding the merger timescales involved. Such data are thus *illustrative* of hierarchical assembly but may not give us the needed *quantitative* verification of mass assembly.

The most striking aspect of the new high quality data is the growing evidence for non-gravitational processes in understanding the physics of the ICM. Whilst temperature profiles obtained with ASCA and BeppoSAX were often discrepant (as discussed by De Grandi), the XMM data gives reasonable support to the assumption of isothermal cores. However, the expected self-similar "scaling laws" between the gas mass, virial radius and X-ray temperature, e.g. in $L_x \propto T^2$ are not obeyed. The steep $L_x - T$ relation implies non-gravitational processes, almost certainly associated with additional heating mechanisms.

Theoretical explanations for the steep $L_x - T$ relation were the subject of almost an entire afternoon session. There was a bewildering number of ideas. Fabio Governato reviewed for us how feedback heating can be incorporated in numerical simulations and showed that a steep relation cannot be easily explained via conventional heating sources such as supernovae. Paolo Tozzi demonstrated that, with an energy of 1-2 kev/baryon, AGN are a promising possibility but it seems there is no obvious evidence for heating from these sources. Peter Thomas from the VIRGO consortium can predict the slope but only at the expense of a strong evolution in the relationship. Uros Seljak gave us an analytical model based on a universal gas profile where departure from self-similarity arises from a temperature-dependent hot gas fraction.

Perhaps that glass of wine at lunchtime was not a good idea but this was a confusing session for me. Unlike observers who compare ASCA/BeppoSAX/XMM temperature profiles, worrying about sensitivity, psf and background differences, when it comes to numerical simulations, I am puzzled that there appears to be no necessity to inter-compare results. Surely this is a first step in convincing anyone of the believability of a particular simulation?

Additional physical complexities in the inner cool cores of clusters were reviewed by Andy Fabian. In the physical environment of a dense cluster core, hot gas is able to cool radiatively in a timescale of less than 10^{9-10} years and a stable flow of cooling gas to the cluster centre is expected. Almost 50% of all X-ray clusters show some signs of the associated temperature gradient implying mass inflow and deposition rates of 10-100 $M_\odot$ yr^{-1}. However, XMM spectra show no sign of the expected Fe XX lines of radiative cooling in the 13-18 Å range. Hans Böhringer also discussed additional temperature diagnostics unaffected by absorption effects. Either the gas is not cooling radiatively or there is perhaps is a balancing effects from a hitherto-unidentified heating mechanism. And there are serious duty cycle issues concerned with preventing cooling beyond 2-3 kev.

As usual, Andy had no shortage of ideas for resolving the puzzle, the most intriguing from my perspective being the hypothesis of a starkly bimodal metallicity distribution which has the desired effect of suppressing the amount of cooling witnessed through the diagnostic iron lines. Like most conference participants I suspect, I have little idea where this additional complexity in the ICM physics is taking us. The Chandra images of the Perseus cluster discussed by Schmidt (with such enthusiasm!) show holes in the X-ray emission where the gas appears hotter and more metal rich. Dynamical complexities were introduced by Ettori in the wealth of data he presented for Abell 1795. A number of speakers introduced magnetic fields associated with radio polarisation (in the case of Abell 2255 discussed by Govoni), radio sources (although as Fabian remarked the gas is not obviously hotter in these regions) or in explaining the survival of cold clouds with associated fronts (Arnaud).

I think we can be optimistic that there is a lot to learn which will assist in the entire interplay between gas cooling, star formation and enrichment. Many years ago at a conference I attended in Cambridge, Andy claimed confidently that cooling flows provided a key mechanism for the formation of giant elliptical galaxies. Although cooling flows remain controversial, Andy can still assert (as Richard Bower emphasised independently) that the physics of gas cooling in cluster cores will tell us much about feedback and star formation: still key ingredients in galaxy formation.

5. Evolution of Galaxy Populations within Dense Environments

On the last day, Pieter van Dokkum surveyed the literature on the role that clusters continue to play in our understanding of galaxy formation and evolution. There is a subtle shift of emphasis however. Rather than being "laboratories of convenience" [8], galaxies in clusters are now being studied alongside equivalent field populations and as likely successors to well-studied Lyman break galaxies at z >2.

Most of the discussion, by van Dokkum, Rosati & Kelson, concentrated on the role of very distant (z >0.5) clusters, largely in terms of differentiating stellar ages and the ages of mass assembly of giant ellipticals. Our large telescopes are being flexed to their limits to secure impressive fundamental plane data (to z =1.2 in 12 hour Keck exposures!) which continues to support the notion that old stellar populations in at least some fraction of the data. That the stars in ellipticals may be older than the assemblies in which they now reside is illustrated in the red mergers seen in the well-studied X-ray luminous cluster

[8] Perhaps there is a better terminology here!

MS1054 (z=0.83). Pieter posed the question of how common such a system might be and gave some new examples of red mergers in other clusters. It is good to see some progress in separating the ages of stars from those of the accumulated mass that makes up giant ellipticals.

The biases introduced not only in selecting galaxies within clusters (by HST morphology, by colour or infrared magnitude) but also by how the clusters themselves were selected (X-ray, optical..) continue to worry me. D. Fadda also reminded us how gas rich and dusty systems may be common even in dense clusters. As remarked earlier, it is hard to know what errors are made by connecting data at different redshifts to delineate an evolutionary picture. One suspects we are reliant either on theoretical modelling (heaven forbid!) or comparative field samples. Beyond z >0.5 this will be some time coming although a number of speakers alluded to the upcoming Keck and VLT spectroscopic surveys.

It is now 20 years since Alan Dressler published his quantitative study of the morphology-density relation in a nearby sample of 55 rich clusters (Dressler 1980). Pieter and Bianca Poggianti urged us to go back and improve the local samples now we have panoramic CCD cameras and multi-object spectrographs so we can be sure of the local fractions (Figure 7). A key issue here is the origin of S0s; the Morphs team (Dressler et al 1997) proposed their recent demise from gas stripping and tidal effects and presented a strong claim for an evolving E/S0 fraction with lookback time. Pieter urged us to be much more cautious in differentiating Es and S0s; with resolved spectroscopy of z >0.5 galaxies feasible on many telescopes, this appears to be a profitable route in conjunction with HST data.

6. Parting Thoughts

This has been a fascinating meeting and we have a lot to be thankful for. Foremost we have amazing observational facilities capable of finding clusters and studying them to great depths in complementary ways. We now can resolve clusters at the arcsec level in X-rays and locate their concentrated masses with S-Z and weak lensing techniques. We can also locate local clusters as byproducts in the comprehensive redshift surveys being undertaken. Instead of viewing these as competing techniques we should exploit them all as complementary probes to check our physical assumptions.

We also have the opportunity to learn new physics in the cluster cores from the paradoxes emerging from the steeper $L_x - T$ relations and the absence of lines of radiative cooling in those with temperature gradients. It should perhaps be no surprise that simple gravitational physics is not enough to explain what we see. The fall-out in our understanding of galaxy and cluster formation from resolving this paradox could be a big one.

As the concluding speaker I want to thank Drs Mardirossian and Mezzetti for all the necessary arrangements here at beautiful Sesto Pusteria and to the Organising Committee for their hard work. We should in particular thank Stefano Borgani for his obvious insight in constructing a thoughtful scientific program and amazing energy he and his colleagues invested in making this a great meeting!

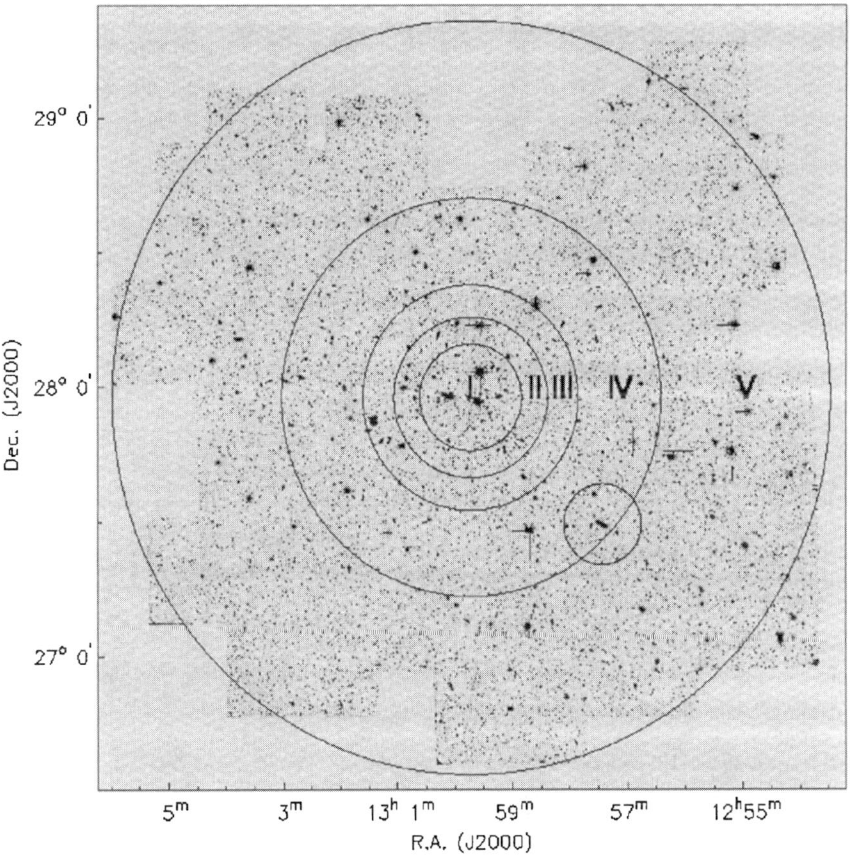

Figure 7. Going back to basics: securing improved morphological and spectroscopic data for galaxies in nearby clusters is important in fixing the low redshift baseline for high z studies. Mosaiced CCD image of the Coma cluster from the study of Beiersbergen et al (2001).

References

Adelberger, K. 2000 in *Clustering at High Redshift*, eds. Mazure, A. et al, ASP Conference Series, 200, p13.

Aragón-Salamanca, A. et al 1998 MNRAS, 297, 427.

Bacon, D.J. et al 2000 MNRAS 318, 625.

Beijersbergen, M. et al 2001 MNRAS in press (astro-ph/0106354).

Borgani, S. 2001 in *Galaxy Clusters and the High Redshift Universe Observed in X-rays*, eds. Neumann, E. et al in press (astro-ph/0106471).

Borgani, S. & Guzzo, L. 2001 Nature 409, 39.

Czoske O., et al. 2001, A&A, in press(astro-ph/0111118)

Daddy, E. et al 2001 A&A 376, 825.

Dressler A. 1980, ApJ, 236, 351.

Dressler A. et al. 1997, ApJ, 490, 577.

Fukugita, M. 2000, in Proc. 30th ICHEP, eds. Lim, C.S. & Yamanata, T., World Sci. p299.

Gioia, I. 2001 in *Chemical Enrichment of the Intracluster Medium* in press (astro-ph/0107452).

Gladders, M. & Yee, H.C. 2001 in *Cosmic Evolution & Galaxy Formation*, ed. Franco, J. et al, ASP Conf. Series 215, p233

Gunn, J.E. & Oke, B. 1975 ApJ, 195, 255.

Kodama T., et al. 2001, ApJ, 562, L9

Kristian, J. et al 1976 ApJ, 205, 688

McCarthy, P. et al 2001 ApJ, 560, L31.

Miller et al 2001 ApJ, 555, 68.

Percival, W. et al 2001 MNRAS 327, 1279.

Smail, I. & Dickinson, M.E. 1995 ApJ, 455, L99.

van Waerbeke, L. et al 2001 A&A 374, 757.

Voit, G.M. 2000 ApJ, 543, 112.

Wittman, D. 2001 ApJ, 557, L89.

POSTERS

(in alphabetic order following the name of the first author)

Tracing Cosmic Evolution with Galaxy Clusters
ASP Conference Series, Vol. 268, 2002
S. Borgani, M. Mezzetti, and R. Valdarnini, eds.

A Dynamical Study of the Hot Galaxy Cluster 1E0657-56

R. Barrena

Instituto de Astrofísica de Canarias, La Laguna, Tenerife, Spain

A. Biviano & M. Ramella

INAF, Osservatorio Astronomico di Trieste, Italy

E.E. Falco

Smithsonian Institution, F.L. Whipple Observatory, AZ, USA

S. Seitz

Universitäts-Sternwarte, Scheinerstrasse 1, Muenchen, Germany

Abstract. Using new spectroscopic observations of 78 member galaxies in 1E0657-56 ($\bar{z} = 0.296$), we investigate the dynamical status of this very X-ray hot cluster. In combination with X-ray and radio data, we suggest that 1E0657-56 is a rich virialized cluster, which is still accreting subclusters from the surrounding large-scale structure.

1. Summary of the main results

1E0657-56 is a medium distant cluster ($\bar{z} = 0.296$), characterized by an extremely high X-ray temperature and luminosity (Tucker et al. 1998). With the aim of deciphering its complex dynamical structure, we have recently determined the redshifts of 78 cluster members, with observations at the ESO NTT. Additional photometric information comes from observations with FORS1 at the ESO VLT.

The projected spatial distribution of (probable) cluster members (see Fig. 1.) selected within the colour-magnitude red sequence $(B - R)$ vs. R, suggests the existence of a subcluster to the West side of the main elongated cluster body. The physical reality of this subcluster is confirmed by the analysis of the galaxy velocities. We compute $\bar{v} = 88710 \pm 75$ km s^{-1} and $\sigma_v = 1260^{+110}_{-101}$ km s^{-1} for the main cluster and $\bar{v} = 89320 \pm 40$ km s^{-1} and $\sigma_v = 213^{+67}_{-52}$ km s^{-1} for the 7 galaxies in the subcluster. The virial masses are $M_{vir,c}(< r_{200}) = 1.24 \times 10^{15} M_{\odot}$ for the main cluster[1], and $M_{vir,c}(< r_{200}) = 0.12 \times 10^{14} M_{\odot}$ for the W subcluster. With these values, following the two-body model of Gregory & Thomson (1984),

[1]Throughout this paper, we use $H_0 = 70$ km s^{-1} Mpc$^{-1}\Omega_0 = 0.3$ and $\Omega_\Lambda = 0.7$

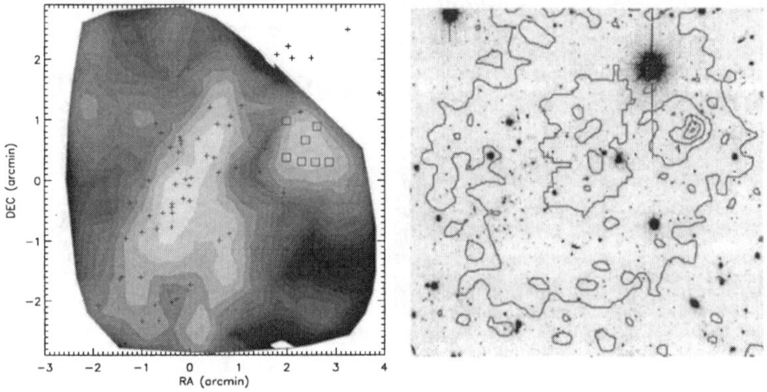

Figure 1. **Left panel.** Bidimensional (2D) projected galaxy density. North is up, East is to the left. Crosses indicate galaxies with measured velocities that belong to the main cluster, squares indicate galaxies with measured velocities that belong to a subcluster. **Right panel.** Contours of ROSAT/HRI image smoothed with a 90″ gaussian filter, superposed onto the R-band FORS/VLT image of 1E0657-056. The field of view is $5' \times 5'$.

we determine that the subcluster is bound to, and currently infalling into the cluster.

The cluster would then be in a pre-merger phase. This is also suggested by the morphology of the X-ray emission (see Fig. 1.), which has a peak between the main cluster and the subcluster. The enhanced X-ray emission in the region between the two structures could be due to the adiabatic compression of the X-ray gas due to the infalling subcluster. The strong X-ray emission from the subcluster is atypical of a small galaxy group (Helsdon et al. 2001). We suggest that the subcluster could be the remnant of a moderately massive cluster, which has been tidally stripped to its current small size by the gravitational field of the cluster.

Consistent with this pre-merger scenario is also the very low fraction of emission-line galaxies, typical of rich, relaxed clusters (Biviano et al. 1997). On the other hand the very presence of a radio-halo (Liang et al. 2000), calls for one or several major merging events that could have accelerated the thermal electrons. This is probably reflected in its very elongated structure (see Fig. 1.), reminiscent of previous accretion events along large-scale structure filaments.

References

Biviano, A., Katgert, P., Mazure, A. et al. 1997, A&A, 321, 84

Gregory, S.A., & Thompson, L.A. 1984, ApJ, 286, 422

Helsdon, S.F., Ponman, T.J., O'Sullivan, E., & Forbes, D.A. 2001, MNRAS, 325, 693

Liang, H., Hunstead, R.W., Birkinshaw, M. & Andreani, P. 2000, ApJ, 544, 686

Tucker, W., Blanco, P., Rappoport, S., et al. 1998, ApJ, 496, L5

Tracing Cosmic Evolution with Galaxy Clusters
ASP Conference Series, Vol. 268, 2002
S. Borgani, M. Mezzetti, and R. Valdarnini, eds.

New 2D Galaxy Systems within PDCS Fields

R. Barrena & E. Mediavilla

Instituto de Astrofísica de Canarias, La Laguna, Tenerife, Spain

M. Ramella, M. Nonino & A. Biviano

INAF, Osservatorio Astronomico di Trieste, Italy

W. Boschin

Univerità degli Studi di Trieste, Dipartimento di Astronomia, Italy

Abstract. We present a sample of 18 galaxy systems identified within two 0.5 × 0.5 square degree subsections of the original V and I galaxy catalogues of the PDCS fields. We identify these clusters using the non-parametric technique VGCF (Ramella et al. 2001). We also identify clusters within new B and R images of the same subsections obtained at the WFC@INT.

1. The BRVI VGCF clusters.

We carried out WFC@INT observations using B and R Harris filters as a part of the GLITP programme (Ramella et al). The two 30x30 arcmin^2 fields are centered on α_{2000}=$00^h29^m00^s$, δ_{2000}=$5°15'00''$ and α_{2000}=$02^h28^m00^s$, δ_{2000}=$1°15'00''$, which are contained in the PDCS 0h+15deg and PDCS 2h+1deg fields respectively (Postman et al. 1996). Our galaxy catalogues reach $B = 24$ and $R = 22$ completness magnitudes, PDCS Catalogues[1] reach $V = 23$ and $I = 21.5$.

We run on each catalogue the Voronoi Galaxy Cluster Finder[2] (VGCF; Ramella et al. 2001). VGCF complements the matched-filter identifications with the identification of irregularly shaped clusters and/or clusters missed by the matched-filter algorithm because of the smoothing procedure.

We identify 18 clusters in more than one band. VGCF also identifies 10 (90%) PDCS clusters. In Table 1 we summarize the identifications in the various bands. Of course, coincident identifications in multiple bands increase the reliability of the detections.

For the detected clusters, we analyze raw colour indexes and perform a preliminary analysis of the colour-magnitude diagrams ($B-R$, R) and ($V-I$, I). We measure colours for galaxies in the red sequence and list them in table 2,

[1] V and I data courtesy of Marc Postman

[2] available at `http://www.ts.astro.it/astro/VoroHome/voronoi.html`

	B	V	R	I	B,R	V,I	B,V,R,I
N_{VGCF}	24	18	21	17	10	12	8
N_{PDCS}	-	11	-	10	-	11	-

Table 1. Total number of cluster identifications and number of coincident identifications in more than one band.

where we also give PDCS matched-filter redshift estimates. There is a generally good qualitative agreement between the trend of colours and estimated redshifts. This fact supports our confidence that the VGCF procedure applied to a multi-band data is actually able to produce a reliable cluster catalogue. In fact, taken at face value, the colour analysis indicates that a fraction up to 75% of our 2D clusters may actually be real physical systems.

Cluster	α (J2000)	δ (J2000)	$\bar{r}$ ($''$)	z_{PDCS}	$B-R$	$V-I$
PDCS 1	00:29:01.2	05:02:21	94	0.6	2.7	?
PDCS 2	00:29:28.9	05:03:36	125	0.4	2.7	?
PDCS 3	00:28:35.0	05:07:24	98	0.6	?	?
PDCS 4	00:29:09.4	05:08:30	175	0.5	?	2.4
PDCS 6	00:29:51.8	05:12 30	145	0.4	?	2.4
PDCS 8	00:28:31.1	05:18:28	86	0.6	2.7	2.3
VGCF 1	00:29:16.1	05:18:11	73	-	2.5	1.7
VGCF 2	00:29:15.3	05:26:53	87	-	?	2.2
VGCF 3	00:28:21.1	05:10:05	92	-	2.2	?
PDCS 20	02:28:32.4	00:57:18	61	0.4	-	1.6
PDCS 22	02:26:56.0	01:05:14	194	0.4	2.4	2.0
PDCS 23	02:27:53.2	01:04:53	191	0.2	1.8	1.1
PDCS 28	02:28:54.4	01:22:39	178	0.3	2.3	-
VGCF 4	02:27:47.3	01:13:48	139	-	2.5	2.3
VGCF 5	02:27:41.3	00:59:46	101	-	2.5	2.0
VGCF 6	02:28:25.7	01:30:11	156	-	2.4	-
VGCF 7	02:28:42.2	01:28:12	74	-	2.1	-
VGCF 8	02:27:19.9	01:05:10	99	-	1.4	1.3

Table 2. Estimated radius and colours of the clusters identified within our fields. Question marks stand for red sequence not detected, dashes indicate no data available.

References

Postman, M., Lubin, L.M., Gunn, J.E., Oke, J.B., Hoessel, J.G., Schneider D.P., & Christensen, J.A.. 1996, AJ, 111, 615

Ramella, M., Boschin, W., Fadda, D., & Nonino, M. 2001, A&A, 368, 776

Tracing Cosmic Evolution with Galaxy Clusters
ASP Conference Series, Vol. 268, 2002
S. Borgani, M. Mezzetti, and R. Valdarnini, eds.

Deep Photographic Cluster Survey — DPCS

Oliver Baumann, Hermann-Josef Röser, Carsten Tschamber

Max-Planck-Institut für Astronomie, Königstuhl 17, D-69117 Heidelberg, Germany

Michael R.S. Hawkins, Harvey MacGillivray

Royal Observatory, Edinburgh, Blackford Hill, Edinburgh, EH9 3HJ, UK

Abstract. A search in three colours (B, R, I) for high-redshifted clusters of galaxies using digitally stacked Schmidt plates is presented. Over an area of 22.7 □ a total of 483 cluster candidates were found, 296 of increased detection significance. For 173 candidates in the full and 63 in the most significant sample the redshift is estimated to be above 0.5.

Clusters of galaxies are ideal laboratories to study evolution of galaxies and formation of structure in the universe. Distant clusters are of special importance, allowing a look back in time. Providing a suitable sample for detailed studies of the evolution of galaxy populations with the VLT was a prime motivation for our survey (see also Scodeggio et al. (1999), ESO Imaging Survey EIS and Postman et al. (1995), Palomar Deep Cluster Survey PDCS).

The survey data base consists of stacked Schmidt plates collected with the UK-Schmidt telescope of survey field 287 (Hawkins & Veron 1993). The plates were scanned with COSMOS and digitally stacked. Blue and red stack images consist of 64 plates each, the I-band stack is based on 30 plates. The resolution of the stacked images is typically $2.2''$. The full field covered has a side-length of $5.3°$, resulting in an image size of $17.5\,\mathrm{k} \times 17.5\,\mathrm{k}$ pixels.

Objects were searched and identified by means of INVENTORY within MIDAS. The result was stored in tables with one entry per object containing all its information in the three colours. Exclusion regions were interactively identified to eliminate bright galaxies, halos and diffraction spikes from bright stars, satellite tracks and plate defects. This exclusion map was consulted during the search for galaxy clusters and the subsequent analysis, reducing the effective search area to 22.7 □. Photometric calibration of the stack was performed at the ESO/MPG 2.2m-telescope with EFOSC II by observing 3 sub-fields together with standard stars at various airmasses. The limiting magnitudes of the stack were derived as $B_J = 24.8^{\mathrm{mag}}$, $R_F = 23.4^{\mathrm{mag}}$ and $I_N = 21.8^{\mathrm{mag}}$ (75% completeness limits). Object counts as a function of magnitude were compared with the literature to check the reliability of the object extraction and calibration algorithm, as was the galaxy-galaxy correlation function. No significant deviations pointing to any problem with the object extraction algorithm were detected.

The cluster search employs a count-in-cell algorithm, analysing density maps of galaxies. From the object tables only pre-selected entries fulfilling the selection criteria for galaxies (useable regions etc.) are used to generate the density map. Cells overlapping with exclusion regions are corrected due to their filling factor. The galaxy density map is smoothed in two steps on small and on large scales. The intermediate scales will contain the signatures of clusters of galaxies. The detection significance of the candidates was estimated using counting statistics imposing a secure and low upper limit for the detection probability of spurious clusters. To generate a reliable catalogue of galaxy clusters the data set is searched in five overlapping magnitude ranges using four cell sizes each. The primary catalogue contains candidates which are found in at least two passes, whereas candidates only found once are in the supplementary list.

Our cluster search algorithm was tested against the adaptive-matched-filter (AMF) algorithm used in simulations for the Sloan-Digital-Sky-Survey, kindly provided by J. Kepner (Kepner et al. 1999), enabling a direct comparison of the two results. The SDSS-simulations were searched in a manner comparable to our real data base (the only difference was a restriction to three instead of five magnitude ranges as the simulations had $z_{\max} = 0.5$). It was evident that the matched-filter algorithm needs some information on the redshift to successfully detect distant clusters. Thus, the new method is superior to the AMF comparing equivalent searches for distant clusters. The small number of spurious detections is acceptable if one keeps in mind the costs of collecting redshift information.

Cluster redshifts were estimated based on the magnitude of the 10th brightest galaxy. Simulations have shown that obviously this estimate is biased towards low values above redshifts of about 0.5. Therefore we expect clusters at redshifts considerably higher than 0.7, the largest redshift estimate, in our candidate list, extending probably well above 1. Our cluster search resulted in a catalogue of 483 cluster candidates (Baumann 2000), 296 of them were detected in several search passes (primary catalogue). A total of 173 have an estimated redshift above 0.5, 63 of them are in the primary catalogue. Contamination with false candidates is estimated to be $< 12\%$ and $< 40\%$ for the main and the supplementary catalogues, respectively. Our algorithm identified 88% of the ACO- and 76% of the APM-clusters within the field. Missing clusters from these lists are due to problematic cases like overlaps or vicinity to excluded regions.

The depth of the I-band stack does not well match the other two colours. We have therefore re-observed the whole area of field 287 with the WFI at the ESO/MPG 2.2m-telescope in the I-band in September 2000 and May 2001 (exposure times 5×300 sec and 5×500 sec, respectively). Median seeing was around $1''$. So in addition to an improved magnitude limit we now also have the advantage of superior morphological information and will repeat the cluster search including the WFI data.

References

Baumann, O. 2000, PhD thesis Universität Heidelberg ,
`http://www.ub.uni-heidelberg.de/archiv/1107`

Hawkins, M.R.S., Veron, P. 1993, MNRAS 260, 202

Kepner, J. et al. 1999, ApJ 517, 78

Postman, M. et al. 1996, AJ 111, 615

Scodeggio, M.et al. 1999, A&AS 137, 83

Tracing Cosmic Evolution with Galaxy Clusters
ASP Conference Series, Vol. 268, 2002
S. Borgani, M. Mezzetti, and R. Valdarnini, eds.

Deprojection of luminosity functions of galaxies in the Coma cluster

M. Beijersbergen, W.E. Schaap and J.M. van der Hulst

Kapteyn Astronomical Institute, P.O. Box 800, 9700 AV Groningen, The Netherlands

Abstract. We present a simple analytic model to deproject 2-d luminosity functions (LF) of cluster galaxies. We apply this model to deproject the LFs of the Coma cluster measured by Beijersbergen et al. 2001. We demonstrate that the shapes of the LFs change after deprojection. The corrections are large enough to change the previously observed trend of increasing faint end slopes with increasing distance to the cluster center. The weighted U, B, and r band slopes of the deprojected LFs show only a weak steepening with increasing projected cluster radius. It is therefore essential to correct LFs for projection effects.

1. Introduction

In the past large field studies of the photometric properties of galaxies in the Coma cluster have been based on photographic plates. More recent studies have been done with CCD detectors which provide higher photometric accuracy. However, a limitation of the use of CCDs has been the relatively small field of view. Consequently, these studies have been limited to very small areas, ranging from ~ 50 arcmin^2 to 1500 arcmin^2, mainly focused on the cluster center. These small fields inhibit to study the dependence of the luminosity function (LF) on the projected distance from the cluster center. Furthermore, lacking the photometric data at larger distances from the core it is not possible to correct a central LF for projection effects. Accordingly, the published LFs of galaxies in the Coma cluster have all been *in projection* and are, therefore, luminosity distributions. It is important to test how well these correspond to the actual deprojected LFs.

In Beijersbergen, Schaap & van der Hulst (2002) we use the survey of Beijersbergen et al. 2001 to deproject 2-d LFs. For the first time deprojected LFs as a function of projected distance from the cluster center are inferred. In addition, we study how the shapes of the LFs are affected by projection.

2. Deprojection of luminosity functions

We use a simple projection scheme to deproject the observed 2-d luminosity distribution of the galaxies in the Coma cluster to the real, 3-d luminosity function. This scheme is based on the assumption that the galaxies are symmetrically distributed in a sphere. This sphere is divided up in a number of concentric shells, in which the galaxy density per magnitude is assumed to be constant. The value of the 3-d luminosity function in a particular shell can then be found by subtracting the contributions from the other shells due to projection.

3. Discussion

Beijersbergen et al. 2001 presented studied changes in the LF as a function of distance to the cluster center. However, they did not correct for the fact that fore- and background galaxies, residing in the outer parts of the cluster, are projected onto the central cluster regions. The question is how much their results are influenced by this effect. Below, we quantify the differences by studying the change in the shapes of the 3-d LFs as function of position in the cluster. The results are compared to the exact same analysis for the 2-d LFs.

3.1. Dependence of luminosity functions on radial distance from the cluster center

One of the main results of Beijersbergen et al. 2001 was the observation of a steepening of the faint end slopes with increasing cluster radius. A straight line fit ($sr+b$), where r is the clustercentric radius, to the weighted U, B and r band slopes as function of radius gives $s = 0.11 \pm 0.02$ Mpc^{-1} for the 2-d LFs. In order to compare this with the situation where all 2-d LFs have been corrected for projection we fitted power laws ($b\ 10^{aM}$) to the faint ends of the 3-d LFs. The LFs were fitted for $M_{\rm r} > -20$, $M_{\rm B} > -20$ and $M_{\rm U} > -19$. A straight line fit to the weighted slopes of the 3-d LFs at each radius gives $s = 0.03 \pm 0.04$ Mpc^{-1}. This is a much weaker trend than derived from the 2-d results. This shows that 2-d LFs derived for relatively small areas should be corrected for projection before one can only reliably interpret their shapes.

4. Conclusions

In Beijersbergen et al. 2002 we present a simple analytic model to deproject 2-d LFs of cluster galaxies. We have shown that these effects affect the shapes of the LFs. Furthermore, the alterations are significant enough as to arrive at different conclusions compared to those inferred from 2-d LFs. The previously derived steepening of the faint end slopes with increasing cluster radius is weakened when projection effects are taken into account. The new results show only a weak increase of faint end slopes with increasing cluster radius. Overall, we have demonstrated that before interpreting shapes of 2-d LFs one needs to correct for projection.

References

Beijersbergen, M., Hoekstra, H., van Dokkum, P.G., van der Hulst J.M. 2001, MNRAS, in preparation

Beijersbergen, M., Schaap, W.E., van der Hulst, J.M. 2002, A&A, in preparation

Tracing Cosmic Evolution with Galaxy Clusters
ASP Conference Series, Vol. 268, 2002
S. Borgani, M. Mezzetti, and R. Valdarnini, eds.

The central X-ray structure of the merging cluster A3921

Elena Belsole, Jean-Luc Sauvageot

Service d'Astrophysique, CEA Saclay, L'Orme des Merisiers Bât 709., F-91191 Gif-sur-Yvette Cedex, France.

Hervé Bourdin

Observatoire de la Côte d'Azur, B.P. 4229, F-06304 Nice Cedex 4, France.

Abstract. The XMM/Newton observation of the merging cluster A3921 is presented. With the high sensitivity and good spatial resolution of the instrument we are able to build a temperature map showing complex behaviour in the region across the merging event and to detect multi temperature structure in the main component which was thought to be in hydrostatic equilibrium.

1. Introduction

Observations and simulations show that merging strongly changes the physical characteristics of the intracluster medium (ICM). The study of this dynamical process would give us new insights into the physical process involved and better constraints in a more general cosmological context. A3921 has been observed by ROSAT and GINGA revealing that a substructure is falling onto the main relaxed cluster from behind with a velocity of 1000 km s^{-1}. The substructure is located at a projected distance of 1.2 Mpc (redshift=0.094; h_o= 0.5) and it has a mass $\sim$ 1/3 of the main component. The XMM/Newton guaranteed time observation of A3921 shows a more complex structure in the centre of the main cluster ruling out the hypothesis of a relaxed main component.

2. Analysis

A3921 has been observed by XMM/Newton on October 2000 as our first guaranteed time target for a total exposure time of 30 ks. We present EPIC/MOS data only as EPIC/pn are still under analysis. The events are corrected for vignetting following the method of Arnaud et al. 2001. As background we used a blank field obtained by adding together several deep exposures after sources are excised.

We subtracted a 2D β model from the EMOS image revealing the merging region on the north-west direction and a central excess. To investigate the physiscal state of the main cluster we extract spectra in concentric truncated (to be far from the merging perturbed region) annuli from the center toward the east. The background subtracted and particle-normalised spectra are fitted with a MEKAL (xspec v11.0) model fixing the N_H to the galactic value and leaving as free parameters the temperature, metallicity and normalisation. The high temperature in the centre excludes the presence of cooling flow. Moreover, the cD galaxy in the centre of the cluster cannot be the main source of this emission because of the high temperature and their relatively large extension of 0.5 Mpc toward north-east. This analysis argues in favour of a multitemperature structure in the centre of the cluster.

We extract 5 energy band images of the central 1 Mpc region, corresponding to the central excess. This approach allows us to better represent the spectral and spatial variation at the same time. From the left to the right (figure 1), the nuclear structure changes with energy suggesting that different physical component are responsible for the localized X-ray emission. We notice the positional shift of the central peak of the emission and the isophotal rotation with energy. The shift of the peak position is of 360 kpc, (2.5 time the EPIC PSF). This mean that at least two sources are present in the nuclear region, however looking at the DSS optical image, no obvious object is founded.

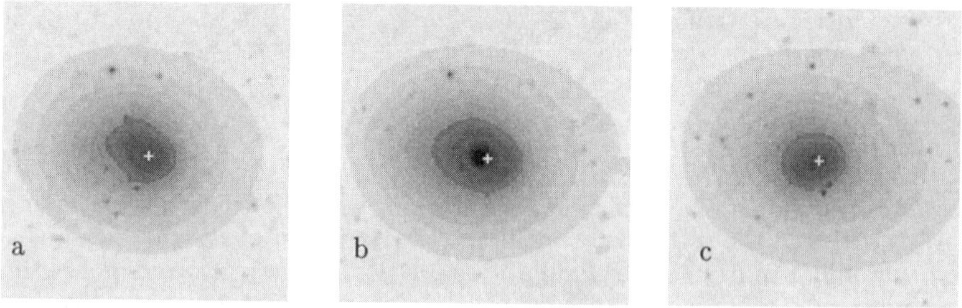

Figure 1. Three of the five energy band images are shown: a) 0.2-0.7 keV; b) 0.75-1.25 keV; c) 4.7-8.0 keV. The cross corresponds to the peak position in the 0.2-0.7 keV image.

3. Discussion

The data presented here show strong evidence of a perturbed structure of the merging cluster A3921. Distortions are evident not only in the region of the falling subcluster in the west but the centre is also far from relaxation. With this preliminary analysis we cannot give a complete interpretation of the temperature structure (this will be confirmed by adding the EPIC/pn data) and the isophotal rotation of the central Mpc, however one interpretation is that the complex structure could be due to a second group that has fallen onto the main cluster in an earlier merger and has not yet the time to relax. To better understand and explain this scenario, dynamical measurements are necessary.

References

Arnaud,M., Neumann, D.M., Aghanim, N., Gastaud, R., Majerowicz, S., Hughes, J.P., 2001, A&A,365, L74

Tracing Cosmic Evolution with Galaxy Clusters
ASP Conference Series, Vol. 268, 2002
S. Borgani, M. Mezzetti, and R. Valdarnini, eds.

Comparing the Field and Cluster Kormendy Relations

Kris Blindert and H.K.C. Yee

Department of Astronomy and Astrophysics, University of Toronto, 60 St. George Street, Toronto, Ontario, Canada, M5S 3H8

Abstract. We compare the R-band Kormendy relations for early-type galaxies in a cluster and in the field at $z \sim 0.2$. The cluster and field samples are selected by similar criteria, and we use data with the same depth and quality, allowing us to make a proper comparison. Preliminary results indicate that the field Kormendy relation *may* be brighter than the cluster relation, but more data is required to confirm the Kormendy relation's dependence on richness.

1. Introduction

The Kormendy relation is a projection of the fundamental plane (FP): a comparison of the Kormendy relation in clusters versus that in the field yields information about the environmental dependence of the structure and evolution of elliptical galaxies. The relatively few existing field samples for the Kormendy relation or FP are typically small ($\sim$ 20 galaxies or less), often over a large redshift range or binned by redshift into very small subsamples (e.g., Treu et al. 2001). Comparisons are nearly always done with cluster data from the literature, introducing many systematic effects (see Ziegler et al. 1999 for a good discussion). These effects are not always well understood, and prevent a reliable comparison. We present preliminary results of a consistent comparison of the Kormendy relation for cluster and field elliptical galaxies.

2. Sample and Analysis

The field sample is drawn from the CNOC2 survey. We make a redshift cut of $0.1 < z < 0.25$, and use CFH12k R-band images from the RCS survey (Gladders & Yee 2001). CFH12k data of Abell 2390 are courtesy of E. Ellingson et al. We use objects within $0.2175 < z < 0.2395$ (virial region, from Carlberg et al. 1996), and retain only those with redshifts from CNOC1 (Yee et al. 1996). For both field and cluster, early-type galaxies are selected using SED class of <0.5 (see Lin et al. 1999).

We use Brown's (1997) *jproto* package to fit ellipses to isophotes. If a residual shows azimuthal structure, we classify the object as 'non-elliptical' (S0,Sa) and remove it from our sample. Surface brightness profiles are fit to a convolution of a de Vaucouleurs law with the local PSF. If the profile is poorly fit, the object is rejected. The final samples consist of: 23 field galaxies, 15 cluster galaxies. Cluster Es are predominantly found inside the virial radius (Carlberg

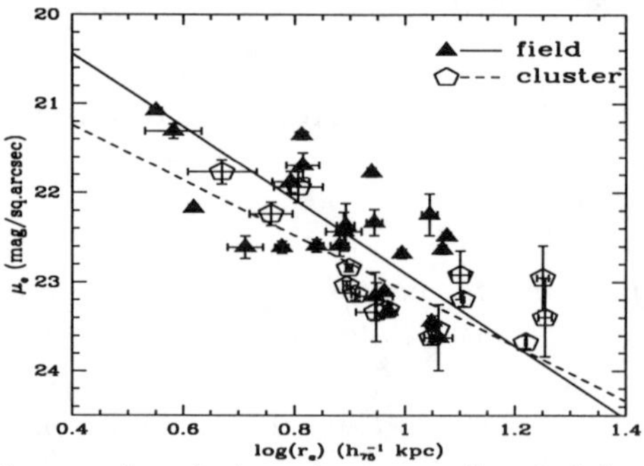

Figure 1. Kormendy relations (mean surface brightness μ_e vs. effective radius r_e) for field and cluster samples.

et al. 1996), and although the redshift selections for the field and cluster are different, the redshift distributions of the final samples are very similar.

3. Preliminary Results

The fraction of SED-selected early-types that we identify as Es is 41% in A2390 and 31% in the field; that is, *the agreement between SED-selection and morphological selection of ellipticals does not depend strongly on environment.* The fraction of all galaxies (i.e., the datasets obtained with the same criteria except for the SED-selection) that we identify as Es is 35% in A2390 and 14% in the field, roughly as expected (Dressler et al. 1997).

Figure 1 shows the Kormendy relations for our field and cluster samples. The best-fit lines differ slightly (1.4σ level). Thus, there may be a dependence of the Kormendy relation on richness: the field Kormendy relation is brighter (i.e., for a given effective radius, the field galaxies have higher surface brightness). Also, the Kormendy relation for clusters is shifted (or stretched) toward larger effective radii; field and cluster galaxies may inhabit different regions along the relation. *However*, to confirm the above suggestions requires more data. Future work will include analysis of the remaining CNOC2 patches, and other clusters at $z \sim 0.2$; this will double both the cluster and field samples, allowing us to make a robust determination of the Kormendy relation's dependence on richness.

References

Brown, J.P. 1997, Ph.D. Thesis, University of Toronto, Toronto, Canada

Carlberg, R.G., et al. 1996, ApJ, 462, 32

Dressler, A., et al. 1997, ApJ, 490, 577

Gladders, M.D., & Yee, H.K.C. 2001, in preparation

Lin, H., et al. 1999, ApJ, 518, 533

Treu, T., et al. 2001, MNRAS, 326, 237

Yee, H.K.C., et al. 1996, ApJS, 102, 289

Ziegler, B. L., et al. 1999, A&A, 346, 13

Tracing Cosmic Evolution with Galaxy Clusters
ASP Conference Series, Vol. 268, 2002
S. Borgani, M. Mezzetti, and R. Valdarnini, eds.

A Serendipitous Cluster Survey in Chandra Archival Data

W. Boschin[1]

Dipartimento di Astronomia, Università di Trieste, via Tiepolo 11, I-34131 Trieste, Italy

Abstract. I present and summarize the status, prediction and initial results of a serendipitous survey I have initiated using Chandra archival data. The main goal of the survey, is to select a sample of distant ($z \geq 0.5$) galaxy clusters in a field of about 5-10 square degrees. Based on an extrapolation of the data of the ROSAT Distant Cluster Survey I expect to identify about 100 galaxy clusters by using an automated procedure based on Voronoi tessellation.

1. Main features of selected Chandra fields

The survey is based on archival Advanced CCD Imaging Spectrometer (ACIS) X-ray data. The Chandra ACIS presents a field of view of 16.9 x 16.9 arcminutes^2 and a size of the instrumental point-spread function (PSF) smaller than 0.5 arcseconds at the aimpoint. This last feature makes the separation between pointlike and extended sources a much easier task than, for instance, in ROSAT PSPC data, where the size of the PSF is at best 25 arcseconds on-axis and so comparable with the typical angular sizes of distant clusters. Selected fields have exposure times larger than 10,000 seconds and high galactic latitude ($|b| \geq 20^o$) to minimize the absorption of X-rays from extragalactic sources by galactic neutral hydrogen. I use the $0.5 - 2$ keV band to detect sources rather than the full ACIS band on which the energy scale is calibrated ($0.3 - 10$ keV). This is done in order to reduce the X-ray background, to minimize the size of the PSF and to maximize sensitivity to distant clusters of galaxies. I reject fields whose targets dominate the ACIS field of view and make impossible the serendipitous detection of other objects. I also exclude fields with special instrument modes (e.g., grating observations with the chips S0-S5).

2. X-ray data analysis

Each field is analized as it becomes available in the archive using uniform X-ray source detection and analysis techniques to produce a well-defined sample of galaxy clusters over the full survey area. Archival data are first elaborated in a standard way (i.e. by filtering good time intervals, removing high background periods, bad columns and pixels, etc) in order to produce an event file

---

[1]e.mail:boschin@ts.astro.it

containing only the detector coordinates of the photons in the selected energy band. This is done by using the software CIAO (Chandra Interactive Analysis of Observations), expressely developed at the Chandra X-ray Center to manage Chandra data. Subsequently, source detection is performed by using the Voronoi Tessellation and Percolation technique (VTP, Ebeling 1993). I decided to use this procedure since it is completely non parametric, i.e. it is equally sensitive to symmetric and irregular sources, and because it detects more faint extended objects than conventional PSF/sliding cell methods (Scharf et al. 1997). The VTP procedure is implemented through an adaptation of a C code written by Ramella et al. (2001) to detect galaxy clusters in optical galaxy catalogs. The procedure is applied to the spatial photon distribution by using different surface brightness thresholds. This allows to distinguish real single sources from those composed of blends of several sources (pointlike or extended) and to reduce uncertainties in source identifications resulting from positive background fluctuations that become grouped with source photons. Finally, the list of deblended sources is passed to an algorithm that estimates their true flux, correcting the raw VTP count rates for the low-surface-brightness emission in the far wings of the sources that has not been detected directly. This is done by adopting the technique proposed by Ebeling et al. (1998). Their technique assumes a well defined surface brightness profile for galaxy clusters, a King profile (Cavaliere & Fusco-Femiano 1976), but since it involves only integral properties of the source spatial photon distribution the results are much more stable to deviations of objects from the assumed profile than in conventional fitting procedures.
In the end, for each source, I provide accurate positions, corrected count rates and fluxes and the spatial extent.

3. Present status and future perspectives

Since the Chandra PSF is a strong function of off-axis angle and energy, I am now carryng out simulations of source detection efficiency as a function of size, off-axis angle, background characteristics, etc. to understand survey limits and minimize false sources while optimizing source detection. When about two dozens of fields are processed, based on an extrapolation of the data of the ROSAT Distant Cluster Survey (Rosati et al. 1998) I expect to identify about 100 galaxy clusters in a field of 5-10 square degrees. The new cluster catalog will complement existing deep cluster catalogs and will contribute to constrain the poorly known faint end of the cluster X-ray luminosity function at high redshifts.

References

Cavaliere, A., & Fusco-Femiano, R. 1976, A&A, 49, 137

Ebeling, H. 1993, Ph.D. Thesis

Ramella, M., Boschin, W., Fadda, D., & Nonino, M. 2001, A&A, 368, 776

Rosati, P., della Ceca, R., Norman, C., & Giacconi, R. 1998, A&A, 331, 41

Scharf, C.A., Jones, L.R., Ebeling, H., Perlman, E., Malkan, M., & Wegner, G. 1997, ApJ, 477, 79

Tracing Cosmic Evolution with Galaxy Clusters
ASP Conference Series, Vol. 268, 2002
S. Borgani, M. Mezzetti, and R. Valdarnini, eds.

The Properties of BCGs in X-Ray Selected Clusters

Sarah Brough

Astrophysics Research Institute, Liverpool John Moores University, England.

Chris A. Collins

Astrophysics Research Institute, Liverpool John Moores University, England.

Doug J. Burke

Harvard-Smithsonian Center for Astrophysics, Cambridge, USA.

Paul D. Lynam

Max-Planck-Institut fur Extraterrestriche Physik, Garching, Germany.

Robert G. Mann

Institute for Astronomy, University of Edinburgh, Edinburgh, England.

Abstract. We present the K-band Hubble diagram for 162 brightest cluster galaxies (BCGs) in X-ray selected clusters, $0.01 < z < 0.83$. The sample incorporates that of Burke, Collins, & Mann (2000) and includes additional infrared data from the 2MASS extended source catalogue. We show that below $z \sim 0.1$ the BCGs show no correlation with their environment, however, above $z \sim 0.1$ BCGs in more X ray luminous clusters are more uniform in their photometric properties. This suggests that there may be two populations of BCGs which have different evolutionary histories.

1. Introduction

Brightest cluster galaxies (BCGs) provide a unique sample with which to study galaxy evolution in a cluster environment. In a sample of 78 BCGs, $0.05 < z < 0.8$, in X-ray selected clusters, Collins & Mann (1998) and Burke, Collins, & Mann (2000) observed a split with cluster properties in their K-band Hubble diagram. This work substantially increases their sample with 84 BCGs below $z \sim 0.1$ from the *ROSAT* All Sky Survey, with K_s-band data from the 2MASS catalogue. The colour correction between the 2MASS K_s-band and the K-band is negligible (Carpenter 2001) and is henceforth neglected. The 2MASS magnitudes have been checked for consistency against data from the 2m UH telescope. A full discussion of this work will be presented in Brough et al., in preparation.

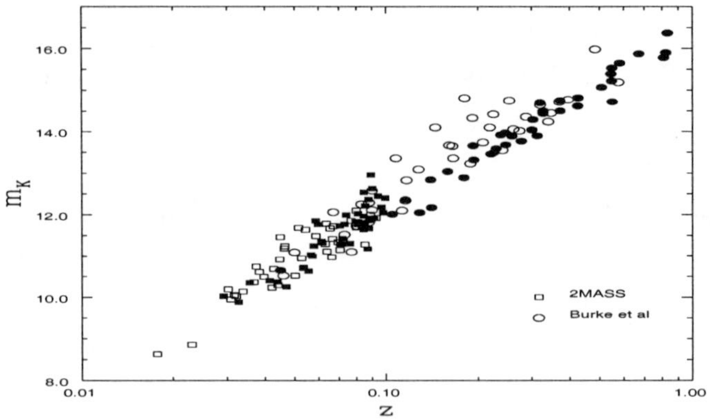

Figure 1. The Hubble diagram. The filled points denote BCGs in clusters with $L_X(0.3 - 3.5keV) > 2.3 \times 10^{44} ergs^{-1}$

2. Results and Conclusions

Fig. 1 shows the K-band Hubble diagram for 162 BCGs. At redshifts beyond $z > 0.1$ BCG magnitudes depend on the X-ray luminosity of the host environment, as suggested by Collins & Mann (1998). In particular, clusters with $L_X(0.3 - 3.5keV) > 2.3 \times 10^{44} ergs^{-1}$ have a scatter of 0.24 mag, compared to those in the low luminosity clusters which have an rms dispersion of 0.5 mag and a Kolmogorov-Smirnov test shows that BCGs in the different environments are different at > 99.9 per cent significance. However, the 97 clusters below z=0.1 indicate that locally BCG magnitudes have a dispersion of 0.3 mag and are uncorrelated with the X-ray luminosity of their host cluster. A KS test shows that they are drawn from the same population at 90 per cent level. These differences suggest that there may be two populations of BCGs with different evolutionary histories: BCGs in less X-ray luminous clusters evolving through a process of mergers and those in more luminous clusters evolving passively. This result has also been seen by Zaritsky et al. in this meeting (astro-ph/0108152).

Acknowledgments

This publication makes use of data products from 2MASS which is a joint project of the University of Massachusetts and the Infrared Processing and Analysis Center/California Institute of Technology, funded by the National Aeronautics and Space Administration and the National Science Foundation.

References

Brough, S. Burke, D.J. Collins, C.A. Lynam, P.D. Mann, R.G. 2001, in prep.

Burke, D.J. Collins, C.A. Mann, R.G. 2000, ApJ, 532, L105

Carpenter, J.M. 2001, AJ, 121, 2851

Collins, C.A. Mann, R.G. 1998, MNRAS, 297, 128

Tracing Cosmic Evolution with Galaxy Clusters
ASP Conference Series, Vol. 268, 2002
S. Borgani, M. Mezzetti, and R. Valdarnini, eds.

The cluster of galaxies Abell 970

Hugo V. Capelato [1], Laerte Sodré Jr.[2], Dominique Proust [3]
Gastão B. Lima Neto [2], Hector Cuevas [2 4]
Hernan Quintana[5] & P. Fouqué [6]

[1] *Divisão de Astrofisica -Instituto de Pesquisas Espaciais, Brazil*
[2] *Instituto Astronômico e Geofísico, Universidade de São Paulo, Brazil*
[3] *Observatoire de Paris - Section de Meudon, France*
[4] *Departamento de Física, Universidad de La Serena, Chile*
[5] *Departamento de Astronomia y Astrofísica, Pontificia Universidad Católica, Chile*
[6] *European Southern Observatory, Santiago, Chile.*

Abstract. We examine the properties of the cluster Abell 970, using a set of 69 new radial velocities. Our analysis indicates that this cluster has a substructure and is out of dynamical equilibrium.

1. Introduction

The X-ray cluster Abell 970 (richness class $R = 0$; B-M III), together with A979, A978 and A993, forms the Sextans supercluster. It is centered on an E/D galaxy which is the the second brightest of the cluster. For this cluster, we have obtained a set of new radial galaxy radial velocities, using the 1.52m ESO telescope at La Silla (Chile), the french 2.0m telescope at Pic du Midi and the 1.93m telescope at Haute-Provence Observatory.

2. The Galaxy Distribution and Kinematics: Substructures

Figure 1 below present adaptative kernel maps of *the projected density* and *the mean radial velocity* (calculated as local kernel averages) of a complete sample of galaxies kinematically linked to the cluster (cf. Sodré et al. for details). These plots cover a region of $\sim 3.1h_{50}^{-1}$ Mpc and are centered on an E/D galaxy which is the the second brightest cluster galaxy ($b_J = 16.57$)

As it can be seen, there is a substructure at NW, near the brightest cluster galaxy, a S0/S type of magnitude $b_J = 16.42$.

The X-ray isophotes as seen by the Einstein IPC detector and confirmed by recent BeppoSax observations (see Lima Neto et al, this volume) are also regular but their peak is not coincident with that of the projected density distribution, being slightly displaced towards the NW substructure associated to the cluster brightest galaxy. Since the relaxation time of the hot gas is much shorter than that of the galaxies, the relative displacement of the peaks may be an evidence of a state of non-equilibrium in the galaxy distribution, as expected

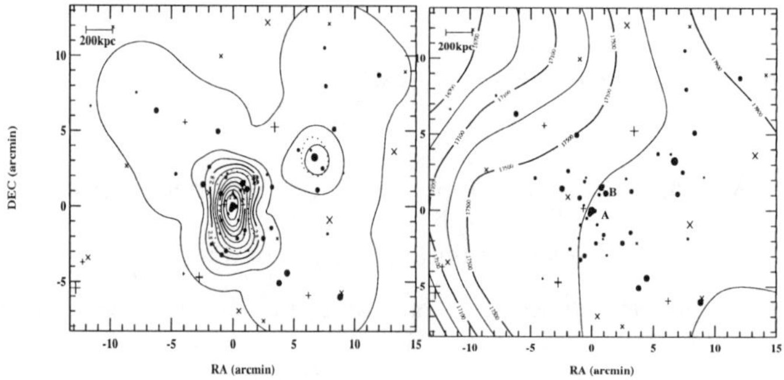

Figure 1. Left: Projected density map of the cluster galaxies brighter than $b_J = 18.9$ (filled circles). Right: the mean velocity adaptative kernel map corresponding to this sample; back/foreground galaxies are denoted by crosses/pluses. The NW group is centered at about (+7,+3) in these maps

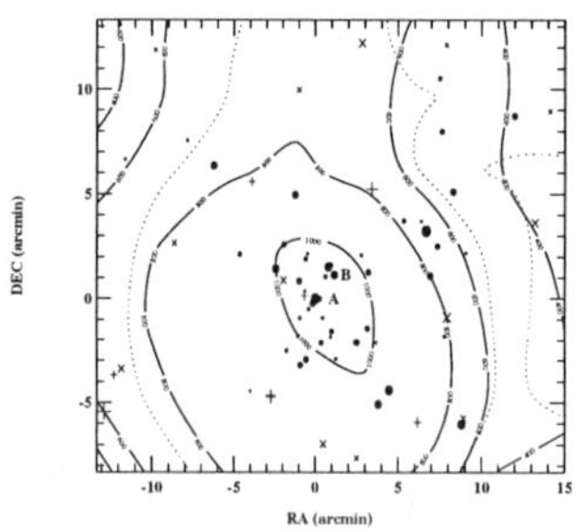

Figure 2. Mean velocity dispersion map for the same sample of Figure 1. Symbols are the same as above. As in Figure 1, coordinates are given as offsets relative to the central E/D galaxy and the contours are given in units of $\mathrm{km\,s^{-1}}$. Notice that the core of the clusters seems dominated by 2 small clumps of galaxies (A and B in these figures). We were not able to disprove the reality of clump B.

if the substructure associated to the brightest galaxy has only recently fallen into the cluster (see also the caption of Figure 2).

The *mean velocity map* (Figure 1) indicates the existence of a velocity gradient across the field, roughly in the E-W direction: **a)** galaxies with $v > 18500\ \mathrm{km\,s^{-1}}$ populate predominantly the north/west-side of the field; **b)** the mean velocity of the NW group detected in the density map is consistent with such a gradient. This gradient is also consistent with the general gradient one would obtain by considering all galaxies of our catalogue with velocities between 12000 and 22000 $\mathrm{km\,s^{-1}}$, suggesting that the cluster may be part of a larger structure running more or less in the E-W direction within, at least, this velocity range. In fact a (smaller) velocity gradient, in this same general direction, is also depicted by the mean redshifts of the supercluster members.

Figure 2 above gives the *mean dispersion velocity* map (also calculated by local kernel averages). It displays a negative radial gradient, which is greater in the E-W direction. This is expected if the cluster has grown through capture of low velocity dispersion groups.

Acknowledgments. We thank the support from FAPESP and CNPq as well as from the USP/COFECUB cooperation and PRONEX.

References

Sodré L., Jr., et al., 2001, A&A in press

Tracing Cosmic Evolution with Galaxy Clusters
ASP Conference Series, Vol. 268, 2002
S. Borgani, M. Mezzetti, and R. Valdarnini, eds.

Testing Cosmological Models with Negative Pressure

Alberto Cappi

Osservatorio Astronomico di Bologna, via Ranzani 1, I-40127 Bologna (Italy)

Abstract. There is now strong evidence that the main contribution to the cosmic energy density is not due to matter, but to a dark component with negative pressure. We briefly discuss the behaviour of different $w < 0$ models, focussing on the Alcock–Paczyński test and VCDM models.

1. The Alcock–Paczyński test

What is the dominant component of the cosmic energy density? The first possible answer is that it is a positive cosmological constant Λ associated to the vacuum energy, with $w \equiv P/\rho c^2 = -1$ (ΛCDM model). Another possibility is "quintessence" (Caldwell et al. 1998), i.e. a scalar field evolving in a potential, coupled to matter through gravitation, spatially inhomogeneous and slowly evolving with time, with $-1 < w < 0$ (QCDM model). Caldwell even suggested a component with $w < -1$; he named it "phantom energy", as in his model there is a negative kinetic term in the Lagrangian of the scalar field, but $w < -1$ can also be obtained if the vacuum energy is due to a quantized free scalar field of low mass (VCDM model, Parker & Raval 2001).

Alcock & Paczyński (1979) showed that in principle the value of Ω_Λ can be inferred from a galaxy redshift survey through a geometric test independent of galaxy evolution: in fact the spatial distribution of galaxies derived using the $D - z$ relation for an Einstein–de Sitter model is artificially distorted if the universe has a Λ component. The problem is that distortions are also induced by peculiar velocities –β distortion ($\beta \equiv \Omega_M^{0.6}/b$) on linear scales and fingers of God on small scales–, but the method might be applied to future redshift surveys, with the simultaneous extraction of Λ and β from anisotropic power–spectrum data. The distortion can be quantified by a flattening factor $F(z)$ relative to the EdS model. At first order, for Q models we have: $F(z) = 1 + (1/4)[1 - \Omega_M - (1 + 3w_Q)\Omega_Q]z + O(z^2)$. In fig.1a–c $F(z)$ is shown for different flat models. There are 3 main systematic effects due to w, shown in fig.1d for flat models with $\Omega_M = 0.3$. With a more negative w we have that: **a)** the amplitude of the maximum anisotropy becomes (slightly) smaller; **b)** the maximum of $F(z)$ shifts towards smaller redshifts; **c)** after the maximum, $F(z)$ decays faster. The amplitude is not very large ($\sim 20\%$), but for the VCDM models the maximum is reached at low z: $z_m \sim 0.32$ if $w_V = -3$. The effect could be detected in on–going galaxy redshift surveys, depending on our ability to disentangle the geometric distortion from the redshift distortion due to peculiar velocities.

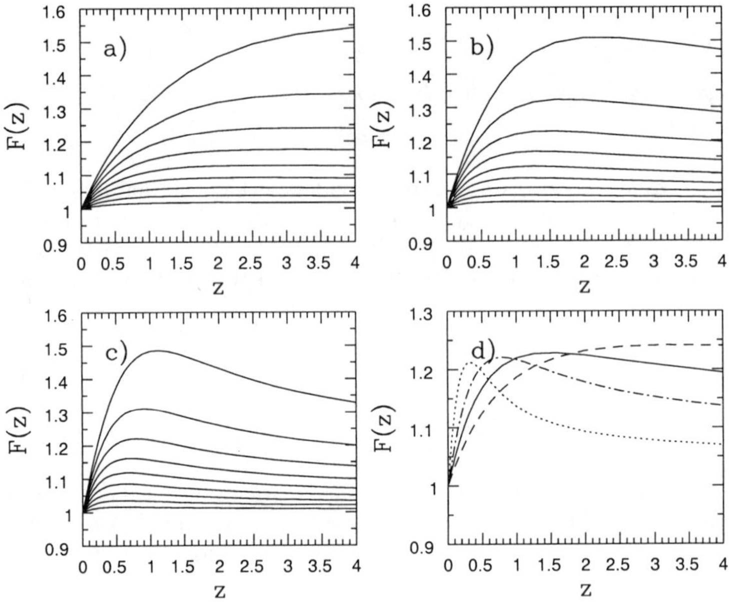

Figure 1. Flattening factor $F(z)$ for flat models; in (a–c), Ω_M varies from 0.1 (top) to $\Omega_M = 0.9$ (bottom), step=0.1. a) QCDM ($\Omega_M+\Omega_Q = 1$, $w_Q = -2/3$); b) ΛCDM ($\Omega_M + \Omega_\Lambda = 1$); c) VCDM ($\Omega_M + \Omega_V = 1$, $w_V = -3/2$); d) 4 flat models with $\Omega_M = 0.3$. Dashed curve: QCDM, $w_Q = -2/3$; solid curve: ΛCDM; dashed–dotted curve: VCDM, $w_V = -3/2$; dotted curve: VCDM, $w_V = -3$.

References

Alcock C., Paczyński B., 1979, Nature 281, 358
Ballinger W.E., Peacock J.A., Heavens A.F., 1996, MNRAS 282, 877
Caldwell R.R., 1999 (astro–ph/9908168)
Caldwell R.R., Dave R., Steinhardt P.J., 1998, Phys.Rev.Lett. 80, 1582
Cappi A., 2001, Astrop. Lett. & Comm., in press (astro–ph/0105382)
Parker L., Raval A., 2001, Phys. Rev. Lett. 86, 749

Tracing Cosmic Evolution with Galaxy Clusters
ASP Conference Series, Vol. 268, 2002
S. Borgani, M. Mezzetti, and R. Valdarnini, eds.

Comparing the Star Formation in the Coma Cluster and the Field

Francisco J. Castander

Yale University, P.O. Box 208101, New Haven, CT 05620-8101, USA & Universidad de Chile, Casilla 36-D, Santiago, Chile

Andes Prize Fellow

Robert C. Nichol, Aronne Merrelli

Carnegie-Mellon University, 5000 Forbes Ave, Pittsburgh, PA 15232, USA

and the SDSS Collaboration

Abstract. We present spectroscopic observations of the Coma Cluster of Galaxies and the local field taken with the Sloan Digital Sky Survey (SDSS) spectroscopic systems. A total of 196 spectra in the Southern part of Coma have been obtained and we compared them to approximately 3000 spectra taken in the field at the same redshifts.

1. Introduction

The Sloan Digital Sky Survey (SDSS; York et al 2000) saw its first spectroscopic light in June 1999. As part of the spectroscopic commissioning the Coma cluster of galaxies was observed. These observations yielded a total of 196 spectra of the brightest galaxy members ($b < 18$) in the Southern part of the Coma cluster (Castander et al 2001). Our observations extend out to a radius of 1.5 deg (1.8 h^{-1} Mpc) slightly larger than the Virial radius.

The SDSS has now obtained over a hundred thousand spectra of galaxies. Among this impressive database, we have selected approximately 3000 spectra that share the same magnitude distribution and redshift range ($0.01 < z < 0.04$) as the Coma galaxies observed.

The SDSS spectrographs deliver a spectral resolution of $R \sim 1800$ across its entire spectral range of 3800 $\rightarrow$ 9100 Å (Uomoto et al 2001, in preparation). The combination of both spectral coverage and spectral resolution allow us to study in detail the star formation mechanisms in the galaxies observed.

Our observations spanning from the Coma cluster core to its Virial radius and the coverage of the NGC4839 subgroup allow us to study the dependency of galaxy properties as a function of cluster environment and compare them to those in the field. In the next section, we present a brief summary of the star formation properties in the Coma cluster compared to the field.

2. Results and Conclusions

- 25% of observed Coma galaxies show signs of star formation, including actively star forming galaxies (20%) and poststarburst galaxies (5%)
- We find no galaxies actively forming stars in the inner $200h^{-1}$ kpc of the cluster.
- Excluding the center, the spatial distribution of active galaxies looks homogeneous, with a slight increase of the fraction of active galaxies with radius.
- No significant increase of star formation is observed around the NGC4839 subgroup.
- There is no significant star formation activity in the cluster regions where the gas density is denser than $\sim 1 \times 10^{-3} h^{1/2}$ cm^{-3}, or one fourth of the central gas density.
- The distribution of star formation rates is similar between active galaxies in Coma and in the field. There is no indication of the cluster medium triggering star formation.
- Most of the emission lines galaxies in Coma have ionization properties similar to HII regions, but there are a few with AGN activity. The ionization properties look slightly different in Coma and the field although they are consistent given the errors.

The Sloan Digital Sky Survey (SDSS) is a joint project of The University of Chicago, Fermilab, the Institute for Advanced Study, the Japan Participation Group, The Johns Hopkins University, the Max-Planck-Institute for Astronomy (MPIA), the Max-Planck-Institute for Astrophysics (MPA), New Mexico State University, Princeton University, the United States Naval Observatory, and the University of Washington. Apache Point Observatory, site of the SDSS telescopes, is operated by the Astrophysical Research Consortium (ARC). Funding for the project has been provided by the Alfred P. Sloan Foundation, the SDSS member institutions, the National Aeronautics and Space Administration, the National Science Foundation, the U.S. Department of Energy, the Japanese Monbukagakusho, and the Max Planck Society. The SDSS Web site is http://www.sdss.org/.

References

Castander, F.J, Nichol, R.C., Merrelli, A. et al, 2001, AJ, 121, 2331

York, D.G., et al, 2000, AJ, 120, 1579

Tracing Cosmic Evolution with Galaxy Clusters
ASP Conference Series, Vol. 268, 2002
S. Borgani, M. Mezzetti, and R. Valdarnini, eds.

Distribution of baryonic and non-baryonic matter in clusters of galaxies

Africa Castillo-Morales
Astrophysics Research Institute, Liverpool John Moores University, England

Sabine Schindler
Astrophysics Research Institute, Liverpool John Moores University, England

Abstract. We present the analysis of baryonic and non-baryonic matter distributions in a sample of eleven nearby clusters (0.03 < z < 0.09) with temperatures between 4.4 and 9.4 keV. These galaxy clusters have been studied in detail using X-ray data and global physical properties have been determined. Correlations between these quantities have been analysed and compared with the results for distant clusters. We found an interesting dependence between the relative gas extent (expressed as the ratio of gas mass fractions at r_{500} and $0.5 \times r_{500}$) and the total cluster mass. The extent of the gas relative to the extent of the dark matter tends to be larger in less massive clusters. This dependence might give us some hints about non-gravitational processes in clusters.

1. Introduction

The study of correlations between different physical properties in clusters of galaxies, as X-ray luminosity, temperature, total mass, gas mass and relative gas extent enable us to draw conclusions about their formation processes, about the thermal history of clusters and hence about cosmological parameters. We selected a sample of nearby clusters of galaxies from ROSAT archive, in which an accurate total mass determination is possible, i.e with relaxed and symmetric morphologies, good temperatures measurements and good surface brightness profiles. To calculate the gas density profiles the standard β-model has been used. All the masses have been calculated in a consistent way and within equivalent volumes for all the clusters assuming isothermality. We analyse relations between different properties and compare them with the relations in the more distant sample (0.3 < z < 1.0) by Schindler 1999 (red circles). We included in the comparison two more distant clusters: RBS797 (z=0.35) (blue circle) analysed with a CHANDRA observation by Schindler et al. 2001 and the cluster RXJ0849+4452 (z=1.26) (green circle) where we have calculated the total and gas mass using the parameters from the CHANDRA data analysis by Stanford et al. 2000. Our preliminary results are presented here. The cosmology used is: $H_0 = 50$ km/s/Mpc and $q_0 = 0.5$.

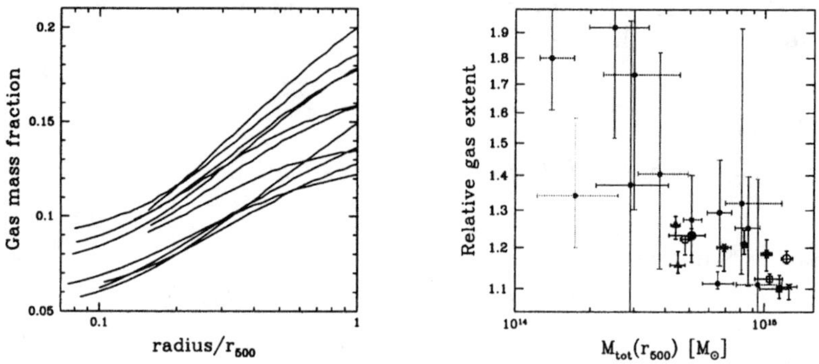

Figure 1. Left: gas mass fraction profiles derived for the nearby cluster sample. Right: Ratio of gas mass fraction at r_{500} and $r_{500}/2$ as a measure for the relative gas extent versus total cluster mass at r_{500}.

2. Results and Discussion

We find inside each cluster the gas mass fraction increases with radius (Fig. 1 left). The gas mass fraction at radius r_{500} is compared with the gas mass fraction at $0.5 \times r_{500}$ in Fig. 1 right (black symbols represent the nearby sample). The ratio of these fractions is a measure of how fast the gas mass fraction is increasing with radius and as such a measure for the extent of the gas distribution with respect to the dark matter distribution. For all the clusters in our sample the relative gas extent $E = \frac{f_{gas}(r_{500})}{f_{gas}(r_{500}/2)}$ is larger than 1. This means that in general the gas distribution is more extended than the dark matter in agreement with results from other authors. This relative gas extent E shows a mild dependence on the total mass. Clusters with larger masses tend to have smaller relative gas extents. Clusters RX J0849+4452 and RBS797 seem to lie on the same trend as the nearby sample. This trend can be explained by the physical processes in the gas, which are assumed to be responsible for the increase of gas mass fraction with radius e.g. energy input by supernovae driven galactic winds. If gas is placed artificially into a model cluster potential in hydrostatic equilibrium the distributions of gas and dark matter have the same slope at radii larger than the core radius, therefore one would expect a priori a ratio $E \approx 1$. It might be that this additional heat input affects low mass clusters more that massive clusters, so that a massive cluster can maintain a ratio $E = 1$ while in the smaller clusters the gas is more extended.

References

Schindler, S. 1999, ApJ, 349, 435

Schindler, S. et al. 2001, A&A, in press, astro-ph/0107504

Stanford, S.A. et al. 2001, ApJ, 552, 504

Tracing Cosmic Evolution with Galaxy Clusters
ASP Conference Series, Vol. 268, 2002
S. Borgani, M. Mezzetti, and R. Valdarnini, eds.

The TexOx Survey of Radio-Selected Galaxy Clusters

Steve Croft, Steve Rawlings

Astrophysics, Physics Department, Keble Road, Oxford OX1 3RH, UK

Gary J. Hill, Pamela L. Gay, Joseph R. Tufts

McDonald Observatory, Department of Astronomy, University of Texas at Austin, RLM 15.308, TX 78712-1083, USA

Abstract. We present some initial results from the TexOx (Texas–Oxford) Cluster (TOC) survey – a new method of selecting distant galaxy clusters. The cosmic evolution of the radio source population suggests that some massive clusters at high redshift will contain several radio-loud AGN. We searched for extreme over-densities at $\sim$ mJy levels in $7' \times 7'$ boxes within the NVSS radio catalogue, covering a large ($\sim$ 1100 deg^2) sky area. We have acquired optical images for $\sim$130 cluster candidates, and followed up a subset of these with the VLA, and with Calar Alto near-IR imaging. Ryle Telescope observations have yielded at least one Sunyaev-Zel'dovich (SZ) detection of a massive $z \sim 1$ system. Spectroscopic follow-up with 8-m class telescopes is in progress.

1. Introduction

Traditional methods of finding clusters, such as searching for peaks in galaxy counts within wide-field optical surveys, become increasingly difficult at high redshift due to contamination by foreground galaxies. The redshift distribution of faint radio AGN means that contamination effects are much reduced for any search for an over-density at redshift $z \sim 1$. The availability of wide-field radio surveys like the 1.4 GHz NRAO VLA Sky Survey (NVSS; Condon et al. 1998), suggests an alternative way of finding clusters which we have been pursuing.

2. The TexOx Cluster Survey and Follow-up Observations

If the NVSS catalogue were gridded into $7' \times 7'$ boxes, we would expect the mean number of sources per box to be $\sim$ 0.8. We created a catalogue of candidate clusters by searching within $7' \times 7'$ boxes, centred on each NVSS source, for an additional four or more sources. In some cases, large low-redshift radio structures are seen by NVSS as several "individual" sources, but these were easily eliminated from our cluster sample by inspection of the NVSS images. The remainder of the fields were our candidate galaxy clusters. At the depth of the NVSS, there is a reasonable probability that the richest clusters will contain several radio-loud AGN, and this probability is not a strong function of redshift out to $z \sim 1$ because of the cosmic evolution of the radio source population.

The cluster candidates were imaged over 30 clear dark nights with the 2.7-m telescope at McDonald Observatory, with typical exposures of 30 – 60 minutes per band in R and I (reaching a limiting sensitivity of $R \sim 24$). We recover some known low-redshift clusters, including several in the Abell catalogue at $z \sim 0.1$, and the $z = 0.37$ cluster around the quasar 3C48. Many of the fields contained optically-obvious clusters near the magnitude limit of our observations. Although any survey using AGN to find clusters will find only a subsample of the total population, the technique has proved very efficient at finding real clusters.

The cluster radio sources are typically associated with galaxies of $R \sim 23$, with the brightest cluster member a magnitude or two brighter, suggesting a typical redshift around 0.5. Importantly, about 25% of TexOx clusters have no radio source identifications to the limits of our 2.7-m observations, suggesting they are at $z > 0.7$. For those fields which we were convinced contained real but distant clusters, we used the Calar Alto 3.5-m telescope to obtain near-IR images. Our collaborators in Cambridge then used the Ryle telescope to try to detect Sunyaev-Zel'dovich (Sunyaev and Zel'dovich, 1972) decrements in some of these. This yielded a detection of an object with a minimum gas mass of around $5 \times 10^{13} M_{\odot}$, in which five of the NVSS sources are identified with host galaxies of $R \approx 24, J \approx 20, K \approx 18$ – presumably a rich cluster at $z \sim 1$ (Cotter et al. 2001; Croft et al. 2001 in prep.). We are using Gemini and the 9.2-m Hobby-Eberly Telescope to get spectroscopic redshifts for this and similar systems.

3. Conclusions

Targeting over-densities in the NVSS provides a powerful way to find high-redshift clusters, with different biases to X-ray, optical or infrared approaches. Our TexOx Cluster (TOC) survey will provide a unique sample of clusters, including some at $z > 1$, which can be used to confirm predictions of Large Scale Structure simulations, to analyse cluster properties, and to study the correlation between cluster dynamics, the intra-cluster gas, and AGN activity.

Acknowledgments. We thank Niv Drory, our collaborator on the Calar Alto programme, and Garret Cotter, Helen Buttery, Rhiju Das, Keith Grainge, William Grainger, Michael Jones, Guy Pooley and Richard Saunders, our collaborators in the Ryle Telescope group. This material is based in part upon work supported by the Texas Advanced Research Program under Grant No. 009658-0710-1999. SC thanks PPARC for the support of a PhD studentship.

References

Condon, J. J., Cotton, W. D., Greisen, E. W., Yin, Q. F., Perley, R. A., Taylor, G. B. & Broderick, J. J. 1998 AJ, 115, 1693

Cotter, G., Buttery, H. J., Rawlings, S., Croft, S., Hill, G. J ., Gay, P. L., Das, R., Drory, N., Grainge, K., Grainger, W. F., Jones, M. E., Pooley, G. G. & Saunders, R. 2001 MNRAS, accepted, astro-ph/0109506

Sunyaev, R. & Zel'dovich, Y. 1972 Comments Astrophys. Space Phys., 4, 173

Tracing Cosmic Evolution with Galaxy Clusters
ASP Conference Series, Vol. 268, 2002
S. Borgani, M. Mezzetti, and R. Valdarnini, eds.

Galaxy morphological segregation in clusters: local vs global conditions

Mariano Domínguez, Hernán Muriel and Diego García Lambas

IATE, Observatorio Astronómico Córdoba - CONICET

Abstract. We study the relative fraction of galaxy morphological types in clusters, as a function of different parameters: local galaxy density, and cluster projected total mass density. Our results suggest that the morphological segregation in clusters is controlled by the local galaxy density in the outskirts. On the other hand, the global projected mass density, shows the strongest correlation with the fraction of morphological types in the central high density region, with a negligible dependence on the local galaxy density.

Our analysis is aimed to provide a better understanding of the relevance of the proposed mechanisms involved in morphological segregation in clusters given their different dependence on gas, mass and galaxy content. The data consists of the clusters originally analyzed by Dressler (1980a) restricted to those objects with determination of its mean temperature.

In order to define the local galaxy density Σ_{Gal} in the position of each galaxy in our sample, we proceed in the same way as proposed by Dressler et al 1997. We compute the area that comprises the ten nearest galaxies around each object and correct for completeness in luminosity and contamination from projections using the same methods described by Whitmore et al 1993.

Based on numerical simulations, Evrard et al. (1996) predict that the average cluster temperature T_X strongly correlates with r_{δ_c}, and propose the following relation:

$$r_{\delta_c}(T_X) = r_{10}(\delta_c) * (T_X/10keV) \tag{1}$$

The normalization $r_{10}(\delta_c)$ corresponds to the radial scale of 10 keV clusters at density contrast δ_c and we adopt a standard overdensity parameters $\delta_c = 500$.

Bartelmann (1996) has derived the projected density profile corresponding to the NFW analytic function for dark matter haloes:

$$\Sigma_{Mass}(x = r/r_{\delta_c}) = \frac{2\rho_s r_{\delta_c}}{x^2 - 1} * \begin{cases} 1 - \frac{2}{\sqrt{x^2-1}} arctanh\sqrt{\frac{x-1}{x+1}}, & \text{if } x > 1 \\ 1 - \frac{2}{\sqrt{1-x^2}} arctanh\sqrt{\frac{1-x}{1+x}}, & \text{if } x < 1 \\ 0, & \text{if } x = 1 \end{cases} \tag{2}$$

We consider the mean cluster temperatures compilation in our sample and we use equations (1) and (2) to correlate projected mass density Σ_{Mass} and the relative fraction of morphological types in clusters. However, given the correlation betwen Σ_{Mass} and Σ_{Gal}, it is important to consider the relevance of each of the parameters for the morphological segregation, ie. is whether one of these parameters is primary or secondary.

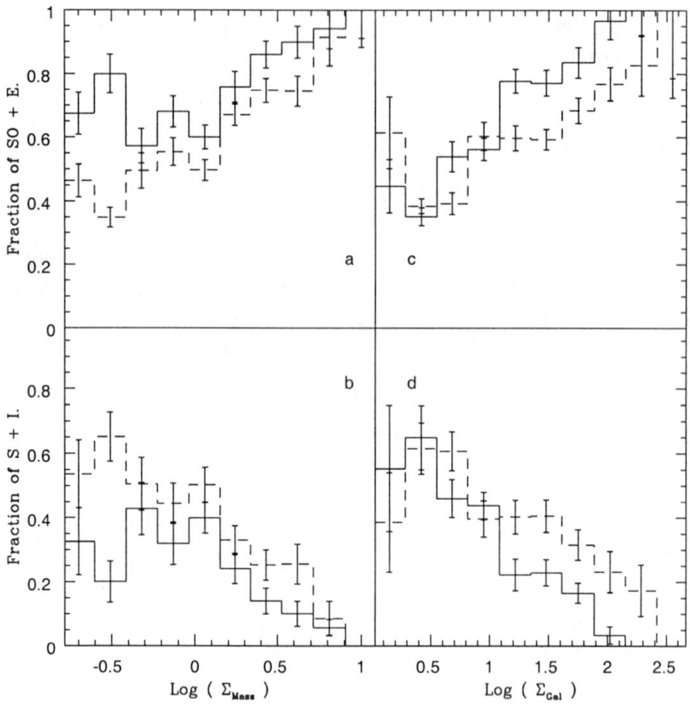

Figure 1. 1a/1b: The fraction of E+S0 and S+I as a function of the proyected mass density for galaxies at high/low (continous/dashed line) projected local galaxy density. 1c/1d: The fraction of E+S0 and S+I as a function of the proyected local galaxy density for galaxies at high/low (continous/dashed line)projected mass density.

The results of our test (see figure 1) strongly suggest that the local galaxy density can not account as a single parameter for the morphological segregation in clusters of galaxies. Nevertheless, Σ_{Gal} is a significat parameter that accounts for the MS in the outskirts of clusters, while Σ_{Mass} provides a primary parameter in the MS of the virialized region of clusters.

References

Bartelmann M. AA. 313, 697.

Dressler A. 1980a. Ap.J Suppl. 424, 564.

Dressler A. 1980b, Ap. J. 236, 351.

Dressler A. et al 1997, Ap. J. 490, 577.

Evrard A.E., Metzler C.A. and Navarro J.F. 1996. Ap.J. 469,494.

Whitmore, B., Gilmore, D. and Jones, C. 1993, Ap.J. 407,489. (WGJ)

Tracing Cosmic Evolution with Galaxy Clusters
ASP Conference Series, Vol. 268, 2002
S. Borgani, M. Mezzetti, and R. Valdarnini, eds.

The Atacama Submillimeter Telescope Experiment (ASTE) and Observations of Galaxy Clusters with Submillimeter Waves

Hajime Ezawa

Nobeyama Radio Observatory, Minamisaku, Nagano 384-1305, Japan

1. The ASTE project

The ASTE (Atamaca Submillimeter Telescope Experiment) is a project to install and operate a 10 m millimeter and submillimeter telescope at Pampa la Bola (4800 m altitude), in Atacama area, Northern Chile. This is a joint project among University of Chile, National Astronomical Observatory of Japan (NAOJ), University of Tokyo, Nagoya university, Osaka Prefecture University, Chiba University, and Toyota National College of Technology. The telescope (Ukita et al. 2000; Ezawa et al. 2000) was first installed at Nobeyama Radio Observatory (NRO) in Japan in 2000 for basic evaluation. The plan is to relocate the telescope to Pampa la Bola, and receive the first light in March 2002. The final design of the telescope is to be equipped with heterodyne receivers, multi-color bolometer at 350, 650, and 850 GHz, and a submillimeter camera for 350 and 650 GHz.

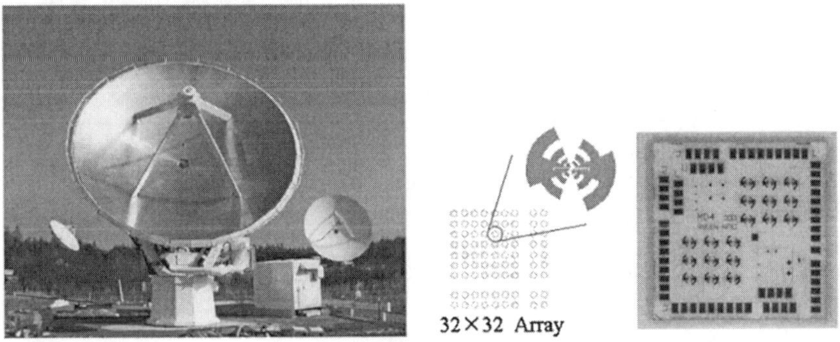

Figure 1. The 10-m sumbillimeter telescope at NRO site (left) and the schematic of the submillimeter camera and its prototype (right).

The submillimeter camera (Matsuo et al. 2001) is a focal plane array of superconducting tunnel junctions utilized as direct photon directors. The camera system will be equipped with two arrays, each for 350 GHz ($\Delta\nu = 40$ GHz) and 650 GHz ($\Delta\nu = 60$ GHz). The final design of the camera will have 32×32 pixels on the focal plane which covers $10' \times 10'$ on the sky.

2. Cluster Observation with ASTE

Sunyaev-Zel'dovich (S-Z) effect is one of the best probe to survey galaxy clusters at high redshift. The large field of view of the submillimeter camera ($10' \times 10'$), enables us to explore wide area on the sky within certain observation time. The estimated number of galaxy clusters observed by ASTE, based on Kitayama et al. (1997), is shown in Figure 2(a), which shows that more than 100 clusters is expected to be detected within a survey for 30 deg^2.

Dust emission from galaxies may contaminate the cluster survey. Since dusts have different SED from S-Z effect (Figure 2(b)), we can distinguish them from cluster S-Z by comparing 350 and 650 GHz results by the ASTE submillimeter camera. The number of galaxies detected within the FOV is estimated to be 8 at 350 GHz and 80 at 650 GHz (see also Takeuchi et al. (2001)). These are negligibly small compared to the number of pixels on the camera (1000).

The bolometer array on board NRO 45 m telescope (NOBA; Kuno et al. 1993) has good sensitivity at 150 GHz where the S-Z decrement becomes maximum. Combining NOBA observation with 350 GHz and 650 GHz results from ASTE, we can explore the hot gas in clusters more effectively, with increased accuracy to filter out false detections caused by background galaxies.

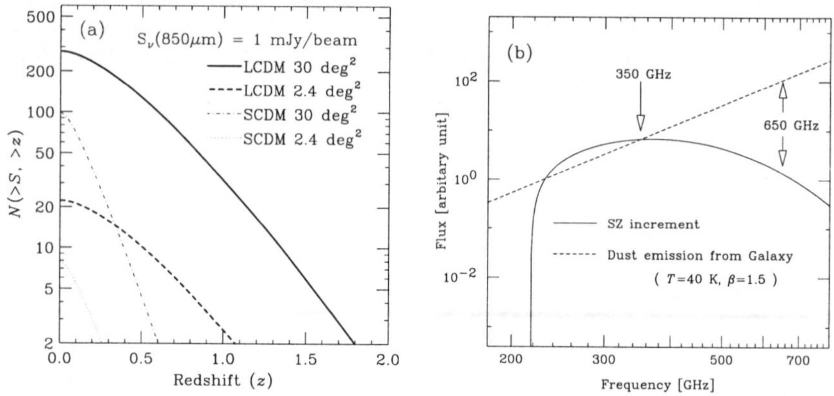

Figure 2. (a) Estimated number of clusters observed by ASTE submillimeter camera, and (b) SED of the S-Z effect and dust emission from typical URFIR galaxies.

References

Ezawa, H., Ishiguro, M., Matsuo et al. 2000, Proc SPIE 4015, 515

Kitayama, T., Sasaki, S., Suto, Y. 1997, PASJ 50, 1

Kuno, N., Matsuo, H. et al. 1993, Int. J. Infrared Millimeter Waves 14, 749

Matsuo, H., et al. 2001, IEEE Trans. Appl. Supercond. 11, 688

Takeuchi, T., Kawabe, R., Kohno, K., Nakanishi, K. et al. 2001, PASP 113, 568

Ukita, N., Kawabe, R., Ishiguri, M. et al. 2000, Proc. SPIE 4015, 177

Tracing Cosmic Evolution with Galaxy Clusters
ASP Conference Series, Vol. 268, 2002
S. Borgani, M. Mezzetti, and R. Valdarnini, eds.

Alignment of Main Axes of Galaxy Clusters

Andreas Faltenbacher, Stefan Gottlöber

Astrophysikalisches Institut Potsdam, An der Sternwarte 16, 14482 Potsdam, Germany

1. Introduction

Within a ΛCDM model ($\Lambda = 0.7$, $\Omega_m = 0.3$) we have investigated the significance and the range of alignment of the major axes of galaxy clusters. To this end we have performed large scale simulations (box 500 h^{-1} Mpc with 256^3 particles of mass $0.89 \times 10^{12} M_\odot$). The 2915 most massive clusters provide the sample under consideration. The mean distance between the clusters is 35 h^{-1} Mpc and the cluster masses range from $2.0 \times 10^{14} M_\odot$ up to $3.3 \times 10^{15} M_\odot$.

2. Intrinsic Cluster Shape

To figure out the intrinsic shape of clusters in this sample we have plotted the ratios between the lengths of the three principal axes of the clusters in two different ways. The axes are ordered by size ($a \geq b \geq c$). The left panel of Figure 1 shows the ratios a/b versus a/c, whereas the right panel shows b/c versus a/c. Assuming a prolate rotational ellipsoid ($a > b$, $b = c$) all points would coincide with the dashed diagonal on the left panel. In the case of a oblate rotational ellipsoid ($a = b$, $a > c$) all data points would coincide with the dashed diagonal plotted in the right panel. Comparing the left to the right panel it clearly turns out that the data points tend to be concentrated rather along the left panel's diagonal than along the right panel's diagonal. Therefore, the clusters have typically prolate shape.

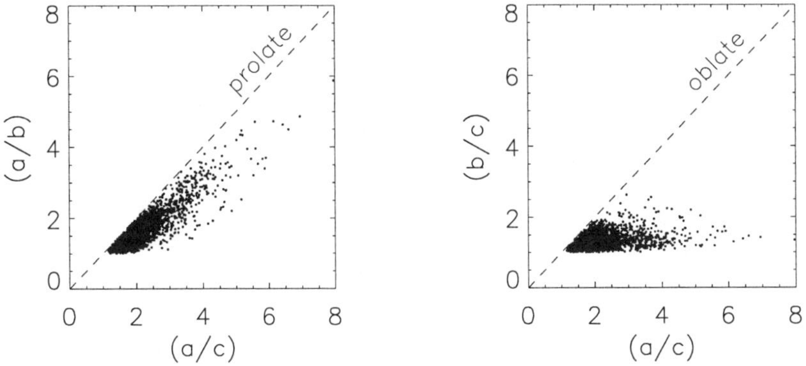

Figure 1. Ratios of the length of the cluster's principal axes.

3. Orientation of Clusters

The question arises how the shape of the clusters is interweaved in the filamentary network of the gravitating matter of the universe. Figure 2 shows the fraction of clusters per bin of cosines of angles ψ between the cluster's major axis and the line pointing to the nearest neighbour. In case of random distribution of major axes all bins would contain 1/10 of clusters (dotted line). The clear signal of alignment decreases with distance. It dissappears for distances above 30 h^{-1} Mpc.

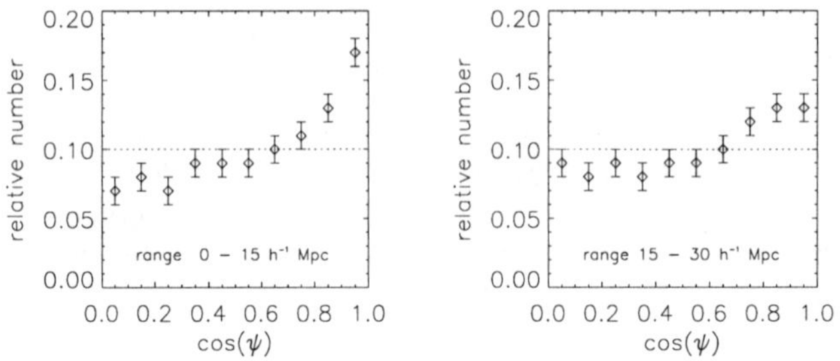

Figure 2. Distributions of the cosine of angles ψ between the cluster's major axis and the line pointing to its nearest neighbour for different ranges of distance: left, $0 - 15$ h^{-1} Mpc, and right, $15 - 30$ h^{-1} Mpc. The dotted line corresponds to a random distribution.

4. Range of Alignment

To explore the range of alignment in more detail we have gathered the data into bins of 10 h^{-1} Mpc width each and calculated the mean value of $\cos(\psi)$ for each of those bins. As Figure 3 reveals there is clear evidence for alignment up to distances of 30 h^{-1} Mpc in a ΛCDM cosmology.

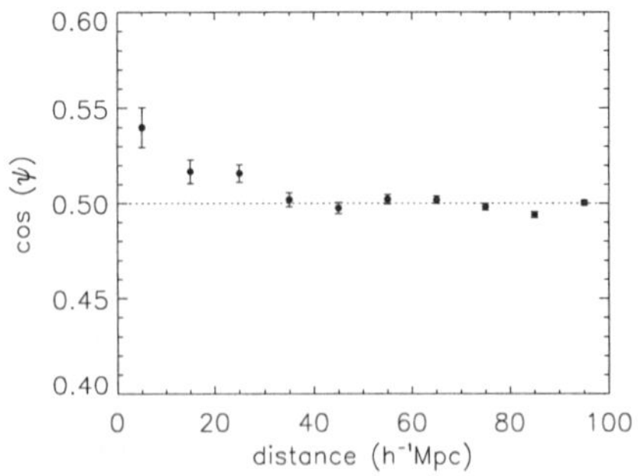

Figure 3. Mean values of $\cos(\psi)$ between the cluster's major axis and the line pointing to the nearest neighbour. The dotted line corresponds to a random distribution.

Tracing Cosmic Evolution with Galaxy Clusters
ASP Conference Series, Vol. 268, 2002
S. Borgani, M. Mezzetti, and R. Valdarnini, eds.

WINGS: a Wide–field Imaging Nearby Galaxy clusters Survey

G.Fasano, D.Bettoni, C.Marmo, E.Pignatelli, B.M.Poggianti

Padova Astronomical Observatory, Vicolo dell'Osservatorio 5, I-35122

M.Moles

CSIC, C/ Serrano 123, E-28006 Madrid

P.Kjærgaard

Copenhagen University Observatory, Juliane Maries Vej 30, DK-2100

Abstract. We present a two-band, wide-field imaging survey of an X–ray selected sample of 78 clusters in the redshift range z=0.03–0.07. The aim of the project is to provide the astronomical community with a complete set of homogeneous, CCD–based, surface photometry and morphological data of the nearby cluster galaxies located within 1.5Mpc from the cluster center.

1. Motivation

In all statistical studies concerning photometric and morphological evolution of galaxies in clusters, the only reference sample available in the nearby universe is the historical one of Dressler (1980) who lists the positions, the visual morphological classifications and the approximate magnitudes of galaxies in 55 clusters at z=0.011–0.066. It is obvious that a local reference sample with quality adequate to the modern technologies is missing and is crucial both for studying the morphological content of nearby clusters in a systematic way and for setting the zero-point for evolutionary studies.

This project is aimed at filling in this gap through a complete CCD surface photometry (in the B and V bands) of galaxies in a well-defined sample of low–z clusters.

2. Cluster sample and status of observations

The growing availability of efficient and reliable Wide–Field Imaging CCD Cameras (WFICs) has opened for the first time the possibility of gathering in a reasonable time a large amount of CCD data on galaxies in nearby clusters. We have exploited this opportunity in both the north (INT, La Palma) and the south (ESO–2.2) hemispheres, by taking observations of clusters selected from an essentially complete X-Ray flux limited sample of Abell clusters (XBACs,

Ebeling et al. 1996) compiled from ROSAT All–Sky Survey data. The redshift range (0.04<z<0.07) is the only selection criterion applied, resulting in a total sample of 78 clusters (36 in the northern hemisphere and 42 in the southern one) over a broad range in X-Ray luminosity. Since now a total of 14 observing nights have been already devoted to this project, resulting in the complete imaging of 61 clusters.

3. Immediate objectives and techniques

The data will be used to produce, for each cluster of the sample, the following outputs:

– a deep (V_{lim} ~22.5; B_{lim} ~23.5) photometric catalog of galaxies containing coordinates, integrated V and B magnitudes, concentration index, rough ellipticity and position angle estimates of each object;

– a surface photometry catalog, relative to a sub-sample of bright/large enough galaxies of the previous, deep list. In this catalog we will include the whole photometric and morphological information extracted from the luminosity and geometrical profiles (color profile, isophotal twisting, disky/boxy c_4 profile, Bulge/Disk decomposition, etc..), as well as the global parameters (effective radius and average surface brightness, total magnitude, Sersic's index, morphological type, etc..) of each galaxy;

The reduction procedures are clearly crucial to achieve these scientific objectives. In particular, automatic pipelines for data reduction and surface photometry are required, together with a fast computing capability.

The deep catalogs will be produced running SExtractor (Bertin and Arnout, 1996) onto the co-added frames in the two filters, whereas the surface photometry catalogs will be produced by using a tool for Galaxy Automatic Surface PHOTometry in wide and/or deep fields (GASPHOT, Pignatelli and Fasano 1999) we are presently developing at the Padova Observatory. It consists of four main tools:

a) STARPROF produces a careful (space varying) representation of the PSF profile;

b) SEXISOPH exploits some capabilities of SExtractor to quickly produce luminosity and geometrical profiles of the galaxies;

c) GALPROF fits the equivalent luminosity profiles of galaxies by using both a Sersic law and a three component ($r^{1/4}$ + exp + PSF) profile convolved with the proper PSF and produces unbiased estimates of total magnitudes, effective radii, Sersic's indices, etc..;

d) MORPHOT exploits some characteristic features of the luminosity and geometrical profiles to estimate the morphological type of individual galaxies.

References

Dressler, A., 1980, ApJS424,565

Ebeling, H., et al., 1996, MNRAS281,799

Bertin, E. & Arnout, S., 1996, A&AS117,393

Pignatelli, E. & Fasano, G., 1999, Ap&SS269,657

Tracing Cosmic Evolution with Galaxy Clusters
ASP Conference Series, Vol. 268, 2002
S. Borgani, M. Mezzetti, and R. Valdarnini, eds.

New evidences of merging for the galaxy cluster Abell 521

C. Ferrari, S. Maurogordato, C. Benoist, E. Slezak

Observatoire de la Côte d'Azur, CERGA, BP 4229, Le Mont-Gros, 06304 Nice Cedex 4, France

M. Arnaud

CEA/DSM/DAPNIA/Service d'Astrophysique, CEA/Saclay, F-91191 Gif sur Yvette Cedex, France

A. Cappi

Osservatorio Astronomico di Bologna, via Ranzani 1, 40127, Bologna, Italy

Abstract. We present new results of an optical analysis of the galaxy cluster A521. Thanks to new data, obtained at the ESO 3.6 m telescope with EFOSC2, we have investigated the dynamical state of this cluster. These new results seem to confirm the hypothesis (Arnaud et al. 2000) that A521 has undergone an old merging along a filament in NE-SW direction, and they show the presence of a galaxy group bound to the BCG that is infalling into the main cluster in the plane of the sky.

1. Introduction

In October 1999 and December 2000 we have obtained new spectra of Abell 521 at the ESO 3.6 m telescope with EFOSC2, and derived a high quality velocity sample of 123 members of the cluster.

ROSAT package applied on the whole dataset gives a high value for velocity dispersion (1342 km/s); moreover, the higher moments of the distribution show some deviations from gaussianity, and typical substructure diagnostics seem to confirm the moderate presence of substructures.

2. Analysis of the sample

We have investigated the existence of substructures in the cluster applying the KMM mixture modeling algorithm to velocity distribution: a partition in three velocity groups comes out, with an estimate of correct allocation rate of 95% (see Fig.2). In particular, we have found a main subgroup (KMM-B), with a still high value of velocity dispersion ($\sim$1000 km/s), and two substructures (KMM-A and KMM-C) at lower and higher radial velocity with respect to the main one.

After having divided the whole catalogue into two subsample, the blue and red ones, it emerges that reddest objects, with low velocity dispersion, seem to

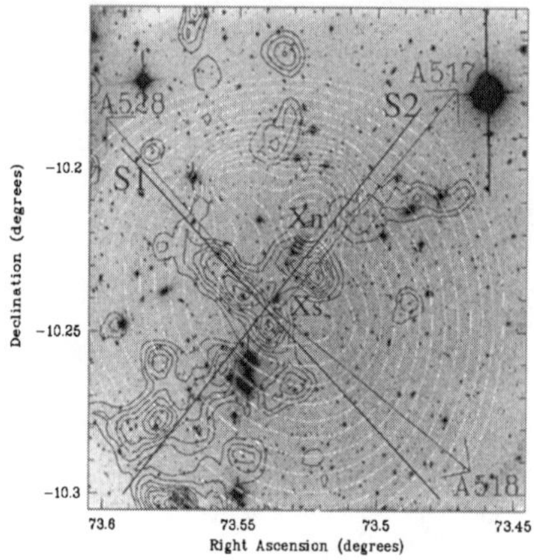

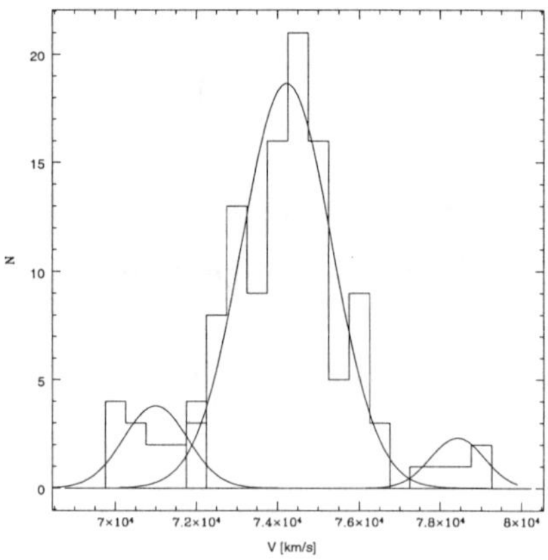

Figure 1. **LEFT:** ROSAT-HRI isocontours are superimposed to the MOS optical image in V-band obtained at CFHT. **RIGHT:** KMM results

trace the main structure of the cluster, whence blue ones (probably late type galaxies) are later captured by cluster potential.

We confirm (Arnaud et al. 2000) the presence of an ancient merging into the line of sight in S1 direction, since this region is characterized by a high value of velocity dispersion and a location close to the whole cluster value; furthermore it contains the greatest part of blue and emission line objects, probably belonging to the impact group that have crossed the core of the target one.

It has been evidenced a region of low velocity dispersion ($\sim$ 400 km/s) as computed from the 10 nearest galaxies around the BCG. This group of galaxies seem to be the optical counterpart of the X-ray group detected in the North of the cluster, corresponding to a local minimum of gas temperature. Since the mean velocity of these galaxies is roughly the same as for the whole cluster, we infer a merger in the plane of the sky.

We have measured radial velocities of all the blobs in the arc-like structure around the BCG: all of them are at the cluster redshift. They probably are smaller objects accreted by the central giant galaxy, which does not show the luminous envelope typical of cD galaxies.

References

Arnaud, M., Maurogordato, S., Slezak, E., Rho, J., 2000, A&A, 355, 461

Benoist, C. et al. 2001, in preparation

Ferrari, C. et al. 2001, in preparation

Maurogordato, S., Proust, D., Beers, T.C., Arnaud, M., Pelloó, R., Cappi, A., Slezak, E., Kriessler, J.R., 2000, A&A, 355, 848

Tracing Cosmic Evolution with Galaxy Clusters
ASP Conference Series, Vol. 268, 2002
S. Borgani, M. Mezzetti, and R. Valdarnini, eds.

Comparison of Subclustering Frequency Determined from 2D and 3D Data

Piotr Flin

Joint Institute for Nuclear Physics, Bogoliubov Laboratory of Theoretical Physics, Dubna, Russia
Pedagogical University, Institute of Physics, ul. Swietokrzyska 15, 25-406, Kielce, Poland

Janusz Krywult

Pedagogical University, Institute of Physics, ul. Swietokrzyska 15, 25-406, Kielce, Poland

Abstract. Frequencies of substructure within Abell clusters determined for the same objects using the wavelet analysis for 2-D and 3-D data are similar.

The frequency of substructure occurrence has been determined by several authors, using both various statistical methods and observational data. Sometimes only the positions of galaxies, i.e. 2-D data, have been used, while some papers have been based on the position and redshifts of member galaxies, i.e. 3-D data. The determined frequency changed from below 30% up to 80% of that for Abell clusters with substructures.

The analysis of previous papers suggests that the differences in the determined substructure occurrences are due to the rather different methods of analysis than to the observational data themselves. Therefore, in order to test this hypothesis, in the present paper a direct comparison of the frequency of substructure occurrence is made for 2-D data and 3-D data for the same Abell clusters.

All 43 galaxy clusters in the present work are the result of the objective algorithms for star/galaxy classification applied to scanned photographic plates. There are three sources of our data, but the photographic material came from 48" Schmidt telescopes. Each galaxy within the radius of 1.5 Mpc ($h = 0.75$) from the cluster centre and with magnitude between m_3 and $m_3 + 3$, where m_3 is the magnitude of the third brightest galaxy, was considered as the cluster member. 33 catalogues of Abell clusters were obtained from DSS using FOCAS package, further 4 catalogues were obtained from scans performed in the Rome Observatory (Trevese et al. 1992; Flin et al. 2000), while the remaining 6 catalogues of clusters came from the COSMOS machine at the Royal Observatory in Edinburgh (Krywult, MacGillivray, & Flin 1999).

The existence of the substructures in galaxy clusters has been checked using wavelet analysis. The wavelet technique is a convolution, on a grid of $N \times N$ pixels, between signal $s(r)$ (in our case, the angular positions of galaxies) and analyzing wavelet function $g(r, a)$. In this work we use two-dimensional radial function called the Mexican Hat (Escalera et al. 1992).

For the analysis presented here, the discrete wavelet was computed on a grid of 256×256 pixels for seven scales increasing from $a = 8$ to 64 (in pixel units), namely $8, 11, 16, 22, 32, 45, 64$ respectively, which ensure the correctness of the Mexican Hat (Daubechies 1990).

We have modelled the significance of the substructuring detected using the Monte Carlo simulations. For each cluster and each scale a, the wavelet analysis was carried out on a set of 1000 structureless distributions of galaxies containing the same number of points as in the true fields.

We assume that a substructure is real if the probability of random fluctuations is less than 0.005. Furthermore, for each scale a only substructures with more than 4 galaxy members in a circle of radius a are considered.

In order to compare our 2-D data with 3-D data, the Girardi et al. (1997) paper served as a source of information on the subclustering in 3-D data. They used wavelet analysis and performed morphological classification of clusters (Escalera et al. 1994), finding bimodal structure for five clusters (A548, A754, A1736, A3526, A3716) and a complex structure in A2151. Moreover, the existence of substructure was detected in A85, A193, A194, A1060, A1367, A1983, A2063, A2877 and A3667.

In our analysis, substructures were detected in clusters: A151, A426, A548, A754, A1060, A1367, A 1736, A1809, A1983, A2052, A2151, A3128, A3395, A3526 and A3667.

Comparison shows that in some clusters with no subclustering in 2-D, the analysis of 3-D data revealed small substructures, both in size and in the number of members (A85, A193, A194, A2063). In a few cases, the 2-D data showed substructures non-existing in the 3-D data (A151, A1809, A2052).

The differences in the substructure existence in the 2-D and the 3-D data do occurr in individual cases. But the frequency of substructure occurrence in 3-D is 31%, while our result, based on the 2-D data, is 34%. At the significance level of 0.05, these numbers are in agreement. So, the frequency of the existence of substructures in Abell clusters determined from 2-D data is a good indicator of subclustering frequency, which allows us to use it for clusters located far away, when the number of galaxies with known redshifts is small.

This work was partially supported by grants KBN/IBM SP/WSPKielce/073 and KBN/AS/BS/052.

References

Daubechies I. 1990, I.E.E.E. Trans. on Information Theory, 36, 961

Escalera, E., Biviano, A, Girardi, M., Mardirossian, F., Mazure, A., Mezzetti, M. 1994, AJ, 423, 539

Escalera, E., Mazure, A. 1992, ApJ, 388, 23

Flin, P., Krywult, J., Trevese, D., Cirimele, G., Hickson, P. 2000, A&AS, 146, 373

Girardi, M., Escalera, E., Fadda, D., Giuricin, G., Mardirossian, F., Mezzetti, M. 1997, ApJ, 482, 41

Krywult, J., MacGillivray, H.T., Flin, P. 1999, A&A, 351, 883

Trevese, D., Flin, P., Migliori, L., Hickson, P., Pittella G. 1992, A&AS, 94, 327

Tracing Cosmic Evolution with Galaxy Clusters
ASP Conference Series, Vol. 268, 2002
S. Borgani, M. Mezzetti, and R. Valdarnini, eds.

The morphology-σ relation in Hickson Compact Groups

Paola Focardi & Birgit Kelm

Dipartimento di Astronomia, Via Ranzani 1, 40127 Bologna, Italy

Abstract. We show that the morphology velocity dispersion relation in HCGs arises from the inclusion within the sample of two different classes of structures displaying characteristics typical of field and embedded systems.

1. Introduction

Due to their high density contrast and low velocity dispersion Compact Groups (CGs) are ideal and unique places in which to investigate the environmental effect on galaxy formation and evolution (Hickson 1997). Visually selected by requiring a minimum of 4 close galaxies per group, HCGs (Hickson 1982) constitute so far the best studied CG sample. Among the 92 confirmed HCGs, 23 do actually include only 3 accordant members. Analysis of a large sample of CGs identified using an automatic algorithm (Focardi & Kelm, 2001) has proven that CG samples selected by compactness include two classes of systems displaying significantly different large scale environmental properties, velocity dispersion and galaxy content. The first class (field CGs) corresponds to isolated structures characterised by a high gas-rich galaxy content and a low σ_r. The second class is associated to embedded structures (embedded CGs) with a low gas-rich galaxy content and a high σ_r. CGs multiplicity appears to be a robust first discriminant between these two classes, as Triplets (Ts) are largely dominant among field CGs and Multiplets (Ms) among embedded ones. Here we argue that the distinction between Ts and Ms may actually also account for the morphology-σ_r correlation found to hold in HCGs (Hickson et al. 1988).

2. The morphology-σ_r correlation in HCGs

In **figure 1** (upper panel) we plot σ_r distributions for HCGs including 3, 4 or more than 4 galaxies respectively. The lower panel in **figure 1** shows distributions for late spirals, early spirals (including S0) and elliptical galaxies in HCGs. A clear correspondence between distributions of galaxy subsamples selected according to multiplicity and morphological type appears to hold. Triplets reflect the behaviour of late spirals, n>4 CGs the behaviour of ellipticals. Similarities are statistically confirmed and indicate that Ts are mainly field systems. With respect to Hickson's analysis, comparing spiral rich (spiral fraction > 0.5) and spiral poor (spiral fraction < 0.5) groups only, our analysis takes into account every galaxy in the HCG sample.

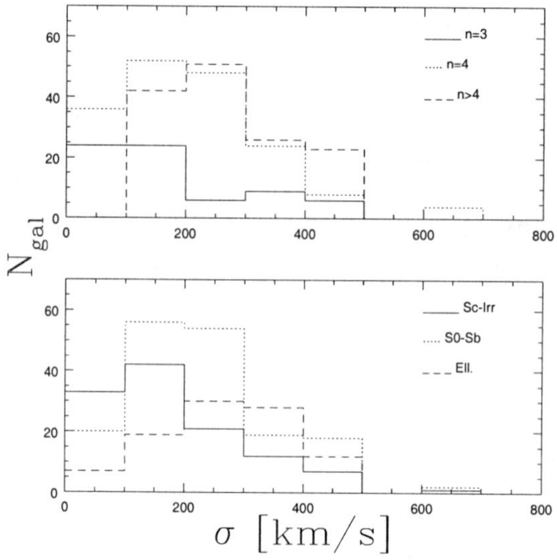

Figure 1. Velocity distribution of galaxies in HCGs when sampling is performed according to multiplicity (upper panel) and to morphological type (lower panel). It emerges that Triplets reflect the distribution of late spirals, Quadruplets the distribution of early spirals and n>4 CGs the distribution of Elliptical galaxies.

3. Discussion

We have shown (**fig. 1**) that the relation between multiplicity and σ_r holds also for HCGs. The analysis of the link between multiplicity and large scale environment in HCGs is more difficult, due to non homogeneous sampling. Rood & Struble (1994) investigated spatial coincidence of nearby HCGs with loose groups and clusters and claimed that only 30% of HCGs are isolated. We find that among isolated structures Ts are 3 times more frequent than higher multiplicity HCGs. HCGs thus appear to include two significantly distinct classes of CGs whose characteristics are responsible for correlation between morphological, dynamical and environmental properties.

References

Focardi P., & Kelm B., 2001, A& A, submitted

Hickson P. 1981, ApJ 255, 382

Hickson P. , Kindl E., & Auman J.R., 1988, ApJ 331,6

Hickson P. 1997, ARAA, 35, 357

Rood H.J., & Struble M.F. 1994, PASP 106, 413

Tracing Cosmic Evolution with Galaxy Clusters
ASP Conference Series, Vol. 268, 2002
S. Borgani, M. Mezzetti, and R. Valdarnini, eds.

ASCA Study of Dynamical Evolution in Nearby Clusters

Tae Furusho

NASA/GSFC, Code 662, Greenbelt, MD 20771, USA

Abstract.
We examine the constraint on the cluster evolution based on the temperature structure combined with the metallicity distribution in 6 nearby clusters of galaxies. Remarkable hot regions in many clusters demonstrate the clusters have undergone strong mergers. Significant gradients of the metallicity profiles found in several systems are shown to trace the galaxy distribution, which suggests that the gas was not substantially mixed after the metal injection period. Combination of the temperature and metallicity distributions enables us to estimate the time history of metal injection, major merger, and quiet periods.

1. Results

Detailed temperature structure and radial profiles of metallicity distribution in the ICM have been revealed for many clusters with ASCA. The clusters of galaxies studied here are 6 nearby ($z < 0.023$) bright clusters: Virgo, Centaurus, A1060, AWM7, Perseus, and Coma. The closeness and brightness of these systems allow ASCA to resolve the internal structures over the whole cluster region with multi-pointing observations. The 2-dimensional temperature maps are reported in Shibata et al. (2001; Virgo), Furusho et al. (2001a; Centaurus, A1060, AWM7, 2001b; Perseus), and Watanabe et al. (1999; Coma). The radial profiles of metallicity distributions are studied by Matsumoto et al. (1996; Virgo), Ikebe et al.(1999; Centaurus), Tamura et al. (1996; A1060), Ezawa et al. (1997; AWM7, 2001; Perseus), and Honda (private communication; Coma).

The temperature variation are divided into 3 classes relatively. A1060 and AWM7 are almost isothermal, Virgo and Centaurus show a significant hot region of a few hundred kpc, and Coma and Perseus show remarkably extended structure over 1 Mpc. Similarly, the observed metallicity profiles can be grouped into 3 types, i.e. almost uniform (A1060 and Coma), large-scale gentle gradient (AWM7 and Perseus), and central concentration (Virgo and Centaurus).

2. Discussion

In Fig. 1, we plotted these data in a 2-dimensional plane of the relative intensity of temperature variation versus the strength of metallicity gradient. Initially, clusters have no structures in metallicity and temperature ($z \gtrsim 10$), and then the central components grow and create pronounced metal concentration during

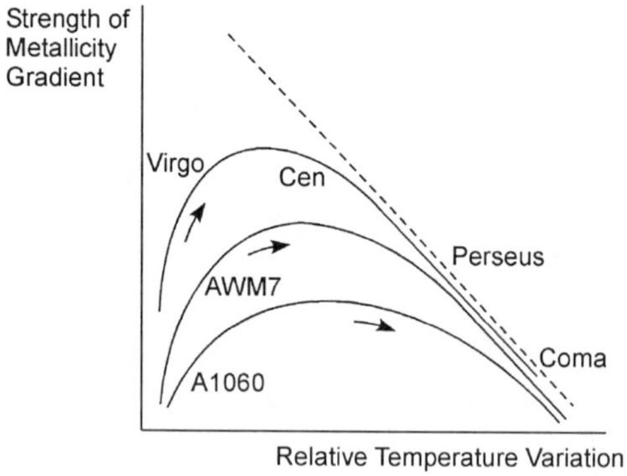

Figure 1. The approximate paths of the cluster evolution implied from temperature and metallicity distributions. The difference among 3 lines roughly corresponds to relative influence of the cD galaxies and frequency of mergers indicated by the dashed line.

the starburst period ($z \sim 3-6$). This causes the upward movement in the plot, which would be steeper if cD galaxies have dominant effect, or would be smaller if cluster centers are disturbed by mergers in a fairly early period. Then, the strong mergers at some later stage tend to disrupt the cluster centers and distribute the metals over a wide region in the ICM. This makes the path downward, and at the same time the development of temperature structures moves the points rightward as indicated by the dashed line in Fig. 1.

The 6 data points suggest that Perseus and Coma are in a fairly evolved stage, and A1060 is in a relatively young stage. These features are broadly consistent with the evolutionary aspects inferred from optical observations, such as the density of galaxies. This scheme is crude and neglecting many factors that would significantly alter the evolution process. However, the general idea based on the combination of the metallicity and temperature distributions would still be useful in describing the history of the cluster evolution.

References

Ezawa, H., Fukazawa, Y., Makishima, K. et al. 1997, ApJ, 490, L33
Ezawa, H., Yamasaki, N. Y., Ohashi, T. et al. 2001, PASJ, 53, 595
Furusho, T., Yamasaki, N.Y., Ohashi, T. et al. 2001a, PASJ, 53, 421
Furusho, T., Yamasaki, N.Y., Ohashi, T. et al. 2001b, ApJsubmitted
Ikebe, Y., Makishima, K., Fukazawa, Y. et al. 1999, ApJ, 525, 58
Matsumoto, H., Koyama, K., Awaki, H. et al. 1996, PASJ, 48, 201
Shibata, R., Matsumoto, H., Yamasaki, N.Y. et al. 2001, ApJ, 549, 228
Tamura, T., Day, C. S., Fukazawa, Y. et al. 1996, PASJ, 48, 671
Watanabe, M., Yamashita, K., Furuzawa, A. et al. 1999, ApJ, 527, 80

Tracing Cosmic Evolution with Galaxy Clusters
ASP Conference Series, Vol. 268, 2002
S. Borgani, M. Mezzetti, and R. Valdarnini, eds.

The Northern Sky Optical Cluster Survey

R. R. Gal and the DPOSS Team
Johns Hopkins University, Bloomberg Center, 3701 San Martin Dr., Baltimore, MD 21218, USA

Abstract. The Northern Sky Optical Cluster Survey is a project to create an objective catalog of galaxy clusters over the entire high–galactic–latitude Northern sky, based on DPOSS data.

1. The Survey

The Northern Sky Optical Cluster Survey is a project to create an objective catalog of galaxy clusters over the entire high–galactic–latitude Northern sky. We use the object catalogs generated from the Digitized Second Palomar Sky Survey (DPOSS) as the basis for this survey. We use a simple adaptive kernel technique to create galaxy density maps from which density peaks are selected. This survey eliminates some of the subjective criteria of past surveys, and utilizes more information (especially colors) than the most similar recent survey, the APM. This contribution presents our cluster detection technique, as well as some initial results for a $\sim 5000\square^\circ$ well-calibrated area in the NGP. The adaptive kernel map of this area is shown in Figure 1. Dark areas represent the highest galaxy density. We find approximately 15,000 candidate galaxy clusters, corresponding to a mean surface density of ~ 3 clusters per square degree, consistent with the detection of r! ic! hness class 0 and higher clusters to $z \sim 0.3$. In addition, we have developed effective photometric redshift and richness estimators for our clusters. Our photometric redshift estimator iteratively utilizes the mean galaxy colors and magnitudes within a fixed physical radius, and is calibrated with nearly 200 spectroscopic cluster redshifts. This estimator is accurate to $\Delta z = 0.028$. Our richness estimator, which counts galaxies in a fixed absolute magnitude range, suggests that most of these cluster candidates have an Abell richness class of $R = 0$ or lower.

In addition to the work presented here, we have obtained extensive follow-up CCD imaging and multislit spectroscopy of many candidates, indicating that we are indeed detecting true systems. Using our catalog, we are measruing the two- and three-dimensional correlation function for clusters out to the largest scales. Comparisons and correlations between X-ray and optically selected catalogs will provide insight into both the physical processes in clusters, as well as the inherent advantages and biases of different techniques.

Figure 1. The adaptive kernel density map of galaxies with $m_r < 19.5$ in nearly 5000 square degrees around the NGP. Dark areas have higher galaxy density. The x-axis is declination ($-2.5°$ to $+70°$) and the y-axis is right ascension ($\sim 142°$ to $\sim 247°$).

Tracing Cosmic Evolution with Galaxy Clusters
ASP Conference Series, Vol. 268, 2002
S. Borgani, M. Mezzetti, and R. Valdarnini, eds.

Reacceleration of relativistic electrons in cooling flows: the case of Perseus Cluster

Myriam Gitti, Gianfranco Brunetti, Giancarlo Setti

Dip. Astronomia Univ. di Bologna, v. Ranzani 1, 40127 Bologna, Italy
Ist. di Radioastronomia del CNR, v. Gobetti 101, 40129 Bologna, Italy

Abstract. Due to radiative losses, the typical ageing time-scale of ultra–relativistic electrons in the ICM is relatively short ($10^7 - 10^8$ yrs), so that the diffuse radio synchrotron emission detected from some clusters of galaxies may support the presence of reacceleration mechanisms in the ICM with an efficiency comparable to the energy loss processes. In particular, the cluster A426 (Perseus) shows a radio mini–halo, which, we argue, can originate by reaccelerating the relativistic particles with the necessary energetics supplied by the cooling flow.

The compression of the gas in a cooling flow is expected to produce a significant increase of the strength of the frozen–in intracluster magnetic field. Depending on the physical conditions prevailing in a cooling flow there are two possibilities: if the turbulent velocity v_T is less than the mean inflow velocity v_F, the inward cooling flow is homogeneous and in a steady state and the magnetic field lines become predominantly **radial** (Soker & Sarazin 1990, *Ap.J.* 348, 73); if $v_T > v_F$, the characterization of the flow as smooth and steady is incorrect and in this case there is an **isotropic** compression of the field (Tribble 1993, *MNRAS* 263, 31). However, in absence of a reacceleration mechanism, the radio emission of the electrons injected at a given time in these strong fields of order of some tens of μG (Ge & Owen 1993, *AJ* 105, 3) would not be observable for more than $\sim 10^8$ yrs. This short lifetime contrasts with the diffuse radio emission observed in the central part of the Perseus cluster, so called *radio mini–halo.* As pointed out by Brunetti et al. (2001 *MNRAS* 320, 365), a reaccelerated relic electron population in cluster radio halos would naturally produce radio synchrotron spectra which steepen with increasing distance from the center. Indeed, this is the case of the Perseus mini–halo (Sijbring 1993, PhD Thesis), so that it might be argued that there is an efficient mechanism for particle acceleration operating in the cooling flow region. In our model we assume that the relativistic electrons are continuosly reaccelerated by *MHD Fermi processes*, which depend on the square of the field like the radiative loss mechanisms.

The mini–halo has a steep spectrum with spectral index $\alpha = 1.4$ between 327 MHz and 609 MHz ($F_\nu \propto \nu^{-\alpha}$) and its extension ($\sim 15'$) is comparable with that of the cooling flow region (cooling radius ~ 210 Kpc). For comparison with the model expectations we have considered the 327 MHz radio image obtained by Sijbring (1993, PhD Thesis) and assumed spherical symmetry. The observational constraints are: *brightness profile* (Fig.1–a), *total spectral index* and *radial spectral steepening* (Fig.1–b). Since we have not investigated the spatial

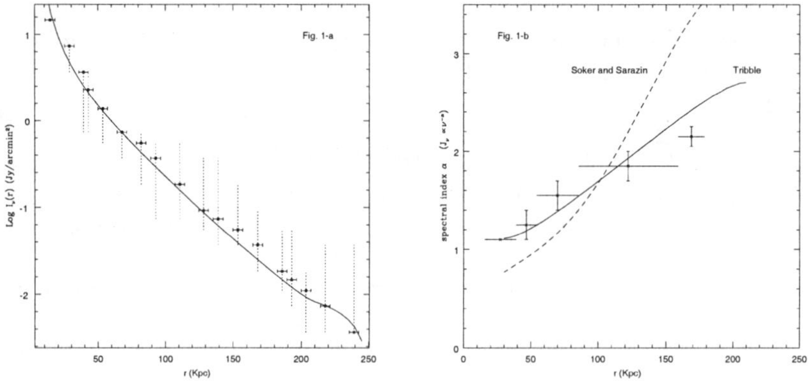

Figure 1. **Fig 1–a**: fit to the surface brightness profile obtained in the case of isotropic field (Tribble's model); vertical errors represent the deviations from the spherical symmetry. **Fig 1–b**: *Solid line*: fit to the spectral steepening obtained in the case of isotropic field (Tribble's model); the data are taken from spectral index distribution map of Sijbring (1993). *Dashed line*: predicted spectral steepening in the case of radial field (Soker & Sarazin's model).

diffusion of the relativistic particles, we assume that the energy density of these particles scales as $\propto x^{-s}$, s being a free parameter and x the distance from the center. The other free parameters in our model are the magnetic field intensity at the cooling radius (B_c) and the characteristic distance between two subsequent peaks of magnetohydrodynamic turbulence (l_c). One of the first indications obtained by our model is that Tribble's case of isotropic compression of the field provides good representations of the data for the parameters in the range: $s = 1.5 \div 2$, $B_c = 1 \div 1.4\mu$G and $l_c = 25 \div 50$ pc, l_c increasing with the strength of the magnetic field. The corresponding expected total spectral index ranges from ~ 1.25 to ~ 1.5. In Fig. 1–a and 1–b (solid line) are plotted fits to the surface brightness profile and to the spectral steepening obtained for representative values of the parameters. On the other hand, Soker & Sarazin's case of radial field appears not to be holding for the Perseus cluster (it provides a too steep slope for the spectral steepening, see dashed line in Fig. 1–b). This indication for the validity of Tribble's model is confirmed by some measurements of magnetic fields in cooling flows (Taylor & Perley 1993, *ApJ* 416, 554).
The possibility to produce an observable radio mini–halo in a cooling flow region depends on the energy density and spectrum of the relativistic electrons at the cooling radius; a mini–halo could form only under favourable conditions, so that their rarity is qualitatively consistent with our model (Gitti et al. in preparation).

Tracing Cosmic Evolution with Galaxy Clusters
ASP Conference Series, Vol. 268, 2002
S. Borgani, M. Mezzetti, and R. Valdarnini, eds.

Statistics of nearby galaxy groups from volume limited catalogues: Algorithm and first results.

Giuseppe Giudice

Dipartimento di Progettazione e Gestione Industriale, University 'Federico II', Naples, Italy; giudice@unina.it

Abstract. Groups of galaxies are found in a set of volume limited samples extract from a magnitude limited catalogue, essencially coming from LEDA database.[1] Some properties of groups are derived.

The catalogue In this paper a catalogue of galaxies obtained from LEDA database and containing both distances and corrected apparent magnitudes is used; it has been implemented with better evaluations of nearby galaxies distances.

LEDA furnishes two evaluations of the distance modulus: one of them is obtained from physical parameters (rotation velocity, de Vaucouleurs luminosity index, see e.g. Garcia, 1993) and the other from a model of the Hubble flow, allowing for Virgocentric infall. First one is given for few galaxies, and second one is given for most galaxies (those with a Virgocentric velocity greater than 500 km s^{-1}). Many nearby galaxies, however, are lacking of distance modulus in LEDA, so we need another source of distances.

Nearby LEDA galaxies lacking of distance modulus have been searched in Tully (1988) and Karachentsev et al. (1999); to those not found there a distance is assigned according to a simple Hubble flow model. Galaxies within 6 degrees from M87 are assigned to Virgo Cluster. No limit in galactic latitude is adopted, so that many galaxies from the Zone of Avoidance are included.

The final catalogue is made of 25878 galaxies; from it eleven samples have been extracted, with distance moduli $\mu \leq 27$, $\mu \leq 28$, and so on, up to $\mu \leq 37$.

The incompleteness of the catalogue has been tested as a function of apparent reduced magnitude and galactic latitude. The first one reveals no incompleteness at lest up to 14.8; the catalogue uses a limit of 15.0, but this is not too significant, given the creation of vulume limited samples. The second one shows an incompleteness starting at $|b| \leq 35°$.

Group-finding procedure In the friends of friends algorithm, here adopted, two galaxies are "friends" if they are less distant (in the sense of "projected distance") than a *distance link* and their velocities differ by less than a *velocity link*. Two galaxies which are "friends" of another one are "friends" also between them. A

[1] I have made use of the LEDA database (http://leda.univ-lyon1.fr), supplied by LEDA Team at the CRAL-Observatoire de Lyon.

galaxy without "friends" is isolated; a set of two friend galaxies is a couple; a group is a set of three or more "friend" galaxies.

In magnitude-limited catalogues, density decreases with distance, so that the links have to be rescaled as a function of distance; in volume-limited ones the density is kept constant within each sample, but decreases when we turn to more distant (more shallow) samples. In either case one needs a link rescaling.

In this study I adopt a distance link proportional to the selection function raised to $\phi^{-1/3}$, but with a tentative superior limit of 1.5 Mpc,[2] and a velocity link proportional to $\phi^{-1/2}$, ϕ being the selection function, without a superior limit; normalization is such that the overdensity be $\delta\rho/\rho = 80$ and $v_L(1000 \text{ km s}^{-1}) = 350$.

Groups at the sample periphery can be 'trimmed' by the sample limits; they can still be submitted to a statistical analysis concerning the grouping properties of the sample but have to be excluded from the internal properties calculation phase.

Main results The ratio of the number of groups with at least three membres to the number of field plus coupled galaxies, and the ratio of the number of galaxies in groups with at least three members to the total number of galaxies present a sistematical trend with the radius of the sample, but are not too distant from the values given in the literature, variying respectively between 0.09 and 0.19 and between 0.35 and 0.63 (see Giuricin et al., 2000 and the references therein).

Crossing time t_{cr} and M/L ratio appear to be quite stable in different samples (median t_{cr} goes from 0.137 to 0.195 Hubble times, and median M_V/L_0 goes from 47.8 to 220.5); on the contrary, dispersion velocity σ_V and inertial radius R_I increase markedly with the sample dimension. I have recalculated groups adopting fixed values of the links; the results show a median velocity dispersion increasing with the samplen diameter and a lack of variation in median inertial radius. A provisional conclusion can be that the variation of velocity dipsrsion is due to the increasing of the samples, in the sense that in a large sample enter many popolated groups (and cluster); The variation of linear dimension (R_I) of groups can instead be partially an artifact.

References

Garcia, A.M. *Détermination et analyse de la distribution en groupes des galaxies de l'univers local*, Thèse de Doctorat, Université Paris VII, 1993.

Giuricin, G., Marinoni, C., Ceriani, L., Pisani, A. 2000, ApJ, 543,178-194

Karachentsev, I., et al. 1999, A&AS, 139, 97

Tully, R. B. 1988, *Nearby Galaxy Catalogue*, Cambridge University Press

[2]This superior limit is well above the average dimension of groups obtained in this study as well as in literature

Tracing Cosmic Evolution with Galaxy Clusters
ASP Conference Series, Vol. 268, 2002
S. Borgani, M. Mezzetti, and R. Valdarnini, eds.

Constraints on $(\Omega_{\rm m}, \Omega_\lambda)$ from strong lensing clusters

Ghislain Golse, Jean-Paul Kneib & Geneviève Soucail

Laboratoire d'Astrophysique, 14, av. E.-Belin, F-31400 Toulouse, France

Abstract. We use three strong lensing clusters to constrain the cosmological parameters $\Omega_{\rm m}$ and Ω_λ. Recent HST observations of galaxy clusters reveal a large number of multiple images, which are predicted to be at different redshifts. We showed in a previous work that if it is possible to measure spectroscopically the redshift of many multiple images then one can constrain $(\Omega_{\rm m}, \Omega_\lambda)$ through ratios of angular diameter distances independently of any external assumptions. Using three strong lensing clusters, our combined results lead to tight constraints.

1. Introduction

Recent work on constraining the cosmological parameters using the CMB and high redshift supernovae seems to converge to a "standard cosmological model" favouring a flat universe with $\Omega_{\rm m} \simeq 0.3$ and $\Omega_\lambda \simeq 0.7$ (Jaffe et al. 2001). However, these results are still uncertain and it is therefore important to explore other independent techniques to constrain these cosmological parameters.

The existence of multiple images – with known redshifts – of the same source allows to calibrate in an absolute way the total cluster mass deduced from the lens model. The great improvement in the mass modelling of cluster-lenses leads to the conclusion that clusters can also be used to constrain the geometry of the universe, through the ratio of angular size distances, $E = D_{\rm LS}/D_{\rm OS}$ (O: observer, L: lens, S: source), which only depends on the redshifts of the lens and sources, as well as the cosmological parameters (Link & Pierce 1998).

2. Method and Application

In a previous work (Golse, Kneib, & Soucail 2001b), we minimized a χ^2 in the source planes to recover some parameters of the lens potential on a grid $(\Omega_{\rm m}, \Omega_\lambda)$, using numerical simulations. We apply this method to 3 strong lensing clusters which show several systems of multiple images with determined redshifts:

• AC114: $z_{\rm L} = 0.312$, $z_1 = 1.691$ (4 images), $z_2 = 1.867$ (3 images), $z_3 = 3.347$ (5 images) (Campusano et al. 2001),

• A2218: $z_{\rm L} = 0.175$, $z_1 = 0.702$ (4 images), $z_2 = 1.034$ (3 images), $z_3 = 2.515$ (3 images) (Ebbels et al. 1998),

• A1689: $z_{\rm L} = 0.184$, $z_1 = 1.834$ (4 images), $z_2 = 4.868$ (2 images) (Golse et al. 2001a).

Fig. 1 shows the confidence levels on the cosmological parameters.

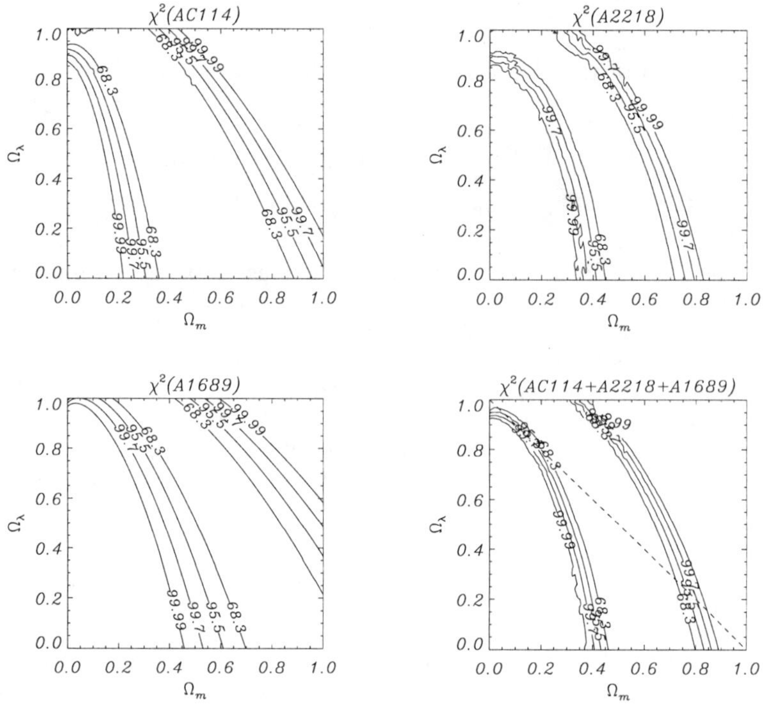

Figure 1. χ^2 confidence levels in the $(\Omega_{\rm m}, \Omega_\lambda)$ plane obtained in the optimisation of the potential of the cluster-lenses AC114, A2218, A1689, and a combined result of the three (dashed line: $\Omega_{\rm m} + \Omega_\lambda = 1$).

3. Conclusion

Combining the constraints from these 3 clusters lead to the Fig. 1 confidence levels. We obtain meaningful constraints compatible with a flat universe ($\Omega_{\rm m} + \Omega_\lambda = 1$). Since the exact degeneracy depends on the different redshift planes involved, results from other cluster lenses will further tighten the error bars. This test can be part of a joint analysis to improve the precision on the parameters.

References

Campusano, L. et al. 2001, astro-ph/0104477, A&Ain press

Ebbels, T. et al. 1998, MNRAS, 295, 75

Golse, G., Castander, F., Soucail, G., Kneib, J.-P., 2001a, in preparation

Golse, G., Kneib, J.-P., & Soucail, G. 2001b, astro-ph/0103500

Jaffe, A. et al. 2001, Phys.Rev.Lett, 86, 3475

Link, R., & Pierce, M. 1998, ApJ, 502, 63

Tracing Cosmic Evolution with Galaxy Clusters
ASP Conference Series, Vol. 268, 2002
S. Borgani, M. Mezzetti, and R. Valdarnini, eds.

Tracing Cosmic Evolution with the Las Campanas Distant Cluster Survey

Anthony H. Gonzalez

Harvard-Smithsonian Center for Astrophysics, 60 Garden Street, Cambridge, MA 02138, USA

Julianne J. Dalcanton

Department of Astronomy, University of Washington, Box 351580, Seattle, WA 98195, USA

Amy E. Nelson

Department of Astronomy and Astrophysics, University of California, Santa Cruz, CA 95064, USA

Luc Simard & Dennis Zaritsky

Steward Observatory, University of Arizona, 933 North Cherry Avenue, Tuscon, AZ 85721, USA

Risa H. Wechsler

Department of Physics, University of California, Santa Cruz, CA 95064

Abstract. The Las Campanas Distant Cluster Survey (LCDCS), which contains over 1000 cluster candidates at $z>0.3$, is a unique sample with which to probe the evolution of both cluster galaxies and the properties of the cluster population. Programs are now underway to utilize the LCDCS for both purposes. We provide a brief overview of these programs, and also present new results for the LCDCS cluster correlation function. Utilizing well-defined, approximately dispersion-limited subsamples, we measure the angular correlation function for clusters at $z\approx0.5$. Spatial correlation lengths are then derived via Limber inversion. We find that the correlation length depends upon mass, as parameterized by the mean cluster separation, in a manner that is consistent with both local observations and CDM predictions for the clustering strength at $z=0.5$.

1. Summary

The past few years have been a period of tremendous progress in detection of distant clusters. As recently as five years ago no single survey contained more than ten cluster candidates at $z>0.5$, but a series of efforts have since led to a several order of magnitude increase in the number of known high-redshift systems (Postman et al. 1996, Scodeggio et al. 1999, Gladders & Yee 2001, Gonzalez et al. 2001). The recently completed Las Campanas Distant Cluster Survey (LCDCS) is our contribution to this field. Clusters in the LCDCS are detected as regions of excess optical surface brightness relative to the background sky, a technique that permits detection of clusters to $z \approx 0.9$ with shallow, drift-scan imaging. The resulting statistical catalog consists of 1073 cluster candidates at $z \gtrsim 0.3$. To enable efficient use of these catalogs we use follow-up imaging and spectroscopy to characterize the properties of the sample, including the contamination rate as well as estimates of the redshifts and relative velocity dispersions of the candidates (see Gonzalez et al. 2001*a* for details).

One application of this sample is determination of the cluster correlation function (Gonzalez et al. 2001*b*). For roughly dispersion-limited subsamples at $z \approx 0.5$ the observed angular correlation functions are well fit by the power law $\omega(\theta) = A_\omega \theta^{1-\gamma}$ with $\gamma = 2.1$. Using the angular correlation data, we derive the spatial correlation length (r_0) as a function of mean intracluster separation (d_c) via the cosmological Limber inversion. We find correlation lengths for the LCDCS that are similar to those observed for local catalogs, and observe a dependence of r_0 upon d_c that is comparable to the results of Croft et al. (1997) for the APM catalog. These results are consistent with theoretical expectations, which predict only modest evolution in the clustering amplitude since $z=0.5$.

Several programs are also being conducted with LCDCS clusters to study the evolution of cluster galaxies . Nelson et al. (2001*a*, *b*) are studying the structural and luminosity evolution of BCG's; Nelson et al. (2001*c*) are studying evolution in the luminosity function and color of the red envelope. Meanwhile, the EDisCS collaboration is conducting a deep 3-color imaging and spectroscopic survey of 20 of the most massive LCDCS candidates at $z=0.5$ and $z=0.8$, with the aim of constraining cluster galaxy morphologies, scaling relations, cluster dynamics, and the evolution of cluster galaxies at these epochs.

References

Croft, R. A. C., Dalton, G. B., Efstathiou, G., Sutherland, W. J., & Maddox, S. J. 1997, 291, 305

Gladders. M., & Yee H. K. C. 2001 (this volume)

Gonzalez, A. H., Zaritsky, D., Dalcanton, J. J., & Nelson, A. E. 2001, astro-ph/0106055

Gonzalez, A. H., Zaritsky, D., & Wechsler, R. H. 2001, submitted to ApJ

Nelson, A. E., Dalcanton, J. J., Simard, L., Zaritsky, D., & Gonzalez, A. H. 2001*a*, submitted to ApJ

Nelson, A. E., Gonzalez, A. H., Zaritsky, D., & Dalcanton, J. J. 2001*b*, submitted to ApJ

Nelson, A. E., Gonzalez, A. H., Zaritsky, D., & Dalcanton, J. J. 2001*c*, astro-ph/0108279

Postman, M., Lubin, L. M., Gunn, J. E., Oke, J. B., Hoessel, J. G., Schneider, D. P., & Christensen, J. A. 1996, AJ, 111, 615

Scodeggio, M. et al. 1999 A&AS, 137, 83

Tracing Cosmic Evolution with Galaxy Clusters
ASP Conference Series, Vol. 268, 2002
S. Borgani, M. Mezzetti, and R. Valdarnini, eds.

Mixed Dimentional Clustering Statistics

Nathan J. Horleston & S. Phillipps

University of Bristol Astrophysics, H.H. Wills Laboratory, Tyndall Avenue, Bristol BS8 1TL

Abstract. Recent work in the literature has made available theoretical predictions of clustering evolution at high redshift for comparison with observational data. The demonstrated dependence of the evolution on cosmological quantities such as Ω_0 and λ_0 makes the comparisons an important consistency test for these parameters. Unfortunately, the accuracy of the observational data available on statistics such as $\xi(r, z)$ is often limited by the small amount of accurate distance information available from high-z galaxy samples. Large amounts of information on positions and magnitudes is, however, usually available. We present a method whereby reliable estimates of clustering statistics can be extracted from such limited information.

1. Clustering Measures

The spatial 2-point correlation function $\xi(y)$ is defined, as usual, with reference to the probability of finding a galaxy in the volume element δV a distance y from a given arbitrary galaxy such that

$$\delta P = n[1 + \xi(y)]\delta V \tag{1}$$

where n is the large scale local number density of galaxies. The observable projection of $\xi(y)$, $w(\theta)$ is defined analogously in terms of positions on the sky. It is possible to estimate ξ from its projection w using Limber's equation. In particular if ξ is a power law then w is also a power law.

$$\xi = By^{-\gamma} \quad w = A\theta^{1-\gamma}$$

The relationship between the amplitudes A and B for any particular sample depends on the distribution in depth of the sample galaxies and is given by

$$B = \frac{A\Gamma(\gamma/2)\left[\int_0^\infty x^2\phi(x)dx\right]^2}{\sqrt{\pi}\Gamma(\gamma - 1/2)\int_0^\infty x^{5-\gamma}\phi^2(x)dx} \tag{2}$$

where $\phi(x)$ is the sample selection function in terms of distance x from the observer. Due to the complexity of this expression it is evident that the uncertainty in B due to the assumed selection function ϕ is almost impossible to quantify. It is clear that, with little or no distance information, w is of little use in investigating the evolution of ξ. What would be desirable, therefore, is a statistic which is directly related to ξ and which can make use of large amounts of data on positions and magnitudes along with limited position data for a few objects thus providing a direct route to $\xi(r,z)$. In the next section we investigate one of these, the projected surface density of galaxies around a galaxy of known distance.

2. Mixed dimentional Statistics

Consider the expected number of galaxies per unit solid angle a distance θ from G

$$E_1(\theta) = N + n\int \xi(y)\phi(r)r^2dr \tag{3}$$

where N is the mean surface density. Since the distance to G is known we can find the number per unit area at G as a function of projected comoving seperation, $s = r_1\theta$, In the case of a power law ξ this can be seen to reduce to

$$\Sigma'(s) = Nr_1^{-2} + n\phi(r_1)BGs^{1-\gamma} \tag{4}$$

where the contant G is a pure function of γ and B is the power law amplitude as before.

If we now note that $n\phi(r_1)$ is the number density of galaxies brighter than some limit, it is apparent that a useful move at this point is to take the same limiting absolute magnitude at r_1, M, for the counts around each centre galaxy G_i. In this case, taking the mean over many centre galaxies,

$$\Sigma(s) = \langle N(r_i)r_1^{-2}\rangle + \Phi(M)BGs^{1-\gamma} \tag{5}$$

where Φ is the galaxies' integral luminosity function. Note that N now depends on r_i as the limiting apparent magnitude for the counts will vary as a function of distance. The first term on the rhs of the above equation can be obtained as the limiting value of Σ at large s, in which case it is appropriate to write

$$\Sigma(s) = \Sigma_\infty + \Phi(M)BGs^{1-\gamma} \tag{6}$$

From this form it becomes relatively straightforward to calculate $\Sigma(s)$ on a galaxy by galaxy basis for objects with measured distances. The simple relationship between Σ and the parameters B and γ then provides a route for the examination of $\xi(z)$.

References

Phillipps, S. 1985, MNRAS, 212, 657

Phillipps, S., Fong, R., Ellis, R.S., Fall, S.M. & MacGillivray, H.T. 1978, MNRAS, 182, 673

Tracing Cosmic Evolution with Galaxy Clusters
ASP Conference Series, Vol. 268, 2002
S. Borgani, M. Mezzetti, and R. Valdarnini, eds.

The Energy Spectrum of Relativistic Electrons in the Coma Galaxy Cluster

Chorng-Yuan Hwang & Ming-Yann Tsay

Graduate Institute of Astronomy, National Central University, Chung-Li, Taiwan; hwangcy@astro.ncu.edu.tw

Abstract.
We present the energy spectrum of relativistic electrons in the Coma Galaxy cluster. The spectrum is derived based on the observed radio, EUV, and hard X-ray data with the hypothesis that the EUV and the hard X-ray excess emission in the Coma cluster is caused by inverse Compton scattering of these relativistic electrons against the cosmic microwave background. The evolution of the relativistic electrons is obtained by solving the diffusion-loss equation assuming magnetic fields of 0.2 & 5 μG.

1. Introduction

EUV observations of the Coma galaxy cluster carried out by the Extreme Ultraviolet Explorer (*EUVE*) showed excess emission compared with a hot thermal plasma model (Lieu et al. 1996; Bowyer et al. 1999). The excess extreme-ultraviolet (EUV) emission detected in the Coma cluster has been explained by the inverse-Compton scattering of relativistic electrons in the cluster against the cosmic microwave background photons (Hwang 1997). With the assumption of an inverse-Compton scattering origin for the EUV excess, the electron spectrum at $\gamma \sim 400$ can be derived.

On the other hand, the Coma galaxy cluster was also found to emit synchrotron radio emission. The electron spectrum around $\gamma \sim 700$ to $\gamma \sim 4 \times 10^4$ can be derived with the radio emission depending on the assuming strength of the magnetic field. These electron might also produce excess emission in the hard X-ray range via the inverse Compton process (Fusco-Femiano et al. 1999; Rephaeli et al. 1999).

We use numerical methods to solve the diffusion-loss equation to derive the evolution and the energy spectra of relativistic electrons in the Coma galaxy cluster. We assume two different values for the intracluster magnetic fields, 0.2 and 5μG. The low magnetic field was implied by the interpretation of an inverse Compton origin for the observed hard X-ray excess emission (Fusco-Femiano et al. 1999; Rephaeli et al. 1999); and the high magnetic field was suggested by Faraday rotation measurements (Feretti et al. 1995).

Our calculations show that the observed relativistic electrons can not be caused by a one-episode injection of relativistic electrons. In particular, we find that it is possible to explain the observed spectrum with continuous injections

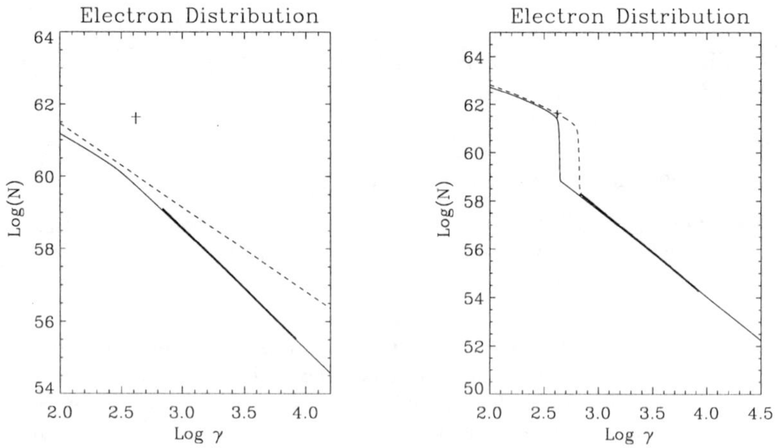

Figure 1. The electron spectra at $B = 5\mu$G. *Left*: It is impossible to fit the EUV and radio data simultaneously with continuous injections of electrons). *Right*: The EUV and radio data can be fitted with an initial major injection plus subsequent continuous minor injections of electrons. The evolution time required to fit the observed EUV and radio data is between 1×10^9 and 1.5×10^9 yr. The plus sign is the distribution of the electrons producing the observed EUV flux and the heavy solid line is the distribution of the electrons producing the synchrotron radio emission.

of electrons only at the low magnetic field; an extra initial major event is required to explain the observed spectrum at the high magnetic field (Figure 1). Furthermore, we find that the relativistic electrons that cause the EUV emission are much older ($\sim 10^9$ yr) than those electrons that radiate the radio emission ($\sim 10^8$ yr). The origin of the energy required for the observed electrons might come from cluster mergers, radio galaxies, and supernovae.

Acknowledgments. The authors acknowledge support from the National Science Council of Taiwan (NSC Grant 90-2112-M-008-040).

References

Bowyer, S., Berghöfer, T. W.,& Korpela, E. J. 1999, ApJ, 526, 592

Feretti, L., Dallacasa, D., Giovannini, G., & Tagliani, A. 1995, A&A, 302, 680

Fusco-Femiano, R., Dal Fiume, D., Feretti, L., Giovannini, G., Grandi, P., Matt, G., Molendi, S., Santangelo, A. 1999, ApJ, 512, L21

Hwang, C.-Y. 1997, *Science*, 278, 1917

R. Lieu, J.P.D. Mittaz, S. Bowyer, J.O. Breen, F.J. Lockman, E.M. Murphy, C.-Y. Hwang 1996, *Science*, 274, 1335

Rephaeli, Y., Gruber, D. & Blanco, P. 1999, ApJ, 511, L21

Tracing Cosmic Evolution with Galaxy Clusters
ASP Conference Series, Vol. 268, 2002
S. Borgani, M. Mezzetti, and R. Valdarnini, eds.

Chandra Observation of the High-Redshift Galaxy Cluster MS 1054−0321

Tesla E. Jeltema, Claude R. Canizares, Mark W. Bautz and Michael R. Malm

Center for Space Research and Department of Physics, Massachusetts Institute of Technology, Cambridge, MA 02139

Megan Donahue

Space Telescope Science Institute, Baltimore, MD 21218

Gordon P. Garmire

The Pennsylvania State University, University Park, PA 16802

Abstract. We summarize the results of the *Chandra* observation of the z = 0.83 cluster MS 1054. We investigated overall cluster properties, cluster substructure, constraints placed on Ω_m, and compared to previous X-ray and optical observations.

1. Introduction

We present the *Chandra* X-ray Observatory observation of MS1054−0321, the highest redshift cluster in the Einstein Extended Medium Sensitivity Survey (EMSS) with z = 0.83. For a full report of this work see Jeltema et al. (2001). This cluster has previously been observed at X-ray wavelengths with *ASCA* and *ROSAT*. Its mass has been estimated from the *ASCA* X-ray temperature of $12.3^{+3.1}_{-2.2}$ keV (Donahue et al. 1998 hereafter D98), from β-model fits to the *ROSAT/HRI* data (Neumann & Arnaud 2000, hereafter NA00), and, at optical wavelengths, from its weak lensing signal (Hoekstra, Franx, & Kuijken 2000; Luppino & Kaiser 1997) and observed velocity dispersion (Tran et al. 1999). All of these methods indicate that MS 1054 is a massive cluster, which in conjunction with its high redshift implies $\Omega_m < 1$. Substructure, an indication that the cluster may not be completely relaxed, was seen in both the *ROSAT/HRI* observation and the weak lensing data (D98; NA00; Hoekstra et al. 2000).

2. MS 1054 with Chandra

Chandra observed MS 1054 with the back-illuminated ACIS-S3 detector on 2000 April 21-22 for 91 ks. We filtered for *ASCA* grade, bad aspect, and high background. Twenty-three sources were identified and removed from the data. The background was estimated from the local background on the ACIS-S3 chip. The net cluster count rate was 0.13 counts s^{-1} in a $2'$ radius region and in the 0.3-7.0

keV band. Spectra were analyzed in the 0.8-7 keV band using XSPEC with standard models. The absorption was fixed at the galactic value of 3.6×10^{20} atoms cm^{-2}, the redshift was fixed at 0.83, and the iron abundance and temperature were free to vary. We find the X-ray temperature of the cluster to be $10.4^{+1.7}_{-1.5}$ keV, lower than, but statistically consistent with, the temperature inferred previously (D98). This temperature agrees well with the observed velocity dispersion and that found from weak lensing (Tran et al. 1999; Hoekstra et al. 2000). We are also able to make the first positive identification of an iron line in this cluster and find a value of 0.26 ± 0.15 for the abundance relative to solar, consistent with early enrichment of the ICM (Mushotzky & Loewenstein 1997). All uncertainties are 90% confidence levels.

We confirm significant substructure in the form of two distinct clumps in the X-ray distribution (D98; NA00). We investigated the spectra of each of these clumps seperately. The eastern clump seems to coincide with the main cluster component. It has a temperature of $10.5^{+3.4}_{-2.1}$ keV, approximately the same as the average spectral temperature for the whole cluster, and it has a position close to the central cD galaxy. The western clump is cooler, with a temperature of $6.7^{+1.7}_{-1.2}$ keV. We found this second peak to have a redshift of z $= 0.84 \pm 0.02$, consistent with a subgroup falling into the cluster. To get an indication of temperature variations in the cluster, we created a map of the hardness ratio. The eastern clump does appear to have a higher hardness ratio and, therefore, a higher temperature than the western clump. There does not appear to be a shock between the two clumps. In their weak lensing study, Hoekstra et al. (2000) find three clumps in the mass distribution of MS 1054. Their central and western clumps seem to correspond to our eastern and western clumps. However, we do not detect the north-eastern weak lensing clump.

Its high X-ray temperature implies that MS 1054 is massive, which along with its high redshift places limits on the mean matter density in the universe, Ω_m. From the overall cluster X-ray temperature we find the virial mass of the cluster to be at least $4.5 \times 10^{14} h^{-1} M_{\odot}$. We confirm the result that the existence of this cluster constrains Ω_m to be less than one (D98; Bahcall & Fan 1998). If $\Omega_m = 1$ and assuming Gaussian initial perturbations, we find the probability of observing MS 1054 in the EMSS is less than $\sim 7 \times 10^{-4}$.

Acknowledgments. This work was funded by contract SAO SV1-61010, NASA contracts NAS8-39073, NAS8-37716, and NAS8-38252, and HST grant GO-6668. MED was also partially supported by NAG5-3257 and NAG5-6236. TEJ would like to acknowledge the support of a NSF fellowship.

References

Bahcall, N.A., & Fan, X. 1998, ApJ, 504, 1

Donahue, M., Voit, G.M., Gioia, I., Luppino, G., Hughes, J.P., & Stocke, J.T. 1998, ApJ, 502, 550 (D98)

Hoekstra, H., Franx, M., & Kuijken, K. 2000, ApJ, 532, 88

Jeltema, T.E., Canizares, C.R., Bautz, M.W., Malm, M.R., Donahue, M., & Garmire G.P. 2001, ApJ, accepted, astro-ph/0107314

Luppino, G.A., & Kaiser, N. 1997, ApJ, 475, 20

Mushotzky, R.F., & Loewenstein, M. 1997, ApJ, 481, L63

Neumann, D.M., & Arnaud, M. 2000, ApJ, 542, 35 (NA00)

Tran, K.V., Kelson, D.D., van Dokkum, P., Franx, M., Illingworth, G.D., & Magee, D. 1999, ApJ, 522, 39

Tracing Cosmic Evolution with Galaxy Clusters
ASP Conference Series, Vol. 268, 2002
S. Borgani, M. Mezzetti, and R. Valdarnini, eds.

Searching for high-redshift clusters with *Planck Surveyor*

Scott T. Kay

Astronomy Centre, University of Sussex, Falmer, Brighton, BN1 9QJ, United Kingdom

Abstract. This poster presents the predicted redshift distribution of galaxy clusters that will be detected by the *Planck Surveyor* satellite, using data drawn from the Hubble Volume simulations. The *Planck* all-sky survey will prove useful in constructing complete samples of clusters at high redshift.

1. Introduction

The redshift distribution of clusters represents a strong test of our underlying cosmological model. Although complete samples of low-redshift clusters are becoming readily available, present knowledge of their distribution at higher redshift ($z \sim 1$) is sparse. Over the coming years, several techniques will be used to find high-redshift clusters, using data at a variety of wavelengths. Perhaps the most appealing method is the Sunyaev-Zel'dovich (SZ; e.g. Sunyaev & Zel'dovich 1980) effect - since the SZ signal is independent of the source redshift, this makes it ideal for searching for high-redshift clusters. The first all-sky survey for SZ sources will be performed by the *Planck Surveyor* satellite, which can then be used in large optical follow-up programmes, in an effort to define complete samples of clusters at high-redshift. In this poster, we present results from using the Hubble Volume simulations (Evrard et al. in preparation) to predict the redshift distribution of the brightest SZ sources. A more detailed study can be found in Kay, Liddle, & Thomas (2001).

2. Redshift distribution of SZ sources

The Hubble Volume simulations give properties of cluster-sized dark matter haloes in lightcones out to $z \sim 4$. Two models are considered for the distribution of intracluster gas, which use the NFW and β model density profiles respectively. Each profile is then convolved with a *Planck* beam function and SZ fluxes are calculated out to an appropriate background value.

Fig. 1 illustrates the redshift distribution of the SZ sources for the two models, in both critical-density (τCDM) and low-density (ΛCDM) cosmologies. A thermal SZ 'flux' limit of $Y > 3 \times 10^{-4}\mathrm{arcmin}^2$ is assumed, and beam full-width-half-maxima of 5 and 8 arcmin are considered, straddling the effective *Planck* beam-size. Integrating the distributions gives the total number of bright SZ sources that will be detected by *Planck*. Both cosmologies predict around 5×10^4 sources, varying by less than a factor of 2 between the different models

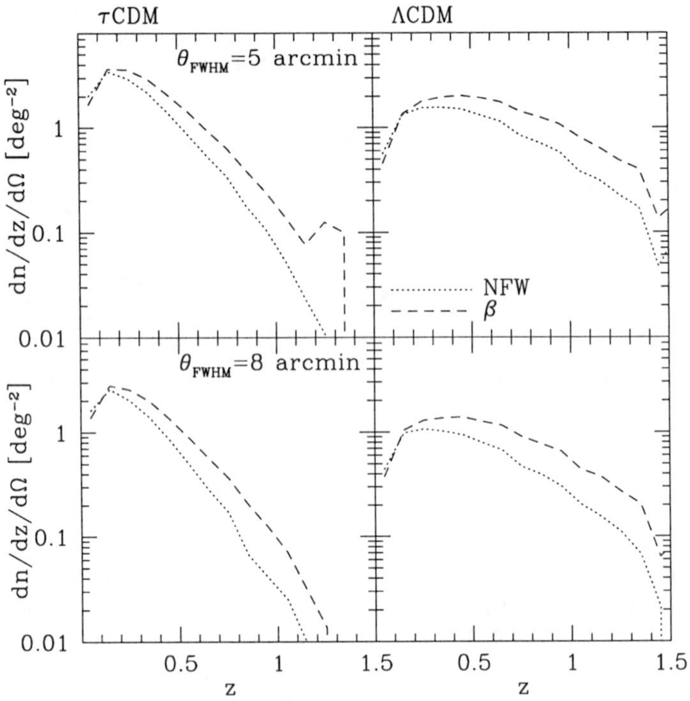

Figure 1. Redshift distributions of bright SZ sources ($Y > 3 \times 10^{-4}\text{arcmin}^2$) in the τCDM and ΛCDM cosmologies.

and beam-sizes. For a given model, the τCDM cosmology predicts a larger number of nearby sources than the ΛCDM cosmology, but fewer sources at high-redshift. This arises due to different rates of structure growth between the two cosmologies (clusters form earlier in the ΛCDM case), as well as the angular diameter distance dependence of the fluxes ($Y \propto d_A^{-2}$), which is larger for a given redshift in the ΛCDM case.

Differences due to the choice of model (NFW or β) are caused by the nature of the gas distribution. In the β model the gas is more extended than in the NFW model, and so predicts more objects above the flux limit. Finally, a larger beam-size smooths the apparent distribution to larger angular scales, reducing the number of detected objects. Indeed, *Planck* will only resolve at most a few per cent of the total number of clusters as extended sources.

References

Kay, S. T., Liddle, A. R., & Thomas, P. A. 2001, MNRAS, 325, 835
Sunyaev, R. A., Zel'dovich, Ya. B. 1980, ARA&A, 18, 537

Tracing Cosmic Evolution with Galaxy Clusters
ASP Conference Series, Vol. 268, 2002
S. Borgani, M. Mezzetti, and R. Valdarnini, eds.

Far-Infrared Studies of Cluster Evolution with SIRTF

Douglas M. Kelly

Steward Observatory, University of Arizona, Tucson AZ 85721, USA

George H. Rieke

Steward Observatory, University of Arizona, Tucson AZ 85721, USA

Abstract. The goal of this program is to study the history of star formation and the evolution of galaxies in cluster environments. Since much of the bolometric luminosity of galaxies emerges at far-infrared wavelengths, this spectral region plays a very important role in galaxy evolution studies. We will measure the far-infrared brightnesses of individual galaxies in a handful of nearby clusters, probing 3–7 magnitudes below L_*. We will measure brightness profiles and total far-infrared fluxes for clusters at redshifts of 0.2-1.27. These observations will let us study how cluster evolution depends on factors such as the gas content, gas infall rate, and galaxy-galaxy interactions.

1. Source Selection

For observations in the far-infrared, the main limiting factors that determine the sensitivity limit of a survey are the background flux due to zodiacal light and Galactic cirrus and confusion from the cosmic far-infrared background. To maximize our sensitivity, we are limiting our source selection to low to moderate background regions, preferably with $n_H < 2.5 \times 10^{20}$ cm^{-2} and at ecliptic latitudes of > 30 degrees. We will detect a large number of background galaxies and an unresolved infrared background, and sorting out these contributions will be an important part of this study. To ensure that we are observing high mass, bright clusters, we impose a source selection based on X-ray luminosity.

Table 1. MIPS Cluster Sensitivity Limits

Wavelength (μm)	Low-z clusters 1σ (μJy)	High-z clusters 1σ (μJy)	L/L_*(Coma) (z=0.023)	L/L_*(A2029) (z=0.077)
24	145	20	0.0005	0.006
70	950	350	0.0015	0.018
160	9000	1500	0.0035	0.040

2. Observing Strategy

The Multiband Imaging Photometer for SIRTF (MIPS) provides very efficient mapping over large areas in its scan-mapping mode. We will use this mode for observing individual galaxies in nearby clusters, with 1 degree long scans and with 5' offsets (one field of view) between scans. With a single pass, we will be far from the ultimate sensitivity limit of MIPS but will nonetheless reach 3-7 magnitudes below L_*(FIR). Deeper guest observer programs could also prove valuable. The high redshift clusters will be observed in a chop-and-step photometry mode. The sensitivity thus achieved should provide an expected signal-to-noise of roughly 100 on the cluster as a whole. After identifying individual cluster members, we plan to measure surface brightness profiles by remapping the data to a resolution of between 150-300 kpc per bin. In a companion program, photometric observations will be made of roughly 20 X-ray selected clusters at redshifts between 0.2 – 0.4. Gravitational lensing is expected to amplify the signal from background galaxies by a factor of 3-6, which should make detectable some galaxies that would normally be below the confusion limit. In addition, the data will allow detection of individual cluster members to L_*(FIR) or below.

3. Goals

- Measure star formation rate as a function of redshift
- Study star formation rate as a function of location in the cluster
- Estimate the dust obscuration as a function of redshift
- Determine the number of obscured star forming galaxies vs redshift
- Measure the evolution in the shape of the far-infrared luminosity function vs redshift
- Study the nature of E+A galaxies
- Measure intracluster dust emission in nearby clusters
- Combine our data with X-ray, UV, visible, near-infrared, and radio data on clusters to study the gas and dust content, masses, star formation rates, and morphologies of cluster galaxies as functions of cluster mass and age
- Compare infrared properties of galaxies in massive, dense clusters with the properties of field galaxies at similar redshifts detected in the SIRTF GTO deep surveys

Tracing Cosmic Evolution with Galaxy Clusters
ASP Conference Series, Vol. 268, 2002
S. Borgani, M. Mezzetti, and R. Valdarnini, eds.

Statistics of galaxies in clusters

Birgit Kelm, Paola Focardi

Dipartimento di Astronomia, Via Ranzani 1, 40127 Bologna, Italy,

Abstract. We performed statistical investigation of galaxy membership in Abell clusters and find the fraction of galaxies that are cluster members to increase roughly from 4% to 8% to 12%, when computed out to 1, 2 and 3 R_A respectively. When distinction between absorption line (A) and emission line (E) galaxies is introduced, cluster membership turns out to be twice as high among A than among E galaxies.

1. Introduction

Rich galaxy cluster contain, by definition (Abell 1958) at least 30 galaxies brighter than m_3+2 within 1 R_{Abell} (1.5 h^{-1} Mpc) of the cluster center. In the literature (Bahcall 1999) the fraction of galaxies in rich clusters is claimed to be ≈5% (within 1 R_A) and to increase with increasing radius. In this study we extend quantitative statistics out to 3 R_A and additionally investigate distributions of galaxies according to their specific spectral properties.

2. Abell Cluster membership

We identify 5 samples, complete in magnitude and volume, in the UZC (Falco et al. 1999) redshift survey, which is 96% complete (to mag=15.5) for Zwicky galaxies. Samples include galaxies brighter than magnitude -18.5, -19, -19.5, -20, -20.5 respectively. Abell clusters above 4000 km/s in the Rood & Struble compilation (1999) constitute the sample to which galaxies are associated. For each sample we plot in **figure 1** (upper left panel) the fraction of galaxies that are cluster members out to 1, 2 and 3 R_A. To avoid severe projection effects we set $\Delta|cz|\leq 1000$ km/s. The fraction of galaxies in the cluster central regions ($r\leq 1R_A$) is ≈ 4% increasing to 8 and 12% at 2 and 3 R_A. The usually quoted fraction (5%) of galaxies associated to clusters is thus confirmed by the present investigation, indicating also that the result is not due to the sampling of galaxies in a specific luminosity range.

3. Abell Cluster membership according to spectral properties

The UZC catalogue lists spectral properties (A=Absorption lines, E=Emission lines, B=Both) for ≈ 80% of the galaxies in our samples. We use spectral information to compute galaxy cluster membership in samples discriminating between E, B and A galaxy respectively. The upper right and the two lower

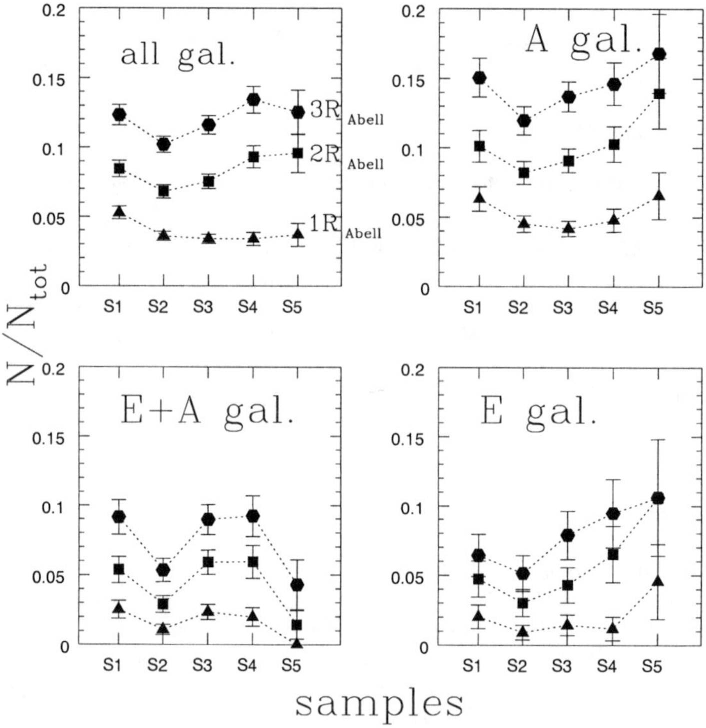

Figure 1. Cumulative fraction of cluster member galaxies out to $1R_A$ (triangles), $2R_A$ (squares) and $3R_A$ (hexagons). Fractions are plotted for 5 different samples (S1 to S5) which are complete in volume for galaxies brighter than Mag -18.5, -19, -19.5, -20, -20.5 respectively. The upper left panel shows fractions for all galaxies, the other panels show fractions for subsamples selected according to spectral properties.

panels in **figure 1** show occurrence of A, B and E galaxies in Abell clusters when normalized to the total UZC population of A, B and E galaxies. As can be seen ≈5-15% (depending on R) of the total population of A galaxies are cluster members versus a fraction of ≈2-8% of E and B galaxies. Once more figures appear not to scatter significantly among samples.

References

Abell, G.O. 1958, ApJS, 3, 211

Bahcall, N.A. 1999, in *Formation of Structures in the Universe* ed A. Dekel & J.P. Ostriker, Cambridge Un. Press, UK

Falco, E.E., Kurtz, M.J., Geller, M.J. et al. 1999, PASP, 111, 438

Struble, M.F., Rood, H.J. 1999, ApJS, 125, 35

Tracing Cosmic Evolution with Galaxy Clusters
ASP Conference Series, Vol. 268, 2002
S. Borgani, M. Mezzetti, and R. Valdarnini, eds.

Non–Gaussian morphology on large scales: Minkowski functionals of the REFLEX clusters of galaxies

M. Kerscher

Sektion Physik, Ludwig–Maximilians–Universität, Theresienstraße 37, D–80333 München, Germany

K. Mecke

Max-Planck-Institut für Metallforschung, Stuttgart, Germany
Fakultät für Physik, Universität Stuttgart, Stuttgart, Germany

P. Schuecker, H. Böhringer

Max-Planck-Institut für Extraterrestrische Physik, Garching, Germany

L. Guzzo, S. De Grandi

Osservatorio Astronomico di Brera, Merate, Italy

C. A. Collins, S. Schindler

Liverpool John Moores University, Liverpool, UK

R. Cruddace

Navel Research Laboratory, Washington DC, USA

1. Methods

The spatial distribution of galaxy clusters poses important constraints on cosmological models. To quantify the large–scale structures traced by the galaxy clusters we have to go beyond the number density. At each cluster we attach a sphere (see Fig. 1). The union set of these spheres is characterized by the

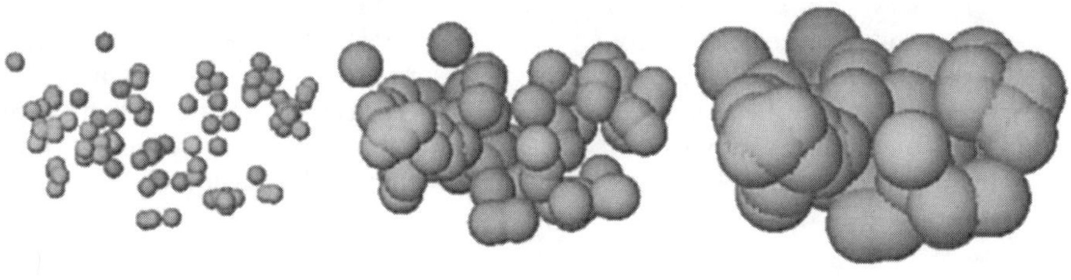

Figure 1. Spheres of varying radius attached to a sample of galaxy cluster from the REFLEX survey.

four Minkowski functionals (Mecke, Buchert, & Wagner 1994): the volume, the surface area, the integral mean curvature and the Euler characteristic. The ra-

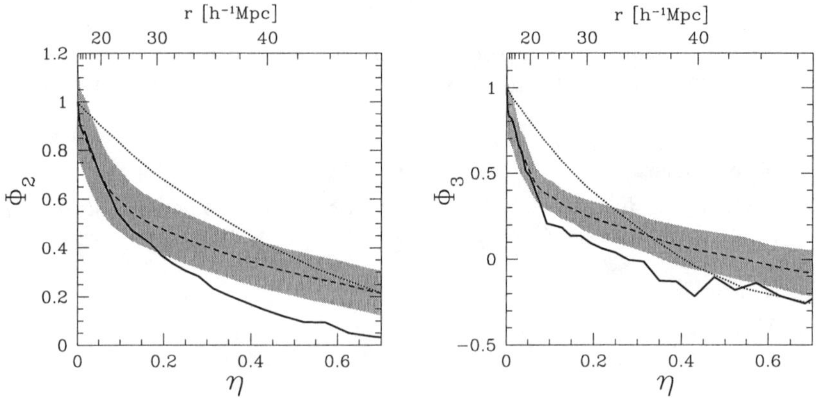

Figure 2. The normalized integral mean curvature Φ_2 and the normalized Euler Characteristic Φ_3 of the REFLEX cluster sample (solid line). The shaded area is determined from simulated Gaussian point distributions with the same two–point correlation function as the REFLEX sample. The dotted line is for Poisson distributed points.

dius of the spheres is used as a diagnostic parameter. Minkowski functionals can be expressed in terms of integrals of n–point correlation functions. Specifically, we derive analytical expressions for a Gaussian point distribution. They can be compared with the Minkowski functionals measured from the REFLEX survey (Kerscher et al. 2001).

2. The Non–Gaussian distribution of REFLEX clusters

We calculate the Minkowski functionals of volume limited subsamples from the REFLEX catalogue of galaxy clusters and compare them with a Gaussian point distributions with the same two–point correlation function as determined from the REFLEX sample. As seen in Fig. 2 the Minkowski functionals of the observed cluster distribution deviate from the Gaussian case. A method to generate Gaussian point sets for a given two–point correlation function allows us to quantify the significance of these deviations. We find that the large–scale distribution of galaxy clusters in the REFLEX survey is incompatible with a Gaussian point distribution at a significance level of 95%. A non–Gaussian cluster distribution could still be compatible with an initially Gaussian mass density field. However, in such a case biasing and/or non–linear structure formation processes are important.

References

Mecke, K. R., Buchert, T., & Wagner, H. 1994, A&A, 288, 697

Kerscher, M., Mecke, K. R., Schuecker, P., Böhringer, H., Guzzo, L., Collins, C.A., Schindler, S., De Grandi, S., & Cruddace, R. 2001, A&A, 377, 1

Tracing Cosmic Evolution with Galaxy Clusters
ASP Conference Series, Vol. 268, 2002
S. Borgani, M. Mezzetti, and R. Valdarnini, eds.

The Alignment Effect of Brightest Cluster Galaxies in the SDSS

R.S.J. Kim[1,2], J. Annis[3], M.A. Strauss[1], R.H. Lupton[1] for the SDSS collaboration

[1]*Princeton University Observatory, Princeton, NJ 08544, USA*

[2]*Department of Physics and Astronomy, The Johns Hopkins University, Baltimore, MD 21218, USA*

[3]*Fermilab, Batavia, IL 60510, USA*

Abstract. We have examined the alignment of Brightest Cluster Galaxies (BCG) with their host clusters, with the Sloan Digital Sky Survey (SDSS). We find that the BCGs are preferentially aligned with the principal axis of their hosts, to a much higher redshift ($z \gtrsim 0.3$) than previously probed ($z \lesssim 0.1$). In particular, the alignment effect is very strong for dominant BCGs, while less dominant BCGs do not show any departure from random alignment. We therefore claim that the alignment process originates from the same process that makes the BCG grow dominant, be it direct mergers in the early stage of cluster formation, or a later process that resembles the galactic cannibalism scenario.

There have been many reports in the past two decades of coherent orientation between galaxies, clusters and large-scale structure (see Djorgovski 1986 for review). In particular, using 44 Abell clusters ($z < 0.1$), Bingelli (1982) reported a tendency for Brightest Cluster Galaxies (BCG) to be aligned with their parent cluster orientations. In this paper, we examine the BCG-cluster alignment effect with 115 clusters detected from the Sloan Digital Sky Survey (York et al. 2000; SDSS) to $z \sim 0.5$. We have develop a automated framework with which we can analyze a much larger set of upcoming SDSS cluster samples; a complete study will be presented in a forthcoming paper (Kim et al. 2001b).

We use 112.5 deg^2 of imaging data that are a part of the SDSS Early Data Release (Stoughton et al. 2001). In order to select a robust cluster sample, we use two independent detection techniques: the Matched Filter (Kim et al. 2001a and references therein) and the MaxBCG technique (Annis et al. 2001). We obtain 211 clusters that are detected by both methods and have robust BCG detections. BCGs are usually conspicuous thanks to their bright luminosities and unique morphologies, however, not all BCGs are as dominant. We quantify this "degree of dominance" (Δm) by the magnitude difference between the BCG and the average magnitude of the second and third brightest member.

The position angle ϕ and axis ratio α of individual BCGs are taken from the outputs of the SDSS photometric pipeline (see York et al. 2000 and references within). We measure the shape and orientation of the clusters by invoking the Voronoi Tessellation cluster finder (Kim et al. 2001a) reversely, retrieving the cluster members with which we compute the inverse-radius weighted moments.

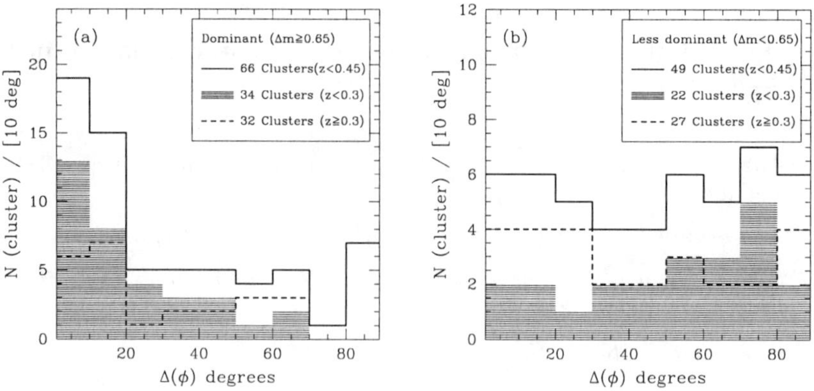

Figure 1. (a) Histogram distribution of the difference in the position angles of the BCG and the cluster, for 66 clusters with dominant BCGs (solid-line), subset of 34 with $z < 0.3$ (shaded) and 32 with $z \geq 0.3$ (dashed). (b) Same as (a) for 49 clusters with less dominant BCGs.

After eliminating cases where the clusters or BCGs are consistent with being round, we finally obtain a set of 115 clusters with well determined parameters.

We divided the cluster sample in two according to their "degree of dominance" measured by the index Δm. The results for 66 clusters with $\Delta m \geq 0.65$ and 49 clusters with $\Delta m < 0.65$ are shown in Fig. 1a and b respectively (solid-lines). The contrast is remarkable: dominant BCGs show striking alignment with their host clusters, whereas we see no departure from randomness for less dominant BCGs. K-S statistics confirm that while the dominant BCGs are aligned at $> 99.999\%$ confidence, the less dominant BCGs are consistent with a random distribution within 1σ. The two samples are further divided in redshift, clusters with $z < 0.3$ (shaded) and $z > 0.3$ (dashed). The low z sample for dominant BCGs are particularly striking, however, we do not draw any conclusions on z evolution, due to the small sample size. Our results support the scenario in which BCG are formed through mergers, however, analyzing a larger sample as a function of z will be crucial to constrain the formation epoch of BCGs.

References

Annis et al. 2001, in preparation

Bingelli, B. 1982, A&A, 107, 338

Djorgovski, S. 1986, *Nearly Normal Galaxies*, p227, Ed. Faber, Springer-Verlag, NY

Kim et al. 2001a, AJ, accepted

Kim et al. 2001b, in preparation

Stoughton et al. 2001, AJ, submitted

York et al. 2000, AJ, 120, 1579

Tracing Cosmic Evolution with Galaxy Clusters
ASP Conference Series, Vol. 268, 2002
S. Borgani, M. Mezzetti, and R. Valdarnini, eds.

Galaxy cluster mass function evolution caused by galaxy mergings

V.M. Kontorovich, S.I. Shelyag

Institute of Radio Astronomy of NAS Ukraine, Chervonopraporna str., 4, 61002, Kharkiv, Ukraine

Abstract. The evolution of the galaxy mass function (MF) both on weak and strong edges is discussed. The mergings are assumed to be responsible for the (observable) weak edge steepness, and MF is adjusted by mass flow and galaxy-number flow on the spectra. We use the Monte-Carlo method with mass depending merging cross section. Recalculation from the current number of galaxies to the own age of cluster and redshift was realized with the kinetic Smoluchowsky equation. Non-monotonic redshift - dependence of cluster richness (and masses of the first-by-brightness galaxies) is obtained. The cD-galaxy formation is retraced. Obtained dependencies conform well to the known galaxy cluster catalogues.

1. Introduction

Development of astronomical observation tools in recent time makes possible to investigate clusters and their evolution on a wide range of distances. Owing to growing of observational data volume the importance of mergings in evolution both galaxies and clusters of galaxies (Dressler, 1984) was confirmed. The role of mergings grows with increasing of distance to the past. In this work the evolution of strong edge of cluster MF, caused by merging of galaxies in cluster, is discussed. The Abell richness and the magnitudes of most massive galaxies of cluster were chosen as evolutionary characteristics for cluster. The data about these magnitudes are contained in the catalogues (Struble & Rood, 1999), (Osborne, 2001). The role of mergings at formation of the spectrum at weak edge of MF was discussed by Kontorovich (1999).

2. The cluster richness evolution in the merging model (The Monte-Carlo method)

The cluster is modeled with using of the Monte-Carlo method. N_0 of particles with initial distribution function localized on small $M \sim M_*$ masses $f(M) = (N_0/M_*)\exp(-M/M_*)$ are merging in pairs with probability $U \propto (M_1 + M_2)(M_1^\beta + M_2^\beta)$ where β is determined by dependence of galaxy mass on its radius $R = C_\beta M^\beta$ (Krivitsky & Kontorovich, 1999). The cluster richness is introduced as the integral of mass function formed as result of merging:

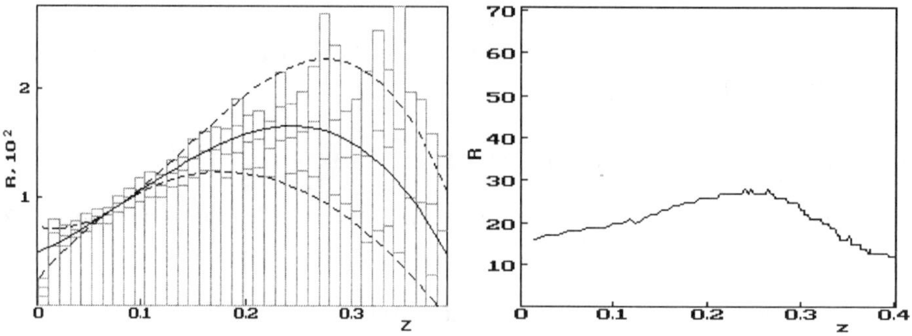

Figure 1. The richness dependence on redshift, obtained from the galaxy cluster catalogues (left). The richness dependence, obtained from the model with $M_* = 2 \cdot 10^{11} M_\odot$, $\beta = 1/3$, $N_0 = 1000$ (right).

$R = \int_{m_3+2}^{m_3} f(M)dM$. The galactic magnitudes m are calculated accordingly to Pogson formula, the mass-luminosity ratio is assumed as independent on masses.

3. Recalculation of the number of merging acts to the redshift

For comparison of received dependences with the observational data it is necessary to execute recalculation of number of merging acts to the time, past from the beginning of merging, and then in redshift. Recalculation is carried out with help of Smoluchowsky equation, describing process of merging: $\partial f(M,t)/\partial t = \int_0^M U(M_1, M - M_1, t) f(M,t) f(M - M_1, t) dM_1 - 2f(M,t) \int_0^\infty U(M_1, M, t) \times$ $\times f(M,t) f(M - M_1, t) dM_1$, where $f(M,t)$ - cluster MF. The Runge-Kutt numerical method is applied for MF calculation. The technique of numerical solution is described by Krivitsky (1995). Using obtained from the solution dependence $N(t)$, where N is the current number of galaxies in cluster, we obtained the average richness dependence on time. Dependence of the richness on redshift (fig. 1, right) was obtained with using of redshift dependence on time for standart cosmological model of flat world ($\Omega = 1$) and of that fact the richness gets it's local minimum on $z \approx 0.4$, accordingly to the observational data (fig. 1, left). Also, the redshift, where process of galaxy merging was began ($z \approx 0.9$), was obtained from the model.

References

Dressler, A. 1984, Ann. Rev. Astron. Astrophys., 22, 185.

Kontorovich, V. 1999, Proceedings of IAU Colloquium 174, 209, 423.

Krivitsky, D. 1995, J. Phys. A: Math. Gen., 28, 2025.

Krivitsky, D., Kontorovich, V. 1999, Astr. & Astrophys. Trans., 18, 43.

Osborne, J. 2001, *ftp.legacy.gsfc.nasa.gov/heasarc/dbase/db_files/abelldb.dat.z*

Struble, M. F., Rood, H. J. 1999, ApJ, 125, 35.

Tracing Cosmic Evolution with Galaxy Clusters
ASP Conference Series, Vol. 268, 2002
S. Borgani, M. Mezzetti, and R. Valdarnini, eds.

Is Abell 970 a Recently Merged Cluster?

Gastão B. Lima Neto[1], Hugo V. Capelato[2] & Laerte Sodré, Jr.[1]

[1]*IAG/USP, av. Miguel Stefano, 4200, São Paulo/S.P., Brazil*

[2]*Div. de Astrofísica INPE/MCT, São José dos Campos/S.P., Brazil*

Abstract. Analysis of new galaxy velocities indicates that Abell 970 may just now be reaching dynamical equilibrium. We report here new results based on observations by BeppoSAX. We were able to obtain a low spatial resolution temperature and abundance profiles. Our results suggest that Abell 970 is a recently ($\sim$ 3–4 Gyr) merged cluster.

1. Introduction

Abell 970 ($z = 0.0587$) was observed in July 2000 by BeppoSAX instruments: the LECS, well calibrated in the [0.1–4.0] keV range, and the MECS 2/3, sensitive in the [1.3–10.5] keV range (Boella et al. 1997), with net exposure times of 33 619 and 97 289 s, respectively. Figure 1 shows the X-ray isocontour superposed on a DSS R image. Note the offset between the X-ray and the galaxy distribution.

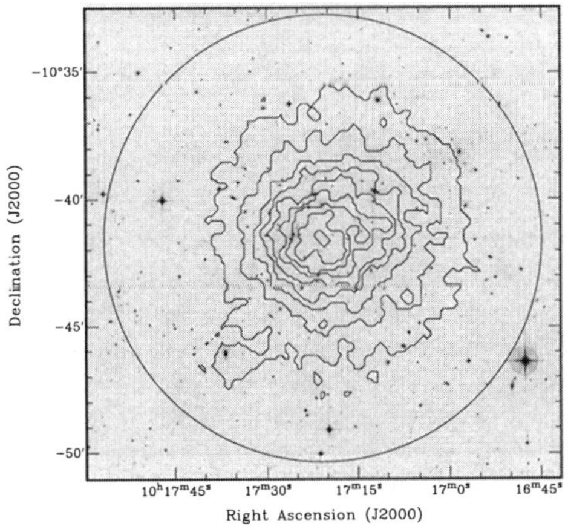

Figure 1. X-ray image obtained with BeppoSAX MECS superposed on a DSS image. The circle indicates the area used in our analysis.

We have defined sub-regions to perform separate spectral analysis: 6 concentric annuli (the radial profile) and 4 regions corresponding to NE and SW halves, and NW and SE halves. These regions were chosen based on the galaxy distribution and their velocities (Sodré et al. 2001, Capelato et al. this volume). For each region we have used a single temperature plasma, the absorbed MEKAL (Liedahl et al. 1995) model under XSPEC v11.

2. Results and conclusions

Fig. 2 shows the temperature radial profile. Notice the strong cooling at $R \sim 6'$.

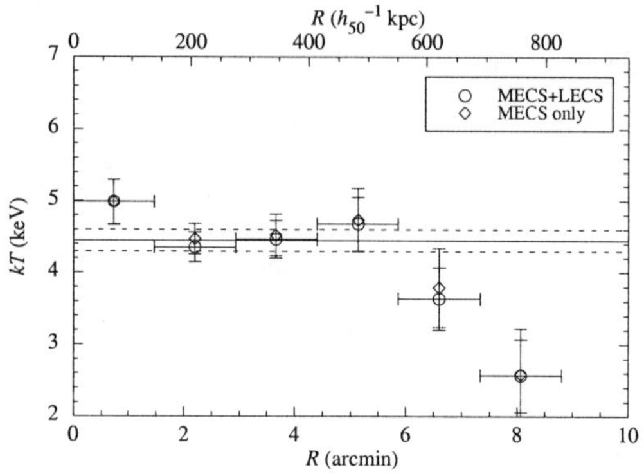

Figure 2. Temperature profile. Error bars are 1σ. The horizontal lines show the mean temperature (full line) and the 1σ error bar (dashed) measured using all data within 8 arcmin ($kT = 4.46^{+0.14}_{-0.15}$).

In the direction NW↔SE the temperature gradient is more than 1σ significant (Fig. 3). The direction of the temperature gradient is about the same as the observed velocity gradient (Sodré et al. 2001, Capelato et al., this volume). Moreover, the galaxy distribution shows the presence of a sub-structure NW of the centre, with a velocity dispersion of 380^{+120}_{-70}km s^{-1}.

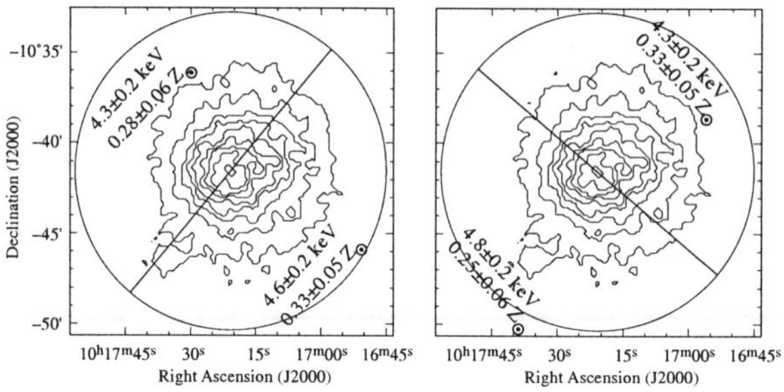

Figure 3. Temperature and metallicity in the regions NE and SW (left) and in the regions NW and SE (right) superposed on the combined MECS 2/3 isocontours. Note the possible temperature gradient in the NW–SE direction.

Although Abell 970 presents rather symmetric X-ray isophotes, the peak of the X-ray distribution do not coincide with the galaxy distribution maximum.

The temperature profile shows a strong gradient at $\sim 6'$. Such a cool region may develop a few Gyrs after a major merging (cf. Schindler & Müller 1993).

The galaxy distribution (spatial and velocity), together with the X-ray isophote off-centre and possible temperature gradient, suggests that Abell 970 had a merger in the recent (3–4 Gyr) past.

Acknowledgments. We thank Sergio Dos Santos and Florence Durret for valuable discussion, and the support from USP/COFECUB cooperation and PRONEX.

References

Schindler S., Müller E., 1993, A&A 272, 137
Sodré L., Jr., et al., 2001, A&A in press

Tracing Cosmic Evolution with Galaxy Clusters
ASP Conference Series, Vol. 268, 2002
S. Borgani, M. Mezzetti, and R. Valdarnini, eds.

The BMW Deep X–Ray Cluster Survey

M. Longhetti, A. Moretti, L. Guzzo, S. Campana, M.R. Panzera, G.Tagliaferri

Oss. Astr. di Brera, via Bianchi 46, 23807 Merate (LC), Italy

I. Dell'Antonio

Physics Department, Brown University, Box 1843, Providence, RI, USA

D. Lazzati

Institute of Astronomy, University of Cambridge, Madingley Road, Cambridge CB3 OHA, UK

Abstract. We describe the main features of the BMW survey of serendipitous X-ray clusters, based on the still unexploited ROSAT-HRI archival observations. The sky coverage, surface density and firts deep optical CCD images of the candidates indicate that this sample can represent an excellent complement to the existing PSPC deep cluster survey and will provide us with a fully indipendent probe of the evolution of the cluster abundance, in addition to significantly increasing the number of clusters known at $z > 0.6$.

1. The BMW survey

The BMW cluster survey is based on the global Brera Multi-scale Wavelet (BMW) project which has analysed 3100 HRI pointings using a wavelet detection algorithm (Lazzati et al. 1999), producing a catalog of $\sim$ 21000 serendipitous sources with measured fluxes and extensions (Campana et al. 1999, Panzera et al. 2001). Cluster candidates are identified as those sources appearing extended at a confidence level $> 99.9\%$ within the well–sampled HRI area between 3 and 15 arcmin off-axis (see Moretti et al. 2001 for details). Figure 1 shows the BMW survey sky coverage as a function of both X–ray flux and source extension (left) and its comparison with that of previous X–ray surveys (right) for a typical faint source with 10–15 arcsec core radius. As can be easily seen, the BMW cluster survey exhibits a good compromise between a fairly large area at intermediate fluxes ($\sim$ 100 sq. deg. around $\sim 10^{-13}$ erg cm^{-2} s^{-1}), and depth (1 sq. deg. at 2.5×10^{-14} erg cm^{-2} s^{-1}).

In 2000, we started a systematic identification campaign of about 150 BMW cluster candidates (limiting flux = 2.5×10^{-14} erg cm^{-2} s^{-1}) at redshift > 0.4 (i.e., non detectable on the DSS images) through optical/NIR imaging on 4 m class telescopes (ESO 3.6m and TNG), and our first results are very encouraging confirming at least 65% of the observed candidates. Important further

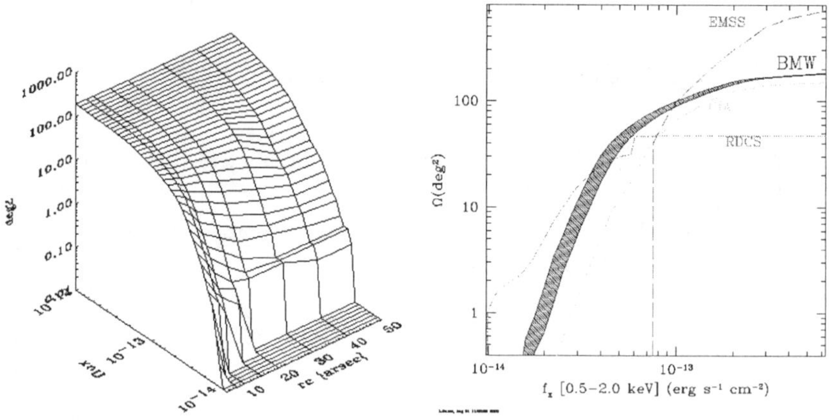

Figure 1. The BMW survey sky coverage (see text).

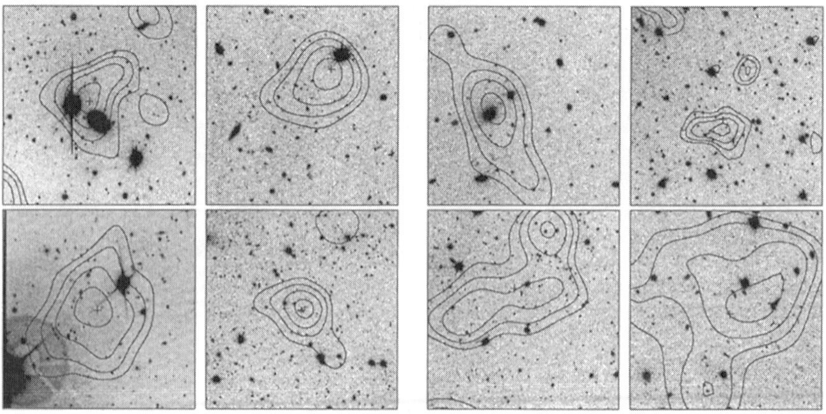

Figure 2. A gallery of a few newly confirmed BMW clusters.

information comes from about $\sim$ 10 clusters found in common with existing PSPC surveys (mostly the RDCS, courtesy of P. Rosati). This set of objects interestingly include clusters at $z = 0.64$, $z = 0.81$ and $z = 0.89$, confirming our ability to peer efficiently into the $z \sim 1$ regime, where only a handful of X-ray clusters are known.

References

Campana, S. et al. 1999, ApJ,524, 423

Lazzati, D. et al. 1999, ApJ, 524, 414

Moretti et al. 2001, astro-ph/010334

Panzera, M.R. et al. 2001, in preparation

Tracing Cosmic Evolution with Galaxy Clusters
ASP Conference Series, Vol. 268, 2002
S. Borgani, M. Mezzetti, and R. Valdarnini, eds.

Wide-Field Multicolour Imaging of Selected *REFLEX* Clusters

P. Lynam, H. Böhringer, P. Schuecker, Th. Reiprich

Max-Planck-Institut für Extraterrestriche Physik, Garching, Germany.

E. Molinari, L. Guzzo, L. Barone

Osservatorio Astronomico di Brera-Merate, Italy.

C. Collins, S. Schindler

Astrophysics Research Institute, Liverpool John Moores University, U.K.

D. Neumann

CEA, DSM, DAPNIA, Services d'Astrophysique, C.E. Saclay, Gif-sur-Yvette, France.

Abstract.

We are engaged in deep, wide-field, BVR photometric imaging of statistically representative samples of *REFLEX* galaxy clusters. This project exploits a new generation of wide-field mosaic imaging arrays. Devices such as the ESO Wide Field Imager (WFI) enable nearby galaxy clusters for the first time to be uniformly sampled in the optical regime, across the whole of their extent. Our unbiased, systematic approach shall provide reliable, homogeneous data sets for subsequent physical studies of these environments. Limited by the field of view of conventional detectors, previous disparate investigations have sampled fewer objects and thus have been susceptible to the possible inclusion of a few extreme constituents.

1. Introduction

The interplay between environment and evolution governs galaxy and cluster behavior at any cosmological epoch. To understand and predict the properties of such structures with look-back time would provide a method to chart the history and probe the future of the Universe. The obvious place to look for evolutionary effects is in the optical galaxy population. X-ray information provides an unbiased reflection of the physical properties of the cluster; the distribution of X-ray emitting gas in a cluster traces its gravitational potential. The most X-ray luminous clusters are correspondingly the most massive (Reiprich & Böhringer 1999). The most comprehensive X-ray cluster sample is the *REFLEX* (Böhringer et al. 2001): a well-controlled sample of over 450 clusters derived from the ROSAT all-sky survey — which remains the premier survey of its kind.

2. Towards the First Statistically Representative Samples

REFLEX pioneers unbiased, *statistically meaningful* investigations linking X-ray and optical properties of clusters. We exploit a new generation of wide-field detectors which enable nearby *bona-fide* galaxy clusters for the first time to be reliably sampled homogeneously in BVR, across their full extent. Clusters within two primary redshift intervals, selected specifically to take maximum advantage of WFI and XMM, are scrutinized. At both epochs, we seek to sample the full range of X-ray luminosity (and therefore cluster mass).

- X-ray flux limit, $F_X \simeq 3 \times 10^{-12}$ erg s^{-1} cm^{-2}.
- Primary redshift intervals, $z \sim 0.14 \rightarrow 0.18$, $z \sim 0.30$, with a supplementary sparse sample at $z \sim 0.45$.
- ~ 50 clusters in each primary redshift interval — 21 so far observed.
- Limiting magnitude: *at least* $B \sim 24.5$ mag. in all cluster fields

Considered separately, each interval presents a comprehensive study of co-eval cluster properties, in which sufficient objects are observed to reduce the skewing effect of anomalous behavior in individual components. Hence the data better reflect the global properties of the Universe and subsequent interpretations shall be statistically robust. Comparison of bulk properties of the ensemble of clusters contained in each distance interval shall probe the recent evolutionary history of the Universe, since the different distances correspond to views of the Universe at different ages.

3. Status

Data acquisition is ongoing. Development of a reduction prescription is at an advanced stage. The remaining task is the creation of a corrective map to recover accurate photometry across the entire field of view. The 0.5° image presented at Sesto and preliminary results for the galaxy luminosity function (Barone et al. 2000), illustrate our capability to manipulate the avalanche of data from WFI, and its potential.

Acknowledgments. Jakob Englhauser (MPE) made his *RGB* software available, through which composite images were produced. Frank Valdes (NOAO) assisted in support of bugs isolated in external IRAF packages. We became aware of the required WFI photometric corrections via correspondence with Fernando Selman (ESO, La Silla) and Jean Manfroid (Liege, Belgium).

References

Barone, L. T., Molinari, E., Böhringer, H. et al. 2000 in Proc. IAP Coll. 'Constructing the Universe with Clusters of Galaxies', Paris, France, July 2000. astro-ph/0009251.

Böhringer, H., Schuecker, P., Guzzo, L., et al. 2001 A&A, 369, 826

Reiprich, Th., Böhringer, H. 1999, Astronomische Nachrichten, 320, 296

Tracing Cosmic Evolution with Galaxy Clusters
ASP Conference Series, Vol. 268, 2002
S. Borgani, M. Mezzetti, and R. Valdarnini, eds.

XMM–NEWTON Observation of Abell 1835 : Temperature, Mass and Gas Mass Fraction Profiles

S. Majerowicz[1], D.M. Neumann[1], T.H. Reiprich[2]

1. *CEA/Saclay, Service d'Astrophysique, L'Orme des Merisiers, Bât. 709, 91191 Gif-sur-Yvette Cedex, France*

2. *Max-Planck Institut für extraterrestrische Physik, PO Box 1312, 85741 Garching, Germany*

Abstract. We present a study of Abell 1835 based on observation of the EPIC cameras aboard XMM–NEWTON. We find a temperature profile up to 1.8 Mpc which is basically flat outside the cooling flow region. Fitting a β-model and assuming hydrostatic equilibrium allow us to determine a gas mass fraction of $0.18\,h_{50}^{-1.5}$ at 70% of the virial radius.

1. Introduction

Abell 1835 is a medium distant (z∼0.25) galaxy cluster which was observed during June 2000 with XMM–NEWTON. We concentrate on EPIC–MOS2 and –PN cameras. The study of Abell 1835 using RGS spectrometer was made by Peterson et al. (2001). We use the remarkable spatial and spectral resolution of XMM–NEWTON to determine the temperature, mass and gas mass fraction profiles with unprecedented accuracy.

2. Imaging

In order to estimate the gas mass, we use the β-model : $S(r) = S_0(1 + (r/r_c)^2)^{-3\beta+0.5}$. Figure 1 shows the cluster in the 0.3 to 3 keV energy band. From the data we calculate S. And its fit give us : $\beta = 0.704 \pm 0.005$ and $r_c = 202.3 \pm 7.1\,\mathrm{kpc}$ (68% confidence level).

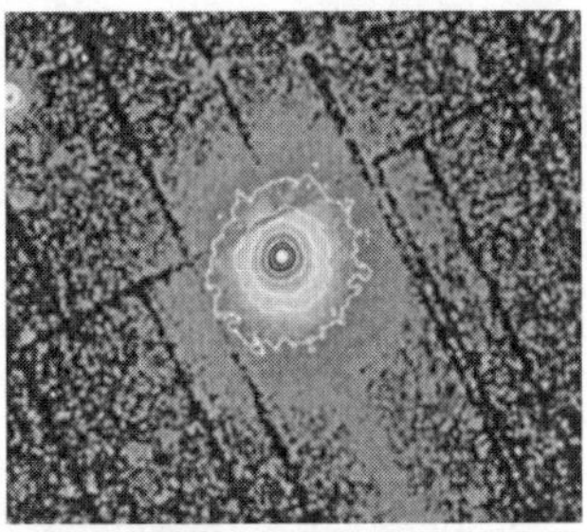

Figure 1. 0.3–3 keV MOS2+PN image of Abell 1835 filtered with a Gaussian.

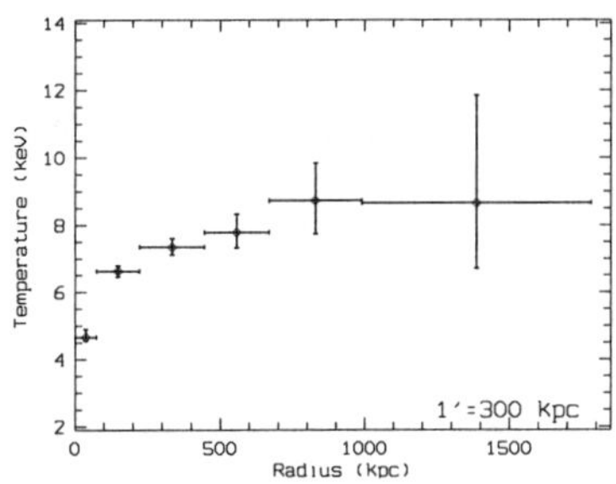

Figure 2. Temperature profile derived from a combined MOS2+PN fit.

3. Temperature profile

All the spectra we use for temperature estimates are vignetting corrected (see Arnaud et al. 2001) and background corrected (see Majerowicz et al. 2001). We fit with a Mekal code associated with a photo electric absorption fixed at the galactic value of $2.24\times10^{20}\,\mathrm{cm}^{-2}$. Figure 2 shows the resulting temperature profile up to 1.8 Mpc from the cluster center. We confirm a temperature drop towards the center of the cluster which is generally associated with a cooling flow (Allen et al 1996). Outside the cooling flow region the temperature profile is flat with a mean value of 8.2 keV.

4. Mass profiles

To calculate the gas mass profile, we use the β-model fit parameters (section 2). Figure 3 shows the corresponding results. The hydrostatic equation and the Monte–Carlo method by Neumann & Böhringer (1995) are used to estimate the gravitational mass (see figure 3). At a radius of 1.7 Mpc ($0.7\,r_{200}$), $\mathrm{M_{gas}} = 2.56 \pm 0.08 \times 10^{14}\,\mathrm{M_\odot}$ and $\mathrm{M_{tot}} = 1.40 \pm 0.21 \times 10^{15}\,\mathrm{M_\odot}$.

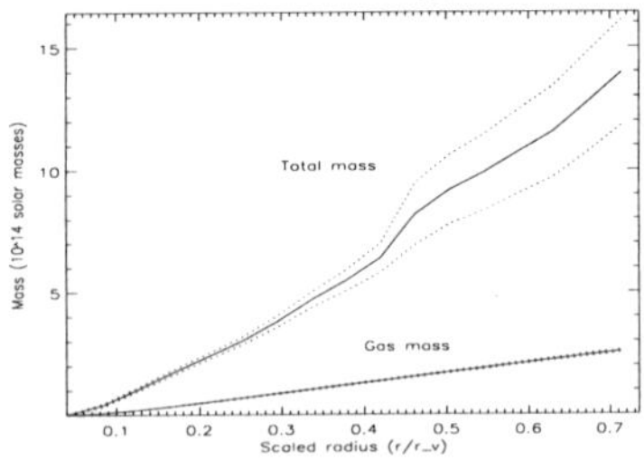

Figure 3. Mass profiles.

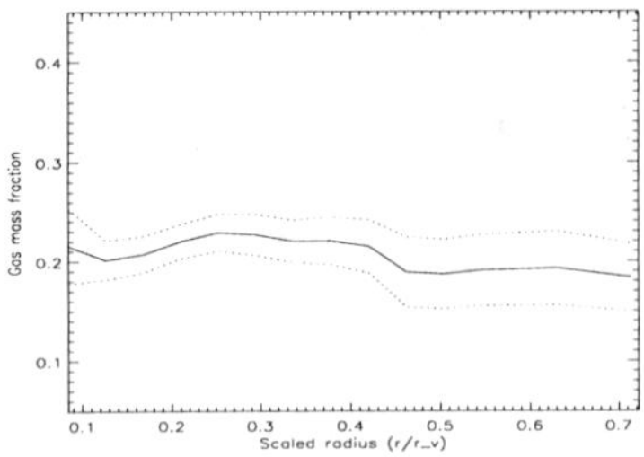

Figure 4. Gas mass fraction profile.

From this two mass calculations, we are able to estimate the gas mass fraction f_b (figure 4) at $0.183 \pm 0.034\,h_{50}^{-1.5}$ (68%). Extrapolating our found f_b throughout the Universe (White et al. 1993) with recent result on Ω_b (Burles et al. 2001) based on primordial nucleosynthesis, we find $\Omega_m = 0.43 \pm 0.10 h_{50}^{-0.5}$ (68%). It is an upper limit since we do not account for other baryon contributions.

References

Allen, S.W., Edge, A.C., Bautz, M.W., Furuzawa, A., Tawara, Y. 1996, MNRAS, 293, 263

Arnaud, M., Neumann, D.M., Aghanim, N., Gastaud, R., Majerowicz, S., Hughes, J.P. 2001, A&A, 365, L80

Burles, S., Nollet, K.M., Turner, M.S. 2001, astro-ph/0010171

Neumann, D.M. & Böhringer, H. 1995, A&A, 301, 865

Majerowicz, S., Neumann, D.M. 2001, Moriond Proceedings

Peterson, J.R., Paerels, F.B.S., Kaastra, J.S., Arnaud, M., Reiprich, T.H., Fabian, A.C. , Mushotzky, R.F., Jernigan, J.G., Sakelliou, I. 2001, A&A, 365, L104

White, S.D.M., Navarro, J.F., Evrard, A.E., Frenk, C.S. 1993, Nature, 366, 429

Tracing Cosmic Evolution with Galaxy Clusters
ASP Conference Series, Vol. 268, 2002
S. Borgani, M. Mezzetti, and R. Valdarnini, eds.

Soft Cores in Late-Type Dwarf and LSB Galaxies from Hα Observations

D. Marchesini[1], E. D'Onghia[1,2], G. Chincarini[2,3], C. Firmani[2], P. Conconi[2], E. Molinari[2], A. Zacchei[4]

1-Università degli Studi di Milano, Via Celoria 16, 20133, Milano, Italy;2-Osservatorio Astronomico di Brera-Merate, Via Bianchi 46, 23807 Merate (LC), Italy;3-Università degli Studi di Milano-Bicocca, Piazza dell'Ateneo Nuovo 1, 20126, Milano, Italy;4-Telescopio Nazionale Galileo (TNG)

Abstract. We present high spatial resolution Hα rotation curves of late-type dwarf and LSB galaxies. From our analysis we find good agreement between our Hα data and the HI observations taken from the literature, concluding that the HI rotation curves for these galaxies suffer very little from beam smearing. A preliminary analysis of our data rules out the CDM model in the inner regions of these galaxies.

1. Introduction

The current scenario for structure formation in the universe is based on the inflationary Cold Dark Matter (CDM) theory. High-resolution N-body simulations predict, for the virialized haloes, density profiles with central cusps with $\rho \propto r^{-1.5}$ when r $\rightarrow$ 0 (Moore et al. 1999), even more cuspy than previous simulation results (Navarro, Frenk, and White 1997, hereafter NFW). These profiles are in conflict with the observations: rotation curves of late-type dwarf and LSB galaxies (DM dominated galaxies) seem to rule out singular halo profiles and are in good agreement with density profiles characterized by soft cores, central regions with constant density (de Blok & McGaugh 1997). On the other hand, van den Bosch & Swaters (2001) claim that these HI rotation curves are affected by beam smearing, concluding that the observed rotation curves are consistent with NFW model. A good spatial resolution (easily achievable with Hα observations) is necessary to explore the galaxy inner regions and rule out beam smearing. We show some preliminary results of our Hα observations for some dwarf and LSB galaxies. This work assumes H_0=75 km s^{-1} Mpc^{-1}.

2. Observations

Observations of late-type dwarf and LSB galaxies were carried out at the 3.6 m Telescopio Nazionale Galileo (TNG) both in photometric and spectroscopic mode using the instrument Dolores (scale = 0.275 Åpxl^{-1}, dispersion = 0.8 Åpxl^{-1}, $\lambda \in (6200, 7800)$ Å). In Table 1 we list the properties of some of the observed galaxies.

Table 1. Properties of the observed galaxies

Name	Type	D (Mpc)	M (mag)	μ_o (mag arcsec^{-1})	h (kpc)
UGC 4325	SA(s)m	10.1	-18.1(R)	21.6(R)	1.7(R)
UGC 4499	SABdm	13.0	-17.8(R)	21.5(R)	2.0(R)
UGC 11861	SABdm	25.1	-20.8(R)	21.4(R)	6.1(R)
LSBC F571-8	Sc	48.0	-17.6(B)	23.9(B)	5.2(B)

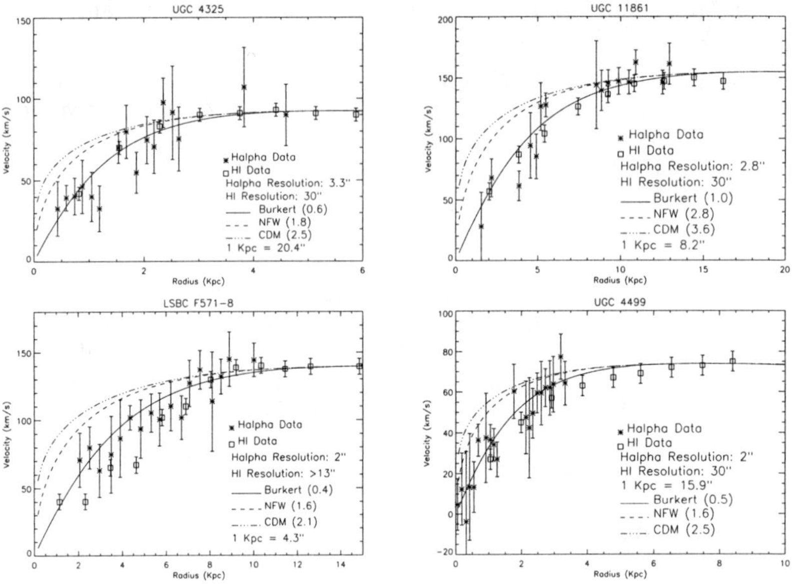

Our Hα data (stars in the figures) are plotted together with HI data (from literature, empty squares). As one can see, the HI data suffer of very little beam smearing. The solid lines are the best fits obtained by using the Burkert profile; the dashed lines are NFW profiles and the dotted-dashed lines are Moore profiles (CDM model). The numbers in parenthesis represent the normalized χ^2 for each fit. For each galaxy is stated the spatial resolution for both the Hα and the HI data. For all the four galaxies, the Burkert profile is the best fit, while the Moore profile does not match data. New observations with high resolution new technology grating (VPHG, Conconi et al. 2001) are being planned.

3. Conclusions

Our preliminary analysis of Hα observations seems to rule out the CDM model in the inner regions of LSB and dwarf galaxies, showing evidence of soft cores.

References

de Blok, W.J.G., McGaugh, S.S. 1997, MNRAS, 290, 533
Moore,B., Guinn,T., Governato,F., Stadel,J., Lake,G. 1999, MNRAS, 310, 1147
Navarro, J.F., Frenk, C.S., White, S.D.M 1997, ApJ, 490, 493
van den Bosch, F.C., Swaters, R.A. 2001, MNRAS, 325, 1017
Conconi, P. et al. 2001 (SPIE Meeting, San Diego, U.S.A.)

Tracing Cosmic Evolution with Galaxy Clusters
ASP Conference Series, Vol. 268, 2002
S. Borgani, M. Mezzetti, and R. Valdarnini, eds.

3D Identification and Reconstruction of z ∼ 1 Clusters: Prospects for the DEEP2 Redshift Survey

C. Marinoni, M. Davis, J. Newman, A. Coil

University of California at Berkeley, Berkeley, CA, 94720-3411, USA

Abstract.
We have developed a geometrical method based on 3D Voronoi polyhedra and Delaunay tessellation for identifying and reconstructing clusters of galaxies in the next generation of deep, flux-limited redshift surveys. We here describe this algorithm and tests of it using mock catalogs that simulate the DEEP2/DEIMOS redshift survey, which will begin observations in the Spring of 2002 and will provide a detailed three dimensional map of the large scale structure up to redshift 1.5.

1. Introduction

Over the past few years there has been considerable work on the detection of two-dimensional galaxy overdensities in wide field imaging surveys for subsequent spectroscopic follow up. However, these systematic searches for clusters, fuelled by the availability of large CCD camera mosaics, are in many cases biased towards high density peaks of the galaxy distribution.

With the next generation of large multi-object spectrographs in their final construction phase, deep redshift surveys (Keck/DEEP2 and VLT/VIRMOS) are well within the horizon. It is now worthwhile to address the more ambitious task of constructing a three dimensional, statistically complete sample of high redshift galaxy clusters. Here we describe a fully automated and objective algorithm for identifying and reconstructing galaxy systems, based upon 3D Voronoi polyhedra and Delaunay triangulation.

2. Method

A Voronoi partition of the space into minimally sized convex polytopes is a natural way to measure packing. A Voronoi polyhedron is the uniquely defined region of space around a galaxy (seed), within which each point is closer to the seed than to any other galaxy. The faces of the Voronoi cell are formed by planes perpendicular to the vectors between a galaxy and its neighbors. The volume inside the polyhedron is inversely proportional to the packing efficiency of its central galaxy; a large cell volume is indicative of an isolated galaxy. The Delaunay triangulation is the geometrical dual of the Voronoi partition and is defined by the tetrahedron whose vertices are the 4 galaxies with the property that the uniquely determined circumscribing sphere does not contain any other galaxy.

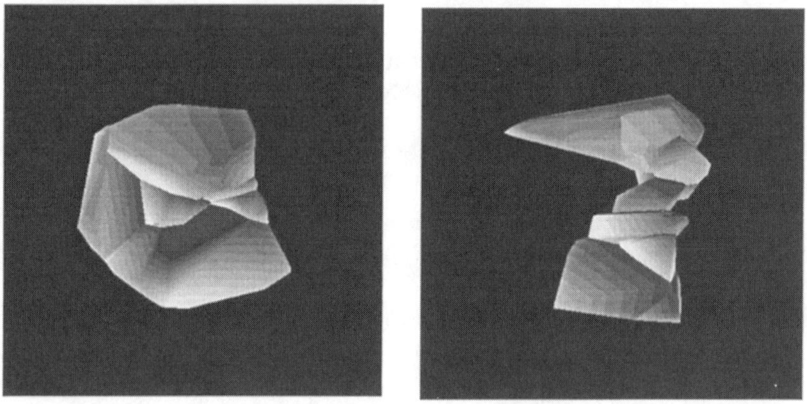

Figure 1. Voronoi reconstruction of a cluster with 10 galaxies in the DEEP2 mock catalog. The Voronoi cells encompassing the cluster are shown in real space (*left*) and in redshift space (*right*).

We thus may identify high-density peaks by selecting as "cluster seeds" the centers of all the Voronoi cells with volume smaller than some threshold, and then use the scale length of the Delaunay mesh to infer the strength of the physical aggregation. We then can define an adaptive cylindrical window in redshift space (elongated in the z direction) with dimensions determined by the local scale factor and process all the Delaunay connected galaxies with a rapidly converging "inclusion-exclusion" logic to identify cluster members.

In our algorithm, the parameters defining the adaptive search window are only weakly dependent on the galaxy density gradient which inevitably occurs in a flux limited survey, since our smoothing procedure immediately identifies regions of enhanced clustering where galaxies are generally more luminous. Moreover the number of cluster interlopers aggregated into a system in a redshift space analysis is minimized and we simultaneously obtain a non-parametric local estimate of the surrounding density environment for each galaxy and a quantitative measure of the distribution of cosmological voids in the survey volume.

3. Results

We have tested our reconstruction scheme using mock catalogs modelling the DEEP2 Redshift Survey (Davis et al. 2000) derived from GIF simulations (Coil et al. 2001). This survey will obtain high quality spectra for $\sim$ 60000 galaxies between z=0.7-1.5 using the Keck 2 Telescope and DEIMOS spectrograph, with the twin goals of studying the evolution of the properties and the large-scale clustering of high redshift galaxies. We used a LCDM model with $\Omega_m = 0.3$, $\Omega_\Lambda = 0.7$, $h = 0.7$ and $\sigma_8 = 0.9$. In order to mimic the selection function of the magnitude limited survey we make a rest-frame B-band cut of 23.4, which at z $\sim$ 1 corresponds to an apparent I-band magnitude with the appropriate K-correction matching the DEEP2 photometric selection criteria ($I_{AB} < 23.5$). In Fig. 2a we present a mock catalog for one DEEP2-like field collapsed along

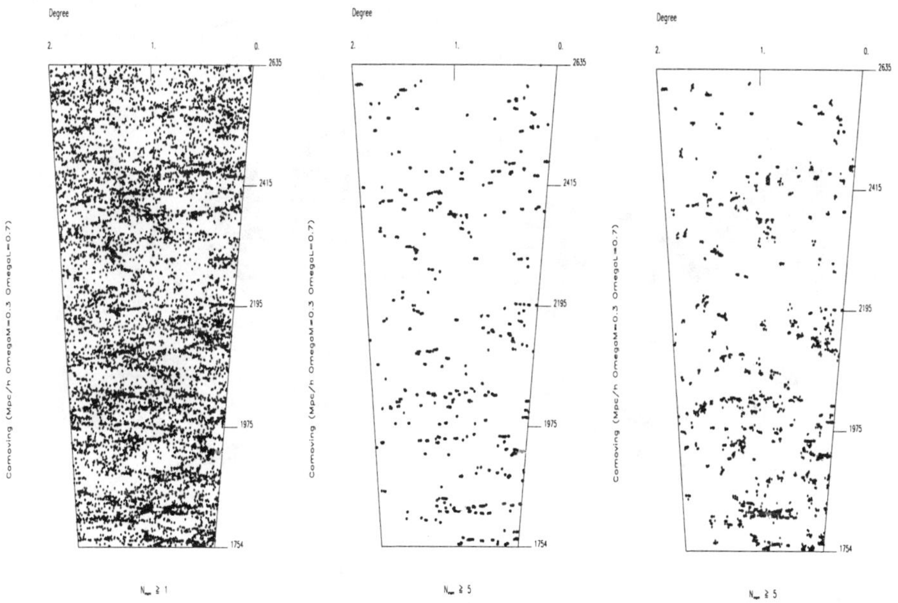

Figure 2. *Left:* One DEEP2 mock catalog in the redshift range z=(0.7,1.2). *Center:* large-scale spatial distribution of clusters with more that 5 members identified in real space using a "Friends of Friends" algorithm. *Right:* Voronoi-Delaunay reconstruction in redshift space.

the smallest axis, covering the redshift range z=0.7-1.2 (a depth which corresponds to selecting the 1200 l/mm grating in the DEIMOS spectrograph) and containing a volume of roughly 10^6 Mpc h^{-3} ($\sim$ 15000 galaxies). We have used six independent mock catalogs to determine how well the algorithm performs in identifying clusters of galaxies.

To determine the "true" distribution of clusters, we have applied in real space (volume-limited simulation) a standard percolation algorithm optimised for selecting virialized objects (mean overdensity $\sim$ 180). The resulting distribution of those clusters with more than 5 members is shown in Fig 2b. Cluster candidates selected by the Voronoi–Delauney algorithm applied in redshift space (flux-limited simulation) are shown in Fig 2c. Note how the large-scale pattern defined by galaxy systems and the associations of clusters in higher order structures reproduce in an unbiased way the underlying real space landscape.

4. Statistical tests

The evolution of the comoving abundance of clusters as a function of their velocity dispersion σ and redshift z is a sensitive function of cosmological parameters (see Newman et al. 2001). The statistical significance and robustness of the clus-

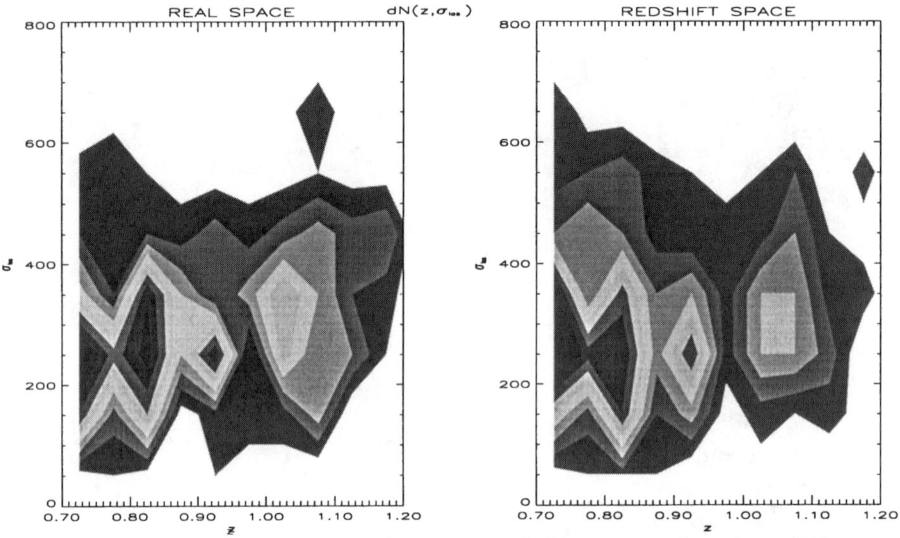

Figure 3. Differential number count of cluster as a function of the redshift z and the projected velocity dispersion σ. The parent distribution of cluster counts identified in real space is shown on the left, while the observed distribution recovered in redshift space with our method is plotted on the right. Density levels, smoothed by 0.05 in z and 50 km/s in σ, are spaced with Δ=2 intervals.

ter abundance test depends critically on an unbiased mapping of the distribution properties of the cluster observables between real to redshift space.

We have investigated the differential and integral distribution functions of the number of reconstructed clusters as a function of their richness, velocity dispersion and redshift. The Kolmogorov-Smirnov statistical test confirms the similarity between the real and reconstructed distributions, and thus the reliability of our algorithm. This can be seen visually in Fig. 4 where we show the two-dimensional distributions of the real and reconstructed clusters in redshift z and projected velocity dispersion σ.

The DEEP2 survey is a collaborative project among astronomers at UC, Caltech and the Univ. of Hawaii. Details of the project can be found at the URL http://astro/berkeley.edu/deep.

References

Coil, A., Davis, M., & Szapudi, I. 2001, PASP, in press, astro-ph/0107541

Davis, M., Newman, J. A., Faber, S. A., & Phillips, A. C., 2000 in Proc. of the ESO/ECF/STSCI workshop on Deep Fields, (Garching, Edition Springer), astro-ph/0012189

Newman, J. A., Marinoni, C., Davis, M., & Coil, A. 2001, PASP, in press

Tracing Cosmic Evolution with Galaxy Clusters
ASP Conference Series, Vol. 268, 2002
S. Borgani, M. Mezzetti, and R. Valdarnini, eds.

The Colour-Magnitude Relation and the Age of Galaxies in the Cluster AC118 at z=0.31

P. Merluzzi, G. Busarello, M. Massarotti

Osservatorio Astronomico di Capodimonte, Napoli, Italy

F. La Barbera

Physics Department, Università Federico II, Napoli, Italy

Abstract. We derive the Colour-Magnitude (CM) relation for the galaxy cluster AC118 at z=0.31 and use this relation to set constraints on the formation epoch and on the age spread of cluster galaxies in the framework of the passive evolution scenario. This work is based on observations at ESO NTT (OAC guaranteed time).

The galaxy population in the core of AC118 was already analyzed by Stanford et al. (1998), who found evidence in favour of the passive evolution scenario. Barger et al. (1996), on the other hand, claimed for evidence of recent bursts of star formation. Here we derive the CM relation in the optical and near infrared (R-K) for a large sample of cluster members (N_{gal}=318) on a wide area (~2.6 Mpc^2, q_0=0.5, H_0=50 Km $s^{-1}Mpc^{-1}$) around the cluster core.

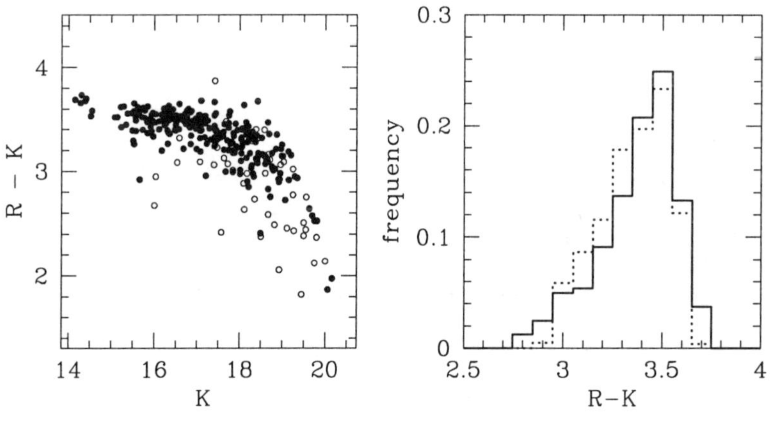

Figure 1. *Left panel*: CM relation of the cluster members selected through photometric redshifts. Filled/open dots mark galaxies described by early–/late–type templates. Typical uncertainties are Δ(R-K)=0.04 mag and ΔK=0.02 mag. *Right panel*: observed R-K colour distribution (solid line) compared with the distribution predicted by the adopted model of galaxy evolution (dotted line).

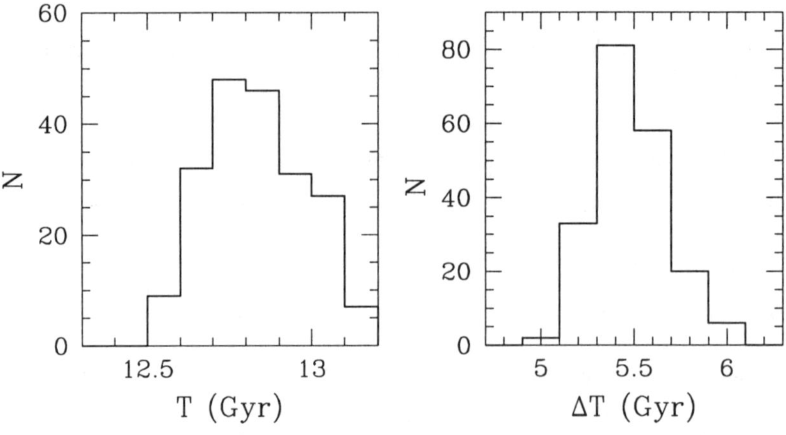

Figure 2. *Left panel*: distribution of the first epoch of star formation as obtained by N=200 sets of simulations of the R-K galaxy colours. *Right panel*: distribution of the age spread.

The sample was selected by a photometric redshift technique which is based on the Spectral Energy Distribution method (e.g. Massarotti, Iovino, & Buzzoni 2001) using the template spectra provided by the code by Bruzual & Charlot (1993).

Figure 1 (left panel) shows the CM relation of the selected cluster members. We analyzed the R-K colour distribution of the galaxies that are described by early–type templates. The observed distribution (solid line in Figure 1) can be recovered by a simple model of galaxy formation (dotted line): stars in galaxies form in short bursts (τ=1 Gyr) randomly occurring in a time interval ($T - \Delta T$, T). T and ΔT are the first epoch of star formation and its duration. For given values of T and ΔT, the expected R-K colour distribution was simulated taking into account the measurement uncertainties on the colours. The best fit estimates of T and ΔT were derived from different sets of simulations. In Figure 2 (left panel) we show the distribution of the first epoch of star formation. In the adopted cosmological model galaxies form the oldest stars very early in the past ($z > 7$).

The spread in star formation times is shown in the right panel of Figure 2. To describe the colours of the bluest cluster galaxies, these objects should have undertaken a recent (T <8 Gyr, z <1) burst of star formation.

References

Barger, A.J., Aragón–Salamanca, A., Ellis, R. et al. 1996, MNRAS, 279, 1

Bruzual, G.A., & Charlot, S. 1993, ApJ, 405, 538

Massarotti, M., Iovino, A., & Buzzoni, A. 2001, A&A, 368, 74

Stanford, S.A., Eisenhardt, P.R.M., & Dickinson, M. 1998, ApJ, 492, 461

Tracing Cosmic Evolution with Galaxy Clusters
ASP Conference Series, Vol. 268, 2002
S. Borgani, M. Mezzetti, and R. Valdarnini, eds.

A Deep Keck Spectral Survey of Cl0024+1654 at $z = 0.4$

Anne J. Metevier and David C. Koo
Department of Astronomy and Astrophysics, University of California, Santa Cruz, CA 95064 USA

Luc Simard
Steward Observatory, University of Arizona, Tucson, AZ, 85721 USA

Gregory D. Wirth
W. M. Keck Observatory, Kamuela HI 96743 USA

Abstract. We have undertaken a Keck spectral survey of galaxies to R$\sim$24 near the central region of Cl0024+1654 at $z = 0.4$. The spectral resolution and slits have been selected to yield kinematics such as the velocity dispersions of early-type systems and rotation curves of disks. The spectra also provide a suite of lines that yield star formation, age, and metallicity information.

1. Introduction

Cl0024 is a rich, intermediate-redshift cluster ideal for galaxy evolution studies. One of the two clusters studied by Butcher and Oemler (1978), Cl0024 was noted for its high fraction of "blue" galaxies relative to local clusters. We have surveyed the central region of this cluster with the LRIS spectrograph to examine the internal kinematics of cluster members as well as the emission signatures of blue disk systems.

2. LRIS Spectra

We have acquired Keck spectra of over 270 objects in the Cl0024 field. In Table 1, we compare our data to other recent spectroscopic surveys of this cluster. Note that our spectral wavelength range covers redshifted [OII] through Hα, and our slits are aligned with disk major axes. We have determined redshifts of nearly 150 cluster members and have confirmed the redshift of the lens source behind this cluster at $z = 1.68$ (Broadhurst et al. 2000). Ellis, Treu, and collaborators are currently undertaking a spectral survey of galaxies in the outer regions of Cl0024. Our data will nicely complement theirs as the majority of the galaxies we have surveyed are concentrated in the cluster center.

Table 1. Our survey compared to recent spectral surveys of Cl0024.

	Dressler et al. 1999	Czoske et al. 2001	**Our Survey**
Survey area	$9.4' \times 7.9'$	$21' \times 25'$	$6.1' \times 10.3'$
N_{tot}	130	687	278
$N_{cluster}$	107	295	145
depth	$r < 22$	$V < 23$	$\mathbf{R < 24.5}$
range (Å)	$3500 - 9800$	$4500 - 8500$	$4500 - 9500$
resolution(″/pix)	$0.27 - 0.59$	$0.31 - 0.44$	**0.215**
resolution(Å/pix)	$2.3 - 5.3$	$3.3 - 5.0$	$\mathbf{1.2 - 1.85}$

3. Efforts Underway

Tully-Fisher study: We have measured the rotation velocities of seven Cl0024 members in a pilot study to charcterize the Tully Fisher relation (TFR) for intermediate-z cluster galaxies (see Metevier et al., also in these proceedings). Our data show no evidence for evolution of the slope or zeropoint of the cluster TFR. However, they do indicate that there may have been more scatter in the past. We intend to strengthen our study with measurements from 15 more cluster spirals which exhibit spatially resolved emission.

Disappearing S0s?: Dressler et al. (1997, MORPHs), in a study of the visually classified morphologies of cluster galaxies, have found evidence that the fraction of S0s in clusters has doubled since $z = 0.5$. However, other studies of distant cluster S0s suggest that they have red stellar populations similar to those of local cluster galaxies whose stellar populations formed before $z = 2$ (Andreon, Davoust, & Heim 1997). We have run a preliminary test of the MORPHs classification of galaxies in the Cl0024 field by measuring the bulge-to-total flux ratios (B/T) of these galaxies via a two-component model (deVaucouleurs bulge + exponential disk). We find that objects classified by MORPHs as late types have low B/T, while objects classified as early types tend to span a wider range of B/T. This result suggests that in distant clusters, disk-dominated systems are being visually classified as bulge-dominated objects (Es and S0s), possibly driving up counts of S0s in distant clusters. We are in the process of running simulations to test the reliability of our models.

References

Andreon, S., Davoust, E. & Heim, T. 1997, A&A, 323, 337

Broadhurst, T., et al. 2000, ApJ, 531, L15

Butcher, H. & Oemler, A., Jr. 1978, ApJ, 219, 18

Czoske, O., et al. 2001, A&A, 372, 391

Dressler, A., et al. 1997, ApJ, 490, 577

Dressler, A., et al. 1999, ApJS, 515, 79

Tracing Cosmic Evolution with Galaxy Clusters
ASP Conference Series, Vol. 268, 2002
S. Borgani, M. Mezzetti, and R. Valdarnini, eds.

The Chemical Evolution of the ICM

A. Moretti and C. Chiosi

Department of Astronomy, Vicolo dell'Osservatorio 2, I-35122 Padova, Italy

L. Portinari

Theoretical Astrophysics Center, Juliane Maries Vej 30, DK-2100 Copenhagen Ø, Denmark

Abstract. We have constructed a toy model to follow the time evolution of the chemical abundances in the Intra Cluster Medium (ICM). The mechanism responsible for the enrichment is assumed to be the galactic wind favoured by supernova explosions.

1. The model

We develop a chemical model for galaxy clusters in analogy with the usual ones for individual galaxies. Just as the ingredients for usual models are (Tinsley, 1980): (a) the stellar yields; (b) the Star Formation Rate; (c) the stellar Initial Mass Function, our model involves: (a') "galactic yields" derived from galactic wind models of ellipticals; (b') a parametric Galactic Formation Rate; (c') a Press-Schechter-like Galactic Initial Mass Function. The model is used to test the response of the predicted metal content and abundance evolution of the ICM to varying input galactic models; in order to constrain model parameters, the resulting luminosity function of cluster galaxies is also calculated. In this first version of the model we have used the Instantaneous Recycling Approximation and we have treated the cluster as a closed system.

2. The galactic yields

To compute the chemical enrichment of the ICM we need an estimation of the galactic yields, which we calculate for two sets of galactic models: a set of models with the variable IMF by Padoan et al. (1997) and a set of models with the classic Salpeter IMF (see Chiosi et al., 1998 and Chiosi, 2000 for details). The comparison between the two sets of models can be seen in Figure 1 (left panel); in particular one can notice that using the variable IMF the fraction of gas re-expelled with the galactic wind represents at least 50% of the total initial proto-galactic mass, on the contrary with the Salpeter IMF more than half of the initial mass remains locked in stars. Results obtained with the variable IMF are sensitive to the redshift of galaxy formation and the above mentioned trend gets more evident at high redshift. Henceforth this set of galactic model predict that a larger fraction of the gas in the ICM is contributed by galactic winds, while in the Salpeter case most of the ICM will be primordial.

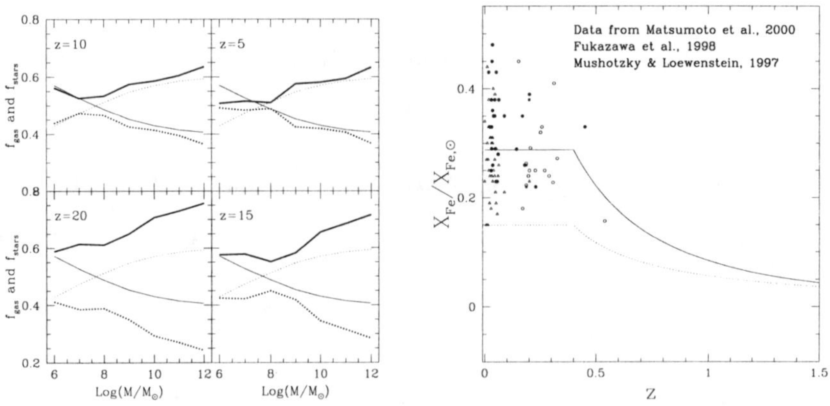

Figure 1. *Left Panel*: Mass fraction in stars left over by the galactic wind (dotted lines), as opposed to the mass fraction in ejected gas (solid lines), for model with the two IMF and for different redshift of formation. Thick lines for variable IMF, thin lines for Salpeter IMF. *Right Panel*: Redshift evolution of the iron abundance: the solid line refers to the variable IMF, the dotted line to the Salpeter IMF

3. The results

Once the Galactic Formation Rate (GFR) of the cluster model is calibrated so that the luminosity function in the B band is reproduced (for details see Moretti et al., in preparation), the observed average iron content is also well reproduced, if the variable IMF is considered (right panel of Figure 1), while in the case of the Salpeter IMF not enough metals are ejected into the ICM. To reproduce the observed luminosity function, the GFR must increase in time until the gas prone to galaxy formation is finished, then it drops. In order to reproduce the faint, steep end of the luminosity function we have introduced also an early, very short burst of galaxy formation, during which the dwarf galaxies are formed. This GFR is reminiscent of a scenario where a high redshift reionization phase is followed by a smoother, Madau–plot like formation rate. Hence we are now developing a more refined model inspired by the observed cosmic star formation history.

References

Chiosi, C. 2000, A&A, 364, 423
Chiosi, C., Bressan, A., Portinari, L. & Tantalo, R. 1998, A&A, 339, 355
Fukazawa, Y., Makishima, K., Tamura, T. et al., 1998, PASJ, 50, 187
Matsumoto, H., Tsuru, T.G., Fukazawa, Y et al., 2000, PASJ, 52, 153
Mushotzky, R.F. and Loewenstein, M. 1997, ApJ, 481, 63
Padoan, P., Nordlund, Å. & Jones, B. 1997, MNRAS, 288, 145
Tinsley, B. M. 1980, Fund. of Cosmic Physics, 5, 287

Tracing Cosmic Evolution with Galaxy Clusters
ASP Conference Series, Vol. 268, 2002
S. Borgani, M. Mezzetti, and R. Valdarnini, eds.

The Radio Butcher-Oemler Effect: The Evolution of the Radio-selected Galaxy Population within Rich Clusters over the Last 5 Gyr

Glenn E. Morrison

Caltech-IPAC and Vanguard Research, Inc., M/S 100-22, Pasadena, CA 91125

Frazer N. Owen

National Radio Astronomy Observatory, P. O. Box O, Socorro, NM 87801

Michael J. Ledlow

Gemini Observatory, Southern Operations Center, AURA, Casilla 603, La Serena Chile

William C. Keel

Dept. of Physics & Astronomy, University of Alabama, Tuscaloosa, AL 35487

John M. Hill

Steward Observatory, University of Arizona, Tucson, AZ 85721

Wolfgang Voges

Max-Planck-Institut für extraterrestrische Physik, Postfach 1312, D-85741 Garching bei München, Germany

Abstract. We performed a deep VLA 20 cm radio continuum survey of 34 extremely rich Abell clusters with $z < 0.41$, down to a constant radio luminosity limit of 2×10^{22} W/Hz (SFR > 5 $M_{\odot}$ yr^{-1}). We discovered that the fraction of radio-selected galaxies in distant rich z $\sim$ 0.4 clusters with luminosities in the range 2×10^{22} W/Hz < L < $10^{22.75}$ W/Hz is higher by a factor of > $6x$ compared to nearby rich clusters. Spectroscopy shows that this population to be made up in part by dusty starbursting galaxies with several optically classified as poststarburst galaxies. The X-ray and optically inferred substructure of some clusters suggests that ongoing cluster-cluster mergers may increase this population regardless of redshift.

1. Introduction

Our research group has performed a radio continuum study of galaxy evolution within dense environments over a cosmologically significant interval of time ($z <$

0.41): a radio study of the well-known Butcher-Oemler (BO) effect (Butcher-Oemler 1984). The BO effect is the enhancement of blue galaxies in the core of rich clusters as a increasing function of redshift. Morrison (1999) and Morrison *et.al.*, (2001a,b) studied the weak radio-selected galaxy ($M_R < -21$) populations associated with rich clusters of galaxies as a function of redshift and $R_{cl} \leq$ 2.5 Mpc from the cluster core. Radio selection of cluster members by their low radio luminosity yields two kinds of objects - weak AGN's and objects with active star formation (SFR $> 5\,M_{\odot}yr^{-1}$, based on our radio luminosity limit of $L_{1.4} > 2 \times 10^{22}$W Hz^{-1}, *i.e.*, uniform sensitivity for all clusters regardless of redshift).

The FIR/radio correlation ties the nonthermal radio emission to the ongoing star formation (Condon 1992). In bursts, radio emission fades faster than blue starlight, so clusters with a large number of radio sources will preferentially be those which have had many recent/ongoing starbursts, perhaps indicating the entry of a new group into the cluster with star formation in its members.

2. Conclusions

- The radio-selected galaxies in clusters are generally associated with two populations: High-luminosity AGN early-type galaxies and low-luminosity dusty starbursting late-type galaxies.
- The fraction of low-luminosity radio galaxies within very-rich clusters has decreased by a factor ~6-10 over the last 5 Gyr.
- Archival HST images of the cluster cores of the four $z = 0.4$ clusters show that 63% of the radio-selected cluster members have disturb morphology indicative of interactions/mergers.
- The increased fraction of faint radio-selected galaxies may be driven by large-scale-structure buildup, *e.g.*, cluster-cluster mergers. This is supported by numerical simulations by Bekki (1999).

Acknowledgments. G.E.M. acknowledges financial support from NRAO predoctoral fellowship program during which most of this work was completed. G.E.M. also gratefully acknowledges the support of JPL (contract #1147-001) and also the support of Vanguard Research, Inc.

References

Bekki, K. 1999, ApJ, 516, L53

Butcher & Oemler, 1984, ApJ, 285, 426.

Condon, J. J. 1992, ARA&A, 30, 575

Morrison, G. E. 1999, PhD Thesis, University of New Mexico

Morrison et al 2001a, submitted to ApJS.

Morrison et al 2001b, submitted to AJ.

Tracing Cosmic Evolution with Galaxy Clusters
ASP Conference Series, Vol. 268, 2002
S. Borgani, M. Mezzetti, and R. Valdarnini, eds.

Galaxy Population in the Cluster of Galaxies around the Radio Galaxy 3C324 at $z = 1.2$

Fumiaki Nakata[1], Masaru Kajisawa[2], Toru Yamada[3], Tadayuki Kodama[1], Kazuhiro Shimasaku[1], Ichi Tanaka[3], Mamoru Doi[4], Hisanori Furusawa[1], Masaru Hamabe[5], Masanori Iye[3], Masahiko Kimura[6], Yutaka Komiyama[7], Satoshi Miyazaki[3], Sadanori Okamura[1], Masami Ouchi[1], Toshiyuki Sasaki[7], Maki Sekigushi[6], Masafumi Yagi[3], and Naoki Yasuda[3]

[1]*Department of Astronomy, School of Science, The University of Tokyo, Bunkyo-ku, Tokyo 113-0033, Japan*
[2]*Astronomical Institute, Tohoku University, Aoba-ku, Sendai, Miyagi 980-8578, Japan*
[3]*National Astronomical Observatory of Japan, 2-21-1 Osawa, Mitaka, Tokyo 181-8588, Japan*
[4]*Institute of Astronomy, The University of Tokyo, Mitaka, Tokyo 181-8566, Japan*
[5]*Department of Mathematical and Physical Science, Faculty of Science, Japan Women's University, 2-8-1 Mejirodai, Bunkyo-ku, Tokyo 112-8681, Japan*
[6]*Institute for Cosmic Ray Research, The University of Tokyo, 5-1-5 Kashiwa-no-Ha, Kashiwa, Chiba 277-8582, Japan*
[7]*Subaru Telescope, National Astronomical Observatory of Japan, 650 North Aohoku Place, Hilo, HI 96720, U.S.A*

Abstract. We study properties of cluster galaxies around the radio galaxy 3C324 at $z = 1.2$ with $BVRIK'$ multi-band imaging data. We apply a photometric-redshift technique to the objects in the 3C324 field, and identify plausible cluster members. With the data, we investigate properties of the cluster galaxies, such as the luminosity function, the color properties, and the blue fraction.

1. Introduction

In the last few years, more than several clusters of galaxies and cluster candidates have been discovered at $z > 1$. Consequently, studies of clusters at $z > 1$ are one of the hottest issues in the present-day astronomy. A spectroscopic observation is extremely time consuming to make a complete sample of member galaxies over a representative field of distant clusters (e.g., 1 Mpc) even with an 8m-class telescope. We try to identify cluster membership of the cluster around 3C324 with a photometric-redshift technique. This approach enables us to investigate galaxy properties of likely cluster members; importantly, not only the galaxies on

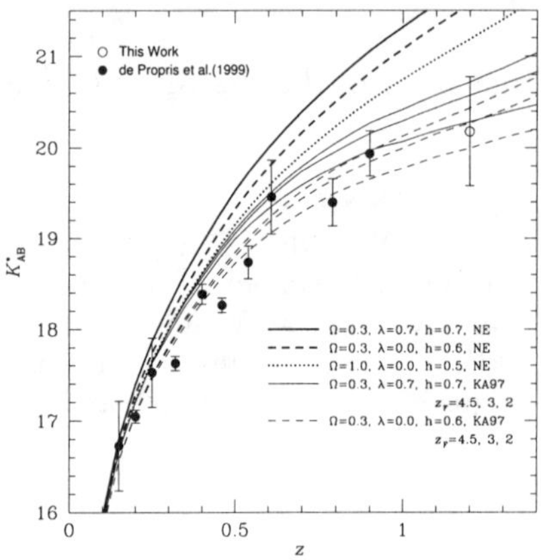

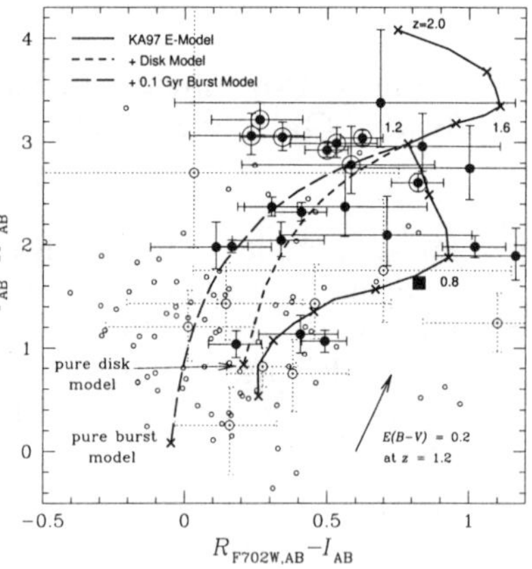

Figure1 (left) : $K^* - z$ Hubble diagram for the cluster $0.1 < z < 0.9$ studied by de Propris et al. (1999; filled circles) and the 3C324 cluster (open circle).
Figure2 (right) : $R_{\mathrm{F702W,AB}} - I_{\mathrm{AB}}$ versus $I_{\mathrm{AB}} - K'_{\mathrm{AB}}$ of the objects detected in the 3C324 field (small open circles : non-members, filled circles : members, large open circles : faint members at $K'_{\mathrm{AB}} > 23$, filled circles marked with a large circle : members on the red sequence). Note that the $(R_{\mathrm{F702W,AB}} - I_{\mathrm{AB}})$ color approximately corresponds to $(U - B)$ color in the rest frame at $z = 1.2$.

the red sequence but also relatively blue galaxies which are otherwise embedded in numerous field populations on a color-magnitude diagram.

2. Results

We identify 35 objects as the plausible cluster members in the 2'×2' field at $K'_{\mathrm{AB}} < 24$. We construct the K'-band luminosity function of the cluster members and fit a Schechter function, and find $K'^{*}_{\mathrm{AB}} = 20.2 \pm 0.6$. Fig.1 compares the K^{*}_{AB} value of the 3C324 cluster with those of lower-redshift clusters. It is found that the K^{*}_{AB} of the 3C324 cluster follows the trend seen in the intermediate-redshift clusters that is consistent with the passive evolution models. On the other hand, most of the giant E/S0 galaxies in this cluster exhibit a significant excess in the rest UV light with respect to the passive evolution model (Fig.2). This result suggests that the massive E/S0 galaxies in this high redshift cluster are still forming stars to some extent, although the possibility of a high star forming activity is ruled out based on the evolution of K^{*}_{AB}. We also derive the blue fraction of this cluster following the definition by Butcher and Oemler (1984), and obtain $f_{\mathrm{B}} = 0.39 \pm 0.28$, which seems to be higher than those for $z < 1$ clusters. This indicates that the star formation activity of this cluster is, on the average, higher than those of lower redshift counterparts.

References

Butcher H., & Oemler A. 1984, ApJ, 285, 426

de Propris R., Stanford S. A., Eisenhardt P. R., Dickinson M., & Elston R. 1999, AJ, 118, 719

Tracing Cosmic Evolution with Galaxy Clusters
ASP Conference Series, Vol. 268, 2002
S. Borgani, M. Mezzetti, and R. Valdarnini, eds.

Cluster Structure in a Large Number of *ROSAT* and *ASCA* Clusters

Naomi Ota

Tokyo Metropolitan University, 1-1 Minami-osawa, Hachiouji, Tokyo 192-0394, Japan

Kazuhisa Mitsuda

ISAS, 3-1-1 Yoshinodai, Sagamihara, Kanagawa 229-8510, Japan

Abstract. We have performed systematic analysis on X-ray structures of eighty distant clusters of galaxies in the redshift range $0.1 < z < 1$. We find that (1) cluster parameters such as temperature, core radius, β do not show significant redshift dependence, (2) the core radius shows a double-peaked distribution in its histogram, and (3) the gas-mass ratio is $0.28 \pm 0.10\, h_{50}^{-3/2}$.

We have analyzed the *ASCA* and *ROSAT* HRI archival data of eighty clusters of galaxies in the redshift range $0.1 < z < 1$, utilizing the isothermal β-model. From the analysis, we obtained the following major results (See Ota 2001 for details). We find no significant redshift dependence in the β-model parameters (β, r_c), the central electron density and the average X-ray temperature (Fig. 1). Among the X-ray parameters, the core radius shows the largest cluster-to-cluster dispersions. The histogram of r_c exhibits significant two peaks at about 60 kpc and 220 kpc (Fig. 2). We find about 20% of morphologically "regular" clusters are better fitted with a double β-model than a single β-model. If we plot the distribution of the two core radii separately (Fig. 2), we find that they remarkably resemble the r_c distribution of the single-β clusters. We also find that the observed luminosity-temperature relation of the clusters is strongly related to the core radius distribution, and can be roughly described as $L_X \propto T_{\rm gas}^{2.9} r_c^{-1}$ (Mitsuda & Ota 2001). From these results, we suggest that there are at least two types of clusters with two different core radii and/or that involve two different length scales in their cluster potentials (Ota & Mitsuda 2001). Finally, from the gas-mass ratio determined with the eighty clusters and $\Omega_{\rm b} = 0.077 \pm 0.0056\, h_{50}^{-2}$ (Burles & Tytler 1998), we obtained $\Omega_{\rm M} = 0.28 \pm 0.10\, h_{50}^{-1/2}$.

References

Burles, S., & Tytler, D. 1998, ApJ, 499, 699

Mitsuda, K., & Ota, N. 2001, in preparation

Mohr, J.J., Mathiesen, B., & Evrard, A.E. 1999, ApJ, 517, 627

Ota, N. 2001, Ph.D. Thesis, University of Tokyo

Ota, N., & Mitsuda, K. 2001, submitted to ApJ

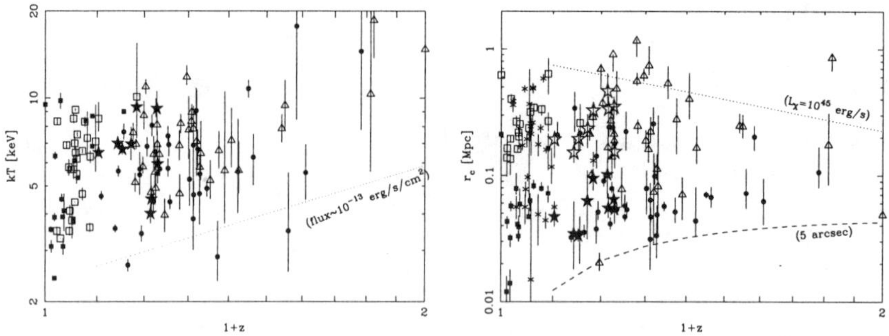

Figure 1. Redshift dependence of the average temperature (left) and the core radius (right). At $z > 0.1$ the results from this work are shown, while at $z < 0.1$ the nearby samples are taken from Mohr et al. (1999). Subgroups of the clusters are denoted with different symbols, the single-β regular and irregular clusters (circles and triangles), the double-β clusters (stars) where the core radii of two components are plotted with open and filled stars in the right panel, the single-β and double-β clusters of the nearby Mohr samples (asterisks and open/filled squares). The selection effects due to the effective areas and the spatial resolution are indicated with the dotted lines and the dashed line, respectively. $H_0 = 50\,\mathrm{km/s/Mpc}$ and $\Omega_\mathrm{M} = 1$ are used throughout the paper.

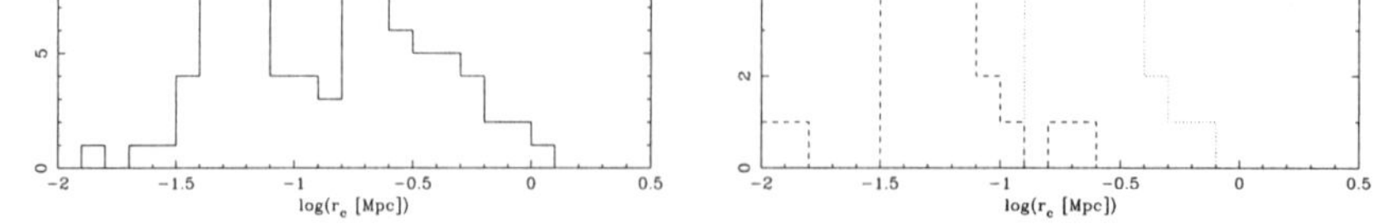

Figure 2. Histograms of the core radius for the single-β clusters (left) and the double-β clusters (right). The nearby Mohr samples and our distant samples are added together.

Tracing Cosmic Evolution with Galaxy Clusters
ASP Conference Series, Vol. 268, 2002
S. Borgani, M. Mezzetti, and R. Valdarnini, eds.

Deep ROSAT-HRI Observation of the cD Galaxy NGC 1399 in the Fornax Cluster: Morphology and Dynamical Status of the X-Ray Halo

M. Paolillo, G. Peres

Università di Palermo, D.S.F.A. - Sez. di Astronomia, Piazza del Parlamento 1, 90134 Palermo, Italy

G. Fabbiano, & D.-W. Kim

Harvard-Smithsonian Center for Astrophysics, High Energy Division, 60 Garden St., Cambridge, Massachusetts 02138

Abstract. We present the results of a deep (167 ks) observation of the NGC 1399/NGC 1404 field obtained from data collected between 1993 and 1996 with the ROSAT High Resolution Imager. We take advantage of the 5" resolution of the HRI to study in detail the structure of the galactic halo and to relate the results to those obtained at larger scales with poorer resolution instruments. We discuss possible non-equilibrium scenarios that may explain the global halo structure. We also find evidence of interactions between the nuclear radio source and the inner gaseous halo.

The study of galactic and cluster gaseous halos has often been based on the assumption of homogeneity and spherical symmetry. We have analyzed a deep ROSAT HRI observation centered on the dominant Fornax cluster galaxy NGC 1399, along with the archival ROSAT PSPC data of the same field. The HRI/PSPC image of the NGC 1399 halo shows significant departures from the above simplified assumptions.

We found an extended and asymmetric halo extending on cluster scales ($r >$ 90 kpc). The HRI image revealed the existence of significant ($> 3\sigma$) filamentary structures and cavities in the galactic halo, due to density fluctuations in the hot gas (Fig.1, left panel).

The halo profile is not consistent with a simple Beta model but suggests the presence of three different components. The combined HRI and PSPC data were fitted with a multi-component bidimensional model, consisting of: (1) a 'central' component (dominating for $r <$ 50") centered on the optical galaxy and whose distribution follows the luminous matter profile, (2) a 'galactic' component (50" $< r \leq$ 400") displaced $1'$ (5.5 kpc) South-West of NGC 1399 and (3) a 'cluster' ellipsoidal component ($r >$ 400") centered $\sim 5.6'$ (31 kpc) North-East of the galaxy. The cooling times of the central and galactic components are smaller than the Hubble time, suggesting the presence of a central cooling flow.

The binding mass estimates, obtained assuming hydrostatic equilibrium, revealed that the dynamics of the central region are dominated by stellar matter while, at radii larger than 2 arcmin, the gravitating mass exceeds the stellar

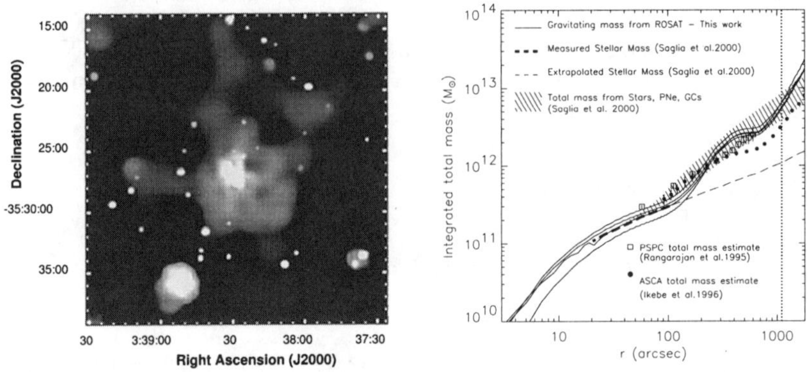

Figure 1. **Left:** Adaptively smoothed X-ray image of NGC 1399 (central peak) and NGC 1404 (SW peak). In the case of NGC 1399 the X-ray emission extends further out than the optical galaxy and reveals the presence filamentary structures and voids. **Right:** Integrated gravitating mass for 4 different temperature profiles (continuous lines). Open squares and filled circles represent previous X-ray determinations. The gravitating mass range derived from optical observations of stars, GCs and PNe, is shown as a shaded region. The contribution of the stellar matter is represented by the dashed lines.

one (Fig.1, right panel). The large scale surface brightness distribution can be explained if the galaxy hosts a large dark halo with different dark matter distributions on galactic and cluster scales. Alternative models that explain the GCs abundance and optical velocity dispersion profiles through tidal interactions with Fornax cluster galaxies are compatible with our data only if such encounters produce a significative flattening of the outer gas distribution. Tidal interactions may also explain the presence of the density fluctuations in the galactic halo.

Ram pressure from the cluster halo is able to account for the decentering of the optical galaxy with respect to the galactic X-ray component if NGC 1399 is moving subsonically (10-100 km s^{-1}) in the cluster potential. Instead, the displacement of the center of the cluster component with respect to the nearest Fornax galaxies may suggest that the cluster is not relaxed and may be undergoing a merger event with a nearby subgroup.

We do not detect the X-ray counterpart of the nuclear radio source and pose an upper limit of $L_X^{3\sigma} = 3.9 \times 10^{39}$ erg s^{-1} on its luminosity in the 0.1-2.4 keV band. Our data also support thermal confinement of the radio emitting plasma. We found X-ray emission enhancements aligned with the radio jets that may be due to shocked gas and X-ray 'holes' and enhancements coincident with the position of the radio lobes. The latter are consistent with a scenario in which the hot gas is displaced by the radio plasma pressure. Alternative models of Inverse Compton scattering of CMB photons fail to account for the observed excesses.

A full discussion of these results can be found in Paolillo et al.(2001, in press in ApJ).

Tracing Cosmic Evolution with Galaxy Clusters
ASP Conference Series, Vol. 268, 2002
S. Borgani, M. Mezzetti, and R. Valdarnini, eds.

The Luminosity Function of Clusters of Galaxies

Ilaria Parolin[1,2], Alberto Moretti[2], Emilio Molinari[2], Guido Chincarini[2,3]

1-Università degli Studi di Milano, Via Celoria 16, 20133, Milano, Italy; 2-Osservatorio Astronomico di Brera-Merate, Via Bianchi 46, 23807 Merate (LC), Italy; 3-Università degli Studi di Milano-Bicocca, Piazza dell'Ateneo Nuovo 1, 20126, Milano, Italy.

Abstract. We present our project about the Luminosity Function of near clusters of galaxies. Further confirmations of the bimodal nature of the LF are found and relations between cluster environment and shape of the function are investigated.

1. The Sample and The Method

In 1994, the Brera Astronomical Observatory (OAB) started a program for optical observations of 14 nearby clusters ($0.02 < z < 0.2$), X-ray selected with the Rosat All-Sky Survey.

Deep observations in g, r and i Gunn colours were performed with the 1.5-m Danish telescope at La Silla (Chile). All frame manipulation was performed using the MIDAS system elaborated by ESO and, in particular, source detection and photometry were performed through the INVENTORY package and our software for extended objects (Moretti et al. 1999).

2. The past

The pioneering work on this sample was published in 1996 by Moretti et al.
The major result of this work is that in the cD cluster A496 the LF is bimodal,(as first noticed in Coma by Biviano et al. 1995; see also Lopez-Cruz et al. 1997), with the early-type population clearly fitted by a Gaussian distribution, while dwarf galaxies obey a Schechter function with a rather steep slope on the faint end. The Luminosity Function depends on the cluster morphology, going from a barely noticeable gap in irregular clusters (e.g. Virgo) to a clear evidence of a gap in evolved clusters (Bauz-Morgan type I).
Moreover, the distribution of the g-r colours of the sequence galaxies as a function of their projected distance from the center of the cluster shows a significant correlation realtive to the population of faint galaxies.
In the years from 1997 to 1999 other clusters were studied: A3395, A85, A4059 (B.M. type I), A3695 (B.M. type II), A119 and A3822 (B.M. type II-III).
The analysis of these clusters yields further confirmation of the bimodal nature of the LF.
Particularly noteworthy is the confirmation of the influence of the level of regularity and the richness of the cluster on the separation of different contributions to the LF (dip in Figure 2) and on the slope of the faint end.

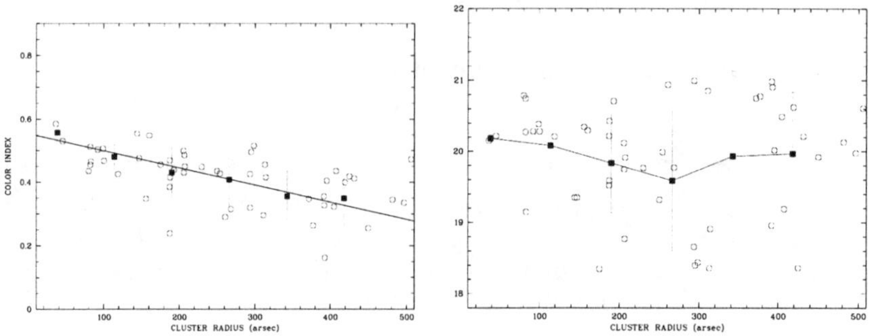

Figure 1. The average colour index of dwarf sequence galaxies shows a gradient from the red to the blue going off the center of the cluster (left panel). This feature cannot be ascribed to the luminosity+colour segregation: the non correlation between cluster radius and mean magnitude of the dwarf sequence galaxies is shown (right panel) - Moretti et al. 1999.

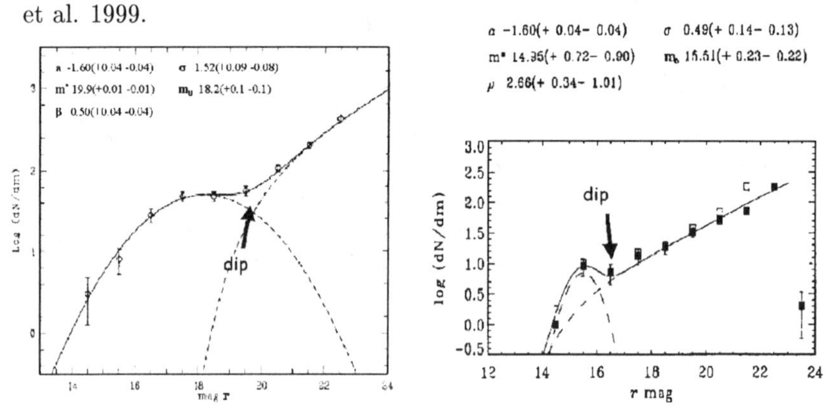

Figure 2. Luminosity Function of A85 (from Ratti, K., masterthesis, 1998; left) and A3822 (from Cozzi, E., masterthesis, 1998; right).

3. The present

The project is reaching its final phase. The remaining group of clusters which is currently being analyzed allows us to study other levels of regularity. In addition to further confirm the composite nature of the LF, we expect results on

- the connection between regularity, richness and evidence for a dip;
- the correlation between the average colour index of dwarf sequence galaxies and their projected distance from the center of the cluster.

References

Biviano, A., Durrett, F., Gerbal, D., Le Fèvre, O., Lobo, C., Mazure, A., Selezak, E. 1995, A&A, 297, 610

Lopez-Cruz, O., Yee, H.K.C., Brown, J.P., Jones, C., Forman, W. 1997, ApJ, 97L, 475

Molinari, E., Chincarini, G., Moretti, A., De Grandi, S. 1998, A&A, 880, 338

Moretti, A., Molinari, E., Chincarini, G., De Grandi, S. 1999, A&AS, 140, 155

Tracing Cosmic Evolution with Galaxy Clusters
ASP Conference Series, Vol. 268, 2002
S. Borgani, M. Mezzetti, and R. Valdarnini, eds.

Recovering the Temperature Structure of Cooling–Flow Clusters

Fabio Pizzolato

Istituto di Fisica Cosmica, CNR, Via Bassini 15, I-20133 Milano, Italy
Dipartimento di Scienze, Universita' dell'Insubria, Via Valleggio 11, I-22100, Como, Italy

Abstract. This paper presents some recent determination of the temperature structure of three cooling–flow clusters of galaxies observed with the instruments on board the X–ray satellite *XMM-Newton*.

1. Introduction

The core of relaxed cluster of galaxies may cool so fast to unbalance the (gross) hydrostatic equilibrium, prompting an inflow of matter called *cooling flow* (see e.g. Fabian 1994).

The expected piling up of hot gas would enhance the core brightness far more than observed, thus part of the falling gas must decouple from the main flow, freezing below X–ray emitting temperatures. If this interpretation is correct, the gas will be multi–phase, and at each radius the gas will not be characterized by one single temperature, but span a wide range of temperatures from $T_{\max}$ down to essentially zero.

In this paper I briefly outline the main results from the observation of three cooling–flow galaxy clusters (A 1795, A 1835 and M87/Virgo) taken with the *EPIC* instruments on board *XMM-Newton*. I show that these results actually points towards a single–phase gas. A more comprehensive discussion may be found in Molendi & Pizzolato 2001.

2. Is multi–phaseness real?

Since multi–phaseness is a cornerstone of cooling–flow models, we investigate whether it is really present, or if the gas is actually single–phase, and its apparent multi–phaseness stems from the conspiracy of a steep temperature gradient and the poor angular resolution of our instruments. If this is the case, our observations mix together neighboring regions with different temperatures, yielding an artifact multi–phaseness.

To pursue our end, we have worked out a spectral model to account for the projection effects and estimate the lowest temperature attained by the gas, and we have compared it to a standard cooling–flow model (Allen 2000), as well as to a reference isothermal model. We have accumulated the spectra of each cluster in concentric ringlets, centered on the peak of emission.

In all our fits, the model which accounts for the projection effects of a single–phase gas is systematically better than the standard cooling flow model. Any radial bin the gas is characterized by a finite range of temperatures.

This is at odds with the prediction of a vanishingly small lower temperature yielded by the cooling flow model, but agrees with Peterson et al. 2001, whose analysis of the data from the *RGS* spectrometer on board *XMM–Newton* show the remarkable absence of emission lines of a plasma below 1 − 2keV.

3. Further developments

If we admit that projection effect do play an essential role in the aspect of cluster of galaxies, it is interesting to reconstruct the actual 3–D shape of the density, temperature and metallicty distributions. To this end we have worked out a new *XSPEC* model for deprojecting the X–ray emission from a spherically symmetric, optically thin cluster. The model is calculated starting from a functional form of the three–dimensional profiles of density, temperature and metallicity (Pizzolato in preparation). With this model we have fitted the spectrum of A 1795, accumulated in four concentric rings.

Although the model is still under development and there are some differences between its predictions and our previous results, the main conclusions are confirmed, in particular the existence of a minimum temperature well above zero.

The role of multi–phaseness calls for a thorough investigation, and the coming into operation of the new generation of X–ray satellites (*XMM–Newton*, *Chandra*) is opening up new possibilities to pursue this end.

References

Allen S. W. 2000, MNRAS, **315**, 269

Fabian A. C., 1994, ARA&A, **32**, 277

Molendi S., Pizzolato F., 2001, ApJ in press

Peterson, J. R., Paerels, F. B. S., Kaastra, J. S., Arnaud, M., Reiprich, T. H., Fabian, A. C., Mushotzky, R. F., Jernigan, J. G., Sakelliou, I. 2001, A&A**365**, L104

Tracing Cosmic Evolution with Galaxy Clusters
ASP Conference Series, Vol. 268, 2002
S. Borgani, M. Mezzetti, and R. Valdarnini, eds.

ISOPHOT Results on Late-Type Virgo Cluster Galaxies

Cristina C. Popescu

The Observatories of the Carnegie Institution of Washington, 813 Santa Barbara Str., Pasadena, 91101 California, USA

Richard J. Tuffs, Heinrich J. Völk

Max Planck Institut für Kernphysik, Saupfercheckweg 1, 69117 Heidelberg, Germany

Daniele Pierini

The University of Toledo, Toledo, Ohio 43606-3390, USA

Barry F. Madore

NASA/IPAC Extragalactic Database, 770 S. Wilson Avenue, Pasadena, California 91125, USA
The Observatories of the Carnegie Institution of Washington, 813 Santa Barbara Str., Pasadena, 91101 California, USA

Abstract. We present statistically significant evidence for the existence of a cold dust component in all late-type (later than S0) galaxies, i.e. spirals, irregulars and Blue Compact Dwarfs (BCDs). This result comes from our analysis of a complete volume and luminosity limited sample of late-type Virgo Cluster galaxies observed with the ISOPHOT instrument (Lemke et al. 1996) on board the ISO satellite.

We present a statistical analysis of the integrated FIR properties of a complete volume- and luminosity-limited sample of Virgo Cluster galaxies ranging in type from S0a to Im and BCD. The sample definition and observational details are given in Tuffs et al. (2001a,b). The sample of 38 galaxies detected in broad-band filters centered on 60, 100 and 170 μm represents the basis for the statistical investigation in this paper.

The flux densities indicate that for most of the galaxies of our sample the SED in the FIR peaks at wavelengths longer than 100 μm and cannot be represented by a single modified blackbody (Planck) function (Popescu et al. 2001). The turn-up in the SED beyond 100 μm is a clear indication of a cold dust component which could not have been detected by IRAS. We fitted the FIR SED with two modified blackbody functions (one representing the "warm dust" component and the other one the "cold dust" component) by constraining one parameter of the fit. We fixed the temperature of the "warm dust" component to be 47 K, on the assumption that this will represent the average dust temperature of the HII regions within these galaxies (for a discussion of the contribution of HII regions and star forming complexes see Popescu et al. 2000, Misiriotis et al. 2001)

Our statistical analysis shows unambiguously the existence of a cold dust component present within all morphological classes of late-type (later than S0) galaxies, e.g. spirals, irregulars and BCDs. Such a cold component previously not seen by IRAS implies the revision of the dust masses in galaxies and of the temperatures of cold dust. The dust masses should be raised by factors of $6 - 13$ from the previous IRAS determinations, with even larger factors for certain individual galaxies. The temperature of the cold dust is found to be typically $8 - 10$ K colder than of the IRAS temperatures, again with individual galaxies having even colder temperatures. A particularly intriguing result of this investigation concerns the masses and temperatures of the cold dust derived for BCD galaxies in our sample. These systems have the highest dust mass surface densities and the coldest dust temperatures. This is not the expected behaviour for galaxies whose FIR emission is dominated by dust heated locally in HII regions, where one would anticipate temperatures of 30 K or more. Two scenarios are proposed to explain the FIR characteristics of BCDs (Popescu et al. 2001).

A good linear correlation is found between the "warm FIR" luminosities and the Hα EW. This is in agreement with the assumptions of our constrained SED fit, that the "warm" component is mainly associated with dust locally heated within star forming complexes. We also found a good but non-linear correlation between the "cold FIR" luminosities and the Hα EW. The correlation itself confirms the predictions of the Popescu et al. (2000) model, where the emission at FIR-sub-mm wavelengths is mainly due to the diffuse UV photons.

A non-linear correlation is found between the "warm" FIR luminosities and the NVSS radio luminosities at 1.4 GHz. We also find a good FIR-radio correlation for the cold dust component, suggesting that the stars mainly responsible for heating the cold dust are the massive progenitors of supernovae whose remnants may be the dominant sources of cosmic ray electrons. This is again in agreement with the model predictions of Popescu et al. (2000). Our findings are the first ones to test the FIR-radio correlation using the FIR luminosities associated with the cold dust component, which was largely missed by IRAS with its lack of wavelength coverage longwards of 100 μm.

References

Lemke, D., Klaas, U., Abolins, J. et al., A&A, 315, L64

Misiriotis, A., Popescu, C.C., Tuffs, R.J., Kylafis, N.D 2001, A&A 372, 775

Popescu, C. C., Misiriotis, A., Kylafis, N. D., Tuffs, R. J., Fischera, J., 2000, A&A, 362, 138

Popescu, C. C., Tuffs, R. J., Völk, H. J., Pierini, D., Madore, F. B. 2001, ApJ submitted

Tuffs, R. J., Popescu, C. C., Pierini, D., Völk, H. J., Hippelein, H., Leech, K., Metcalfe, L., Heinrichsen, I., Xu, C. 2001a, this volume

Tuffs, R. J., Popescu, C. C., Pierini, D., Völk, H. J., Hippelein, H., Leech, K., Metcalfe, L., Heinrichsen, I., Xu, C. 2001b, ApJS in press

Tracing Cosmic Evolution with Galaxy Clusters
ASP Conference Series, Vol. 268, 2002
S. Borgani, M. Mezzetti, and R. Valdarnini, eds.

An XMM-Newton Mosaic Observation of A2163

G.W. Pratt, M. Arnaud

CEA/Saclay Service d'Astrophysique, L'Orme des Merisiers, Bât. 709, 91191 Gif-sur-Yvette Cedex, France

N. Aghanim

IAS-CNRS, Université Paris Sud, Bât. 121, 91405 Orsay Cedex, France

Abstract. Preliminary results from an XMM-Newton observation of the massive, medium-distant cluster A2163 are presented.

One useful application of the increased spatial and spectral sensitivity of the most recent generation of X-ray satellites is in probing the mass distribution of galaxy clusters. Accurate mass estimates are needed to further the use of clusters as cosmological probes and as calibrators for numerical simulation. A2163 is an ideal candidate for such studies as it is medium distant ($z = 0.2$), and massive and thus represents the upper end of the cluster mass scale. XMM-Newton observed A2163 in a mosaic of 5 pointings in 2000 Aug: here we present preliminary radial temperature and mass profiles. We assume $\Omega = 1$, $H_0 = 50$ km s^{-1} Mpc^{-1}.

After generating calibrated event files with SASv5.1, we corrected for vignetting using the photon weighting method of Arnaud et al. (2001). For background we used the blank-sky event files accumulated by D. Lumb. The background subtraction procedure is described in Pratt et al. (2001) – this method accounts for all the background components including the soft X-ray diffuse emission local to A2163. We show the background-subtracted 0.5-2.0 keV MOS image in Figure 1; we use only events within 12' of the centre for each pointing.

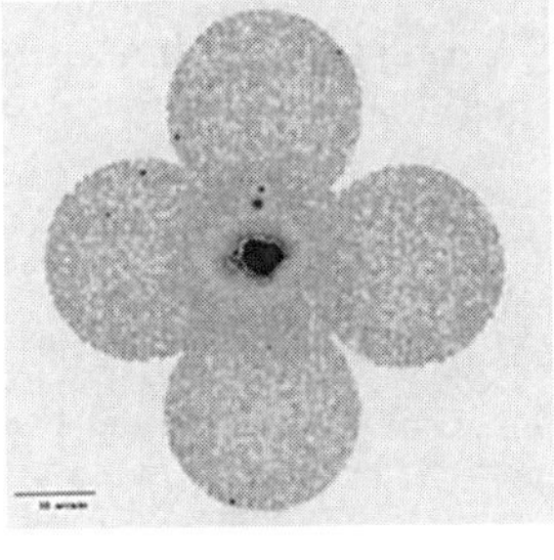

Figure 1. 0.5–2 keV MOS (Gaussian filtered) image of A2163

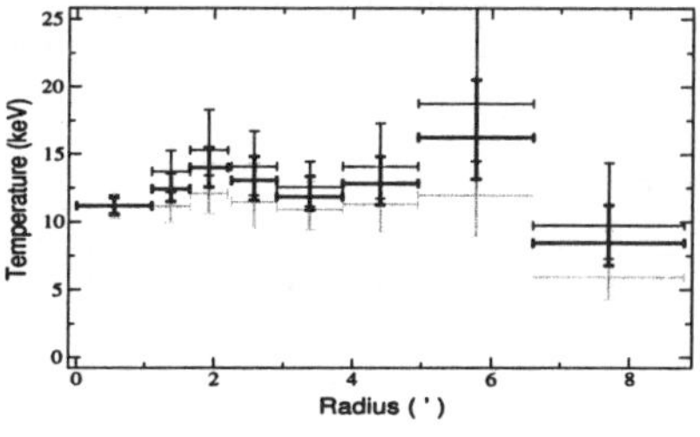

Figure 2. MOS (dark grey), pn (light grey) and combined MOS+pn (black) temperature profiles

We extracted spectra in annular regions centred on the peak of the emission from the cluster, and fitted these spectra with an absorbed isothermal plasma model in the 0.6-10.0 keV band using a fixed absorption of $n_H = 1.65 \times 10^{21}$

cm^{-2}, as found in the combined *ROSAT/GINGA* study by Elbaz et al. (1995a). The radial temperatue profile for each instrument, and the combined MOS+pn temperature profile, are shown in Figure 2. We show also the abundance profile (MOS only) in Figure 3. Using the combined temperature profile, we then produced a preliminary mass profile assuming hydrostatic equilibrium and using the Monte-Carlo method of Neumann & Böhringer (1995). The radii and masses were converted into units of the radius at density contrast 200 using the normalisations provided in Neumann & Arnaud (1999). We show the resulting mass profile of A2163 in Figure 4.

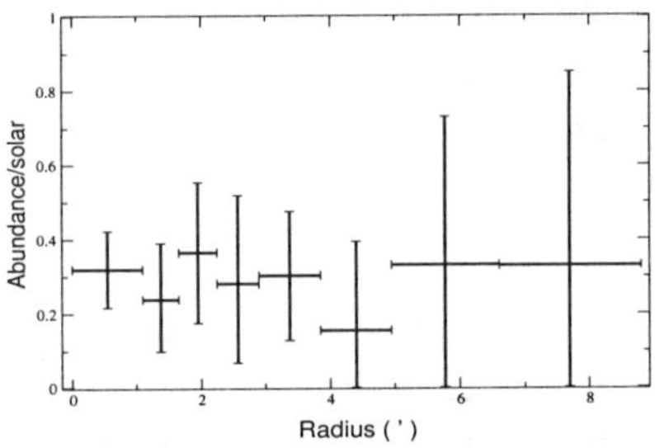

Figure 3. Abundance profile of A2163 MOS only

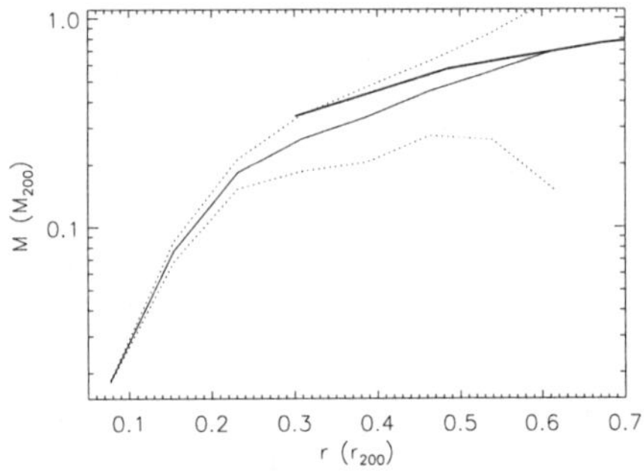

Figure 4. Mass profile of A2163, compared with the (scaled) prediction from the numerical simulations of Evrard et al. (1996)

The XMM-Newton mosaic observation has allowed us to derive an radial temperature profile out to $\sim 0.7 r_{200}$. MOS and pn temperatures agree within their respective errors, but pn temperatures are generally lower; notice though that the MOS and pn profiles have the same form. The profile is isothermal out to $\sim 0.5 r_{200}$, after which there is evidence for a decline. The abundance profile is flat, although the errors are large (a consequence of the high temperature of this cluster). The total mass found, $M(r \leq 2.2 \text{ Mpc}) = 2.8 - 3.0 \times 10^{15}\ M_{\odot}$.
The resulting gravitational mass profile agrees well with numerical simulations, particularly in the outer regions. Temperature substructure such as that found by e.g. Markevitch & Vikhlinin (2001) and Bourdin et al. (2001) may account for the divergence of the (observed and numerical) profiles in the inner regions. A more detailed discussion will be presented in a forthcoming paper.

References

Arnaud, M. et al., 2001, A&A, 365, L80

Bourdin, H., et al. 2001, In XXXVI Rencontres de Moriond: Galaxy Clusters and the High-Redshift Universe, ed. D.M. Neumann, F. Durret and J. Trân Thanh Van, in press

Elbaz, D., Arnaud, M., & Böhringer, H., 1995, A&A, 293, 337

Evrard, A.E., Metzler, C.A., Navarro, J.F., 1996, ApJ, 469, 494

Markevitch, M., & Vikhlinin, V., 2001, ApJ submitted: `fr.arXiv.org/abs/astro-ph/0105093`

Neumann, D., & Böhringer, H., 1995, A&A, 301, 865

Pratt, G.W., Arnaud, M., Aghanim, N., 2001, In XXXVI Rencontres de Moriond: Galaxy Clusters and the High-Redshift Universe, ed. D.M. Neumann, F. Durret and J. Trân Thanh Van, in press

Tracing Cosmic Evolution with Galaxy Clusters
ASP Conference Series, Vol. 268, 2002
S. Borgani, M. Mezzetti, and R. Valdarnini, eds.

VIMOS IFU as a Tool to Observe the Cores of High z Clusters

D. Rizzo, on behalf of the VIRMOS Consortium

Brera Astronomical Observatory, Via Bianchi 46, I-23807 Merate (LC), Italy

Abstract. Research on the core of medium and high z clusters of galaxies can derive great benefits from integral field spectroscopy, and a key role in this respect will be played by the Integral Field Unit (IFU) being developed by the VIRMOS Consortium as part of the VIMOS spectrograph. After a brief technical outline of the instrument, some of the problems to be addressed by data reduction techniques are described, and the scientific issues to which VIMOS IFU is likely to give its major contribution are pointed out.

1. Technical overview

The VIMOS Integral Field Unit (hereafter IFU) allows to obtain one spectrum for each resolution element of its field of view, thanks to a square array of microlenses, each one coupled to an optical fiber. Technical details about VIMOS IFU can be found in Prieto et al. (2000). See also Le Fèvre et al. (1998).

VIMOS IFU can be used in various configurations, outlined in Table 1.

Table 1. IFU configurations

Field of view	Spatial resolution	Spectral resolution	Number of fibers
$54'' \times 54''$	$0\farcs67$	250	80×80
$27'' \times 27''$	$0\farcs33$	250	80×80
$27'' \times 27''$	$0\farcs67$	700, 2500	40×40
$13'' \times 13''$	$0\farcs33$	700, 2500	40×40

2. Reducing IFU data

IFU data reduction presents new challenges and requires to set up new data reduction techniques with respect to "traditional" MOS data.

Two main aspects that have to be taken into account for the data reduction of every IFU are the high density of spectra on the detector, causing some degree of crosstalk (overlapping of light from neighbouring spectra) and the sky subtraction, demanding special care, since the sky must be evaluated using different fibers for different observations. For VIMOS IFU in particular a third aspect is the high number of spectra collected in a single exposure (up to 6400 spectra on four 2K×4K CCDs), requiring at least a partial automatization of the reduction process.

Once data have been reduced, exploiting their scientific content is even more challenging. For this reason a European network has been set up, formed by all the European institutes involved in IFU instrument development, with the purpose to provide the European astronomical community with powerful 3D data analysis tools.

3. Doing science with IFU

The fact that an IFU can obtain spectra for all the objects in the field of view at once makes it much more efficient than a MOS in very crowded field, where complete sampling is required. This is especially true in the case of high-z clusters, where the numbers of interlopers is roughly 90% of all the field objects.

This can have great advantages regarding spectroscopic confirmation of clusters and dynamical studies of cluster cores, where just one IFU observation (or a few ones, if using the smaller size and higher resolution configuration) could obtain the same result as several MOS exposures. It is clear that medium and high-z clusters, being of a smaller angular size, are ideal targets for IFU observations, as shown by Figure 1.

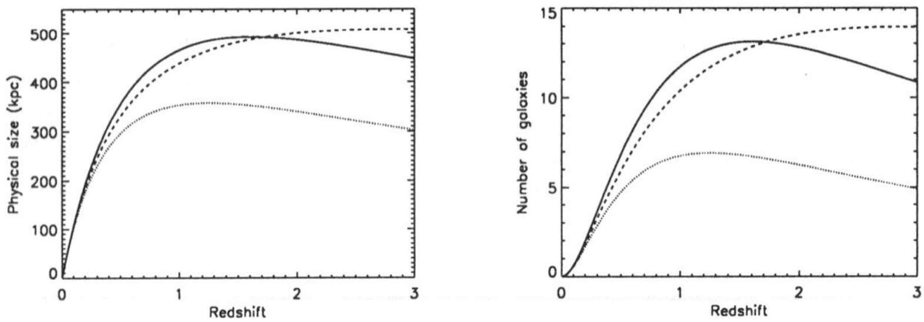

Figure 1. Left: physical size of VIMOS IFU field (54″×54″). Right: estimate of the number of galaxies in the field, with $m_3 < m < m_3 + 2$, belonging to a Coma-like cluster moved at different redshifts (computed using data from Biviano et al. 1995). Solid line: $\Omega_M = 0.3, \Omega_\Lambda = 0.7$; dashed line: $\Omega_M = 0.1, \Omega_\Lambda = 0$; dotted line: $\Omega_M = 1.0, \Omega_\Lambda = 0$.

Systematic IFU observations of cluster cores will be part of the VIMOS-VLT Deep Survey (Le Fèvre et al. 2000).

References

Biviano, A., et al. 1995, A&A 297, 610
Le Fèvre, O., et al. 1998, Proc. SPIE 3355, 8
Le Fèvre, O., et al. 2000, *astro-ph*/0101034
Prieto, J.P., et al. 2000, Proc. SPIE 4008, 510

Tracing Cosmic Evolution with Galaxy Clusters
ASP Conference Series, Vol. 268, 2002
S. Borgani, M. Mezzetti, and R. Valdarnini, eds.

The Origin of the Density Profiles of Halos

Eduard Salvador-Solé & Alberto Manrique

Dpt. d'Astronomia i Meteorologia. Universitat de Barcelona. Martí i Franquès 1, E-08028 Barcelona. Spain

Abstract. Numerical simulations have shown that the density profile of halos has a universal shape, but its origin remains unknown. We solve this problem by taking into account the combined effects of major and minor mergers.

1. Introduction

In the last 30 years, many efforts has been dedicated to determining the dark-matter halo structure and to understanding how it is set. From the theoretical viewpoint there are two approaches: the *spherical infall* model (based on smooth accretion onto isolated seeds) which yields halos growing inside-out, and the modeling of the effects of *violent relaxation* at major mergers which yields halos with a new structure dependent on boundary conditions. None of these approaches is fully satisfactory since, in hierarchical scenarios, major mergers break the spherical assumption of infall, and minor mergers generate unknown varying boundary conditions. From the empirical viewpoint, recent high-resolution N-body simulations have shown that the spherically averaged density profiles have a universal form (Navarro, Frenk & White 1997). In this paper we explain the halo structure in the framework of hierarchical gravitational clustering.

2. A New More Accurate Theoretical Approach

In hierarchical scenarios, dark-matter halos evolve through a sequence of major mergers and smooth accretion (the cumulative effect of minor mergers). Major mergers are rare binary events producing violent relaxation and, hence, spherical objects with new profiles independent of their past history. On the other hand, smooth accretion can be modeled through spherical infall, causing the halo to grow inside-out during this stage. The extended PS model (Lacey & Cole 1993) can be modified to distinguish major and minor mergers (Salvador-Solé, Solanes & Manrique 1998; Raig, González-Casado & Salvador-Solé 2001) through a relative increase mass threshold. This modified PS model provides analytical expressions for several growth rates, in particular for the mass accretion rate, $r_{\rm a}(M,t)$, which allows one to determine the track $M(t)$ followed by accreting halos by solving the differential equation $\mathrm{dM/dt} = \mathrm{r_a[M(t), t]}$.

After its last major merger at $t_{\rm i}$ a spherical halo with mass $M_{\rm i}$ and radius $R_{\rm i}$ grows inside-out by retaining the profile within the instantaneous radius (the radius within which the mean density of the halo is 200 $\rho_{\rm crit}$). Thus, the accreted

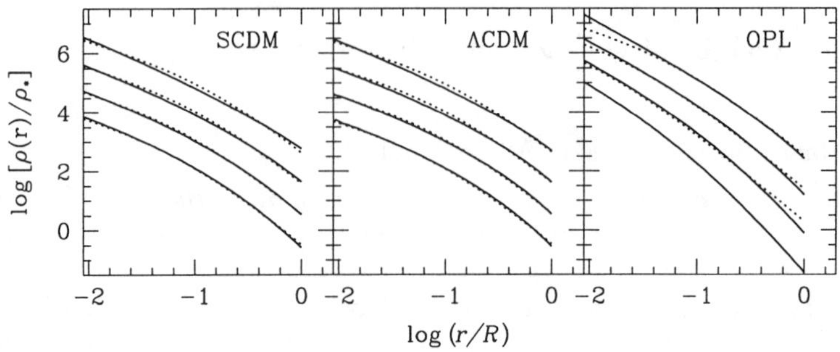

Figure 1. Predicted (full lines) and empirical (dotted lines) density profiles for current halos of several masses ($10^{-2}M_*$, $10^{-1}M_*$, M_*, 10 M_* from bottom to top). M_* is the mass scale at which primordial density fluctuations leave the linear regime, and ρ_* stands for $\rho_{\rm crit}M_*/M$

mass, $M(t) - M_{\rm i}$ is the integral of $4\pi r^2\rho(r)\,{\rm d}r$ from $R_{\rm i}$ to $R(t)$. By differentiating this expression and taking into account the definition of the instantaneous radius

$$R(t) = \left[\frac{3\,M(t)}{800\pi\rho_{\rm crit}(t)}\right]^{1/3} \tag{1}$$

we are led to

$$\rho(t) = 200\,\rho_{\rm crit}\left\{1 - \frac{M(t)}{tr_{\rm a}[M(t),t]}\frac{{\rm d}\ln\rho_{\rm crit}}{{\rm d}\ln{\rm t}}\right\}^{-1} \tag{2}$$

with $M(t)$ the accretion track with $M(t_{\rm i}) = M_{\rm i}$. Equations (1) and (2) define in a parametric form the outer profile for $r \geq R_{\rm i}$. The inner profile, $r < R_{\rm i}$, is determined by violent relaxation with boundary conditions imposed by accretion at $t_{\rm i}$. The continuity condition implies that the inner and outer profiles (and its derivatives) must coincide at $R_{\rm i}$. Since both $M(t)$ and $\rho_{\rm crit}(t)$ are analytical, continuously differentiable functions, the Taylor series expansion of the density profile at $R_{\rm i}$ allows the estimation of its value at $r < R_{\rm i}$ with arbitrary accuracy.

In Figure 1 we plot the density profiles predicted for current halos of four different masses in three cosmologies: the standard CDM model (SCDM), a flat, $\Lambda = 0.75$, CDM model (ΛCDM), and an open, $\Omega_0 = 0.1$, power-law model with index $n = -1$ (OPL). For comparison we show the corresponding empirical profiles obtained by Navarro et al. (1997). As can be seen the agreement is very good. In Manrique, Salvador-Solé & Raig (submitted to MNRAS) further details are given on the halo profiles obtained through the modified PS model.

References

Lacey, C., & Cole S. 1993, MNRAS, 262, 627.

Navarro, J.F., Frenk, C.S., & White S.D.M. 1997, ApJ, 490, 493

Raig, A., González-Casado, G., & Salvador-Solé, E. 2001, MNRAS, in press

Salvador-Solé, E., Solanes J.M., & Manrique, A. 1998, ApJ, 499, 542

Tracing Cosmic Evolution with Galaxy Clusters
ASP Conference Series, Vol. 268, 2002
S. Borgani, M. Mezzetti, and R. Valdarnini, eds.

The Structure of the Virgo Cluster from HI Data

Teresa Sanchis[1], José M. Solanes[2], and E. Salvador-Solé[1]

[1]*Departament d'Astronomia i Meteorologia, Universitat de Barcelona. Av. Diagonal 647; E-08028 Barcelona, Spain*
[2]*Dept. d'Enginyeria Informàtica i Matemàtiques, Universitat Rovira i Virgili. Av. dels Països Catalans, 26; E-43007 Tarragona, Spain*

Abstract. We present the first results of an exhaustive study of the neutral hydrogen content of Virgo cluster spirals that is aimed to investigate the most plausible causes of HI removal and to gain insight on the dynamical history of this system.

1. Motivation and Data Sources

The Virgo cluster of galaxies is the nearest, fairly rich cluster where studies of the environmental effects on galaxy evolution can be carried out with greatest detail. A large fraction of the spiral population of this cluster exhibits significant HI deficiencies (e.g., Giovanelli & Haynes 1985; Solanes et al. 2001), doubtless linked to the interaction of the fragile gaseous disks with the intracluster medium. We are now beginning an exhaustive investigation of the 3D spatial distribution and kinematics of a large sample of Virgo spirals with the aim of unveiling the physical process(es) behind the HI removal. Our study should also provide a better understanding of the dynamical history of this complex cluster, which in turn may have important implications on theories of structure formation and the extragalactic distance scale.

The primary source for the neutral hydrogen contents is the Arecibo General Catalog, a private database maintained by R. Giovanelli and M.P. Haynes at the Cornell University. Galaxy distances are extracted from surveys based on the Tully-Fisher (TF) method by Ekholm et al. (2000), Federspiel, Tammann, & Sandage (1998), Fouqué et al. (1990), Gavazzi et al. (1999), Kraan-Korteweg, Cameron, & Tammann (1988), Mould, Aaronson, & Huchra (1980), Pierce & Tully (1988), and Yasuda, Fukugita, & Okamura (1997). All the TF datasets have been homogenized by a regression analysis.

2. Preliminary Results

The distribution of the spiral galaxies in the Virgo cluster region shows a very large depth as it goes from ~ 10 to ~ 50 Mpc (see Figure 1). The filamentary structure of the galaxy distribution in the tree dimensional space results in an axis ratio of 100:25:30 in the Cartesian coordinates with the x-axis lying almost along the line of sight. Interestingly, we have found the elongation to be more marked on the west side of the cluster, where the radial coordinates of the galaxies reach values twice larger than the easternmost objects.

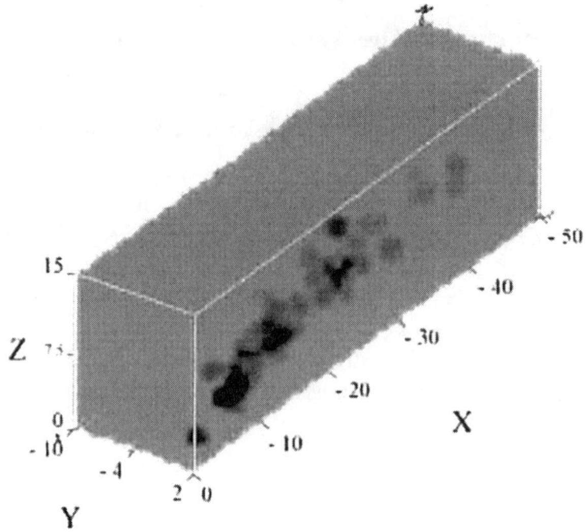

Figure 1. Voxel projection of the 3D distribution of HI deficiency in rectangular equatorial coordinates. Shade intensity increases with HI deficiency. The xy plane is taken to be $\delta = 0$, with the x- and y-axis pointing to RA=0 and 6 hrs, respectively, and the z-axis to the north. Distances are in Mpc.

The spatial distribution of HI deficiency mimics the elongation seen in the spiral distribution, although slightly shifted towards shorter distances. Most of the highly HI-deficient galaxies concentrate in the distance range from 10 to 20 Mpc, with a second region of significant HI deficiency at 25-30 Mpc involving objects associated with the M49/W'/W bridge. We also observe a strong coherence in the line-of-sight velocities of the galaxies, especially in front of ($d_{TF} < 15$ Mpc) and behind ($d_{TF} > 25$ Mpc) the cluster core (M87) suggestive of radial motions. There is no correlation, however, between the degree of HI deficiency and the observed radial velocity.

References

Ekholm T., et al. 2000, A&A, 355, 835

Federspiel, M., Tammann, G.A., & Sandage, A. 1998, ApJ, 495, 115

Fouqué, P. et al. 1990, ApJ, 349, 1

Gavazzi, G., et al. 1999, MNRAS, 304, 595

Giovanelli, R., & Haynes, M.P. 1985, ApJ, 292, 404

Kraan-Korteweg, R.C., Cameron, L.M., & Tammann, G.A. 1988, ApJ, 331, 620

Mould, J., Aaronson, M., & Huchra, J. 1980, ApJ, 238, 458

Pierce, M.J., & Tully, R.B. 1988, ApJ, 330, 579

Solanes, J.M., et al. 2001, ApJ, 548, 97

Yasuda, N., Fukugita, M., & Okamura, S. 1997, ApJS, 108, 417

Tracing Cosmic Evolution with Galaxy Clusters
ASP Conference Series, Vol. 268, 2002
S. Borgani, M. Mezzetti, and R. Valdarnini, eds.

The Nature of The Dwarf Population in Abell 868

R.M. Smith

Dept. of Physics and Astronomy, Univ. of Wales Cardiff, P.O. Box 913, Cardiff, CF24 3YB, U.K.,

P.J. Boyce(1), S. Phillipps, (1), J. Bryn Jones(2), S. P. Driver(3), W.J. Couch(4)

(1)University of Bristol, Tyndall Avenue, Bristol, BS8 1TL, U.K. (2)Univ. of Nottingham, University Park, Nottingham, NG7 2RD, U.K., (3)Univ. of St. Andrews, North Haugh, St. Andrews, KY16 9SS, U.K., (4)Univ. of New South Wales, Sydney, NSW 2052, Australia

Abstract. We have observed the colour distribution of the galaxy populations within the rich Abell cluster A868. These observations, together with recent theoretical models of cluster evolution, provide a natural explanation for the dwarf-density vs environment relations found in many other studies.

1. Observations and Results

Abell 868 is a richness class 3 cluster at a redshift of 0.154. The presence of a central giant cD galaxy, the richness and the apparent association with an X-ray source suggest it is not simply the projection of large-scale structure along the line of sight, which Valotto et al. (2001) have shown can lead to erroneous determinations of the Luminosity Function (LF). Deep B and R band images of the central 0.9 Mpc x 0.9 Mpc region have now been obtained to derive broad morphological information about the dwarf population in this cluster. Similar observations were taken of a nearby off-cluster field for statistical subtraction of the background.

The background-subtracted B-R colour distributions clearly showed three populations - one with a peak at about $\mathrm{B-R} \sim 1.1$, which is the colour expected for dIrrs at the redshift of Abell 868, a red tail with $\mathrm{B-R} > 2.5$ which can be associated with background giant ellipticals, and a central peak of dEs. The variation of the LF with population is shown in Figure 1. The steep faint-end slope clearly arises primarily from the faint blue dwarf population. Background ellipticals have little influence on the shape of the LF whilst the dE LF is approximately flat. Figure 1 also shows the radial distribution of the dEs and dIrrs. Although the ratio of dEs to dIrrs remains approximately constant with radius, there is an absence of dIrrs in the central $0.12\mathrm{Mpc}^2$.

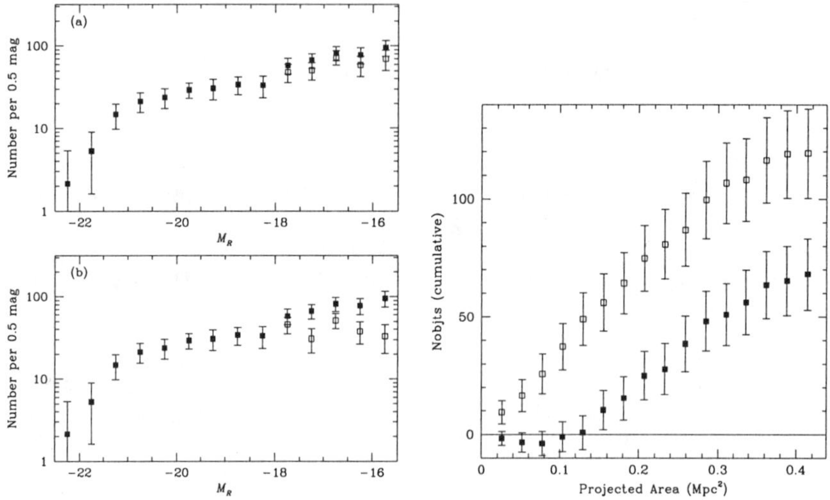

Figure 1. Left -(a) - Variation of LF with population. Solid squares - all galaxies, open squares - bakground Es excluded. (b) - solid squares as (a), open squares - Es and dIrrs excluded. Right - Open squares - Excess of dEs over background, Solid squares - Excess of dIrrs over background. Both as a cumulative function of cluster projected area.

2. Discussion and Conclusions

The apparent absence of any dIrrs in the central region of the cluster is consistent with the 'harrassment' model of Moore et al. (1996,1998). Such tidal processes would certainly fade the dIrrs in the central regions of A868, leaving behind a faint dE remnant.The similar radial distribution of the dEs and dIrrs suggests that they have also formed part of the cluster for similar timescales. These observations, together with the 'harrassment' models, provide possible explanations for several other observed properties of cluster LFs. With dIrrs providing the steep slope, their destruction would lead to a flattening of the slope with cluster evolution - a result in agreement with the observations of Driver et al. (1998). In addition, with dIrr destruction a function of local density, less dense regions/clusters should have steeper faint-end slopes - again in agreement with many observations (e.g. Phillipps et al. 1998).

References

Driver, S.P., Couch, W.J. and Phillipps, S. 1998,MNRAS, 301, 369

Moore, B., Katz, N., Lake, G., Dressler, A. and Oemler, A. 1996, Nature, 379, 613

Moore, B., Lake, G. and Katz, N. 1998, ApJ, 495, 139

Phillipps, S., Driver, S.P., Couch, W.J. and Smith, R.M. 1998, ApJ, 498, 119

Valotto, C.A., Moore, B. and Lambas, D.G. 2001, ApJ, 546, 157

Tracing Cosmic Evolution with Galaxy Clusters
ASP Conference Series, Vol. 268, 2002
S. Borgani, M. Mezzetti, and R. Valdarnini, eds.

PCA of Cluster Surface Brightness Profiles

L. Spurgeon, S. Dos Santos, G.C. Stewart

Department of Physics and Astronomy University of Leicester LE1 7RH

Abstract. Self-similarity of cluster surface brightness (SB) profiles is a very important as, any non-gravitational energy release into the ICM would be expected to cause departures. Studies so far have used analytical fits to the data, which can introduce spurious parameter correlations. In this paper, we present initial results of an investigation into properties of 17 clusters. After standard data reduction, we use a principal component analysis (PCA) to extract information from the SB profiles. The degree of self-similarity amongst the clusters, the minimum number of parameters needed to describe a profile and a comparison with analytic models (eg β-model) are discussed.

1. Principal Component Analysis -Method and Initial results

Recent work on the SB of rich clusters has shown they are self-similar on large scales (Mohr et al. 1999, Vikhlinin et al. 1999) and outlined a correlation between asymptotic slope *beta* with core radius (r_c) (Neumann and Arnaud 1999). These studies used 1 or 2 β-model components to represent radial profiles. An intrinsic correlation, inherent to the fitting procedure, particularly between β and r_c, casts doubt on physical correlations found in these studies. A non-parametric method eg PCA does not have this problem. We present preliminary results of a PCA of 17 low-z ROSAT PSPC rich ($T > 2keV$) clusters.

The cluster data were retrieved from LEDAS (Univ. of Leicester) and was reduced using Snowden's software for extended sources. First and second-order moments of the counts distribution were used to find the emission center, ellipticity ϵ and position angle θ. For each pixel, a distance from the center was computed correcting for ϵ and θ . Figure 1a is an example of a raw SB profile. A median sliding filter was applied to the data to obtain a median SB profile (MSBP), which was divided by the exposure time then rescaled accounting for cosmological dimming and cluster temperature (using ASCA cooling-flow corrected temperatures). The radius was rescaled by the characteristic radius R500, taken from numerical simulations (Evrard et al.1996). Figure 1b shows the set of rescaled MSBP. Each MSBP was fitted with cubic splines to produce the SB value at equally spaced radii ($\delta r = 0.01$). The data points from the whole sample form a cloud of points in a multidimensional k-space. The PCA first finds the direction in k-space where the points are most elongated by least-squares minimization of the variance. This defines the first principal component (PC1) and the data are collapsed along this axis leaving a (k-1)-space without this source of variation. This is repeated for other components.

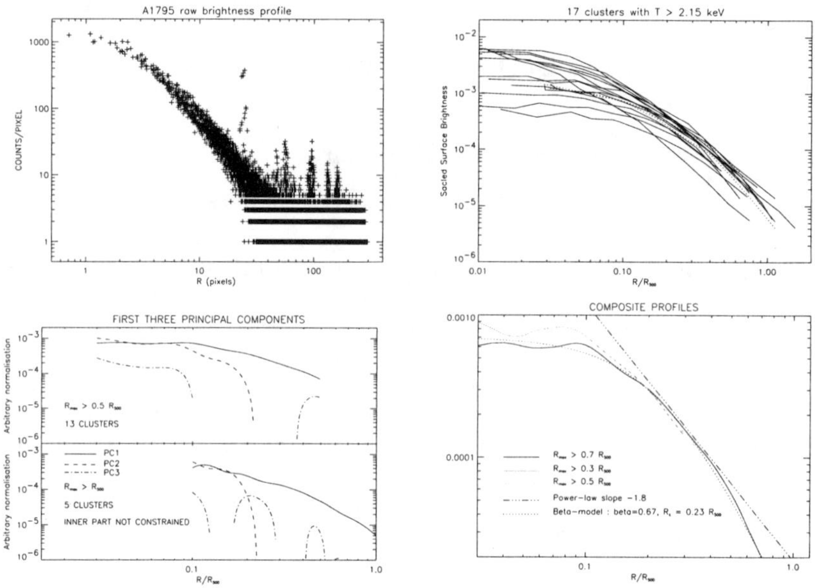

Figure 1. a (top left): Raw brightness profile of A1795. b (top right): Scaled median profiles. c (lower left): First three PC's of two subsets of the sample. d (lower right): Composite profiles for different subsets obtained by the sum of the first 3 PC, weighted by the variance percentage they represent.

Figure 1c shows the first 3 PC's in two subsets. They account for more than 98% of the sample variance. PC1, explains $\sim 80 - 85\%$ of the variance, while only $\sim 10 - 15\%$ is attributed to PC2. Physical interpretations of the PCs have, PC1 corresponding to the large radius profile shape, PC2 to the core radius and part of PC3 to the central cooling-flow component. A composite profile can be constructed by summing the 3 PCs, weighted by the percentage represented (see figure 1d). The whole data set is consistent with 2-3 PC's which describe the cluster data ($\gtrsim 98\%$ variance) and a unique asymptotic profile shape with a slope steepening from -1.8 to -2.5 (similar to a $\beta = 2/3$ profile) There is a unique $r_c \approx 0.2 - 0.3 r_{500}$ (the constant extent of PC2) Work is underway to using a larger sample, extending to groups and high-z clusters and to look at physical correlations between PCs normalisation.

References

Evrard, A. E., Metzler, C. A., & Navarro, J. F. 1996, ApJ, 469, 494

Mohr, J. J., Mathieson, B., & Evrard, A. E. 1999, ApJ, 517, 627

Neumann, D. M., & Arnaud, M. 1999, A&A, 348.711

Vikhlinin, A., Forman., & W. Jones, C. 1999, ApJ, 525, 47

Tracing Cosmic Evolution with Galaxy Clusters
ASP Conference Series, Vol. 268, 2002
S. Borgani, M. Mezzetti, and R. Valdarnini, eds.

Indices for SSPs with Enhanced Mix of α-Elements

Rosaria Tantalo

Department of Astronomy, University of Padua,
Vicolo dell'Osservatorio 2, 35122 – PADOVA Italy

Abstract. In this poster line strength indices for new Single Stellar Populations (SSP) based on the recent stellar models with chemical compositions enhanced in the α-elements (Salasnich et al. 2000) are presented. The SSP indices have been calculated taking into account different methods to correct the *Standard-Worthey Indices* for the real chemical composition of α-enhanced isochrones.

1. Introduction

Over the years, much effort has been spent to interpret main properties of Ellipticals galaxies such as the age, metallicity, and chemical properties. To this aim, Tantalo et al. (1998) (TCB98) have developed the so-called Δ-Method to get the metallicity, age, and *enhancement-degree* for a sample of galaxies. Since the isochrones adopted by TCB98 were with solar partition of elements, in this study by using isochrones based on α-enhanced stellar models (see Salasnich et al. 2000 for all details on these models) and calibrations of indices in which the effect of different $[\alpha/\mathrm{Fe}]$ ratios are simultaneously taken into account, we seek to improve upon a point of weakness in the previous analysis.

2. How the different mixture of α-elements will affect the indices ?

In presence of a certain amount of enhancement in the α-elements one has to suitably modify the relationship between the total metallicity Z and the iron content $[Fe/H]$. This can be expressed by the following relation in which Γ is the *Degree of Enhancement*

$$\left[\frac{Fe}{H}\right] = \lg\left(\frac{Z}{\mathrm{Z}_{\odot}}\right) - \lg\left(\frac{X}{\mathrm{X}_{\odot}}\right) + \Gamma \qquad (1)$$

For solar metallicity and solar-scaled isochrones ($\Gamma = 0.0$), [Fe/H] is $= 0.0$, whereas in the case of α-enhanced isochrones with solar metallicity the real abundance is [Fe/H]$= -0.3557$.

In the analysis below, we have calculated two different sets of SSPs indices following two different procedures for including the effects of the enhancement of α-elements on the line strength indices. **BC**: by using the empirical calibration for Mg_2 given by Borges et al. (1995) as in TCB98. **TC**: by using the index corrections according eq.(6) of Trager et al. (2000).

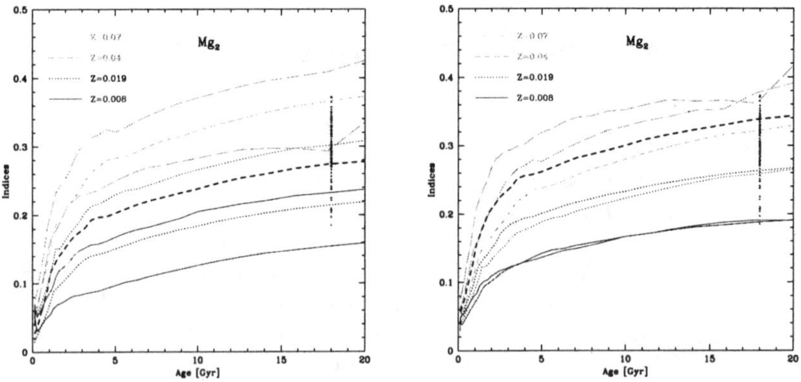

Figure 1. Both panels shown the Mg_2 index as function of age for solar-scaled and α-enhanced abundances, *thin* and *thick lines* respectively calculated with **TC** (left) and **BC** (right) method. *Solid*, *dotted*, *dashed* and *long-dashed* lines are for Z=0.008, Z=0.019, Z=0.040 and Z=0.070 respectively. *Full-triangles* visualizes the range of the Trager et al. (2000) data sample.

3. Conclusions

Fig.1 shows, the results for the Mg_2 index corrected according **TC** and **BC** method in left-panel and right-panel respectively (see Tantalo et al. 2001, for more details). In both cases indices tend to flatten out at old ages (say above 3Gyr), whereas they are strongly depend on metallicity and enhancement.

The major drawback of the new indices is their *age-degeneracy* beyond 3Gyr. Therefore removing the classical age-metallicity degeneracy is a cumbersome affair. A preliminary determination of the age by means of Δ-Method of TCB98 does not lead either to a unique solution or acceptable ages (too old).

Despite the above *age-degeneracy*, in Tantalo et al. (2001) using a minimum-distance procedure, we have estimated the age, metallicity , and Γ for a sample of galaxies. The ranges spanned by age, metallicity, and Γ are now compatible with estimates derived from other, independent methods (Trager et al. 2000).

References

Borges, C.A., Idiart, T.P., de Freitas-Pacheco, J.A. & Thevein, F. 1995, ApJ, 110, 2408

Salasnich, B., Girardi, L., Weiss, A. & Chiosi, C. 2000, A&A, 361,1023

Tantalo, R., Chiosi, C., & Bressan, A. 1998, in A&A, 333, 419 (TCB98)

Tantalo, R. & Chiosi, C. 2001, submitted

Trager, S.C., Faber, S.M., Worthey, G. & González, J.J. 2000, ApJ, 119, 164

Tracing Cosmic Evolution with Galaxy Clusters
ASP Conference Series, Vol. 268, 2002
S. Borgani, M. Mezzetti, and R. Valdarnini, eds.

Cluster Galaxy Populations from $z = 0.33 - 0.83$: Characterizing Evolution with CL1358+62

Kim-Vy Tran and Garth D. Illingworth

Department of Astronomy and Astrophysics, University of California, Santa Cruz, CA 95064

Marijn Franx

Leiden Observatory, P.O. Box 9513, NL-2300 RA Leiden, Netherlands

Abstract. We combine HST WFPC mosaics of clusters at multiple redshifts with Keck LRIS spectroscopy to characterize different galaxy populations and how they evolve. By combining quantitative measurements of structural parameters with velocity dispersions, line indices, and morphological types, we attempt to establish links between ellipticals, S0's, E+A's, and later type galaxies. Here we focus on galaxies in CL1358+62, a massive cluster at $z = 0.33$.

1. Introduction

Using both high resolution imaging from HST WFPC2 and spectroscopy from LRIS on Keck, we study the cluster galaxy populations in CL1358+62 using 142 confirmed members. The imaging provides colors, morphological type, and quantitative structural parameters such as bulge/total fraction, bulge and disk scale lengths, half-light radius, galaxy asymmetry, and ellipticity. The spectroscopy provides membership, velocity dispersions (~masses), and line indices for measuring recent star formation, e.g. Hδ and [OII]. With our complementary approach, we can place constraints on the evolution of cluster early-type (E/S0) and post-starburst (E+A) galaxies in CL1358+62.

2. Elliptical vs. S0: A Matter of Viewing Angle?

Jorgensen & Franx (1994) suggest that elliptical and S0 galaxies ($M_B \geq -22$) are of the same galaxy class where E's are face-on members and S0's the more edge-on ones, *i.e.* these are the same galaxies where viewing angle is the primary discriminator. If the CL1358+62 sample has ellipticals that are mainly round and S0's that are elongated while both share a similar bulge fraction (B/T) distribution, this would support a common parent population for both. We show the ellipticity distributions of E's and S0's in Fig. 1 (left). Results from the Kolmogorov-Smirnov tests of the B/T and ellipticity distributions for this sample supports a common parent population for ellipticals and S0's.

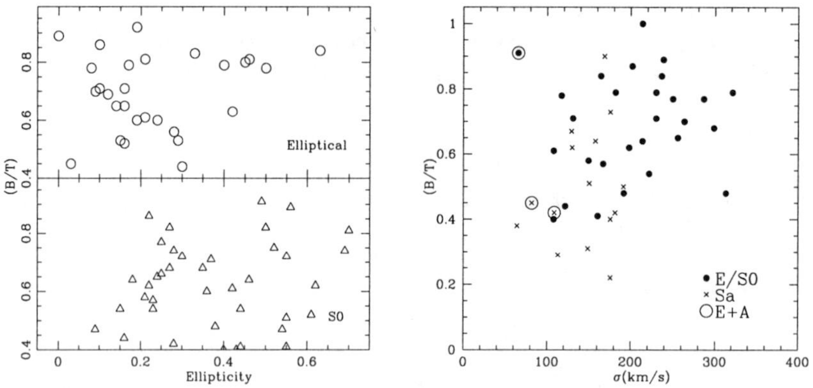

Figure 1. *Left:* Bulge fraction vs. bulge ellipticity for ellipticals and S0's; we show only the most bulge-dominated galaxies ($B/T \geq 0.4$). The B/T distributions of E's and S0's are indistinguishable while their ellipticity distributions do not share a common parent population (> 99% confidence). *Right:* Velocity dispersion (Kelson et al. 2000) vs. B/T for a subset of E/S0, Sa, and E+A members. As the E+A's in CL1358+62 do not share the same velocity dispersion distribution as the E/S0 population, they cannot evolve into massive E/S0 galaxies.

3. E+A Galaxies: What Do They Become?

E+A galaxies are still a mystery (Dressler et al. 1997). Spectroscopically defined as galaxies that formed stars within the recent past (< 1.5 Gyr; Balogh et al. 1999), they may be the transition between late type spirals and early type systems. If E+A's are a snapshot in the transformation of the former to the latter, they are a fundamental phase in galaxy evolution.

To test if E+A's can evolve into the bright, massive early type cluster galaxies observed nearby, we examine the bulge fraction, bulge scale length, and velocity dispersion of E+A's in CL1358+62. Their scale lengths and velocity dispersions (Fig. 1, right) show a population different from E/S0 galaxies. The E+A's in CL1358+62 may evolve into early type systems but they cannot become massive ones.

References

Balogh, M. L., Morris, S. L., Yee, H. K. C., Carlberg, R. G., & Elligson, E. 1999, ApJ, 527, 54

Dressler, A., Oemler, A., Couch, W. J., Smail, I., Ellis, R. S., Barger, A., Butcher, H., Poggianti, B. M., & Sharples, R. M. 1997, ApJ, 490, 577

Jorgensen, I. & Franx, M. 1994, ApJ, 433, 533

Kelson, D. D., Illingworth, G. D., van Dokku, P. G., & Franx, M. 2000, ApJ, 531, 159

Tracing Cosmic Evolution with Galaxy Clusters
ASP Conference Series, Vol. 268, 2002
S. Borgani, M. Mezzetti, and R. Valdarnini, eds.

ISOPHOT Observations of Late-Type Virgo Cluster Galaxies

Richard J. Tuffs

Max Planck Institut für Kernphysik, Saupfercheckweg 1, 69117 Heidelberg, Germany

Cristina C. Popescu

The Observatories of the Carnegie Institution of Washington, 813 Santa Barbara Str., Pasadena, 91101 California, USA

Heinrich J. Völk

Max Planck Institut für Kernphysik

Daniele Pierini

The University of Toledo, Toledo, Ohio 43606-3390, USA

Hans Hippelein

Max Planck Institut für Kernphysik, Königstühl 1, Heidelberg, Germany

Kieron Leech, Leo Metcalfe

ISO Data Centre, Villafranca del Castillo, P.O. Box 50727, 28080 Madrid, Spain

Ingolf Heinrichsen & Cong Xu

IPAC, 770 S. Wilson Avenue, Pasadena, California 91125, USA

Abstract. We present deep diffraction-limited maps in the FIR of a complete luminosity- and volume-limited sample of 63 late-type Virgo cluster galaxies. The ISOPHOT instrument on board the Infrared Space Observatory (ISO) was used to achieve limiting sensitivities over an order of magnitude deeper than IRAS in the 60 and 100 micron bands and to reach the confusion limit at 170 micron. The maps provide an unbiased probe of the FIR spectral energy distributions of gas rich galaxies in the local universe spanning a broad range in star-formation activity and morphological types, including dwarf systems.

63 member galaxies of the Virgo cluster with Hubble type later than S0 have been observed using the ISOPHOT instrument (Lemke et al. 1996) on board ISO in band-passes centred on 60, 100 and 170 μm. The 63 galaxies belong to two contrasting luminosity- *and* volume- limited subsamples selected from the Virgo Cluster Catalogue (Binggeli, Sandage & Tammann 1985; Binggeli,

Popescu & Tammann 1993) on the basis of projected position and integrated B-band magnitude $B_{\rm T}$. 39 of the observed galaxies belong to the "cluster periphery" subsample comprised of all 44 VCC cluster member galaxies brighter than $B_{\rm T}$ 16.8 seen towards the main cluster but with angular separations of greater than 4 degrees from the position of maximum projected galaxy density. These are thought to be predominantly composed of systems freshly falling in from the field. The "cluster core" subsample is comprised of all objects brighter than $B_{\rm T}$ 14.5 with projected positions within 2 degrees of M87. Due to the faint completeness level of the VCC ($B_{\rm T} \sim 18$) the observed objects should be representative of a wide range in star formation activity, including relatively quiescent galaxies. The observed sample includes also a variety of morphological types, from early type spirals to irregulars and BCDs.

Rapid oversampled scans covering the entire optical extent of each target down to the 25.5 mag arcsec^{-2} B-band isophote and adjacent background directions were made using ISOPHOT's focal plane chopper in conjunction with a spacecraft raster. For the first time, data taken in this mode could be corrected for the complex non-linear response to the illumination history in each observation. This allowed robust integrated photometry and structural information to be extracted down to the confusion limit in the 170 micron band and to the sensitivity limits imposed by low-level glitching in the 60 and 100 micron bands. The photometry in the 60 and 100 μm bands was found to be well correlated with the corresponding IRAS measurements for the 33 of the 63 galaxies detected by IRAS, with relative gains ISO/IRAS = 0.95 and 0.82 at 60 and 100 μm respectively.

The averaged 3 σ upper limits (of galaxies with pointlike source appearance) were 43, 33 and 58 mJy (at 60, 100 and 170 μm). 54 galaxies (85.7%) were detected at least at one wavelength, and 40 galaxies (63.5%) were detected at all three wavelengths. The highest detection rate (85.7%) was in the 170 μm band. From the observed 63 galaxies, 21 (33.3%) presented evidence for a nuclear and a disk component at least at one wavelength and 4 (6.3%) galaxies presented evidence for a multi-component morphology at least at one wavelength.

A full account of the observations, sample and data analysis, together with plots of surface brightness profiles and tabulated photometry is given by Tuffs et al. (2001). A statistical analysis of the integrated FIR properties of the sample is given by Popescu et al. (2001a,b).

References

Lemke, D., Klaas, U., Abolins, J. et al., 1996 A&A, 315, L64

Binggeli, B., Sandage, A. & Tammann, G. A. 1985, AJ, 90, 1681

Binggeli, B., Popescu, C. C., Tammann, G. A. 1993, A&AS, 98, 275

Tuffs, R. J., Popescu, C. C., Pierini, D., Völk, H. J., Hippelein, H., Leech, K., Metcalfe, L., Heinrichsen, I., Xu, C. 2001b, ApJS in press

Popescu, C. C., Tuffs, R.J., Völk, H. J., Pierini, D., Madore, B.F. 2001a, this volume

Popescu, C. C., Tuffs, R. J., Völk, H. J., Pierini, D., Madore, B.F, 2001b, ApJ submitted

Tracing Cosmic Evolution with Galaxy Clusters
ASP Conference Series, Vol. 268, 2002
S. Borgani, M. Mezzetti, and R. Valdarnini, eds.

Detection of Non-Random Patterns in Large-Scale Structure

Riccardo Valdarnini

SISSA Via Beirut 2-4 34014 TRIESTE

Abstract. A new method for analyzing the morphological features of point patterns is presented. The method is taken from the study of molecular liquids, where it has been introduced for making a statistical description of anisotropic distributions. The statistical approach is based on the spherical harmonic expansion of angular correlations.

1. THE METHOD

In Valdarnini (2001) I propose an alternative method for analyzing the clustering morphology which is a generalization of the 2-point function $\xi(r)$ (Peebles & Hauser 1974) and is based on the spherical harmonic analysis. The method is drawn from molecular dynamics simulations, where it has been introduced for studying orientational order of supercooled liquids and metallic glasses (Wang & Stroud 1991). Let us consider a system of N_p particles. The $i-th$ particle has coordinates $\vec{r}_i$, in an arbitrary reference frame. For a specified cutoff radius R_c all the particles such that $|\vec{r}_i - \vec{r}_j| < R_c$ are neighbors of i. The line joining i to one of the j is termed a *bond*. The angular coordinates of the vector $\vec{\Delta}_{ji} \equiv \vec{r}_j - \vec{r}_i$ are θ_j, ϕ_j and the quantity

$$Q_{lm}(\vec{r}_i) = \sum_{j \neq i} Y_{lm}(\theta_j, \phi_j), \tag{1}$$

is the coefficient of the spherical harmonic expansion of the angular density of the bonds associated with the particle i. In Eq. 1, and hereafter, summation is understood over all particles j of the distribution such that $|\vec{r}_i - \vec{r}_j| < R_c$.

The coefficients $Q_{lm}(\vec{r}_i)$ are defined as the bond-orientational order parameters and they can be drastically changed by a rotation of the reference systems. A natural quantity to consider , which is rotation invariant, is

$$Q_l(\vec{r}_i) = \sqrt{\frac{4\pi}{2l+1} \sum_{m=-l}^{m=l} Q^{\star}_{lm}(\vec{r}_i) Q_{lm}(\vec{r}_i)}. \tag{2}$$

Using the addition theorem for the spherical harmonics, the expression simplifies to

$$Q_l(\vec{r}_i) = \sqrt{\sum_j \sum_k P_l(\gamma_{jk})}, \tag{3}$$

where P_l is the Legendre polynomial,

$$\gamma_{jk} \equiv cos(\theta_{jk}) = \vec{\Delta}_{ji} \cdot \vec{\Delta}_{ki}/(|\vec{\Delta}_{ji}||\vec{\Delta}_{ki}|)$$

is the angle between two bonds and the summations in Eq. 3 are then independent of the chosen frame. An useful quantity is the auto-correlation $G_l(r)$ of the coefficients $Q_l(\vec{r}_i)$. The function G_l is defined as follows: for all of the M_p pairs (i, k), such that $|\vec{r}_i - \vec{r}_k| = r \pm \Delta r$, where Δr is the thickness of the radial bin, then $G_l(r)$ is the sum over all of these pairs

$$G_l(r) = \frac{1}{M_p} \sum_i \sum_k \frac{4\pi}{2l+1} \sum_{m=-l}^{m=l} Q^{\star}_{lm}(\vec{r}_i) Q_{lm}(\vec{r}_i + \vec{r}). \quad (4)$$

This equation can be greatly simplified: let j be the set of neighbors of the particle i and p that of the particle k, which satisfy $|\vec{r}_j - \vec{r}_i| < R_c$ and $|\vec{r}_p - \vec{r}_k| < R_c$. Then the summation becomes

$$G_l(r) = \frac{1}{M_p} \sum_i \sum_k \sum_j \sum_p P_l(\Gamma_{jp}), \quad (5)$$

with Γ_{jp} being the angle between $\vec{\Delta}_{ji}$ and $\vec{\Delta}_{pk}$. The summation over the pairs is $\sum_i \sum_k$, with the sum over the particles k only for those particles with $|\vec{r}_i - \vec{r}_k| = r \pm \Delta r$.

The effectiveness of the statistical analysis in quantifying clustering morphology is studied (Valdarnini 2001) by applying the statistical estimator G_l to point distributions produced by an ensemble of cosmological $N-$ body simulations with a CDM spectrum. The results shown that the statistical method defined by the function $G_l(r)$ can be used to analyze the clustering morphology produced by gravitational clustering in a quantitative way. The function $G_l(r)$ describes anisotropies in the clustering distribution by measuring the degree of correlation between the angular densities as seen from two different observers separated by r. $G_l(r)$ can then be considered a statistical measure of clustering patterns, with different scales probed by varying the input parameters l and R_c. With large redshift surveys becoming available in the next few years, the proposed statistical method appears as a promising tool for analyzing patterns in the galaxy distribution.

References

Peebles, P.J.E. & Hauser, M.C. 1974, ApJS 28, 19

Valdarnini, R. 2001, AA 366, 376

Wang, Z.Q. & Stroud, D. 1991, J. Chem. Phys. 94, 3896

Tracing Cosmic Evolution with Galaxy Clusters
ASP Conference Series, Vol. 268, 2002
S. Borgani, M. Mezzetti, and R. Valdarnini, eds.

A Dynamical Modeling Technique for Clusters of Galaxies

Emmanuel Van Hese & Herwig Dejonghe

Universiteit Gent, Sterrenkundig Observatorium, Krijgslaan 281, B-9000 Gent, Belgium

Abstract. We present a flexible technique to construct the full distribution function of the member galaxies of rich clusters, in order to (1) get information on the orbital structure of the clusters, and (2) constrain the amount of dark matter. As an example, we apply this method to Abell 119.

The internal structure and dynamics of galaxy clusters provides important information about their formation and evolution. Present observations are now making it possible to search for the phase space distribution function $F(\mathbf{r}, \mathbf{v})$ (DF in short) of the galaxies in clusters. We developped a quadratic programming technique (Dejonghe 1989) to construct the DF in a given potential as a linear combination of basis functions, by minimizing a χ^2-like function. This provides us with a broad range of distributions, from which the best fit is obtained in an essentially non-parametric manner.

As we use only the projected moments, there remains a degeneracy in the solutions, but in this way we do show the range of DFs and potential forms that is consistent with these data.

When we limit ourselves to spherical equilibrium models, the DF $F(E, L)$ can be expressed as a function of the energy and the angular momentum. The galaxies are incorporated in a dark halo, for which we will adopt the well-known profile suggested by Navarro, Frenk & White (1997):

$$\rho_{\mathrm{NFW}}(r) \sim \frac{1}{(r/r_s)(1 + r/r_s)^2}, \tag{1}$$

with r_s a scale radius. Since the corresponding mass $M(r)$ diverges, we use instead a measure for the virialized mass which is commonly denoted by M_{200}. The two quantities r_s and M_{200} will serve as input parameters to fix the potential and search for the best DF. The basis functions are essentially of the form $(E - E_0)^\alpha L^{2\beta}$, making the calculation of the projected moments $\nu_p(r_p)$ and $\sigma_p(r_p)$ fairly straightforward, while the form of the DF is very general and in essence non-parametric.

As an example, we discuss Abell 119, which is a rich cluster at redshift $z = 0.045$, observed by Fabricant et al. (1993) and in the ENACS survey (Katgert et al. 1996). In total, we compiled a list of positions and velocities for 151 galaxies. With an adaptive kernel technique (Pisani 1993), we selected 125 member galaxies, which were then binned to yield the projected moments.

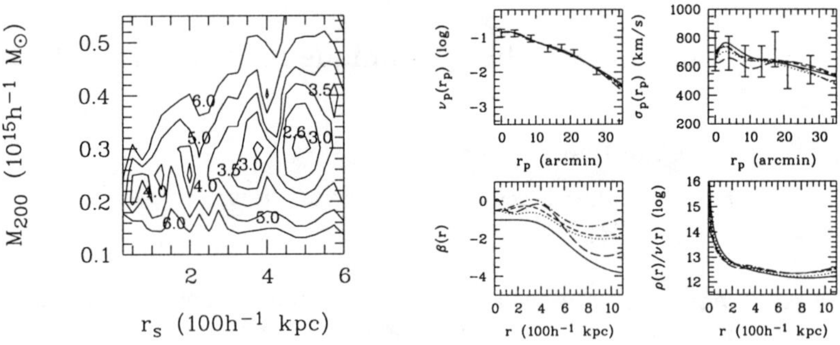

Figure 1. A119: (left) χ^2 of the best-fit models. (right) Proj. moments, anisotropy and mass-to-light ratio for five models.

Figure 1 (left) represents a contour plot of the χ^2 of the best fits (with five basis functions) for each value of the parameters (taking $H_0 = 100h^{-1}$ km/s/Mpc). The degeneracy is obvious: a wide range of scale radii is possible, while M_{200} increases mildly with r_s. This is also visible in figure 1 (right), where some profiles of five models are shown. All models fit the data well, have marginally decreasing mass-to-light ratios and decreasing anisotropies $\beta(r) = 1 - \sigma_t^2(r)/\sigma_r^2(r)$. The latter indicates that tangential orbits at larger radii are necessary to explain the shallow projected dispersion.

In this way we have modeled six clusters (A85, A119, A496, A2670, A3128 and A3158), from which we can draw some general conclusions. In every case, the mass-to-light ratio varies little, explaining the contour diagrams and the orbits: the larger r_s, the less concentrated the dark matter, and the more total mass is needed to support the galaxy motions in the central regions. The relation between r_s and M_{200} will be steeper as the profiles get steeper. Also, shallower profiles require tangential orbits to produce the large projected velocities at large distances from the center, while steeper profiles lead to more isotropic orbits.

We will be able to construct an ensemble cluster out of our clusters; this should allow modeling with more accuracy, higher-order moments could impose additional constraints on the parameters, and it will be possible to investigate the orbits of separate galaxy types.

References

Dejonghe, H. 1989, ApJ, 343, 113

Fabricant, D., et al. 1993, AJ, 105, 788

Katgert, P. et al. 1996, A&A, 310, 8

Navarro, J., Frenk, C. & White, S. 1997, ApJ, 490, 493

Pisani, A. 1993, MNRAS, 265, 706

van der Marel, R. et al. 2000, AJ, 119, 2038

Tracing Cosmic Evolution with Galaxy Clusters
ASP Conference Series, Vol. 268, 2002
S. Borgani, M. Mezzetti, and R. Valdarnini, eds.

Radio Emission from the A3571 Cluster Complex in the Shapley Concentration

T. Venturi[1], S. Bardelli[2], M. Zagaria[3], I. Prandoni[3], R. Morganti[4]

[1] *Istituto di Radioastronomia, CNR, Via Gobetti 101, I-40129 Bologna*
[2] *Osservatorio Astronomico di Bologna, Via Ranzani 1, I-40126 Bologna*
[3] *Università di Bologna, Via Ranzani 1, I-40126 Bologna*
[4] *NFRA, Postbus 2, AA7990 Dwingeloo, The Netherlands*

Abstract. We summarise the most relevant results of a 22 cm radio survey of the A3571 cluster complex, located in the central region of the Shapley Concentration, and propose that this complex represents the final stage of a cluster merger event.

1. The A3571 Cluster Complex and the Radio Observations

The cluster region dominated by A3571 is one of the three main cluster complexes located at the centre of the Shapley Concentration. The other two are dominated by the two clusters A3528 and A3558 respectively, and have been extensively studied from radio wavelengths to X–ray energies (see also Bardelli et al., these proceedings). The A3571 complex includes also the two Abell clusters A3572 and A3575. A3571 is dominated by a large cD galaxy, it is bright and hot in the X–ray band, with a gas temperature $T_{gas} \sim 7.6$ KeV (Nevalainen et al. 2000, ApJ 536, 73). Very little is known on the other two clusters in the chain.

The whole A3571 cluster region was imaged at 22 cm with the Australia Telescope Compact Array (ATCA), using the mosaic technique. The Compact Array operated in the 6.0C configuration, which yields a resolution of $\sim 11'' \times 5''$. The extent of the area covered by these observations is 2.4 deg. The rms in the final mosaicked image is $\sigma \sim 0.13$ mJy b^{-1}. We placed a flux density detection limit of 0.65 mJy b^{-1}.

We note that the results on the 22 cm survey of the A3571 cluster complex are part of a larger project, whose aim is to investigate the role of cluster mergers on the radio properties of galaxies in clusters. A more complete description of the present work can be found in Venturi et al. (2001, submitted to A&A).

2. Results

We detected 124 sources with $S_{22cm} \geq 0.65$ mJy, and identified 36 of them (i.e. $\sim 29\%$) with optical counterparts from the COSMOS/UKST Southern Object Catalogue. Among the optical counterparts, six are located at the distance of the Shapley Concentration. We point out that the spectroscopic information in this region is rather poor, and nine more radio sources are associated with

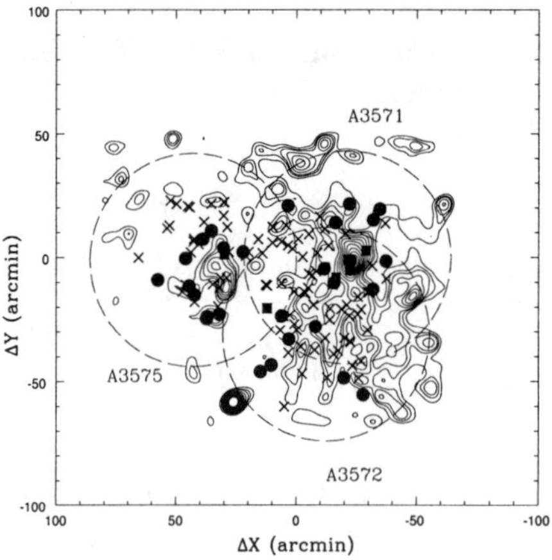

Figure 1. Location of the radio sources overlaid on the distribution of the optical galaxies. Filled squares are the radio galaxies in the A3571 complex; filled circles are the radio galaxies without redshif information; crosses are the unidentified radio sources.

candidate Shapley galaxies. In Figure 1 we show the isodensity contours of the optical galaxies in the A3571 complex, with the distribution of the radio sources superposed.

Comparison of the number of detected radio galaxies in the A3571 complex with the expectations from the radio luminosity function for cluster ellipticals made by Ledlow & Owen (1996, AJ 112, 9) suggests a deficit of radio galaxies over the whole complex. However, if we limit our analysis only to the dominant cluster A3571, the number of radio galaxies detected is in agreement with the expectations.

The radio power of the radio galaxies in the A3571 complex, and of the other candidates most likely located at the same distance, is low, i.e. $\log P_{22} \leq 22.5$ (W Hz^{-1}). This, coupled with the fact that the radio galaxies are unresolved with our observations, suggests that starburst could be partly responsible for the radio emission.

Finally, the radio source counts in the A3571 region are in very good agreement with the background (Prandoni et al. 2001, A&A 365, 392) despite the fact that the density of the optical galaxies is on average a factor of 1.5 higher than the local background density. The same result was obtained also for the A3528 and A3558 complexes (Venturi et al. 2001, MNRAS 324, 1131).

The results of our 22 cm radio survey, coupled with the optical and X–ray properties of the A3571 cluster complex, led us to suggest the possibility that this complex represents the final stage of a merger event, where A3571 is the resulting cluster after virialisation. A more detailed discussion is given in Venturi et al. (2001, A&A, submitted).

Tracing Cosmic Evolution with Galaxy Clusters
ASP Conference Series, Vol. 268, 2002
S. Borgani, M. Mezzetti, and R. Valdarnini, eds.

How to Pick up Interesting Clusters in X-Ray Surveys

Pedro T. P. Viana

Centro de Astrofísica da UP, R. das Estrelas, 4150-762 Porto, Portugal

Andrew R. Liddle

Astronomy Centre, Univ. of Sussex, Falmer, Brighton BN1 9QJ, UK

A. Kathy Romer

Physics Dep., Carnegie Mellon Univ., Pittsburgh PA15213, U.S.A.

Robert G. Mann

Inst. for Astronomy, Univ. of Edinburgh, Edinburgh EH9 9HJ, UK

Abstract.
We briefly show how X-ray fluxes and apparent temperatures of galaxy clusters can be combined to constrain their redshifts.

The *Xcs* is a galaxy cluster survey based on serendipitous detections in *XMM-Newton* images, and it is expected to contain over two thousand clusters with sufficient photon counts for their X-ray temperatures to be accurately estimated without further observations. It will allow the determination of the redshift evolution of the galaxy cluster temperature function beyond $z = 1$, thus providing a means for estimating the value of the density parameter Ω_0 (Romer et al. 2001). For this to be possible, we need to know the redshifts of the clusters in the *Xcs*, both to place them correctly in the evolutionary sequence, and because the redshifts are required to convert the apparent X-ray temperatures, Tapp = T/(1+z), obtained by fitting the shape of the thermal bremsstrahlung spectrum for $z = 0$, into the actual cluster X-ray temperatures. Clearly, optical follow-up to obtain spectroscopic redshifts for all the clusters in the *Xcs* will represent a substantial task.

In addition to the spectrum of the photons received, the main cluster X-ray observables are the angular extent and the flux. In Liddle et al. (2001) we found that at fixed apparent temperature the later depends much more strongly on the cluster redshift than the former. In order to ascertain how useful the X-ray flux is as a redshift estimator, even in the presence of scatter in the luminosity - temperature relation and uncertainties in the apparent temperature estimates, we then carried out Monte Carlo simulations of catalogues corresponding to three years of *Xcs* data. We assumed a cosmological constant dominated Universe with $\Omega_0 = 0.3$. The intrinsic scatter in the luminosity distribution is estimated from N-body hydrodynamical simulations as being 20 per cent at 1σ (O. Muan-

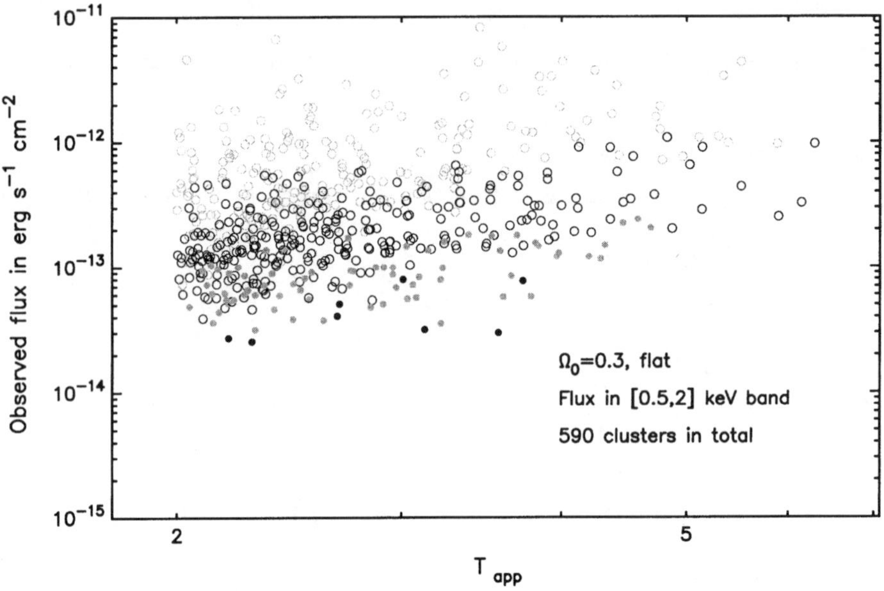

Figure 1. Simulated *Xcs* data in the X-ray flux versus apparent temperature plane. The clusters are divided into four redshift intervals: $0 < z < 0.3$, cyan/grey open circles; $0.3 < z < 0.6$, black open circles; $0.6 < z < 0.9$, cyan/grey points; $z > 0.9$, black points.

wong, private communication). The likely temperature estimation errors were calculated using XSPEC simulations to be about 20 per cent (Liddle et al. 2001).

In Figure 1 we can see that the segregation of clusters with redshift remains quite good. For example, drawing a line between the points (2,10^{-13}) and (7,10^{-12}) clearly differentiates between local clusters, with $z < 0.3$, which lie above the line, and clusters with $z > 0.6$ which lie below the line; it reduces the sample to 44 per cent of its original size without losing any of the $z > 0.6$ clusters. A more stringent cut a factor of two lower in flux can create a sample of mostly high-redshift clusters; it reduces the sample to 117 clusters, including 65 of the 78 clusters above $z = 0.6$. Overall, we find that an estimated redshift with 33 per cent 1σ uncertainty at high redshift should be readily achievable, though the absolute scale relating flux and apparent temperature to redshift will require calibration against real data. Most likely we have been quite conservative in the assessment of temperature errors, so the actual accuracy of the estimation may considerably exceed this in practice.

References

Liddle A.R., Viana P.T.P., Romer A.K., Mann R.G. 2001, MNRAS, 325, 875
Romer A.K., Viana P.T.P., Liddle A.R., Mann R.G. 2001, ApJ, 547, 594

Tracing Cosmic Evolution with Galaxy Clusters
ASP Conference Series, Vol. 268, 2002
S. Borgani, M. Mezzetti, and R. Valdarnini, eds.

The L_x - σ Relation For Poor Groups of Galaxies

Marc E. Zimer

Max Planck Institute for Astronomy, Heidelberg

John S. Mulchaey

Carnegie Observatories, Pasadena

Ann I. Zabludoff

Steward Observatory, University of Arizona, Tucson

Abstract. We use a sample of 20 poor groups of galaxies to study the low L_x tail of the L_x - σ relationship. We have obtained redshifts for fainter members and deep X-ray imaging of these groups. We find that the L_x - σ relationship derived for the rich clusters in the sample of Mushotzky & Scharf (1997) is consistent with the data for our 20 poor groups. However, it is not possible to reach strong conclusions about differences in the L_x - σ relationship for groups and clusters in general at the present time because of significant differences between various cluster samples.

1. Introduction

Poor groups of galaxies, such as the Local Group are the most common environment for galaxies. ROSAT observations indicate that approximately half of all nearby groups of galaxies contain a diffuse, X-ray emitting 'intragroup medium' analogous to the intracluster medium in rich clusters (Mulchaey 2000). Recently it has been realized that the intragroup medium may hold important clues into the formation and evolution of groups (see for example, Davé et al, these proceedings). Numerical simulations indicate that in the absence of non-gravitational heating, the relationships between X-ray luminosity (L_x), X-ray temperature (T) and optical velocity dispersion (σ) are expected to be similar for groups and clusters. The true nature of the scaling relationships for groups and clusters has proved to be a very controversial topic. For example, different authors have reached vastly different conclusions about the relationship between the optical velocity dispersion σ and the X-ray luminosity L_x. Previous work has found either a similar (Zabludoff & Mulchaey 1998) or a more shallow (Helsdon & Ponman 2000; Mahdavi et al. 2000) relation for groups than for clusters. Several explanations for a shallower group slope have been proposed including the possibility that individual galaxy halos contribute significantly to the X-ray luminosity of groups (Mahdavi et al. 2000).

Most studies of the L_x-σ relationship for groups have been based on velocity dispersions determined from only 3 or 4 of the brightest group members. Simulations indicate that when small numbers of galaxies are used, the velocity dispersion does not trace the group mass (Davé et al., these proceedings). We have embarked on a study to measure velocities for more group members and therefore obtain accurate dispersions for a large number of groups.

2. The L_x - σ Plot

Our sample contains 20 groups of galaxies observed with fiber spectrographs at the 2.5m telescope at Las Campanas Observatory and the 3.5m WIYN telescope (Zabludoff & Mulchaey 1998; Zimer et al. 2001). The X-ray properties of these groups have been derived from ROSAT PSPC data (Mulchaey et al. 2001). Fourteen of these groups are detected in the ROSAT images. From the fiber spectroscopy, we measure velocities for typically $\sim$ 15–50 group members. These numbers are high enough to derive reliable velocity dispersions (Davé et al., this proceedings). Figure 1 shows the resulting L_x-σ relationship we derive (data points), with various other fits to the L_x-σ relationship found in the literature. A summary of the various published L_x-σ relationships is also given in Table 1. The *upper left* panel in Figure 1 shows the best fit to our cluster and group data using a parametric bootstrap technique (Lubin & Bahcall 1993):

$$log\ L_x = (4.39 \pm 0.27)\ log\ \sigma + (31.29 \pm 0.80) \quad (1)$$

while for the cluster data alone (Mushotzky & Scharf 1997), we derive:

$$log\ L_x = (3.67 \pm 0.51)\ log\ \sigma + (33.46 \pm 1.54). \quad (2)$$

Within the errors, these fits are in agreement with each other and with the earlier results of Mulchaey & Zabludoff (1998) (see Table 1). Given the small number of groups in our sample, we have not attempted to fit the group data alone. However, a visual inspection of Figure 1 suggests the group + cluster fit adequately describes the groups. In the *upper right* panel we compare our data with fits in Xue & Wu (2000). Their group + cluster fit is consistent with our result within the errors. The *lower left* panel shows the best fit slopes for groups only from Mahdavi et al. (2000) and Helsdon & Ponman (2000). It is immediately obvious that the Mahdavi et al. (2000) fit is inconsistent with our data and cannot be correct for groups in general. The Helsdon & Ponman (2000) group fit is consistent with our group + cluster fit. However, Helsdon & Ponman (2000) claim to find a significant difference between the L_x-σ relation for groups and the relation for clusters. Despite finding similar slopes for groups, we reach a different conclusion than Helsdon & Ponman (2000) because there are significant differences in the cluster samples used in each survey. Helsdon & Ponman (2000) use the White et al. (1997) cluster sample (see the *lower right* panel in Figure 1), which has a considerably steeper L_X-σ relation than the Mushotzky & Scharf (1997) cluster sample we use (White et al. 1997 : $L_x \propto \sigma^{6.38}$, Mushotzky & Scharf 1997 : $L_x \propto \sigma^{3.67}$). The reason for the differences in these cluster samples is not apparent. It is clear, however, that these differences must be understood before any strong conclusions can be made regarding differences in the L_x-σ relationship of groups and clusters.

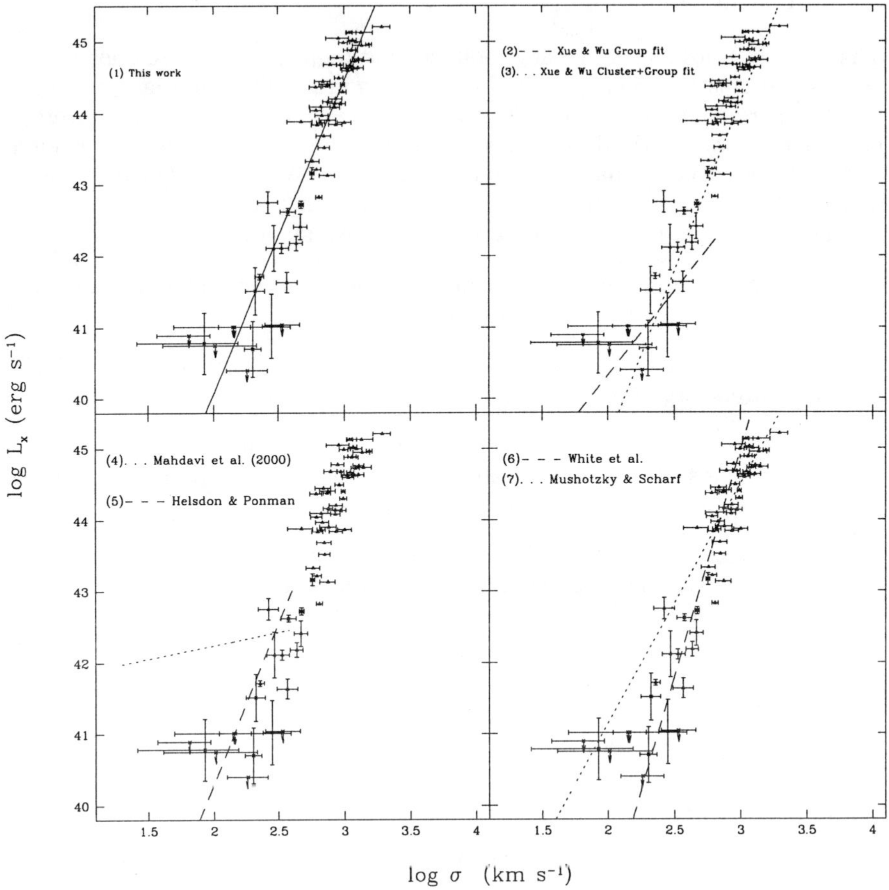

Figure 1. Logarithm of velocity dispersion vs. logarithm of X-ray luminosity for the X-ray detected groups and clusters. The non-detected groups are plotted using the upper limits on L_x (arrows)

Table 1. The L_x - σ Relation fits

	Helsdon & Ponman 2000	Mahdavi et al. 2000	Xue & Wu 2000
groups only	4.7±0.9 (5)	0.37±0.3 (4)	2.35±0.21 (2)
clusters only	6.38±0.46 (6)	3.90±0.10	5.30±0.21
clusters + groups	/	/	4.75±0.18 (3)

	Zabludoff & Mulchaey 1998	This work	
groups only	/	/	
clusters only	/	3.67±0.51 (7)	
clusters + groups	4.29±0.37	4.39±0.27 (1)	

3. Conclusion

There is still considerable debate about the nature of the L_x - σ relationship for groups and clusters. Some of the discrepancies in the literature may be due to poorly determined σ's and/or L_x's. In many studies, the σ's are quite uncertain because they were derived from only the three or four brightest group members. The L_x measurements may also be uncertain because of the low signal-to-noise of most ROSAT images and contamination from other sources.

From our fiber spectroscopy program we have found:

- low-mass groups have many members if the luminosity function is sampled down to sufficiently low luminosities.
- robust velocity dispersions for groups require at least $\sim$ 10–15 velocity measurements.
- strong conclusions about differences in the L_x - σ relationship for groups and clusters cannot be reached at the present time because of differences in the various cluster samples.
- the uncertainties in the various scaling relationships suggest that it may be somewhat premature at this time to draw strong conclusions about the need for additional heating mechanisms in groups.

References

Davé, R., Katz, N., & Weinberg, D., this proceedings.
Helsdon, S.F., & Ponman, T.J. 2000, MNRAS, 315, 356
Lubin, L.M., & Bahcall, N.A. 1993, ApJ 415, L17
Mahdavi, A., Boehringer, H., Geller, M.J., & Ramella, M. 2000, ApJ, 534, 114
Mulchaey, J.S. 2000, ARA&A, 38, 289
Mulchaey, J.S., Zabludoff, A.I., & Zimer, M.E. 2001, ApJL in Prep
Mulchaey, J.S., Davis, D.S., Mushotzky, R.F., & Burstein, D. 2001, in Prep.
Mushotzky, R.F., & Scharf, C.A. 1997, ApJ, 482, L13
White, D.A., Jones, C., & Forman, W. 1997, MNRAS, 292, 419
Xue, Y., & Wu, X. 2000, ApJ, 538, 65
Zabludoff, A.I., & Mulchaey, J.S. 1998, ApJ, 496, 39
Zimer, M.E., Mulchaey, J.S., & Zabludoff, A.I. 2001, ApJ, in Prep

Tracing Cosmic Evolution with Galaxy Clusters
ASP Conference Series, Vol. 268, 2002
S. Borgani, M. Mezzetti, and R. Valdarnini, eds.

Large-Scale Distribution and Spectral Properties of Galaxies in the Shapley Concentration

E.Zucca[1], S.Bardelli[1], A.Baldi[2]

[1] *Osservatorio Astronomico di Bologna, via Ranzani 1, I-40127 Bologna*

[2] *Istituto di Fisica Cosmica "Occhialini", via Bassini 15, I-20133 Milano*

Abstract. The Shapley Concentration is the richest supercluster known and therefore it can be used to test cosmological models and to study the influence of extreme environments on cluster dynamics. We present some results based on a survey of $\sim$ 2000 spectra in this region.

We are carrying on a long term, multiwavelength study of the central part of the Shapley Concentration (see also the contribution of Bardelli et al., these proceedings); here we report some results from the analysis of an optical redshift survey of $\sim$ 2000 galaxies both in clusters and in the intercluster regions.

• **Large-scale distribution** (Bardelli et al. 2000)
The average velocity of the observed intercluster galaxies in this supercluster appears to be a function of the (α, δ) position, and can be fitted by a plane in the three-dimensional space (α, δ, v): the distribution of the galaxy distances around the best fit plane is described by a Gaussian with $\sigma = 3.8$ h^{-1} Mpc .
We estimate the total overdensity in galaxies, the mass and the dynamical state of this structure, discussing the effect of considering a bias between the galaxy distribution and the underlying matter. The estimated total overdensity in galaxies is $(N/\bar{N}) \sim 5.2$ on a scale of 15.5 h^{-1} Mpc for the Shapley Concentration. If light traces the mass distribution, the corresponding mass is 2.3×10^{16} Ω h^{-1} $M_{\odot}$. The dynamical analysis reveals that, if light traces mass and $\Omega = 1$, the Shapley Concentration already reached its turnaround radius and started to collapse: the final collapse will happen in $\sim 3 \times 10^9$ h^{-1} yrs.
We find an indication that the value of the bias between clusters and galaxies in the Shapley Concentration is higher than that reported in literature, confirming the impression that this supercluster is very rich in clusters.
Finally from the comparison with some theoretical scenarios, we find that the Shapley Concentration is consistent with the predictions of the models with more power on large scale (such as Open and ΛCDM), while it is inconsistent with the standard CDM normalized to the cluster abundance.

• **Spectral properties** (Bardelli et al. 2001; Baldi et al. 2001).
The large amount of spectra in our sample allows a uniform spectral classification of galaxies in the Shapley Concentration, in order to investigate the morphology–density relation in a wide range of densities, from the rich cluster cores to the supercluster environment, in this dynamically active region. Moreover, these results were compared with galaxies from the ESP survey, which represent the reference value for the "true" field [$(N/\bar{N}) \sim 1$].
Using a PCA classification technique, we found that all spectra can be well re-

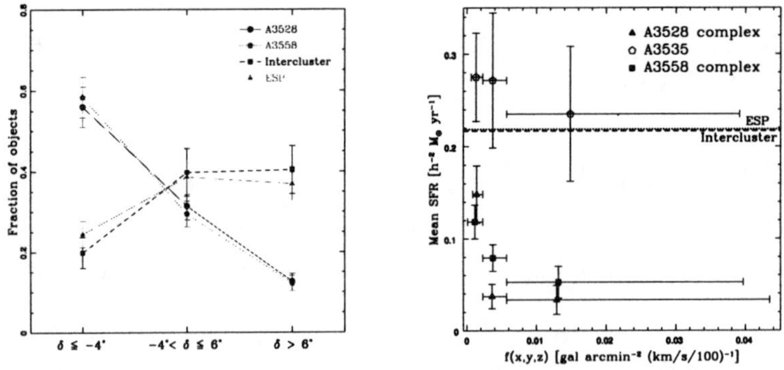

Figure 1. Left: Fraction of galaxies of different types in the various samples; higher values of δ mean later spectral types. Right: Mean star formation rate (derived from the [OII] equivalent width) as a function of local density (derived with an adaptive kernel method).

constructed using only three components, representing a) an early-type galaxy spectrum, b) a spectrum with very prominent blue continuum and emission lines, and c) an emission line dominated spectrum. This classification can be parametrized using two parameters: δ, which represents the contribution of the blue over the red part of the spectrum, and θ, which represents the importance of the emission lines. In the $\delta - \theta$ plane the galaxies follow a well defined sequence, and it results that δ is enough to discriminate among the various types. This spectral classification, together with the [OII] equivalent widths and the star formation rates, has been used to study the properties of galaxies at different densities: cluster, intercluster (i.e. galaxies in the supercluster but outside clusters) and field environment.

No significant differences are present between samples at low density regimes (i.e. intercluster and field galaxies): this result, although not completely unexpected, has been proved here for the first time. Cluster galaxies, instead, not only have values significantly different from the field ones, but also show a dependence on the local density. Moreover, a well defined morphology-density relation is present in the cluster complexes, although these structures are known to be involved in major merging events: the existence of this relation is not obvious a priori in merging clusters, because the dynamical phenomena could have shuffled galaxies from different local densities.

Finally we analyzed the mean star formation rate as a function of the density, finding again a decreasing trend, suggesting that the star formation is depressed.

References

Baldi A., Bardelli S., Zucca E., 2001, MNRAS 324, 509

Bardelli S., Zucca E., Zamorani G., et al., 2000, MNRAS 312, 540

Bardelli S., Zucca E., Baldi A., 2001, MNRAS 320, 387

Author Index

Aghanim, N., 433
Allen, S.W., 195
Andreani, P., 86
Annis, J., 395
Arnaud, M., 157,203,363,433

Baker, J.C., 15
Baldi, A., 463
Balogh, M., 253
Bardelli, S., 143,455,463
Barkhouse, W.A., 289
Barone, L., 403
Barr, J., 15
Barrena, R., 329,331
Bartelmann, M., 86
Bartlett, J.G., 101
Basilakos, S., 81,94
Baumann, O., 333
Bautz, M.W., 385
Beijersbergen, M., 335
Belsole, E., 337
Benoist, C., 363
Best, P., 19
Bettoni, D., 361
Biviano, A., 127,139,329,331
Blindert, K., 339
Böhringer, H., 69,77,187,393,403
Borgani, S., 31,35
Boschin, W., 331,341
Bourdin, H., 337
Boyce, P.J., 441
Bower, R., 253,257,301
Bremer, M.N., 15
Bridges, T.J., 285
Briel, U.G., 203
Brough, S., 343
Brunetti, G., 373
Bryn Jones, J., 441
Burke, D.J., 305,343
Busarello, G., 293,413

Campana, S., 401
Canizares, C.R., 385
Capaccioli, M., 293
Capelato, H.V., 345,399
Cappi, A., 347,363
Carlstrom, J.E., 113
Carter, D., 285
Castander, F.J., 349
Castillo–Morales, A., 199,351
Cavaliere, A., 240
Chincarini, G., 151,407,427
Chiosi, C., 417
Churazov, E., 187
Coil, A., 409
Collins, C.A., 43,69,77,305,343,393,403
Conconi, P., 407
Couch, W.J., 441
Croft, S., 353
Cruddace, R., 393
Cuevas, H., 345

Dalcanton, J.J., 47,379
Dallacasa, D., 207
Davé, R., 236
Davis, M., 409
Dawson, K., 113
De Filippis, E., 199
De Grandi, S., 143,393
Dejonghe, H., 453
Dell'Antonio, I., 401
Doi, M., 421
Dominguez, M., 355
Donahue, M., 385
D'Onghia, E., 151,407
Dos Santos, S., 443
Dressler, A., 277
Driver, S.P., 441
Duc, P.-A., 297

Ebeling, H., 113
Eisenhardt, P., 35
Ellis, R.S., 277,311
Ettori, S., 143
Ezawa, H., 357

Fabbiano, G., 425
Fabian, A.C., 177,195
Fadda, D., 297
Fairley, B.W., 305
Falco, E., 329
Faltenbacher, A., 359
Fasano, G., 361
Ferrari, C., 363

Feretti, L., 207
Firmani, C., 151,407
Flin, P., 365
Focardi, P., 367,391
Fouqué, P., 345
Franx, M., 447
Furusawa, H., 421
Furusho, T., 369

Gal, R.R., 371
Gardini, A., 220
Garmire, G.P., 385
Gay, P.L., 353
Giovannini, G., 207
Girardi, M., 147
Gitti, M., 373
Giudice, G., 375
Giuricin, G., 147
Golse, G., 377
Gonzalez, A.H., 47,379
Gottlöber, S., 359
Governato, F., 213
Govoni, F., 207
Grego, L., 113
Guzzo, L., 69,77,393,401,403

Hamabe, M., 421
Hawkins, M.R.S., 333
Heinrichsen, I., 449
Henry, J.P., 203
Hernquist, L., 236
Hill, G.J., 353
Hill, J.M., 419
Hippelein, H., 449
Holden, B., 35
Holzapfel, W.L., 113
Horleston, N.J., 381
Hwang, C.-Y., 383

Ikebe, Y., 169,187
Illingworth, G.D., 447
Iye, M., 421

Jeltema, T.E., 385
Jones, L.R., 305
Jones, M.E., 121
Joy, M., 113

Kajisawa, M., 421
Katgert, P., 139
Katz, N., 236
Kay, S.T., 387
Keel, W.C., 419
Kelly, D.M., 389
Kelm, B., 367,391
Kerscher, M., 393
Kim, D.-W., 425
Kim, R.S.J., 395
Kimura, M., 421
Kjaergaard, P., 361
Kneib, J.-P., 277,377
Kneissl, R., 121
Kodama, T., 301,421
Kolokotronis, V., 94
Komiyama, Y., 421
Kontorovich, V.M., 397
Koo, D.C., 173,415
Krywult, J., 365
Kurk, J.D., 23,27

La Barbera, F., 293,413
Lambas, D.G., 355
Lapi, A., 240
LaRoque, S., 113
Lazzati, D., 401
Ledlow, M.J., 419
Leech, K., 449
Lehnert, M.D., 15,19
Liddle, A.R., 43,457
Lima Neto, G.B., 345,399
Limboz, F., 147
Longhetti, M., 401
López-Cruz, O., 289
Lupton, R.H., 395
Lynam, P.D., 343,403

MacGillavray, H., 333
Madore, B.F., 431
Majerowicz, S., 405
Malm, M.R., 385
Mann, R.G., 43,343,457
Manrique, A., 437
Manzato, P., 147
Marchesini, D., 151,407
Marinoni, C., 409
Marmo, C., 361
Massarotti, M., 293,413
Matarrese, S., 86
Matsushita, K., 187
Maurogordato, D., 363
Mediavilla, E., 331

Mecke, K., 393
Menci, N., 240
Merluzzi, P., 293,413
Merrelli, A., 349
Metcalfe, L., 449
Metevier, A., 173,415
Mezzetti, M., 147
Miley, G., 19,23,27
Mitsuda, K., 423
Miyazaki, S., 421
Mo, H., 86
Mobasher, B., 285
Molendi, S., 143
Moles, M., 361
Molinari, E., 403,407,427
Moretti, Alberto, 401,427
Moretti, Alessia, 417
Morganti, R., 455
Morrison G.E., 419
Moscardini, L., 86
Muanwong, O., 228
Mulchaey, J.S., 459
Muriel, H., 355

Nagai, D., 113
Nakata, F., 301,421
Natarajan, P., 277
Nelson, A.E., 47,379
Neumann, D., 203,403,405
Newman, J., 409
Nichol, R.C., 43,57,305,349
Nonino, M., 331

Oemler, A., 277
Okamura, S., 301,421
Ota, N., 423
Ouchi, M., 421
Owen, F.N., 419

Panzera, M.R., 23,27,401
Paolillo, M., 425
Parolin, I., 427
Pentericci, L., 23,27
Peres, G., 425
Phillips, S., 381,441
Pierini, D., 431,449
Pierpaoli, E., 90
Pignatelli, E., 361
Pizzolato, F., 429
Plionis, M., 81,94

Poggianti, B.M., 285,361
Popescu, C.C., 431,449
Portinari, L., 417
Postman, M., 3
Prandoni, I., 455
Pratt, G.W., 433
Proust, D., 345

Quilis, V., 253
Quintana, H., 345

Ramella, M., 329,331
Rawlings, S., 353
Reese, E.D., 113
Reiprich, T., 403,405
Rieke, G.H., 389
Rizzo, D., 435
Romer, A.K., 43,305,457
Rosati, P., 31,35
Röser, H.-J., 333
Röttgering, H., 19,23,27

Salvador–Solé, E., 437,439
Sanchis, T., 439
Sasaki, T., 421
Sauvageot, J.-L., 337
Schaap, W.E., 335
Schindler, S., 199,351,393,403
Schmidt, R.W., 195
Schuecker, P., 69,77,393,403
Schwope, A., 199
Scott, D., 90
Sekigushi, M., 421
Setti, G., 373
Seitz, S., 329
Shelyag, S.I., 397
Shimasaku, K., 421
Simard, L., 47,173,379,415
Slezak, E., 363
Smail, I.R., 277,301
Smith, R.M., 441
Sodré, L.Jr., 345,399
Solanes, J.M., 439
Soucail, G., 377
Spurgeon, L., 443
Squires, G., 35
Stanford, S.A., 35
Stewart, G.C., 443
Strauss, M.A., 395

Tagliaferri, G., 401

Tanaka, I., 421
Tantalo, R., 445
Taylor, G.B., 207
Theureau, G., 293
Thomas, P., 224
Thomas, T., 139
Tozzi, P., 35,248
Tran, K.V., 447
Treu, T., 277
Trivedi, P., 277
Tsay, M.-Y., 383
Tschamber, C., 333
Tuffs, R.J., 431,449
Tufts, J.R., 353

Valdarnini, R., 232,451
van Dokkum, P.G., 265
van der Hulst, J.M., 335
Van Hese, E., 453
Venemans, B.P., 23,27
Venturi, T., 143,455
Viana, P.T.P., 43,457
Voges, W., 419
Völk, H.J., 431,449

Wake, D.A., 305
Wambsganss, J., 199
Wechsler, R.H., 379
Weinberg, D., 236
White, M., 90
Wirth, G.D., 415

Xu, C., 449

Yagi, M., 421
Yamada, T., 421
Yasuda, N., 421
Yee, H.K.C., 289,339

Zabludoff, A.I., 459
Zacchei, A., 407
Zagaria, M., 455
Zaritski, D., 47,379
Zimer, M.E., 459
Zucca, E., 143,463

ASTRONOMICAL SOCIETY OF THE PACIFIC
CONFERENCE SERIES

and

INTERNATIONAL ASTRONOMICAL UNION
VOLUMES

Published
by

The Astronomical Society of the Pacific
(ASP)

ASP CONFERENCE SERIES VOLUMES

Published by the Astronomical Society of the Pacific

PUBLISHED: 1988 (* asterisk means OUT OF STOCK)

Vol. CS -1 PROGRESS AND OPPORTUNITIES IN SOUTHERN HEMISPHERE OPTICAL ASTRONOMY: CTIO 25TH Anniversary Symposium
eds. V. M. Blanco and M. M. Phillips
ISBN 0-937707-18-X

Vol. CS-2 PROCEEDINGS OF A WORKSHOP ON OPTICAL SURVEYS FOR QUASARS
eds. Patrick S. Osmer, Alain C. Porter, Richard F. Green, and Craig B. Foltz
ISBN 0-937707-19-8

Vol. CS-3 FIBER OPTICS IN ASTRONOMY
ed. Samuel C. Barden
ISBN 0-937707-20-1

Vol. CS-4 THE EXTRAGALACTIC DISTANCE SCALE:
Proceedings of the ASP 100th Anniversary Symposium
eds. Sidney van den Bergh and Christopher J. Pritchet
ISBN 0-937707-21-X

Vol. CS-5 THE MINNESOTA LECTURES ON CLUSTERS OF GALAXIES AND LARGE-SCALE STRUCTURE
ed. John M. Dickey
ISBN 0-937707-22-8

PUBLISHED: 1989

Vol. CS-6 SYNTHESIS IMAGING IN RADIO ASTRONOMY: A Collection of Lectures from the Third NRAO Synthesis Imaging Summer School
eds. Richard A. Perley, Frederic R. Schwab, and Alan H. Bridle
ISBN 0-937707-23-6

PUBLISHED: 1990

Vol. CS-7 PROPERTIES OF HOT LUMINOUS STARS: Boulder-Munich Workshop
ed. Catharine D. Garmany
ISBN 0-937707-24-4

Vol. CS-8* CCDs IN ASTRONOMY
ed. George H. Jacoby
ISBN 0-937707-25-2

Vol. CS-9 COOL STARS, STELLAR SYSTEMS, AND THE SUN: Sixth Cambridge Workshop
ed. George Wallerstein
ISBN 0-937707-27-9

Vol. CS-10* EVOLUTION OF THE UNIVERSE OF GALAXIES:
Edwin Hubble Centennial Symposium
ed. Richard G. Kron
ISBN 0-937707-28-7

Vol. CS-11 CONFRONTATION BETWEEN STELLAR PULSATION AND EVOLUTION
eds. Carla Cacciari and Gisella Clementini
ISBN 0-937707-30-9

Vol. CS-12 THE EVOLUTION OF THE INTERSTELLAR MEDIUM
ed. Leo Blitz
ISBN 0-937707-31-7

PUBLISHED: 1991

Vol. CS-13 THE FORMATION AND EVOLUTION OF STAR CLUSTERS
ed. Kenneth Janes
ISBN 0-937707-32-5

ASP CONFERENCE SERIES VOLUMES

Published by the Astronomical Society of the Pacific

PUBLISHED: 1991 (* asterisk means OUT OF STOCK)

Vol. CS-14 ASTROPHYSICS WITH INFRARED ARRAYS
ed. Richard Elston
ISBN 0-937707-33-3

Vol. CS-15 LARGE-SCALE STRUCTURES AND PECULIAR MOTIONS IN THE UNIVERSE
eds. David W. Latham and L. A. Nicolaci da Costa
ISBN 0-937707-34-1

Vol. CS-16 Proceedings of the 3rd Haystack Observatory Conference on ATOMS, IONS, AND MOLECULES: NEW RESULTS IN SPECTRAL LINE ASTROPHYSICS
eds. Aubrey D. Haschick and Paul T. P. Ho
ISBN 0-937707-35-X

Vol. CS-17 LIGHT POLLUTION, RADIO INTERFERENCE, AND SPACE DEBRIS
ed. David L. Crawford
ISBN 0-937707-36-8

Vol. CS-18 THE INTERPRETATION OF MODERN SYNTHESIS OBSERVATIONS OF SPIRAL GALAXIES
eds. Nebojsa Duric and Patrick C. Crane
ISBN 0-937707-37-6

Vol. CS-19 RADIO INTERFEROMETRY: THEORY, TECHNIQUES, AND APPLICATIONS, IAU Colloquium 131
eds. T. J. Cornwell and R. A. Perley
ISBN 0-937707-38-4

Vol. CS-20 FRONTIERS OF STELLAR EVOLUTION:
50th Anniversary McDonald Observatory (1939-1989)
ed. David L. Lambert
ISBN 0-937707-39-2

Vol. CS-21 THE SPACE DISTRIBUTION OF QUASARS
ed . David Crampton
ISBN 0-937707-40-6

PUBLISHED: 1992

Vol. CS-22 NONISOTROPIC AND VARIABLE OUTFLOWS FROM STARS
eds. Laurent Drissen, Claus Leitherer, and Antonella Nota
ISBN 0-937707-41-4

Vol CS-23 ASTRONOMICAL CCD OBSERVING AND REDUCTION TECHNIQUES
ed. Steve B. Howell
ISBN 0-937707-42-4

Vol. CS-24 COSMOLOGY AND LARGE-SCALE STRUCTURE IN THE UNIVERSE
ed. Reinaldo R. de Carvalho
ISBN 0-937707-43-0

Vol. CS-25 ASTRONOMICAL DATA ANALYSIS, SOFTWARE AND SYSTEMS I - (ADASS I)
eds. Diana M. Worrall, Chris Biemesderfer, and Jeannette Barnes
ISBN 0-937707-44-9

Vol. CS-26 COOL STARS, STELLAR SYSTEMS, AND THE SUN:
Seventh Cambridge Workshop
eds. Mark S. Giampapa and Jay A. Bookbinder
ISBN 0-937707-45-7

Vol. CS-27 THE SOLAR CYCLE: Proceedings of the
National Solar Observatory/Sacramento Peak 12th Summer Workshop
ed. Karen L. Harvey
ISBN 0-937707-46-5

ASP CONFERENCE SERIES VOLUMES

Published by the Astronomical Society of the Pacific

PUBLISHED: 1992 (asterisk means OUT OF STOCK)

Vol. CS-28 AUTOMATED TELESCOPES FOR PHOTOMETRY AND IMAGING
eds. Saul J. Adelman, Robert J. Dukes, Jr., and Carol J. Adelman
ISBN 0-937707-47-3

Vol. CS-29 Viña del Mar Workshop on CATACLYSMIC VARIABLE STARS
ed. Nikolaus Vogt
ISBN 0-937707-48-1

Vol. CS-30 VARIABLE STARS AND GALAXIES
ed. Brian Warner
ISBN 0-937707-49-X

Vol. CS-31 RELATIONSHIPS BETWEEN ACTIVE GALACTIC NUCLEI AND STARBURST GALAXIES
ed. Alexei V. Filippenko
ISBN 0-937707-50-3

Vol. CS-32 COMPLEMENTARY APPROACHES TO DOUBLE AND MULTIPLE STAR RESEARCH, IAU Colloquium 135
eds. Harold A. McAlister and William I. Hartkopf
ISBN 0-937707-51-1

Vol. CS-33 RESEARCH AMATEUR ASTRONOMY
ed. Stephen J. Edberg
ISBN 0-937707-52-X

Vol. CS-34 ROBOTIC TELESCOPES IN THE 1990's
ed. Alexei V. Filippenko
ISBN 0-937707-53-8

PUBLISHED: 1993

Vol. CS-35* MASSIVE STARS: THEIR LIVES IN THE INTERSTELLAR MEDIUM
eds. Joseph P. Cassinelli and Edward B. Churchwell
ISBN 0-937707-54-6

Vol. CS-36 PLANETS AROUND PULSARS
ed. J. A. Phillips, S. E. Thorsett, and S. R. Kulkarni
ISBN 0-937707-55-4

Vol. CS-37 FIBER OPTICS IN ASTRONOMY II
ed. Peter M. Gray
ISBN 0-937707-56-2

Vol. CS-38 NEW FRONTIERS IN BINARY STAR RESEARCH: Pacific Rim Colloquium
eds. K. C. Leung and I.-S. Nha
ISBN 0-937707-57-0

Vol. CS-39 THE MINNESOTA LECTURES ON THE STRUCTURE AND DYNAMICS OF THE MILKY WAY
ed. Roberta M. Humphreys
ISBN 0-937707-58-9

Vol. CS-40 INSIDE THE STARS, IAU Colloquium 137
eds. Werner W. Weiss and Annie Baglin
ISBN 0-937707-59-7

Vol. CS-41 ASTRONOMICAL INFRARED SPECTROSCOPY: FUTURE OBSERVATIONAL DIRECTIONS
ed. Sun Kwok
ISBN 0-937707-60-0

ASP CONFERENCE SERIES VOLUMES

Published by the Astronomical Society of the Pacific

PUBLISHED: 1993 (* asterisk means OUT OF STOCK)

Vol. CS-42 GONG 1992: SEISMIC INVESTIGATION OF THE SUN AND STARS
ed. Timothy M. Brown
ISBN 0-937707-61-9

Vol. CS-43 SKY SURVEYS: PROTOSTARS TO PROTOGALAXIES
ed. B. T. Soifer
ISBN 0-937707-62-7

Vol. CS-44 PECULIAR VERSUS NORMAL PHENOMENA IN A-TYPE AND RELATED STARS, IAU Colloquium 138
eds. M. M. Dworetsky, F. Castelli, and R. Faraggiana
ISBN 0-937707-63-5

Vol. CS-45 LUMINOUS HIGH-LATITUDE STARS
ed. Dimitar D. Sasselov
ISBN 0-937707-64-3

Vol. CS-46 THE MAGNETIC AND VELOCITY FIELDS OF SOLAR ACTIVE REGIONS, IAU Colloquium 141
eds. Harold Zirin, Guoxiang Ai, and Haimin Wang
ISBN 0-937707-65-1

Vol. CS-47 THIRD DECENNIAL US-USSR CONFERENCE ON SETI -- Santa Cruz, California, USA
ed. G. Seth Shostak
ISBN 0-937707-66-X

Vol. CS-48 THE GLOBULAR CLUSTER-GALAXY CONNECTION
eds. Graeme H. Smith and Jean P. Brodie
ISBN 0-937707-67-8

Vol. CS-49 GALAXY EVOLUTION: THE MILKY WAY PERSPECTIVE
ed. Steven R. Majewski
ISBN 0-937707-68-6

Vol. CS-50 STRUCTURE AND DYNAMICS OF GLOBULAR CLUSTERS
eds. S. G. Djorgovski and G. Meylan
ISBN 0-937707-69-4

Vol. CS-51 OBSERVATIONAL COSMOLOGY
eds. Guido Chincarini, Angela Iovino, Tommaso Maccacaro, and Dario Maccagni
ISBN 0-937707-70-8

Vol. CS-52 ASTRONOMICAL DATA ANALYSIS SOFTWARE AND SYSTEMS II - (ADASS II)
eds. R. J. Hanisch, R. J. V. Brissenden, and Jeannette Barnes
ISBN 0-937707-71-6

Vol. CS-53 BLUE STRAGGLERS
ed. Rex A. Saffer
ISBN 0-937707-72-4

PUBLISHED: 1994

Vol. CS-54* THE FIRST STROMLO SYMPOSIUM: THE PHYSICS OF ACTIVE GALAXIES
eds. Geoffrey V. Bicknell, Michael A. Dopita, and Peter J. Quinn
ISBN 0-937707-73-2

Vol. CS-55 OPTICAL ASTRONOMY FROM THE EARTH AND MOON
eds. Diane M. Pyper and Ronald J. Angione
ISBN 0-937707-74-0

Vol. CS-56 INTERACTING BINARY STARS
ed. Allen W. Shafter
ISBN 0-937707-75-9

ASP CONFERENCE SERIES VOLUMES

Published by the Astronomical Society of the Pacific

PUBLISHED: 1994 (* asterisk means OUT OF STOCK)

Vol. CS-57	STELLAR AND CIRCUMSTELLAR ASTROPHYSICS eds. George Wallerstein and Alberto Noriega-Crespo ISBN 0-937707-76-7
Vol. CS-58*	THE FIRST SYMPOSIUM ON THE INFRARED CIRRUS AND DIFFUSE INTERSTELLAR CLOUDS eds. Roc M. Cutri and William B. Latter ISBN 0-937707-77-5
Vol. CS-59	ASTRONOMY WITH MILLIMETER AND SUBMILLIMETER WAVE INTERFEROMETRY, IAU Colloquium 140 eds. M. Ishiguro and Wm. J. Welch ISBN 0-937707-78-3
Vol. CS-60	THE MK PROCESS AT 50 YEARS: A POWERFUL TOOL FOR ASTROPHYSICAL INSIGHT, A Workshop of the Vatican Observatory --Tucson, Arizona, USA eds. C. J. Corbally, R. O. Gray, and R. F. Garrison ISBN 0-937707-79-1
Vol. CS-61	ASTRONOMICAL DATA ANALYSIS SOFTWARE AND SYSTEMS III - (ADASS III) eds. Dennis R. Crabtree, R. J. Hanisch, and Jeannette Barnes ISBN 0-937707-80-5
Vol. CS-62	THE NATURE AND EVOLUTIONARY STATUS OF HERBIG Ae/Be STARS eds. Pik Sin Thé, Mario R. Pérez, and Ed P. J. van den Heuvel ISBN 0-9837707-81-3
Vol. CS-63	SEVENTY-FIVE YEARS OF HIRAYAMA ASTEROID FAMILIES: THE ROLE OF COLLISIONS IN THE SOLAR SYSTEM HISTORY eds. Yoshihide Kozai, Richard P. Binzel, and Tomohiro Hirayama ISBN 0-937707-82-1
Vol. CS-64*	COOL STARS, STELLAR SYSTEMS, AND THE SUN: Eighth Cambridge Workshop ed. Jean-Pierre Caillault ISBN 0-937707-83-X
Vol. CS-65*	CLOUDS, CORES, AND LOW MASS STARS: The Fourth Haystack Observatory Conference eds. Dan P. Clemens and Richard Barvainis ISBN 0-937707-84-8
Vol. CS-66*	PHYSICS OF THE GASEOUS AND STELLAR DISKS OF THE GALAXY ed. Ivan R. King ISBN 0-937707-85-6
Vol. CS-67	UNVEILING LARGE-SCALE STRUCTURES BEHIND THE MILKY WAY eds. C. Balkowski and R. C. Kraan-Korteweg ISBN 0-937707-86-4
Vol. CS-68*	SOLAR ACTIVE REGION EVOLUTION: COMPARING MODELS WITH OBSERVATIONS eds. K. S. Balasubramaniam and George W. Simon ISBN 0-937707-87-2
Vol. CS-69	REVERBERATION MAPPING OF THE BROAD-LINE REGION IN ACTIVE GALACTIC NUCLEI eds. P. M. Gondhalekar, K. Horne, and B. M. Peterson ISBN 0-937707-88-0
Vol. CS-70*	GROUPS OF GALAXIES eds. Otto-G. Richter and Kirk Borne ISBN 0-937707-89-9

ASP CONFERENCE SERIES VOLUMES

Published by the Astronomical Society of the Pacific

PUBLISHED: 1995 (* asterisk means OUT OF STOCK)

Vol. CS-71 TRIDIMENSIONAL OPTICAL SPECTROSCOPIC METHODS IN ASTROPHYSICS, IAU Colloquium 149
eds. Georges Comte and Michel Marcelin
ISBN 0-937707-90-2

Vol. CS-72 MILLISECOND PULSARS: A DECADE OF SURPRISE
eds. A. S Fruchter, M. Tavani, and D. C. Backer
ISBN 0-937707-91-0

Vol. CS-73 AIRBORNE ASTRONOMY SYMPOSIUM ON THE GALACTIC ECOSYSTEM: FROM GAS TO STARS TO DUST
eds. Michael R. Haas, Jacqueline A. Davidson, and Edwin F. Erickson
ISBN 0-937707-92-9

Vol. CS-74 PROGRESS IN THE SEARCH FOR EXTRATERRESTRIAL LIFE: 1993 Bioastronomy Symposium
ed. G. Seth Shostak
ISBN 0-937707-93-7

Vol. CS-75 MULTI-FEED SYSTEMS FOR RADIO TELESCOPES
eds. Darrel T. Emerson and John M. Payne
ISBN 0-937707-94-5

Vol. CS-76 GONG '94: HELIO- AND ASTERO-SEISMOLOGY FROM THE EARTH AND SPACE
eds. Roger K. Ulrich, Edward J. Rhodes, Jr., and Werner Däppen
ISBN 0-937707-95-3

Vol. CS-77 ASTRONOMICAL DATA ANALYSIS SOFTWARE AND SYSTEMS IV - (ADASS IV)
eds. R. A. Shaw, H. E. Payne, and J. J. E. Hayes
ISBN 0-937707-96-1

Vol. CS-78 ASTROPHYSICAL APPLICATIONS OF POWERFUL NEW DATABASES: Joint Discussion No. 16 of the 22nd General Assembly of the IAU
eds. S. J. Adelman and W. L. Wiese
ISBN 0-937707-97-X

Vol. CS-79* ROBOTIC TELESCOPES: CURRENT CAPABILITIES, PRESENT DEVELOPMENTS, AND FUTURE PROSPECTS FOR AUTOMATED ASTRONOMY
eds. Gregory W. Henry and Joel A. Eaton
ISBN 0-937707-98-8

Vol. CS-80* THE PHYSICS OF THE INTERSTELLAR MEDIUM AND INTERGALACTIC MEDIUM
eds. A. Ferrara, C. F. McKee, C. Heiles, and P. R. Shapiro
ISBN 0-937707-99-6

Vol. CS-81 LABORATORY AND ASTRONOMICAL HIGH RESOLUTION SPECTRA
eds. A. J. Sauval, R. Blomme, and N. Grevesse
ISBN 1-886733-01-5

Vol. CS-82* VERY LONG BASELINE INTERFEROMETRY AND THE VLBA
eds. J. A. Zensus, P. J. Diamond, and P. J. Napier
ISBN 1-886733-02-3

Vol. CS-83* ASTROPHYSICAL APPLICATIONS OF STELLAR PULSATION, IAU Colloquium 155
eds. R. S. Stobie and P. A. Whitelock
ISBN 1-886733-03-1

ATLAS INFRARED ATLAS OF THE ARCTURUS SPECTRUM, 0.9 - 5.3 μm
eds. Kenneth Hinkle, Lloyd Wallace, and William Livingston
ISBN: 1-886733-04-X

ASP CONFERENCE SERIES VOLUMES

Published by the Astronomical Society of the Pacific

PUBLISHED: 1995 (* asterisk means OUT OF STOCK)

Vol. CS-84 THE FUTURE UTILIZATION OF SCHMIDT TELESCOPES, IAU Colloquium 148
eds. Jessica Chapman, Russell Cannon, Sandra Harrison, and Bambang Hidayat
ISBN 1-886733-05-8

Vol. CS-85* CAPE WORKSHOP ON MAGNETIC CATACLYSMIC VARIABLES
eds. D. A. H. Buckley and B. Warner
ISBN 1-886733-06-6

Vol. CS-86 FRESH VIEWS OF ELLIPTICAL GALAXIES
eds. Alberto Buzzoni, Alvio Renzini, and Alfonso Serrano
ISBN 1-886733-07-4

PUBLISHED: 1996

Vol. CS-87 NEW OBSERVING MODES FOR THE NEXT CENTURY
eds. Todd Boroson, John Davies, and Ian Robson
ISBN 1-886733-08-2

Vol. CS-88* CLUSTERS, LENSING, AND THE FUTURE OF THE UNIVERSE
eds. Virginia Trimble and Andreas Reisenegger
ISBN 1-886733-09-0

Vol. CS-89 ASTRONOMY EDUCATION: CURRENT DEVELOPMENTS, FUTURE COORDINATION
ed. John R. Percy
ISBN 1-886733-10-4

Vol. CS-90 THE ORIGINS, EVOLUTION, AND DESTINIES OF BINARY STARS IN CLUSTERS
eds. E. F. Milone and J. -C. Mermilliod
ISBN 1-886733-11-2

Vol. CS-91 BARRED GALAXIES, IAU Colloquium 157
eds. R. Buta, D. A. Crocker, and B. G. Elmegreen
ISBN 1-886733-12-0

Vol. CS-92* FORMATION OF THE GALACTIC HALO .... INSIDE AND OUT
eds. Heather L. Morrison and Ata Sarajedini
ISBN 1-886733-13-9

Vol. CS-93 RADIO EMISSION FROM THE STARS AND THE SUN
eds. A. R. Taylor and J. M. Paredes
ISBN 1-886733-14-7

Vol. CS-94 MAPPING, MEASURING, AND MODELING THE UNIVERSE
eds. Peter Coles, Vicent J. Martinez, and Maria-Jesus Pons-Borderia
ISBN 1-886733-15-5

Vol. CS-95 SOLAR DRIVERS OF INTERPLANETARY AND TERRESTRIAL DISTURBANCES: Proceedings of 16th International Workshop National Solar Observatory/Sacramento Peak
eds. K. S. Balasubramaniam, Stephen L. Keil, and Raymond N. Smartt
ISBN 1-886733-16-3

Vol. CS-96 HYDROGEN-DEFICIENT STARS
eds. C. S. Jeffery and U. Heber
ISBN 1-886733-17-1

Vol. CS-97 POLARIMETRY OF THE INTERSTELLAR MEDIUM
eds. W. G. Roberge and D. C. B. Whittet
ISBN 1-886733-18-X

ASP CONFERENCE SERIES VOLUMES

Published by the Astronomical Society of the Pacific

PUBLISHED: 1996 (* asterisk means OUT OF STOCK)

Vol. CS-98 FROM STARS TO GALAXIES: THE IMPACT OF STELLAR PHYSICS ON GALAXY EVOLUTION
eds. Claus Leitherer, Uta Fritze-von Alvensleben, and John Huchra
ISBN 1-886733-19-8

Vol. CS-99 COSMIC ABUNDANCES:
Proceedings of the 6th Annual October Astrophysics Conference
eds. Stephen S. Holt and George Sonneborn
ISBN 1-886733-20-1

Vol. CS-100 ENERGY TRANSPORT IN RADIO GALAXIES AND QUASARS
eds. P. E. Hardee, A. H. Bridle, and J. A. Zensus
ISBN 1-886733-21-X

Vol. CS-101 ASTRONOMICAL DATA ANALYSIS SOFTWARE AND SYSTEMS V – (ADASS V)
eds. George H. Jacoby and Jeannette Barnes
ISBN 1080-7926

Vol. CS-102 THE GALACTIC CENTER, 4th ESO/CTIO Workshop
ed. Roland Gredel
ISBN 1-886733-22-8

Vol. CS-103 THE PHYSICS OF LINERS IN VIEW OF RECENT OBSERVATIONS
eds. M. Eracleous, A. Koratkar, C. Leitherer, and L. Ho
ISBN 1-886733-23-6

Vol. CS-104 PHYSICS, CHEMISTRY, AND DYNAMICS OF INTERPLANETARY DUST, IAU Colloquium 150
eds. Bo Å. S. Gustafson and Martha S. Hanner
ISBN 1-886733-24-4

Vol. CS-105 PULSARS: PROBLEMS AND PROGRESS, IAU Colloquium 160
ed. S. Johnston, M. A. Walker, and M. Bailes
ISBN 1-886733-25-2

Vol. CS-106 THE MINNESOTA LECTURES ON EXTRAGALACTIC NEUTRAL HYDROGEN
ed. Evan D. Skillman
ISBN 1-886733-26-0

Vol. CS-107 COMPLETING THE INVENTORY OF THE SOLAR SYSTEM:
A Symposium held in conjunction with the 106th Annual Meeting of the ASP
eds. Terrence W. Rettig and Joseph M. Hahn
ISBN 1-886733-27-9

Vol. CS-108 M.A.S.S. -- MODEL ATMOSPHERES AND SPECTRUM SYNTHESIS:
5th Vienna - Workshop
eds. Saul J. Adelman, Friedrich Kupka, and Werner W. Weiss
ISBN 1-886733-28-7

Vol. CS-109 COOL STARS, STELLAR SYSTEMS, AND THE SUN: Ninth Cambridge Workshop
eds. Roberto Pallavicini and Andrea K. Dupree
ISBN 1-886733-29-5

Vol. CS-110 BLAZAR CONTINUUM VARIABILITY
eds. H. R. Miller, J. R. Webb, and J. C. Noble
ISBN 1-886733-30-9

Vol. CS-111 MAGNETIC RECONNECTION IN THE SOLAR ATMOSPHERE:
Proceedings of a Yohkoh Conference
eds. R. D. Bentley and J. T. Mariska
ISBN 1-886733-31-7

ASP CONFERENCE SERIES VOLUMES

Published by the Astronomical Society of the Pacific

PUBLISHED: 1996 (* asterisk means OUT OF STOCK)

Vol. CS-112 THE HISTORY OF THE MILKY WAY AND ITS SATELLITE SYSTEM
eds. Andreas Burkert, Dieter H. Hartmann, and Steven R. Majewski
ISBN 1-886733-32-5

PUBLISHED: 1997

Vol. CS-113 EMISSION LINES IN ACTIVE GALAXIES: NEW METHODS AND TECHNIQUES, IAU Colloquium 159
eds. B. M. Peterson, F.-Z. Cheng, and A. S. Wilson
ISBN 1-886733-33-3

Vol. CS-114 YOUNG GALAXIES AND QSO ABSORPTION-LINE SYSTEMS
eds. Sueli M. Viegas, Ruth Gruenwald, and Reinaldo R. de Carvalho
ISBN 1-886733-34-1

Vol. CS-115 GALACTIC CLUSTER COOLING FLOWS
ed. Noam Soker
ISBN 1-886733-35-X

Vol. CS-116 THE SECOND STROMLO SYMPOSIUM:
THE NATURE OF ELLIPTICAL GALAXIES
eds. M. Arnaboldi, G. S. Da Costa, and P. Saha
ISBN 1-886733-36-8

Vol. CS-117 DARK AND VISIBLE MATTER IN GALAXIES
eds. Massimo Persic and Paolo Salucci
ISBN-1-886733-37-6

Vol. CS-118 FIRST ADVANCES IN SOLAR PHYSICS EUROCONFERENCE:
ADVANCES IN THE PHYSICS OF SUNSPOTS
eds. B. Schmieder. J. C. del Toro Iniesta, and M. Vázquez
ISBN 1-886733-38-4

Vol. CS-119 PLANETS BEYOND THE SOLAR SYSTEM
AND THE NEXT GENERATION OF SPACE MISSIONS
ed. David R. Soderblom
ISBN 1-886733-39-2

Vol. CS-120 LUMINOUS BLUE VARIABLES: MASSIVE STARS IN TRANSITION
eds. Antonella Nota and Henny J. G. L. M. Lamers
ISBN 1-886733-40-6

Vol. CS-121 ACCRETION PHENOMENA AND RELATED OUTFLOWS, IAU Colloquium 163
eds. D. T. Wickramasinghe, G. V. Bicknell, and L. Ferrario
ISBN 1-886733-41-4

Vol. CS-122 FROM STARDUST TO PLANETESIMALS:
Symposium held as part of the 108th Annual Meeting of the ASP
eds. Yvonne J. Pendleton and A. G. G. M. Tielens
ISBN 1-886733-42-2

Vol. CS-123 THE 12th 'KINGSTON MEETING': COMPUTATIONAL ASTROPHYSICS
eds. David A. Clarke and Michael J. West
ISBN 1-886733-43-0

Vol. CS-124 DIFFUSE INFRARED RADIATION AND THE IRTS
eds. Haruyuki Okuda, Toshio Matsumoto, and Thomas Roellig
ISBN 1-886733-44-9

Vol. CS-125 ASTRONOMICAL DATA ANALYSIS SOFTWARE AND SYSTEMS VI
eds. Gareth Hunt and H. E. Payne
ISBN 1-886733-45-7

ASP CONFERENCE SERIES VOLUMES

Published by the Astronomical Society of the Pacific

PUBLISHED: 1997 (* asterisk means OUT OF STOCK)

Vol. CS-126 FROM QUANTUM FLUCTUATIONS TO COSMOLOGICAL STRUCTURES
eds. David Valls-Gabaud, Martin A. Hendry, Paolo Molaro, and Khalil Chamcham
ISBN 1-886733-46-5

Vol. CS-127 PROPER MOTIONS AND GALACTIC ASTRONOMY
ed. Roberta M. Humphreys
ISBN 1-886733-47-3

Vol. CS-128 MASS EJECTION FROM AGN (Active Galactic Nuclei)
eds. N. Arav, I. Shlosman, and R. J. Weymann
ISBN 1-886733-48-1

Vol. CS-129 THE GEORGE GAMOW SYMPOSIUM
eds. E. Harper, W. C. Parke, and G. D. Anderson
ISBN 1-886733-49-X

Vol. CS-130 THE THIRD PACIFIC RIM CONFERENCE ON RECENT DEVELOPMENT ON BINARY STAR RESEARCH
eds. Kam-Ching Leung
ISBN 1-886733-50-3

PUBLISHED: 1998

Vol. CS-131 BOULDER-MUNICH II: PROPERTIES OF HOT, LUMINOUS STARS
ed. Ian D. Howarth
ISBN 1-886733-51-1

Vol. CS-132 STAR FORMATION WITH THE INFRARED SPACE OBSERVATORY (ISO)
eds. João L. Yun and René Liseau
ISBN 1-886733-52-X

Vol. CS-133 SCIENCE WITH THE NGST (Next Generation Space Telescope)
eds. Eric P. Smith and Anuradha Koratkar
ISBN 1-886733-53-8

Vol. CS-134 BROWN DWARFS AND EXTRASOLAR PLANETS
eds. Rafael Rebolo, Eduardo L. Martin, and Maria Rosa Zapatero Osorio
ISBN 1-886733-54-6

Vol. CS-135 A HALF CENTURY OF STELLAR PULSATION INTERPRETATIONS: A TRIBUTE TO ARTHUR N. COX
eds. P. A. Bradley and J. A. Guzik
ISBN 1-886733-55-4

Vol. CS-136 GALACTIC HALOS: A UC SANTA CRUZ WORKSHOP
ed. Dennis Zaritsky
ISBN 1-886733-56-2

Vol. CS-137 WILD STARS IN THE OLD WEST: PROCEEDINGS OF THE 13th NORTH AMERICAN WORKSHOP ON CATACLYSMIC VARIABLES AND RELATED OBJECTS
eds. S. Howell, E. Kuulkers, and C. Woodward
ISBN 1-886733-57-0

Vol. CS-138 1997 PACIFIC RIM CONFERENCE ON STELLAR ASTROPHYSICS
eds. Kwing Lam Chan, K. S. Cheng, and H. P. Singh
ISBN 1-886733-58-9

Vol. CS-139 PRESERVING THE ASTRONOMICAL WINDOWS: Proceedings of Joint Discussion No. 5 of the 23rd General Assembly of the IAU
eds. Syuzo Isobe and Tomohiro Hirayama
ISBN 1-886733-59-7

ASP CONFERENCE SERIES VOLUMES

Published by the Astronomical Society of the Pacific

PUBLISHED: 1998 (* asterisk means OUT OF STOCK)

Vol. CS-140 SYNOPTIC SOLAR PHYSICS --18th NSO/Sacramento Peak Summer Workshop
eds. K. S. Balasubramaniam, J. W. Harvey, and D. M. Rabin
ISBN 1-886733-60-0

Vol. CS-141 ASTROPHYSICS FROM ANTARCTICA:
A Symposium held as a part of the 109th Annual Meeting of the ASP
eds. Giles Novak and Randall H. Landsberg
ISBN 1-886733-61-9

Vol. CS-142 THE STELLAR INITIAL MASS FUNCTION: 38th Herstmonceux Conference
eds. Gerry Gilmore and Debbie Howell
ISBN 1-886733-62-7

Vol. CS-143* THE SCIENTIFIC IMPACT OF THE GODDARD HIGH RESOLUTION SPECTROGRAPH (GHRS)
eds. John C. Brandt, Thomas B. Ake III, and Carolyn Collins Petersen
ISBN 1-886733-63-5

Vol. CS-144 RADIO EMISSION FROM GALACTIC AND EXTRAGALACTIC COMPACT SOURCES, IAU Colloquium 164
eds. J. Anton Zensus, G. B. Taylor, and J. M. Wrobel
ISBN 1-886733-64-3

Vol. CS-145 ASTRONOMICAL DATA ANALYSIS SOFTWARE AND SYSTEMS VII – (ADASS VII)
eds. Rudolf Albrecht, Richard N. Hook, and Howard A. Bushouse
ISBN 1-886733-65-1

Vol. CS-146 THE YOUNG UNIVERSE GALAXY FORMATION AND EVOLUTION AT INTERMEDIATE AND HIGH REDSHIFT
eds. S. D'Odorico, A. Fontana, and E. Giallongo
ISBN 1-886733-66-X

Vol. CS-147 ABUNDANCE PROFILES: DIAGNOSTIC TOOLS FOR GALAXY HISTORY
eds. Daniel Friedli, Mike Edmunds, Carmelle Robert, and Laurent Drissen
ISBN 1-886733-67-8

Vol. CS-148 ORIGINS
eds. Charles E. Woodward, J. Michael Shull, and Harley A. Thronson, Jr.
ISBN 1-886733-68-6

Vol. CS-149 SOLAR SYSTEM FORMATION AND EVOLUTION
eds. D. Lazzaro, R. Vieira Martins, S. Ferraz-Mello, J. Fernández, and C. Beaugé
ISBN 1-886733-69-4

Vol. CS-150 NEW PERSPECTIVES ON SOLAR PROMINENCES, IAU Colloquium 167
eds. David Webb, David Rust, and Brigitte Schmieder
ISBN 1-886733-70-8

Vol. CS-151 COSMIC MICROWAVE BACKGROUND AND LARGE SCALE STRUCTURES OF THE UNIVERSE
eds. Yong-Ik Byun and Kin-Wang Ng
ISBN 1-886733-71-6

Vol. CS-152 FIBER OPTICS IN ASTRONOMY III
eds. S. Arribas, E. Mediavilla, and F. Watson
ISBN 1-886733-72-4

Vol. CS-153 LIBRARY AND INFORMATION SERVICES IN ASTRONOMY III -- (LISA III)
eds. Uta Grothkopf, Heinz Andernach, Sarah Stevens-Rayburn, and Monique Gomez
ISBN 1-886733-73-2

ASP CONFERENCE SERIES VOLUMES

Published by the Astronomical Society of the Pacific

PUBLISHED: 1998 (* asterisk means OUT OF STOCK)

Vol. CS-154 — COOL STARS, STELLAR SYSTEMS AND THE SUN: Tenth Cambridge Workshop
eds. Robert A. Donahue and Jay A. Bookbinder
ISBN 1-886733-74-0

Vol. CS-155 — SECOND ADVANCES IN SOLAR PHYSICS EUROCONFERENCE: THREE-DIMENSIONAL STRUCTURE OF SOLAR ACTIVE REGIONS
eds. Costas E. Alissandrakis and Brigitte Schmieder
ISBN 1-886733-75-9

PUBLISHED: 1999

Vol. CS-156 — HIGHLY REDSHIFTED RADIO LINES
eds. C. L. Carilli, S. J. E. Radford, K. M. Menten, and G. I. Langston
ISBN 1-886733-76-7

Vol. CS-157 — ANNAPOLIS WORKSHOP ON MAGNETIC CATACLYSMIC VARIABLES
eds. Coel Hellier and Koji Mukai
ISBN 1-886733-77-5

Vol. CS-158 — SOLAR AND STELLAR ACTIVITY: SIMILARITIES AND DIFFERENCES
eds. C. J. Butler and J. G. Doyle
ISBN 1-886733-78-3

Vol. CS-159 — BL LAC PHENOMENON
eds. Leo O. Takalo and Aimo Sillanpää
ISBN 1-886733-79-1

Vol. CS-160 — ASTROPHYSICAL DISCS: An EC Summer School
eds. J. A. Sellwood and Jeremy Goodman
ISBN 1-886733-80-5

Vol. CS-161 — HIGH ENERGY PROCESSES IN ACCRETING BLACK HOLES
eds. Juri Poutanen and Roland Svensson
ISBN 1-886733-81-3

Vol. CS-162 — QUASARS AND COSMOLOGY
eds. Gary Ferland and Jack Baldwin
ISBN 1-886733-83-X

Vol. CS-163 — STAR FORMATION IN EARLY-TYPE GALAXIES
eds. Jordi Cepa and Patricia Carral
ISBN 1-886733-84-8

Vol. CS-164 — ULTRAVIOLET–OPTICAL SPACE ASTRONOMY BEYOND HST
eds. Jon A. Morse, J. Michael Shull, and Anne L. Kinney
ISBN 1-886733-85-6

Vol. CS-165 — THE THIRD STROMLO SYMPOSIUM: THE GALACTIC HALO
eds. Brad K. Gibson, Tim S. Axelrod, and Mary E. Putman
ISBN 1-886733-86-4

Vol. CS-166 — STROMLO WORKSHOP ON HIGH-VELOCITY CLOUDS
eds. Brad K. Gibson and Mary E. Putman
ISBN 1-886733-87-2

Vol. CS-167 — HARMONIZING COSMIC DISTANCE SCALES IN A POST-HIPPARCOS ERA
eds. Daniel Egret and André Heck
ISBN 1-886733-88-0

Vol. CS-168 — NEW PERSPECTIVES ON THE INTERSTELLAR MEDIUM
eds. A. R. Taylor, T. L. Landecker, and G. Joncas
ISBN 1-886733-89-9

ASP CONFERENCE SERIES VOLUMES

Published by the Astronomical Society of the Pacific

PUBLISHED: 1999 (* asterisk means OUT OF STOCK)

Vol. CS-169 11th EUROPEAN WORKSHOP ON WHITE DWARFS
eds. J.-E. Solheim and E. G. Meištas
ISBN 1-886733-91-0

Vol. CS-170 THE LOW SURFACE BRIGHTNESS UNIVERSE, IAU Colloquium 171
eds. J. I. Davies, C. Impey, and S. Phillipps
ISBN 1-886733-92-9

Vol. CS-171 LiBeB, COSMIC RAYS, AND RELATED X- AND GAMMA-RAYS
eds. Reuven Ramaty, Elisabeth Vangioni-Flam, Michel Cassé, and Keith Olive
ISBN 1-886733-93-7

Vol. CS-172 ASTRONOMICAL DATA ANALYSIS SOFTWARE AND SYSTEMS VIII
eds. David M. Mehringer, Raymond L. Plante, and Douglas A. Roberts
ISBN 1-886733-94-5

Vol. CS-173 THEORY AND TESTS OF CONVECTION IN STELLAR STRUCTURE: First Granada Workshop
ed. Álvaro Giménez, Edward F. Guinan, and Benjamín Montesinos
ISBN 1-886733-95-3

Vol. CS-174 CATCHING THE PERFECT WAVE: ADAPTIVE OPTICS AND INTERFEROMETRY IN THE 21st CENTURY,
A Symposium held as a part of the 110th Annual Meeting of the ASP
eds. Sergio R. Restaino, William Junor, and Nebojsa Duric
ISBN 1-886733-96-1

Vol. CS-175 STRUCTURE AND KINEMATICS OF QUASAR BROAD LINE REGIONS
eds. C. M. Gaskell, W. N. Brandt, M. Dietrich, D. Dultzin-Hacyan, and M. Eracleous
ISBN 1-886733-97-X

Vol. CS-176 OBSERVATIONAL COSMOLOGY: THE DEVELOPMENT OF GALAXY SYSTEMS
eds. Giuliano Giuricin, Marino Mezzetti, and Paolo Salucci
ISBN 1-58381-000-5

Vol. CS-177 ASTROPHYSICS WITH INFRARED SURVEYS: A Prelude to SIRTF
eds. Michael D. Bicay, Chas A. Beichman, Roc M. Cutri, and Barry F. Madore
ISBN 1-58381-001-3

Vol. CS-178 STELLAR DYNAMOS: NONLINEARITY AND CHAOTIC FLOWS
eds. Manuel Núñez and Antonio Ferriz-Mas
ISBN 1-58381-002-1

Vol. CS-179 ETA CARINAE AT THE MILLENNIUM
eds. Jon A. Morse, Roberta M. Humphreys, and Augusto Damineli
ISBN 1-58381-003-X

Vol. CS-180 SYNTHESIS IMAGING IN RADIO ASTRONOMY II
eds. G. B. Taylor, C. L. Carilli, and R. A. Perley
ISBN 1-58381-005-6

Vol. CS-181 MICROWAVE FOREGROUNDS
eds. Angelica de Oliveira-Costa and Max Tegmark
ISBN 1-58381-006-4

Vol. CS-182 GALAXY DYNAMICS: A Rutgers Symposium
eds. David Merritt, J. A. Sellwood, and Monica Valluri
ISBN 1-58381-007-2

Vol. CS-183 HIGH RESOLUTION SOLAR PHYSICS: THEORY, OBSERVATIONS, AND TECHNIQUES
eds. T. R. Rimmele, K. S. Balasubramaniam, and R. R. Radick
ISBN 1-58381-009-9

ASP CONFERENCE SERIES VOLUMES

Published by the Astronomical Society of the Pacific

PUBLISHED: 1999 (* asterisk means OUT OF STOCK)

Vol. CS-184 THIRD ADVANCES IN SOLAR PHYSICS EUROCONFERENCE: MAGNETIC FIELDS AND OSCILLATIONS
eds. B. Schmieder, A. Hofmann, and J. Staude
ISBN 1-58381-010-2

Vol. CS-185 PRECISE STELLAR RADIAL VELOCITIES, IAU Colloquium 170
eds. J. B. Hearnshaw and C. D. Scarfe
ISBN 1-58381-011-0

Vol. CS-186 THE CENTRAL PARSECS OF THE GALAXY
eds. Heino Falcke, Angela Cotera, Wolfgang J. Duschl, Fulvio Melia, and Marcia J. Rieke
ISBN 1-58381-012-9

Vol. CS-187 THE EVOLUTION OF GALAXIES ON COSMOLOGICAL TIMESCALES
eds. J. E. Beckman and T. J. Mahoney
ISBN 1-58381-013-7

Vol. CS-188 OPTICAL AND INFRARED SPECTROSCOPY OF CIRCUMSTELLAR MATTER
eds. Eike W. Guenther, Bringfried Stecklum, and Sylvio Klose
ISBN 1-58381-014-5

Vol. CS-189 CCD PRECISION PHOTOMETRY WORKSHOP
eds. Eric R. Craine, Roy A. Tucker, and Jeannette Barnes
ISBN 1-58381-015-3

Vol. CS-190 GAMMA-RAY BURSTS: THE FIRST THREE MINUTES
eds. Juri Poutanen and Roland Svensson
ISBN 1-58381-016-1

Vol. CS-191 PHOTOMETRIC REDSHIFTS AND HIGH REDSHIFT GALAXIES
eds. Ray J. Weymann, Lisa J. Storrie-Lombardi, Marcin Sawicki, and Robert J. Brunner
ISBN 1-58381-017-X

Vol. CS-192 SPECTROPHOTOMETRIC DATING OF STARS AND GALAXIES
ed. I. Hubeny, S. R. Heap, and R. H. Cornett
ISBN 1-58381-018-8

Vol. CS-193 THE HY-REDSHIFT UNIVERSE: GALAXY FORMATION AND EVOLUTION AT HIGH REDSHIFT
eds. Andrew J. Bunker and Wil J. M. van Breugel
ISBN 1-58381-019-6

Vol. CS-194 WORKING ON THE FRINGE: OPTICAL AND IR INTERFEROMETRY FROM GROUND AND SPACE
eds. Stephen Unwin and Robert Stachnik
ISBN 1-58381-020-X

PUBLISHED: 2000

Vol. CS-195 IMAGING THE UNIVERSE IN THREE DIMENSIONS: Astrophysics with Advanced Multi-Wavelength Imaging Devices
eds. W. van Breugel and J. Bland-Hawthorn
ISBN 1-58381-022-6

Vol. CS-196 THERMAL EMISSION SPECTROSCOPY AND ANALYSIS OF DUST, DISKS, AND REGOLITHS
eds. Michael L. Sitko, Ann L. Sprague, and David K. Lynch
ISBN: 1-58381-023-4

Vol. CS-197 XVth IAP MEETING DYNAMICS OF GALAXIES: FROM THE EARLY UNIVERSE TO THE PRESENT
eds. F. Combes, G. A. Mamon, and V. Charmandaris
ISBN: 1-58381-24-2

ASP CONFERENCE SERIES VOLUMES

Published by the Astronomical Society of the Pacific

PUBLISHED: 2000 (* asterisk means OUT OF STOCK)

Vol. CS-198 EUROCONFERENCE ON "STELLAR CLUSTERS AND ASSOCIATIONS: CONVECTION, ROTATION, AND DYNAMOS"
eds. R. Pallavicini, G. Micela, and S. Sciortino
ISBN: 1-58381-25-0

Vol. CS-199 ASYMMETRICAL PLANETARY NEBULAE II: FROM ORIGINS TO MICROSTRUCTURES
eds. J. H. Kastner, N. Soker, and S. Rappaport
ISBN: 1-58381-026-9

Vol. CS-200 CLUSTERING AT HIGH REDSHIFT
eds. A. Mazure, O. Le Fèvre, and V. Le Brun
ISBN: 1-58381-027-7

Vol. CS-201 COSMIC FLOWS 1999: TOWARDS AN UNDERSTANDING OF LARGE-SCALE STRUCTURES
eds. Stéphane Courteau, Michael A. Strauss, and Jeffrey A. Willick
ISBN: 1-58381-028-5

Vol. CS-202* PULSAR ASTRONOMY – 2000 AND BEYOND, IAU Colloquium 177
eds. M. Kramer, N. Wex, and R. Wielebinski
ISBN: 1-58381-029-3

Vol. CS-203 THE IMPACT OF LARGE-SCALE SURVEYS ON PULSATING STAR RESEARCH, IAU Colloquium 176
eds. L. Szabados and D. W. Kurtz
ISBN: 1-58381-030-7

Vol. CS-204 THERMAL AND IONIZATION ASPECTS OF FLOWS FROM HOT STARS: OBSERVATIONS AND THEORY
eds. Henny J. G. L. M. Lamers and Arved Sapar
ISBN: 1-58381-031-5

Vol. CS-205 THE LAST TOTAL SOLAR ECLIPSE OF THE MILLENNIUM IN TURKEY
eds. W. C. Livingston and A. Özgüç
ISBN: 1-58381-032-3

Vol. CS-206 HIGH ENERGY SOLAR PHYSICS – *ANTICIPATING HESSI*
eds. Reuven Ramaty and Natalie Mandzhavidze
ISBN: 1-58381-033-1

Vol. CS-207 NGST SCIENCE AND TECHNOLOGY EXPOSITION
eds. Eric P. Smith and Knox S. Long
ISBN: 1-58381-036-6

ATLAS VISIBLE AND NEAR INFRARED ATLAS OF THE ARCTURUS SPECTRUM 3727-9300 Å
eds. Kenneth Hinkle, Lloyd Wallace, Jeff Valenti, and Dianne Harmer
ISBN: 1-58381-037-4

Vol. CS-208 POLAR MOTION: HISTORICAL AND SCIENTIFIC PROBLEMS, IAU Colloquium 178
eds. Steven Dick, Dennis McCarthy, and Brian Luzum
ISBN: 1-58381-039-0

Vol. CS-209 SMALL GALAXY GROUPS, IAU Colloquium 174
eds. Mauri J. Valtonen and Chris Flynn
ISBN: 1-58381-040-4

Vol. CS-210 DELTA SCUTI AND RELATED STARS: Reference Handbook and Proceedings of the 6th Vienna Workshop in Astrophysics
eds. Michel Breger and Michael Houston Montgomery
ISBN: 1-58381-043-9

ASP CONFERENCE SERIES VOLUMES

Published by the Astronomical Society of the Pacific

PUBLISHED: 2000 (* asterisk means OUT OF STOCK)

Vol. CS-211 MASSIVE STELLAR CLUSTERS
eds. Ariane Lançon and Christian M. Boily
ISBN: 1-58381-042-0

Vol. CS-212 FROM GIANT PLANETS TO COOL STARS
eds. Caitlin A. Griffith and Mark S. Marley
ISBN: 1-58381-041-2

Vol. CS-213 BIOASTRONOMY '99: A NEW ERA IN BIOASTRONOMY
eds. Guillermo A. Lemarchand and Karen J. Meech
ISBN: 1-58381-044-7

Vol. CS-214 THE Be PHENOMENON IN EARLY-TYPE STARS, IAU Colloquium 175
eds. Myron A. Smith, Huib F. Henrichs and Juan Fabregat
ISBN: 1-58381-045-5

Vol. CS-215 COSMIC EVOLUTION AND GALAXY FORMATION:
STRUCTURE, INTERACTIONS AND FEEDBACK
The 3rd Guillermo Haro Astrophysics Conference
eds. José Franco, Elena Terlevich, Omar López-Cruz, and Itziar Aretxaga
ISBN: 1-58381-046-3

Vol. CS-216 ASTRONOMICAL DATA ANALYSIS SOFTWARE AND SYSTEMS IX
eds. Nadine Manset, Christian Veillet, and Dennis Crabtree
ISBN: 1-58381-047-1 ISSN: 1080-7926

Vol. CS-217 IMAGING AT RADIO THROUGH SUBMILLIMETER WAVELENGTHS
eds. Jeffrey G. Mangum and Simon J. E. Radford
ISBN: 1-58381-049-8

Vol. CS-218 MAPPING THE HIDDEN UNIVERSE: THE UNIVERSE BEHIND THE MILKY WAY
THE UNIVERSE IN HI
eds. Renée C. Kraan-Korteweg, Patricia A. Henning, and Heinz Andernach
ISBN: 1-58381-050-1

Vol. CS-219 DISKS, PLANETESIMALS, AND PLANETS
eds. F. Garzón, C. Eiroa, D. de Winter, and T. J. Mahoney
ISBN: 1-58381-051-X

Vol. CS-220 AMATEUR - PROFESSIONAL PARTNERSHIPS IN ASTRONOMY:
The 111th Annual Meeting of the ASP
eds. John R. Percy and Joseph B. Wilson
ISBN: 1-58381-052-8

Vol. CS-221 STARS, GAS AND DUST IN GALAXIES: EXPLORING THE LINKS
eds. Danielle Alloin, Knut Olsen, and Gaspar Galaz
ISBN: 1-58381-053-6

PUBLISHED: 2001

Vol. CS-222 THE PHYSICS OF GALAXY FORMATION
eds. M. Umemura and H. Susa
ISBN: 1-58381-054-4

Vol. CS-223 COOL STARS, STELLAR SYSTEMS AND THE SUN:
Eleventh Cambridge Workshop
eds. Ramón J. García López, Rafael Rebolo, and María Zapatero Osorio
ISBN: 1-58381-056-0

Vol. CS-224 PROBING THE PHYSICS OF ACTIVE GALACTIC NUCLEI
BY MULTIWAVELENGTH MONITORING
eds. Bradley M. Peterson, Ronald S. Polidan, and Richard W. Pogge
ISBN: 1-58381-055-2

ASP CONFERENCE SERIES VOLUMES

Published by the Astronomical Society of the Pacific

PUBLISHED: 2001 (* asterisk means OUT OF STOCK)

Vol. CS-225 VIRTUAL OBSERVATORIES OF THE FUTURE
eds. Robert J. Brunner, S. George Djorgovski, and Alex S. Szalay
ISBN: 1-58381-057-9

Vol. CS-226 12th EUROPEAN CONFERENCE ON WHITE DWARFS
eds. J. L. Provencal, H. L. Shipman, J. MacDonald, and S. Goodchild
ISBN: 1-58381-058-7

Vol. CS-227 BLAZAR DEMOGRAPHICS AND PHYSICS
eds. Paolo Padovani and C. Megan Urry
ISBN: 1-58381-059-5

Vol. CS-228 DYNAMICS OF STAR CLUSTERS AND THE MILKY WAY
eds. S. Deiters, B. Fuchs, A. Just, R. Spurzem, and R. Wielen
ISBN: 1-58381-060-9

Vol. CS-229 EVOLUTION OF BINARY AND MULTIPLE STAR SYSTEMS
A Meeting in Celebration of Peter Eggleton's 60th Birthday
eds. Ph. Podsiadlowski, S. Rappaport, A. R. King, F. D'Antona, and L. Burderi
IBSN: 1-58381-061-7

Vol. CS-230 GALAXY DISKS AND DISK GALAXIES
eds. Jose G. Funes, S. J. and Enrico Maria Corsini
ISBN: 1-58381-063-3

Vol. CS-231 TETONS 4: GALACTIC STRUCTURE, STARS, AND THE INTERSTELLAR MEDIUM
eds. Charles E. Woodward, Michael D. Bicay, and J. Michael Shull
ISBN: 1-58381-064-1

Vol. CS-232 THE NEW ERA OF WIDE FIELD ASTRONOMY
eds. Roger Clowes, Andrew Adamson, and Gordon Bromage
ISBN: 1-58381-065-X

Vol. CS-233 P CYGNI 2000: 400 YEARS OF PROGRESS
eds. Mart de Groot and Christiaan Sterken
ISBN: 1-58381-070-6

Vol. CS-234 X-RAY ASTRONOMY 2000
eds. R. Giacconi, S. Serio, and L. Stella
ISBN: 1-58381-071-4

Vol. CS-235 SCIENCE WITH THE ATACAMA LARGE MILLIMETER ARRAY (ALMA)
ed. Alwyn Wootten
ISBN: 1-58381-072-2

Vol. CS-236 ADVANCED SOLAR POLARIMETRY: THEORY, OBSERVATION, AND INSTRUMENTATION, The 20th Sacramento Peak Summer Workshop
ed. M. Sigwarth
ISBN: 1-58381-073-0

Vol. CS-237 GRAVITATIONAL LENSING: RECENT PROGRESS AND FUTURE GOALS
eds. Tereasa G. Brainerd and Christopher S. Kochanek
ISBN: 1-58381-074-9

Vol. CS-238 ASTRONOMICAL DATA ANALYSIS SOFTWARE AND SYSTEMS X
eds. F. R. Harnden, Jr., Francis A. Primini, and Harry E. Payne
ISBN: 1-58381-075-7

Vol. CS-239 MICROLENSING 2000: A NEW ERA OF MICROLENSING ASTROPHYSICS
ed. John Menzies and Penny D. Sackett
ISBN: 1-58381-076-5

ASP CONFERENCE SERIES VOLUMES

Published by the Astronomical Society of the Pacific

PUBLISHED: 2001 (* asterisk means OUT OF STOCK)

Vol. CS-240 GAS AND GALAXY EVOLUTION,
A Conference in Honor of the 20th Anniversary of the VLA
eds. J. E. Hibbard, M. P. Rupen, and J. H. van Gorkom
ISBN: 1-58381-077-3

Vol. CS-241 CS-241 THE 7TH TAIPEI ASTROPHYSICS WORKSHOP ON
COSMIC RAYS IN THE UNIVERSE
ed. Chung-Ming Ko
ISBN: 1-58381-079-X

Vol. CS-242 ETA CARINAE AND OTHER MYSTERIOUS STARS:
THE HIDDEN OPPORTUNITIES OF EMISSION SPECTROSCOPY
eds. Theodore R. Gull, Sveneric Johannson, and Kris Davidson
ISBN: 1-58381-080-3

Vol. CS-243 FROM DARKNESS TO LIGHT:
ORIGIN AND EVOLUTION OF YOUNG STELLAR CLUSTERS
eds. Thierry Montmerle and Philippe André
ISBN: 1-58381-081-1

Vol. CS-244 YOUNG STARS NEAR EARTH: PROGRESS AND PROSPECTS
eds. Ray Jayawardhana and Thomas P. Greene
ISBN: 1-58381-082-X

Vol. CS-245 ASTROPHYSICAL AGES AND TIME SCALES
eds. Ted von Hippel, Chris Simpson, and Nadine Manset
ISBN: 1-58381-083-8

Vol. CS-246 SMALL TELESCOPE ASTRONOMY ON GLOBAL SCALES, IAU Colloquium 183
eds. Wen-Ping Chen, Claudia Lemme, and Bohdan Paczyński
ISBN: 1-58381-084-6

Vol. CS-247 SPECTROSCOPIC CHALLENGES OF PHOTOIONIZED PLASMAS
eds. Gary Ferland and Daniel Wolf Savin
ISBN: 1-58381-085-4

Vol. CS-248 MAGNETIC FIELDS ACROSS THE HERTZSPRUNG-RUSSELL DIAGRAM
eds. G. Mathys, S. K. Solanki, and D. T. Wickramasinghe
ISBN: 1-58381-088-9

Vol. CS-249 THE CENTRAL KILOPARSEC OF STARBURSTS AND AGN:
THE LA PALMA CONNECTION
eds. J. H. Knapen, J. E. Beckman, I. Shlosman, and T. J. Mahoney
ISBN: 1-58381-089-7

Vol. CS-250 PARTICLES AND FIELDS IN RADIO GALAXIES CONFERENCE
eds. Robert A. Laing and Katherine M. Blundell
ISBN: 1-58381-090-0

Vol. CS-251 NEW CENTURY OF X-RAY ASTRONOMY
eds. H. Inoue and H. Kunieda
ISBN: 1-58381-091-9

Vol. CS-252 HISTORICAL DEVELOPMENT OF MODERN COSMOLOGY
eds. Vicent J. Martínez, Virginia Trimble, and María Jesús Pons-Bordería
ISBN: 1-58381-092-7

PUBLISHED: 2002

Vol. CS-253 CHEMICAL ENRICHMENT OF INTRACLUSTER AND INTERGALACTIC MEDIUM
eds. Roberto Fusco-Femiano and Francesca Matteucci
ISBN: 1-58381-093-5

ASP CONFERENCE SERIES VOLUMES

Published by the Astronomical Society of the Pacific

PUBLISHED: 2002 (* asterisk means OUT OF STOCK)

Vol. CS-254 EXTRAGALACTIC GAS AT LOW REDSHIFT
eds. John S. Mulchaey and John T. Stocke
ISBN: 1-58381-094-3

Vol. CS-255 MASS OUTFLOW IN ACTIVE GALACTIC NUCLEI: NEW PERSPECTIVES
eds. D. M. Crenshaw, S. B. Kraemer, and I. M. George
ISBN: 1-58381-095-1

Vol. CS-256 OBSERVATIONAL ASPECTS OF PULSATING B AND A STARS
eds. Christiaan Sterken and Donald W. Kurtz
ISBN: 1-58381-096-X

Vol. CS-257 AMiBA 2001: HIGH-Z CLUSTERS, MISSING BARYONS, AND CMB POLARIZATION
eds. Lin-Wen Chen, Chung-Pei Ma, Kin-Wang Ng, and Ue-Li Pen
ISBN: 1-58381-097-8

Vol. CS-258 ISSUES IN UNIFICATION OF ACTIVE GALACTIC NUCLEI
eds. Roberto Maiolino, Alessandro Marconi, and Neil Nagar
ISBN: 1-58381-098-6

Vol. CS-259 RADIAL AND NONRADIAL PULSATIONS AS PROBES OF STELLAR PHYSICS
eds. Conny Aerts, Timothy R. Bedding, and Jørgen Christensen-Dalsgaard
ISBN: 1-58381-099-4

Vol. CS-260 INTERACTING WINDS FROM MASSIVE STARS
eds. Anthony F. J. Moffat and Nicole St-Louis
ISBN: 1-58381-100-1

Vol. CS-261 THE PHYSICS OF CATACLYSMIC VARIABLES AND RELATED OBJECTS
eds. B. T. Gänsicke, K. Beuermann, and K. Reinsch
ISBN: 1-58381-101-X

Vol. CS-262 THE HIGH ENERGY UNIVERSE AT SHARP FOCUS: CHANDRA SCIENCE
The 113th Annual Meeting of the ASP
eds. Eric M. Schlegel and Saeqa Dil Vrtilek
ISBN: 1-58381-102-8

Vol. CS-263 STELLAR COLLISIONS, MERGERS AND THEIR CONSEQUENCES
ed. Michael M. Shara
ISBN: 1-58381-103-6

Vol. CS-264 CONTINUING THE CHALLENGE OF EUV ASTRONOMY: CURRENT ANALYSIS AND PROSPECTS FOR THE FUTURE
eds. Steve B. Howell, Jean Dupuis, Daniel Golombek, Frederick M. Walter, and Jennifer Cullison
ISBN: 1-58381-104-4

Vol. CS-265 ω CENTAURI, A UNIQUE WINDOW INTO ASTROPHYSICS
eds. Floor van Leeuwen, Joanne D. Hughes, and Giampaolo Piotto
ISBN: 1-58381-105-2

Vol. CS-266 ASTRONOMICAL SITE EVALUATION IN THE VISIBLE AND RADIO RANGE
eds. J. Vernin, Z. Benkhaldoun, and C. Muñoz-Tuñón
ISBN: 1-58381-106-0

Vol. CS-267 HOT STAR WORKSHOP III: THE EARLIEST STAGES OF MASSIVE STAR BIRTH
ed. Paul A. Crowther
ISBN: 1-58381-107-9

Vol. CS-268 TRACING COSMIC EVOLUTION WITH GALAXY CLUSTERS
eds. Stefano Borgani, Marino Mezzetti, and Riccardo Valdarnini
ISBN: 1-58381-108-7

A LISTING OF IAU VOLUMES MAY BE FOUND ON THE NEXT PAGE

INTERNATIONAL ASTRONOMICAL UNION (IAU) VOLUMES

Published by the Astronomical Society of the Pacific

PUBLISHED: 1999 (* asterisk means OUT OF STOCK)

Vol. No. 190 NEW VIEWS OF THE MAGELLANIC CLOUDS
eds. You-Hua Chu, Nicholas B. Suntzeff, James E. Hesser, and David A. Bohlender
ISBN: 1-58381-021-8

Vol. No. 191 ASYMPTOTIC GIANT BRANCH STARS
eds. T. Le Bertre, A. Lèbre, and C. Waelkens
ISBN: 1-886733-90-2

Vol. No. 192 THE STELLAR CONTENT OF LOCAL GROUP GALAXIES
eds. Patricia Whitelock and Russell Cannon
ISBN: 1-886733-82-1

Vol. No. 193 WOLF-RAYET PHENOMENA IN MASSIVE STARS AND STARBURST GALAXIES
eds. Karel A. van der Hucht, Gloria Koenigsberger, and Philippe R. J. Eenens
ISBN: 1-58381-004-8

Vol. No. 194 ACTIVE GALACTIC NUCLEI AND RELATED PHENOMENA
eds. Yervant Terzian, Daniel Weedman, and Edward Khachikian
ISBN: 1-58381-008-0

PUBLISHED: 2000

Vol. XXIVA TRANSACTIONS OF THE INTERNATIONAL ASTRONOMICAL UNION REPORTS ON ASTRONOMY 1996-1999
ed. Johannes Andersen
ISBN: 1-58381-035-8

Vol. No. 195 HIGHLY ENERGETIC PHYSICAL PROCESSES AND MECHANISMS FOR EMISSION FROM ASTROPHYSICAL PLASMAS
eds. P. C. H. Martens, S. Tsuruta, and M. A. Weber
ISBN: 1-58381-038-2

Vol. No. 197 ASTROCHEMISTRY: FROM MOLECULAR CLOUDS TO PLANETARY SYSTEMS
eds. Y. C. Minh and E. F. van Dishoeck
ISBN: 1-58381-034-X

Vol. No. 198 THE LIGHT ELEMENTS AND THEIR EVOLUTION
eds. L. da Silva, M. Spite, and J. R. de Medeiros
ISBN: 1-58381-048-X

PUBLISHED: 2001

IAU SPS ASTRONOMY FOR DEVELOPING COUNTRIES
Special Session of the XXIV General Assembly of the IAU
ed. Alan H. Batten
ISBN: 1-58381-067-6

Vol. No. 196 PRESERVING THE ASTRONOMICAL SKY
eds. R. J. Cohen and W. T. Sullivan, III
ISBN: 1-58381-078-1

Vol. No. 200 THE FORMATION OF BINARY STARS
eds. Hans Zinnecker and Robert D. Mathieu
ISBN: 1-58381-068-4

Vol. No. 203 RECENT INSIGHTS INTO THE PHYSICS 0F THE SUN AND HELIOSPHERE: HIGHLIGHTS FROM SOHO AND OTHER SPACE MISSIONS
eds. Pål Brekke, Bernhard Fleck, and Joseph B. Gurman
ISBN: 1-58381-069-2

Vol. No. 204 THE EXTRAGALACTIC INFRARED BACKGROUND AND ITS COSMOLOGICAL IMPLICATIONS
eds. Martin Harwit and Michael G. Hauser
ISBN: 1-58381-062-5

INTERNATIONAL ASTRONOMICAL UNION (IAU) VOLUMES

Published by the Astronomical Society of the Pacific

PUBLISHED: 2001 (* asterisk means OUT OF STOCK)

Vol. No. 205 GALAXIES AND THEIR CONSTITUENTS AT THE HIGHEST ANGULAR RESOLUTIONS
eds. Richard T. Schilizzi, Stuart N. Vogel, Francesco Paresce, and Martin S. Elvis
ISBN: 1-58381-066-8

Vol. XXIVB TRANSACTIONS OF THE INTERNATIONAL ASTRONOMICAL UNION REPORTS ON ASTRONOMY
ed. Hans Rickman
ISBN: 1-58381-087-0

PUBLISHED: 2002

Vol. No. 12 HIGHLIGHTS OF ASTRONOMY
ed. Hans Rickman
ISBN: 1-58381-086-2

Complete lists of proceedings of past IAU Meetings are maintained at the IAU Web site at the URL: http://www.iau.org/publicat.html

Volumes 32 - 189 in the IAU Symposia Series may be ordered from:

Kluwer Academic Publishers
P. O. Box 117
NL 3300 AA Dordrecht
The Netherlands

Kluwer@wKap.com

All book orders or inquiries concerning ASP or IAU volumes listed should be directed to the:

The Astronomical Society of the Pacific Conference Series
390 Ashton Avenue
San Francisco CA 94112-1722 USA

Phone: 415-337-2126
Fax: 415-337-5205

E-mail: catalog@astrosociety.org
Web Site: http://www.astrosociety.org